PLANT TAXONOMY,
PHYTOGEOGRAPHY AND RELATED SUBJECTS

THE

DAVIS & HEDGE FESTSCHRIFT

PLANT TAXONOMY,
PHYTOGEOGRAPHY AND RELATED SUBJECTS

THE

DAVIS & HEDGE FESTSCHRIFT

commemorating
the seventieth birthday of Peter Hadland Davis
and
the sixtieth birthday of Ian Charleson Hedge

edited by

KIT TAN

assisted by

R.R. Mill and T.S. Elias

EDINBURGH
University Press

22 George Square, Edinburgh

Set in Lasercomp Times Roman and
printed in Great Britain by
The Alden Press, London & Oxford

British Library Cataloguing
in Publication Data
Tan, Kit
The Davis & Hedge festschrift:
plant taxonomy, phytogeography
and related subjects.
1. Plants. Taxonomy
I. Title II. Mill, R.R
III. Elias, S. IV. Davis, Peter
V. Hedge, Ian
581′.012

ISBN 0 85224 638 2

CONTENTS

FOREWORD

It gives me very great pleasure indeed to have the opportunity of writing a foreword to this volume, recognising as it does the work of two of the outstanding Edinburgh botanists of the latter half of this century. Peter Davis, by his extensive collections and by his work on the Flora of Turkey, provided the foundation for the Royal Botanic Garden's current expertise on southwest Asian botany; Ian Hedge built on this, ensuring that the research was not confined to Turkey, and developing the outstanding herbarium resources that drew specialists to Edinburgh.

Both have had lifetime involvement with Edinburgh, being students in the Department of Botany of the University, and carrying out their subsequent professional work at or in association with the Royal Botanic Garden. Peter Davis – 'PD' to many generations of Edinburgh University taxonomy students – had a University appointment even before the completion of his Ph.D. and then moved through the academic ranks to that of Professor; moreover, despite the move of the rest of the Botany Department in 1961 to King's Buildings, the Science campus of the University, he and the 'Taxonomy Section', which later became the 'Flora of Turkey Unit', remained associated with the Herbarium, Library and living plant resources of the RBG. As one who had the privilege of attending his lectures when he was still 'Mr Davis', I remember with gratitude the all too rare excitement that both his enthusiasm and the content of his lectures engendered. Few other lecturers at that time used as their main reference works books 'hot off the press', and that this should be the case in taxonomy with its rich and still relevant older literature was all the more notable. That the author of the *Flora of Turkey* introduced biometrics and biosystematics into Edinburgh's taxonomy teaching should not surprise anyone who remembers the outstanding influence and success of his 1963 text *Principles of Angiosperm Taxonomy* (co-authored with V.H. Heywood).

Ian Hedge is, of course, still in full career and this volume recognises his continuing contributions to botany as well as his achievements to date. In addition to his expertise on families such as the Cruciferae, Labiatae, and Umbelliferae, recognised through his extensive publications, his service as Curator of the Herbarium of the RBG Edinburgh for some twenty years was particularly notable for the network of international contacts that he developed particularly on the floristics of southwest Asia.

This volume in many ways reflects the results of these international links. Botanists from twenty-five institutes in fifteen countries have presented papers that commemorate the outstanding contributions of Peter Davis and Ian Hedge; they do this by highlighting current taxonomic problems, especially in the floras of those regions upon which these botanists' research has concentrated. Indeed, although not designed as such, the work is in a very real sense an addendum to their floristic work, bringing together papers dealing with a diversity of plants, but most with the common theme of systematics and phytogeography in the Mediterranean and southwest Asian regions. The book will, I am sure, provide a valuable insight into current systematic work on the plants of these regions that goes beyond the information incorporated in the floristic studies that have made Peter Davis and Ian Hedge so well recognised.

In addition to acknowledging the achievements of these two botanists by contributing to this 'Festschrift', I am sure that all the authors join me in wishing Peter Davis and Ian Hedge many more happy and enjoyable years in botany as in all the other strands of life.

JOHN MCNEILL
18 July 1988

ACKNOWLEDGMENTS

I acknowledge with gratitude the following persons for giving of their time to review the manuscripts submitted:

Rupert C. Barneby
B.L. Burtt
Gertrud Dahlgren
Roger F.O. Kemp
Jennifer M. Lamond
John McNeill
David G. Mann
and Paddy J.B. Woods

I am especially grateful to Robert R. Mill and Thomas S. Elias for their constructive criticism and assistance in editing several papers.

KIT TAN

PETER HADLAND DAVIS: BIOGRAPHICAL SKETCH; LIST OF PUBLICATIONS

KIT TAN

Royal Botanic Garden, Edinburgh, EH3 5LR, Scotland

AND

THOMAS S. ELIAS

Rancho Santa Ana Botanic Garden, Claremont, California 91711

Peter Davis approaching his seventieth birthday. Photo: by courtesy of Alex George, CSIRO, Canberra, Australia.

PROFESSOR PETER HADLAND DAVIS
B.Sc., Ph. D., D.Sc., F.L.S., F.R.S.E.

Born in Weston-Super-Mare on 18 June 1918, Peter Davis was educated first at Nash House, Burnham-on-Sea and subsequently at Bradfield College and Maiden Erleigh, Reading. In 1937 he was apprenticed to Will Ingwersen's Alpine Plant Nursery at East Grinstead and it was during that time that his early interest in plants was nurtured, eventually to become an all-consuming passion which he continues to sustain in his retirement. In 1938 he began botanizing as an amateur in the Middle East, including Turkey, but his work was interrupted by the outbreak of the Second World War in 1939. During the period from the beginning of hostilities until 1945 he served in the Armed Forces eventually spending the last two years of his service on special duties in Cairo. On release from His Majesty's Forces he came to the University of Edinburgh to study botany and graduated with a First Class Honours B.Sc. in the summer of 1949. This was the beginning of a distinguished academic career spent wholly in the Botany Department, broken up by numerous collecting trips to various parts of the world. Appointed Lecturer in Botany in 1950 he began a programme of research which led him and his research team inexorably forward to the completion of the *Flora of Turkey and the East Aegean Islands*. However, between the initiation and completion of this major Flora lay much research and a colourful life. His Ph.D. on the 'Taxonomy of Middle East Flora' was awarded in 1952 and through the 1950s he travelled widely collecting and exploring in Kurdistan, Russia, and various regions of the Middle East. In 1958 he was awarded the Cuthbert Peek Medal by the Royal Geographical Society for exploration in remote regions of Kurdistan.

In 1961 he began in earnest the preparation of the *Flora of Turkey* having secured a grant from the Science Research Council, then known as D.S.I.R. It is particularly significant that the S.R.C. (subsequently S.E.R.C.) supported this research until the completion of the main Flora in 1985. Indeed, this work is one of the jewels in the crown of research supported by the Biological Sciences Committee, representing as it does a major and distinguished piece of scholarship. Through the 1960s and 1970s he was promoted to Senior Lecturer, then Reader and, finally, to a Personal Chair in Plant Taxonomy. His work outside the University has been widely recognised by the award of the Symposium Medal at the International Symposium on the Problems of Balkan Flora and Vegetation, Certificate of Merit from Turkish Minister of State for outstanding contributions in the field of science, The OPTIMA Silver Medal, Gold Medal of the Linnean Society, by his appointment as an Honorary Associate of the Royal Botanic Garden, Edinburgh and by his election to the Royal Society of Edinburgh. In 1980 he was awarded an S.E.R.C. Replacement Fellowship which enabled him to spend all of his time on research and the completion of the Flora.

In addition to a life dedicated to research he has also made a substantial contribution to the teaching of plant taxonomy both directly and by the books he has written. He has always given of his time freely to help students and colleagues. His contributions to Plant Taxonomy are considerable and he is recognised as one of the most distinguished and energetic Plant Taxonomists of his time. Apart from his many other publications, his enduring achievement is undoubtedly the study of the *Flora of Turkey and the East Aegean Islands* which is now complete. This substantial work in nine volumes (1965–1985) plus a Supplement (1988) has been received with great acclaim and is one of the most important taxonomic achievements this century.

Currently Emeritus Professor of Plant Taxonomy at the University of Edinburgh, Professor Davis was awarded the Neill Prize for the 40th Biennial Period (1985–87) by the Royal Society of Edinburgh in recognition of his distinguished contribution to Plant Taxonomy and as a field botanist.

He is a collector of paintings and ceramics and has prepared several catalogues of the 'Peter Davis Collection of Modern Paintings and Drawings' for the Scottish National Gallery of Modern Art and for the 'Wemyss Ware' exhibition of the Scottish Arts Council. Surrounded by his superb collection of Wemyss Ware, in his retirement he will have a little more time to spend on travelling and his other cultural pursuits. We affectionately wish him well.

Reproduced with kind permission of the University Senatus, University of Edinburgh, with additions by Dr Kit Tan.

LIST OF PUBLICATIONS

Davis, P.H. 1933. The Cheddar Pink. The Sunday Observer 12 November 1933: 13.

——. 1936a. The alpine flora of Zermatt. Bull. Alp. Gard. Soc. 4: 348–353.

——. 1936b. The butterfly garden. Good Gardening September 1936: 11.

——. 1937. Unusual rock plants. Plants that are beautiful, but easy to grow. Bull. Alp. Gard. Soc. 5: 158–164.

——. 1938. A collector in Crete. Bull. Alp. Gard. Soc. 5: 385–405.

——. 1939a. Plant hunting in the Aegean. Good Gardening June 1939: 12, 38.

——. 1939b. Some plants from the eastern Mediterranean. Bull. Alp. Gard. Soc. 7: 25–64.

——. 1939c. *Omphalodes luciliae*. Gardening Illustr. 61(3142): 339.

——. 1939d. Flowers in a Cretan gorge. My Garden 16(62): 207–213.

——. 1939e. Alpine flowers of Baba Dagh. My Garden 16(63): 323–329.

——. 1940a. Plant hunters in Persia (Part 1). My Garden 20(78): 131–136.

——. 1940b. Plant hunters in Persia (Part 2). My Garden 20(79): 257–263.

——. 1940c. Plant hunters in Persia (Part 3). My Garden 20(80): 377–383.

——. 1944. Through the Antilebanon. J. Roy. Hort. Soc. 69(1): 7–13.

——. 1946a. Oncocyclus Irises in the Levant. J. Roy. Hort. Soc. 71: 93–97.

——. 1946b. Mediterranean days (Part 1). My Garden 33(151): 19–24.

——. 1946c. Mediterranean days (Part 2). My Garden 33(152): 169–172.

——. 1946d. Mediterranean days (Part 3). My Garden 33(153): 269–272.

——. 1947a. On the rocks. Bull. Alp. Gard. Soc. 15: 37–53.

——. 1947b. Plants and experiences in the eastern Mediterranean. J. Roy. Hort. Soc. 72: 13–17.

——. 1947c. Two colours. J. Roy. Hort. Soc. 72: 487.

——. 1947d. Fritillaries in the eastern Mediterranean. R.H.S. Lily Year Book 11: 141–144.

——. 1947e. Two new chasmophytes from the Antilebanon. Kew Bull. 1947 (1): 64–66.

Zohary, M., and ——. 1947. New plants from the Nearer East. Kew Bull. 1947 (1): 87–94.

Davis, P.H. 1949a. *Ranunculus asiaticus*. Gardening Illustr. 1 April 1949: 199.
——. 1949b. Additamenta ad floram Anatolicae, I. Kew Bull. 1949: 393–426.
——. 1949c. A journey in south-west Anatolia. Part I. J. Roy. Hort. Soc. 74: 104–115.
——. 1949d. A journey in south-west Anatolia. Part II. J. Roy. Hort. Soc. 74: 154–164.
——. 1950a. *Zeugandra iranica*. Hooker's Ic. Pl. 35: t. 3497.
——. 1950b. *Tracheliopsis antilibanotica*. Hooker's Ic. Pl. 35: t. 3498.
——. 1950c. Can we increase the vigour of difficult species? J. Roy. Hort. Soc. 75: 110.
——. 1951a. Cliff vegetation in the eastern Mediterranean. J. Ecol. 39: 63–93.
——. 1951b. The Taurus revisited, 1949. J. Roy. Hort. Soc. 76: 54–63.
——. 1951c. Additamenta ad floram Anatolicae, II. Kew Bull. 1951: 63–121.
——. 1951d. Three new species from the nearer East. Notes Roy. Bot. Gard. Edinb. 21: 46–51.
——. 1952a. Additamenta ad floram Anatolicae, III. Notes Roy. Bot. Gard. Edinb. 21: 61–98.
——. 1952b. Botanical expedition to Turkey. Edinb. Univ. Gazette 2 (November 1952): 7–8.
——. 1953a. Notes on the summer flora of the Aegean. Notes Roy. Bot. Gard. Edinb. 21: 101–142.
——. 1953b. The vegetation of the deserts near Cairo. J. Ecol. 41: 157–173.
——. 1954a. *Primula apennina*. Bot. Mag. 170: t. 224.
——. 1954b. Turkish Kurdistan. The Scotsman, Dec. 30 and 31.
——. 1955a. Mediterranean itineraries. Notes Roy. Bot. Gard. Edinb. 21: 271–278.
——. 1955b. Turkish itineraries. Notes Roy. Bot. Gard. Edinb. 22: 1–7.
——. 1955c. Karpathos: stronghold of traditions. The Scotsman, 11 June.
——, and J. Roberts. 1955. Materials for a Flora of Turkey: I. — Geraniaceae. Notes Roy. Bot. Gard. Edinb. 22: 9–27.
——. 1956a. A new Turkish Primula. Quart. Bull. Alp. Gard. Soc. 24(1): 73–77.
——. 1956b. Fourteen new species from Turkey. Notes Roy. Bot. Gard. Edinb. 22: 65–84.
——. 1956c. Lake Van and Turkish Kurdistan: a botanical journey. Geogr. J. 122(2): 156–165.
——. 1956d. Turkish odyssey. Geogr. J. 122(4): 495–497.
——. 1957a. The spring flora of the Turkish Riviera. J. Roy. Hort. Soc. 82: 165–173.
——. 1957b. Materials for a Flora of Turkey: *II: *Linum* Linn. Notes Roy. Bot. Gard. Edinb. 22: 135–161.
——. 1957c. New Turkish species of *Tunica*, *Velezia* and *Potentilla*. Notes Roy. Bot. Gard. Edinb. 22: 163–171.
——. 1957d. Additions to the flora of Eigg (V.C. 104). Trans. Bot. Soc. Edinb. 37: 111–113.
——. 1958a. Turkish itineraries: II. Notes Roy. Bot. Gard. Edinb. 22: 583–586.
——. 1958b. Old and new place names used in studies on the Turkish flora. Notes Roy. Bot. Gard. Edinb. 22: 587–591.
——. P.K. Chowdhuri, and M. Hossain. 1958. Materials for a Flora of Turkey: III. Ranunculaceae: I. Notes Roy. Bot. Gard. Edinb. 22: 403–425.
——. 1960a. Materials for a Flora of Turkey, IV. Ranunculaceae: II. Notes Roy. Bot. Gard. Edinb. 23: 103–161.
——. 1960b. [Determinations of Haradjian's materials of *Ceratocephalus*, p. 138, *Ranunculus*, pp. 138–143, and *Erodium*, pp. 273–276.] *In* K.H. Rechinger, Zur Flora von Syrien, Libanon und den angrenzendenden türkischen Gebieten. . Reliquiae Samuelssoniae, VI. Ark. Bot. 5(1): 1–488.
——. 1961a. Hints for hard-pressed collectors. Watsonia 4: 283–289.

——. 1961b. *Muscari macrocarpum* Sweet. J. Roy. Hort. Soc. 86: 492–494.
——, and M.S. Khan. 1961. *Aristolochia* in the Near East. Notes Roy. Bot. Gard. Edinb. 23: 515–546.
——. 1962a. Materials for a Flora of Turkey: VI. Notes Roy. Bot. Gard. Edinb. 24: 19–33.
——. 1962b. The distribution of *Epilobium* in Turkey. Notes Roy. Bot. Gard. Edinb. 24: 198–203.
——. 1962c. *Rhodothamnus sessilifolius* P.H. Davis. Hooker's Ic. Pl. 36: t. 3575.
——. 1964. Materials for a Flora of Turkey: VIII. Cruciferae I: *Isatis*. Notes Roy. Bot. Gard. Edinb. 26: 11–25.
——, and M.S. Khan. 1964. Two new Aristolochias from Turkey. Notes Roy. Bot. Gard. Edinb. 25: 67–69.
Coode, M.J.E., and P.H. Davis. 1964. A neglected Mediterranean *Fumana*. Notes Roy. Bot. Gard. Edinb. 26: 27–34.
Davis, P.H., J. Cullen, M.J.E. Coode, and I.C. Hedge. 1965. Materials for a Flora of Turkey: X. Notes Roy. Bot. Gard. Edinb. 26: 165–206.
——. 1966. A decorative Scottish pottery. Country Life (Feb. 3): 248–250.
——. 1967. Materials for a Flora of Turkey: XVI. Geraniaceae, Linaceae. Notes Roy. Bot. Gard. Edinb. 28: 35–38.
——, and D.C. Stuart. 1967. Three new species of *Muscari*. R.H.S. Lily Year Book 30: 123–125.
——. 1968a. A new *Cerastium* from the Amanus (Turkey). Notes Roy. Bot. Gard. Edinb. 28: 233–234.
——. 1968b. A new Turkish lily. Notes Roy. Bot. Gard. Edinb. 28: 235–236.
——. 1969. Materials for a Flora of Turkey: XIX. Leguminosae — Vicieae. Notes Roy. Bot. Gard. Edinb. 29: 311–320.
——, and D.M. Henderson. 1969. A revision of Turkish lilies. R.H.S. Lily Year Book 33: 212–223.
——, and B.M.G. Jones. 1969. Turkish itineraries: IV. Notes Roy. Bot. Gard. Edinb. 29: 203–209.
——. 1970. *Geranium* sect. *Tuberosa*, revision and evolutionary interpretation. Israel J. Bot. 19: 91–113.
——. 1971. Distribution patterns in Anatolia with particular reference to endemism, pp. 15–26. *In* P.H. Davis, P.C. Harper, and I.C. Hedge [eds.], Plant Life of South-west Asia. Botanical Society of Edinburgh, Edinburgh.
——, and I.C. Hedge. 1971a. Some retrospective comments, pp. 329–331. *In* P.H. Davis, P.C. Harper, and I.C. Hedge [eds.], Plant Life of South-west Asia. Botanical Society of Edinburgh, Edinburgh.
——, and ——. 1971b. Floristic links between NW Africa and SW Asia. Ann. Naturhistor. Mus. Wien 75: 43–57.
——. 1972. *Campanula myrtifolia*. Bot. Mag. n.s. 179: t. 637.
—— (compiler). 1974. Materials for a Flora of Turkey: XXX. Compositae I. Notes Roy. Bot. Gard. Edinb. 33: 207–264.
—— (compiler). 1975. Materials for a Flora of Turkey: XXXI. Compositae II. Notes Roy Bot. Gard. Edinb. 33: 409–435.
——, and R. Bhattacharjee. 1974. Two new species from Iraq. Notes Roy. Bot. Gard. Edinb. 33: 293–294.
——, and I.C. Hedge. 1975. The Flora of Turkey: past, present and future. Candollea 30: 331–351.
——. 1976a. Turkey: present state of floristic knowledge, pp. 93–113. *In* La flore du bassin méditerranéen: essai de systématique synthétique. Coll. Internat. CNRS 235.
——. 1976b. North African and other itineraries: 1961–1975. Notes Roy. Bot. Gard. Edinb. 35: 61–68.
——, and M.S. Khan. 1976. A new *Aristolochia* from SW Turkey. Notes Roy. Bot. Gard. Edinb. 35: 319–320.
—— (compiler). 1977. Materials for a Flora of Turkey: XXXV: Boraginaceae,

Gentianaceae and Solanaceae. Notes Roy. Bot. Gard. Edinb. 35: 297–314.
——. 1978a. The moving staircase: a discussion on taxonomic rank and affinity. Notes Roy. Bot. Gard. Edinb. 36: 325–340.
—— (compiler). 1978b. Materials for a Flora of Turkey: XXXVI: Primulaceae and Scrophulariaceae. Notes Roy. Bot. Gard. Edinb. 36: 1–22.
——. 1979a. Towards a supplement for the Flora of Turkey. Webbia 34: 135–141.
——. 1979b. The Flora of Turkey. Bull. Univ. Edinb. 15(9): 4–5.
——, and J.R. Edmondson. 1979. Flora of Turkey: a floristic bibliography. Notes Roy. Bot. Gard. Edinb. 37: 273–283.
——, and F. Sorger. 1979. A new *Campanula* from Anatolia. Notes Roy. Bot. Gard. Edinb. 37: 265.
——. 1980a. New species from Turkey, Arabia and Morocco. Notes Roy. Bot. Gard. Edinb. 38: 443–446.
—— (compiler). 1980b. Materials for a Flora of Turkey XXXVII: Labiatae, Plumbaginaceae, Plantaginaceae. Notes Roy. Bot. Gard. Edinb. 38: 23–64.
——, and D.C. Stuart. 1980. *Muscari* Miller, pp. 46–49. *In* T.G. Tutin, V.H. Heywood, N.A. Burges, D.M. Moore, D.H. Valentine, S.M. Walters, and D.A. Webb [eds.], Flora Europaea vol. 5: Alismataceae to Orchidaceae (Monocotyledones). Cambridge University Press, Cambridge.
Robson, N.K.B., and P.H. Davis. 1980. A new Turkish species of *Hypericum* sect. *Arthrophyllum*. Notes Roy. Bot. Gard. Edinb. 38: 103–104.
Davis, P.H., and B. Mathew. 1982. *Muscari massayanum* Grunert. The Garden (R.H.S.) 107(8): 335–336.
——, and F. Sorger. 1982. Two new Anatolian species of *Consolida*. Notes Roy. Bot. Gard. Edinb. 40: 89–91.
—— (compiler). 1983. Materials for a Flora of Turkey: XXXVIII. Araceae, Dioscoreaceae and Liliaceae. Notes Roy. Bot. Gard. Edinb. 41: 45–54.
——. 1985. Why is the flora of Turkey interesting and important? Kew Mag. 2: 357–367.
——, and Kit Tan. 1986. Two new Delphiniums from Anatolia. Notes Roy. Bot. Gard. Edinb. 43: 267–268.

FLORA OF TURKEY AND THE EAST AEGEAN ISLANDS

Davis, P.H. [ed., assisted by M.J.E. Coode and J. Cullen]. 1965. Volume 1. Edinburgh University Press, Edinburgh, 567 p.
—— [ed., assisted by M.J.E. Coode and J. Cullen]. 1967. Volume 2. Edinburgh University Press, Edinburgh. 581 p.
—— [ed., assisted by D.F. Chamberlain and V.A. Matthews]. 1970. Volume 3. Edinburgh University Press, Edinburgh. 628 p.
—— [ed., assisted by D.F. Chamberlain and V.A. Matthews]. 1972. Volume 4. Edinburgh University Press, Edinburgh. 657 p.
—— [ed., assisted by F.K. Kupicha, V.A. Matthews and B.S. Parris]. 1975. Volume 5. Edinburgh University Press, Edinburgh. 890 p.
—— [ed., assisted by J.R. Edmondson, R.R. Mill and B.S. Parris]. 1979 (as 1978). Volume 6. Edinburgh University Press, Edinburgh. 825 p.
—— [ed., assisted by J.R. Edmondson, R.R. Mill and Kit Tan]. 1982. Volume 7. Edinburgh University Press, Edinburgh. 947 p.
—— [ed., assisted by R.R. Mill and Kit Tan]. 1984. Volume 8. Edinburgh University Press, Edinburgh. 632 p.
—— [ed., assisted by R.R. Mill and Kit Tan]. 1985. Volume 9. Edinburgh University Press, Edinburgh. 724 p.
——, R.R. Mill, and Kit Tan [eds.]. 1988. Volume 10 (Supplement). Edinburgh University Press, Edinburgh. 590 p.

OTHER BOOKS

Davis, P.H., and V.H. Heywood. 1963. Principles of angiosperm taxonomy. Oliver & Boyd, Edinburgh. 556 p.

——, and J. Cullen. 1965. The identification of flowering plant families. Oliver & Boyd, Edinburgh. 122 p.

——, P.C. Harper, and I.C. Hedge [eds.]. 1971. Plant life of South-west Asia. Botanical Society of Edinburgh, Edinburgh. 335 p.

——, and J. Cullen. 1979. The identification of flowering plant families. 2nd edition. Cambridge University Press, Cambridge. 113 p.

——, and R. Rankine. 1986. Wemyss Ware: a decorative Scottish pottery. Scottish Academic Press, Edinburgh and London. 168 p.

Stearn, W.T., and P.H. Davis. 1984. Peonies of Greece: a taxonomic and historical survey of the genus *Paeonia* in Greece. Goulandris Museum, Kifissia. 126 p.

BOOK REVIEWS

Davis, P.H. 1967. Flora of Iraq, vols. I and II [Review]. New Phytol. 66: 827–828.

——. 1979. Flowering plants of the world [Review]. Notes Roy. Bot. Gard. Edinb. 37: 339–340.

——. 1980. Flora Palaestina part 3 [Review]. Notes Roy. Bot. Gard. Edinb. 38: 22.

——, and I.C. Hedge. 1980. A synopsis of the Middle East flora. Notes Roy. Bot. Gard. Edinb. 38: 486.

BIARUM DAVISII

NIKI GOULANDRIS

The Goulandris Natural History Museum, 13, Levidou Str., Kifissia, Greece

AND

KIT TAN

Royal Botanic Garden, Edinburgh, EH3 5LR, Scotland

Biarum davisii Turrill in Gard. Chron. ser. 3, 104: 437, f. 190 (1938).

Tuber ± hemispherical, c. 2–3 cm high. Leaves 3–8, hysteranthous, lamina elliptic-oblong or -obovate, 1.5–6.5 cm × 4.5–30 mm, obtuse to subacute at apex, base shortly cuneate, attenuate into petiole; lateral veins 3–9 per side; margin slighlty undulate. Scape 2–4 cm. Spathe with tube very narrowly ellipsoid, 2–3 cm, pale yellow-green or whitish, suffused purplish-brown; lamina lanceolate, strongly cucullate and curved forwards, 2.5–5 cm long, pale green with faint purplish mottling outside and purple-spotted towards apex; rim of throat dark brown to yellow. Spadix very slender, 4.5–7.5 cm, slightly shorter than spathe, female inflorescence very short (c. 1 mm at most), c. 10-flowered, separated from the male inflorescence by a sterile zone 7–10 mm long bearing no or very few sterile flowers; male inflorescence cylindrical, 8–16 mm. Appendix lacking odour, 3–5 cm × 0.5–2 mm, pale yellowish green towards base, passing gradually upwards to purplish-brown, and dark red to purplish-brown near apex; uppermost part suffused olive-green. Berries (Crete; not seen in Turkish material) 6–7 × 3.5–5 mm. Flowering October (Turkey)–November (Cretan material cult, Kew).

1. Leaves 1.5–2.5 cm × 4.5–9 mm, with 3–5 lateral veins; spathe lamina 2.5–3 cm; spadix 4.5–5.5 cm, with male inflorescence 8–10 mm (Crete) . . subsp. **davisii**
1. Leaves 4–6.5 × 15–30 mm, with 7–9 lateral veins per side; spathe lamina 4–5 cm. spadix 6.5–7.5 cm with male inflorescence 15–16 mm (W Anatolia). . . . . subsp. **marmarisensis** Boyce

Biarum davisii was first discovered in April 1938 by Peter Davis when he as only 19. It was growing amidst rocks and on semi-scree above and northeast of Sphakia in western Crete. Dried material of leaves and fruits and living tubers were presented to Kew and, later, pickled flowering material. The plants collected flowered in October and from these, Peter Davis prepared the illustrations of his new species, the very first of hundreds currently bearing his name or described by him.

The plant has been referred to by several workers as perhaps the most attractive little aroid in Europe, resembling in the words of Peter himself, "a dear little penguin!"

Mrs Niki Goulandris has also kindly sponsored the colour printing of the original illustrations produced by Peter Davis in 1938 — Kit Tan.

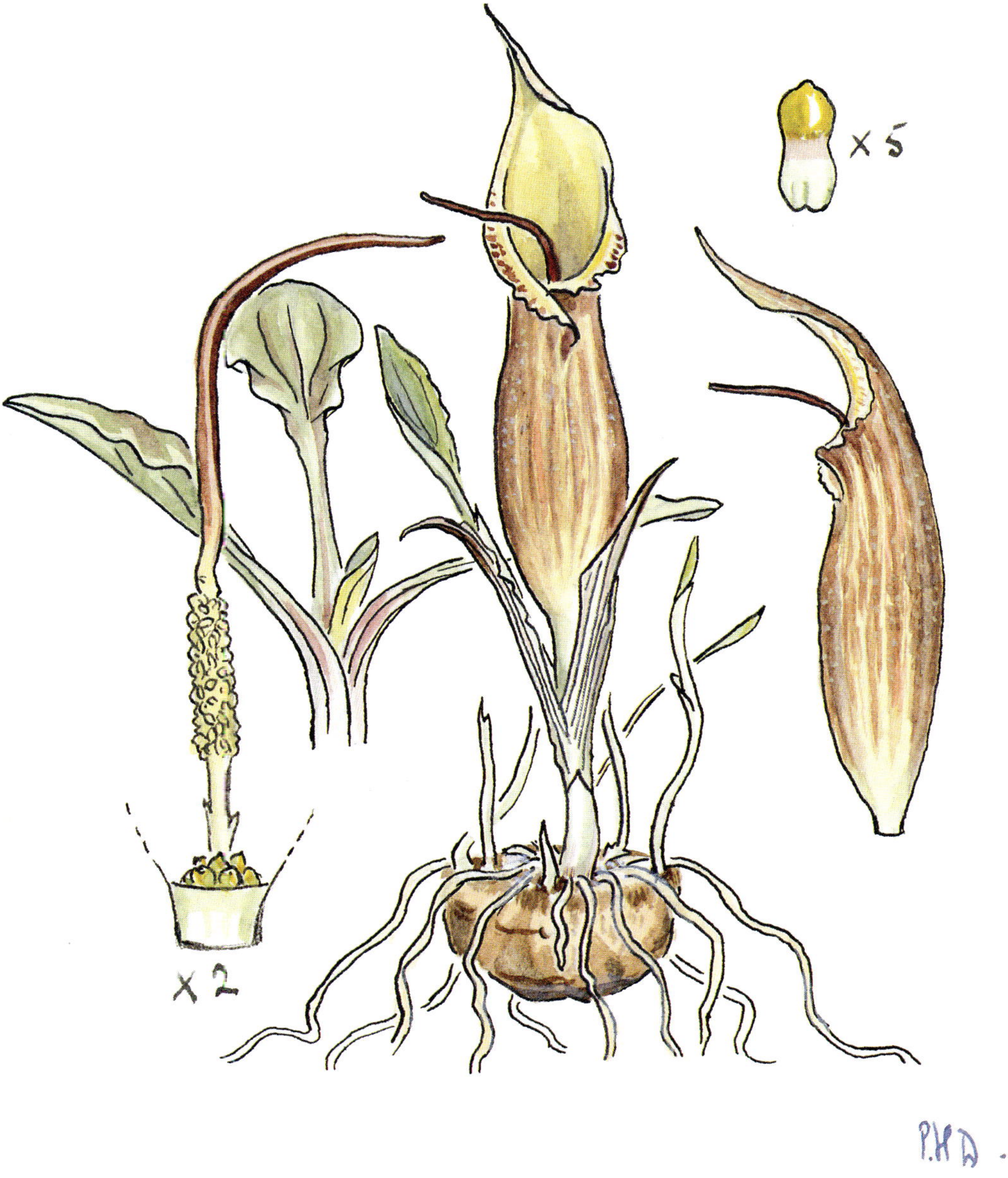

Biarum davisii: drawn by Peter Davis at the age of 19.

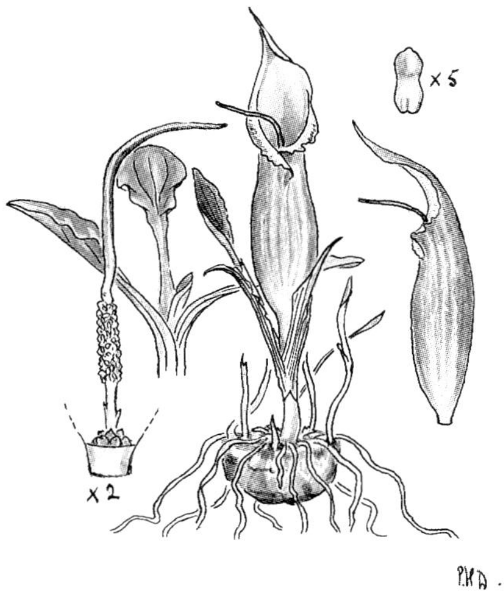
×5
×2
P.H.D.

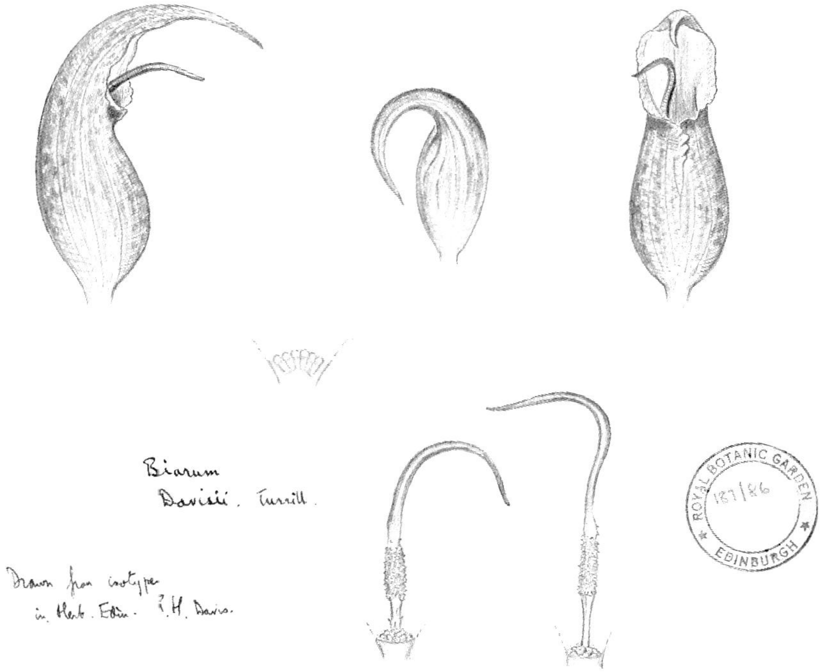
Biarum
Davisii. Turrill.
Drawn from isotype
in Herb. Edin. P.H. Davis.
ROYAL BOTANIC GARDEN
187/86
EDINBURGH

IAN CHARLESON HEDGE: BIOGRAPHICAL SKETCH; LIST OF PUBLICATIONS

ROBERT R. MILL AND KIT TAN

Royal Botanic Garden, Edinburgh, EH3 5LR, Scotland

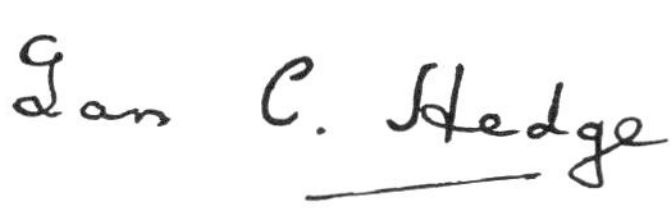

Photo: by courtesy of Ken Grant, Royal Botanic Garden, Edinburgh.

IAN CHARLESON HEDGE B. Sc.

Ian Charleson Hedge was born in Edinburgh on 18 August 1928. He joined the staff of the Royal Botanic Garden, Edinburgh in 1951 and did his National Service in the Intelligence Corps from 1951–1953. In the mid-1960s he assumed the duties of Curator of the Herbarium, a post which he held very ably until early 1988 when he stepped down to concentrate on research. From 1967–1977 he was the editor of *Notes from the Royal Botanic Garden, Edinburgh* and was also responsible for editing many other Garden publications.

His botanical research interests include S.W. Asiatic botany, and the taxonomy of Cruciferae, Labiatae and Umbelliferae; he is the widely respected and foremost authority on *Salvia* of the Old World. Since 1970 he has been a consultant for *Flora of Pakistan* and since 1984 has also served on the Editorial Board of *Flora of Thailand.* Ian is a valued and close collaborator to the *Flora of Turkey* Unit and has also contributed greatly to the *Flora of Iraq, Flora of Cyprus* and especially to K.H. Rechinger's *Flora Iranica* project. He has, in recent years, co-supervised several postgraduate students at the University of Edinburgh.

He has travelled widely in the Mediterranean and S.W. Asia, and has botanized in Turkey (several months in 1957, with Peter Davis) and Afghanistan (1962, with Per Wendelbo; 1969, with Wendelbo and Lars Ekberg). His favourite leisure activities are fishing and golf. He is also very fond of pigs.

LIST OF PUBLICATIONS

Hedge, I.C. 1957. Studies in East Mediterranean species of *Salvia.* Notes Roy. Bot. Gard. Edinb. 22: 173–188.

——. 1958a. The nomenclature of *Salvia acetabulosa.* Notes Roy. Bot. Gard. Edinb. 22: 427–428.

——. 1958b. A new species of *Oxytropis* from Kurdistan. Notes Roy. Bot. Gard. Edinb. 22: 429–430.

——. 1959a. A new *Arabis* from Turkey Notes Roy. Bot. Gard. Edinb. 23: 23–24.

——. 1959b. Studies in Eastern Mediterranean species of *Salvia*: II. Notes Roy. Bot. Gard. Edinb. 23: 47–69.

——. 1960a. Two remarkable new Salvias from Afghanistan. Notes Roy. Bot. Gard. Edinb. 23: 163–165.

——. 1960b. Studies in Eastern Mediterranean species of Salvia: III. Notes Roy. Bot. Gard. Edinb. 23: 167–171.

——. 1960c. The status of *Thylacodraba* O.E. Schulz. Notes Roy. Bot. Gard. Edinb. 23: 173–174.

——. 1960d. Notes on some cultivated species of *Salvia.* J. Roy. Hort. Soc. 85: 451–454.

——. 1961a. *Salvia glutinosa* L. and *S. nubicola* Sweet — Two previously confused species. Notes Roy. Bot. Gard. Edinb. 23: 205–208.

——. 1961b. Some remarks on the perennial species of *Ziziphora* with special reference to those in Turkey. Notes Roy. Bot. Gard. Edinb. 23: 209–221.

——. 1961c. Materials for a Flora of Turkey: V. Notes Roy. Bot. Gard. Edinb. 23: 547–558.

——. 1961d. Studies in Eastern Mediterranean species of *Salvia*: IV. Notes Roy. Bot. Gard. Edinb. 23: 559–567.

——. 1961e. Iranian plants collected by Per Wendelbo in 1959: IV. *Labiatae.* Årbok Univ. Bergen Mat.-Naturv. ser. 1961, No. 7, 19 p.

——. 1962. *Nepeta fissa* and the species allied to it. Notes Roy. Bot. Gard. Edinb. 24: 51–71.

——. 1963a. A new species of *Lamium* from Nepal. Notes Roy. Bot. Gard. Edinb. 25: 49–50.

——. 1963b. Notes on Nábělek's Labiatae from Turkey, Iran and Iraq. Notes Roy. Bot. Gard. 25: 51–54.

——. 1963c. Notes on giant asphodels in Afghanistan. J. Roy. Hort. Soc. 88: 402–406.

——. 1963d. Iranian plants collected by Per Wendelbo in 1959: *Cruciferae.* Årbok Univ. Bergen Mat.-Naturv. ser. 1963, No. 13. 10 p.

——, and R. Hendrych. 1964. Three new species of *Thesium* from Kurdistan and Samos. Preslia 36: 240–244.

——, and J.M. Lamond. 1964. A guide to the Turkish genera of Umbelliferae. Notes Roy. Bot. Gard. Edinb. 25: 171–177.

——, and P. Wendelbo. 1964a. *Acanthophyllum xanthoporphyranthum* sp. nov. from W. Afghanistan. Bot. Not. 117: 315–317.

——, and ——. 1964b. Studies in the flora of Afghanistan: I. Årbok Univ. Bergen Mat.-Naturv. ser. 1963, No. 18, 56 p.

Davis, P.H., J. Cullen, M.J.E. Coode, and I.C. Hedge. 1965. Materials for a Flora of Turkey: X. Notes Roy. Bot. Gard. Edinb. 26: 165–201.

Hedge, I.C. 1966a. Materials for a Flora of Turkey XI: A new species of *Saponaria.* Notes Roy. Bot. Gard. Edinb. 26: 353.

——. 1966b. A new *Nepeta* from W. Nepal. Notes Roy. Bot. Gard. Edinb. 26: 355–356.

——. 1966c. Studies in the Flora of Afghanistan III: An account of *Salvia.* Notes Roy. Bot. Gard. Edinb. 26: 407–425.

——. 1967a. The specimens of Paul Dietrich Giseke in the Edinburgh Herbarium. Notes Roy. Bot. Gard. Edinb. 28: 73–86.

——. 1967b. Studies in the Flora of Afghanistan: VI. Labiatae A-Lal. Notes Roy. Bot. Gard. Edinb. 27: 149–173.

——. 1968a. Studies in the Flora of Afghanistan, VIII: Labiatae: Conclusions and key to genera. Notes Roy. Bot. Gard. Edinb. 28: 163–172.

——. 1968b. *Physocardamum*: a new genus of Cruciferae from Turkey. Notes Roy. Bot. Gard. Edinb. 28: 293–296.

——, and N. Feinbrun. 1968. *Satureia thymbrifolia* Hedge et Feinbrun *sp. nov.* Israel J. Bot. 17: 213–216.

——, and J.M. Lamond. 1968. Studies in the Flora of Afghanistan, VII. Labiatae: Lam-end. Notes Roy. Bot. Gard. Edinb. 28: 89–161.

Henderson, D.M., H. Prentice, and I.C. Hedge. 1968. Pollen morphology of *Salvia* and some related genera. Grana Palynologica 8(1): 70–85.

Hedge, I.C. 1969a. Two new species of *Lamium* from Nepal. Notes Roy. Bot. Gard. Edinb. 29: 29–31.

——. 1969b. Materials for a Flora of Turkey XVII: *Onobrychis* and *Hedysarum.* Notes Roy. Bot. Gard. Edinb. 29: 33–37.

——. 1969c. *Elburzia* — a new genus of Cruciferae from Iran. Notes Roy. Bot. Gard. Edinb. 29: 181–184.

Breckle, S.-W., W. Frey, and I.C. Hedge. 1969. Botanical literature of Afghanistan. Notes Roy. Bot. Gard. Edinb. 357–371.

Hedge, I.C. 1970. Observations on the mucilage of *Salvia* fruits. Notes Roy. Bot. Gard. Edinb. 30: 79–95.

Burtt, B.L., I.C. Hedge, and P.F. Stevens. 1970. A taxonomic critique of recent numerical studies in Ericales and *Salvia.* Notes Roy. Bot. Gard. Edinb. 30: 141–158.

Hedge, I.C., and J.M. Lamond. 1970. Index of collectors in the Edinburgh Herbarium. Her Majesty's Stationery Office, Edinburgh. 147 p.

——, and P. Wendelbo. 1970. Some remarks on endemism in Afghanistan. Israel J. Bot. 19: 401–417.

Bokhari, M.H., and I.C. Hedge. 1971. Observations on the tribe *Meriandreae* of the Labiatae. Notes Roy. Bot. Gard. Edinb. 31: 53–67.

Davis, P.H., P.C. Harper, and I.C. Hedge [eds.]. 1971. Plant life of South-West Asia. Botanical Society of Edinburgh, Edinburgh. 335 p.

——, and I.C. Hedge. 1971a. Some retrospective comments, pp. 329–331. *In* P.H. Davis, P.C. Harper and I.C. Hedge [eds.], Plant life of South-West Asia. Botanical Society of Edinburgh, Edinburgh.

——, and ——. 1971b. Floristic links between NW Africa and SW Asia. Ann. Naturhistor. Mus. Wien 75: 43–57.

Hedge, I.C., J.M. Lamond and V.A. Matthews. 1971. Materials for a Flora of Turkey XXV: some new species of Umbelliferae. Notes Roy. Bot. Gard. Edinb. 31: 75–79.

——. 1972a. *Salvia* in Madagascar. Notes Roy. Bot. Gard. Edinb. 32: 1–11.

——. 1972b. The pollination mechanism of *Aeollanthus njassae*. Notes Roy. Bot. Gard. Edinb. 32: 45–48.

——, and P. Wendelbo. 1972. Studies in the Flora of Afghanistan XIII: Various new taxa and records. Notes Roy. Bot. Gard. Edinb. 31: 331–350.

——. 1973. Umbelliferae in 1672 and 1972. Notes Roy. Bot. Gard. Edinb. 32: 151–158.

——, and J.M. Lamond. 1973. A review of the tribe *Echinophoreae* (Umbelliferae). Notes Roy. Bot. Gard. Edinb. 32: 167–188.

Danin, A., and I.C. Hedge. 1973. Contributions to the flora of Sinai I: new and confused taxa. Notes Roy. Bot. Gard. Edinb. 32: 259–271.

Hedge, I.C. 1974a. A revision of *Salvia* in Africa including Madagascar and the Canary Islands. Notes Roy. Bot. Gard. Edinb. 33: 1–121.

—. 1974b. A further note on *Salvia tetrodonta*. Notes Roy. Bot. Gard. Edinb. 33: 295–299.

——. 1974c. *Echinops*, in P.H. Davis, Materials for a Flora of Turkey XXX: Compositae I. Notes Roy. Bot. Gard. Edinb. 33: 237.

Breckle, S.-W., W. Frey, and I.C. Hedge. 1975. Botanical literature of Afghanistan: Supplement I. Notes Roy. Bot. Gard. Edinb. 33: 503–521.

Davis, P.H., and I.C. Hedge. 1975. The flora of Turkey: past, present and future. Candollea 30: 263–283.

Bokhari, M.H., and I.C. Hedge. 1975. Anatomical characters in *Capparis spinosa* and its allies. Notes Roy. Bot. Gard. Edinb. 34: 231–240.

Hedge, I.C. 1976. A systematic and geographical survey of the Old World Cruciferae, pp. 1–45. *In* J. Vaughan, A.J. MacLeod and B.M.G. Jones [eds.], The biology and chemistry of the Cruciferae. Academic Press, London & New York.

——, and M.H. Bokhari. 1976. *Zhumeria* (Labiatae): anatomy, taxonomy and affinities. Iran. J. Bot. 1: 1–10.

——, and A.G. Miller. 1977. New and interesting taxa from NE Tropical Africa. Notes Roy. Bot. Gard. Edinb. 35: 179–193.

Bokhari, M.H., and I.C. Hedge. 1977. Anatomical observations on a desert group of *Salvia* species. Notes Roy. Bot. Gard. Edinb. 35: 377–389.

Hedge, I.C. 1978. *Parentucellia*, in P.H. Davis (compiler), Materials for a Flora of Turkey XXXV. Notes Roy. Bot. Gard. Edinb. 36: 10–11.

——, and J.M. Lamond. 1978. Further notes on Umbelliferae — *Echinophoreae*. Notes Roy. Bot. Gard. Edinb. 37: 61–66.

——, and P. Wendelbo. 1978. Patterns of distribution and endemism in Iran. Notes Roy. Bot. Gard. Edinb. 36: 441–464.

——. 1980. *Salvia*, in P.H. Davis (compiler), Materials for a Flora of Turkey XXXVII. Notes Roy. Bot. Gard. Edinb. 38: 47–49.

——, A. Khaer, and O. Malver. 1980. *Dipterygium* — Cruciferae or Capparaceae? Notes Roy. Bot. Gard. Edinb. 38: 247–250.

——, and J.M. Lamond. 1980a. *Nepeta*, in P.H. Davis (compiler), Materials for a Flora of Turkey XXXVII. Notes Roy. Bot. Gard. Edinb. 38: 41–46.

——, and ——. 1980b. Notes on Umbelliferae: some Asiatic *Scandiceae*. Notes Roy. Bot. Gard. Edinb. 38: 251–257.

——. 1982. Studies in the Flora of Arabia II: some new and interesting species of Labiatae. Notes Roy. Bot. Gard. Edinb. 40: 63–73.

Miller, A.G., I.C. Hedge, and R.A. King. 1982. Studies in the flora of Arabia I: a botanical bibliography of the Arabian peninsula. Notes Roy. Bot. Gard. Edinb. 40: 43–61.

Hedge, I.C. 1983. Two new monotypic genera of Labiatae. Notes Roy. Bot. Gard. Edinb. 41: 115–121.

——, and R.A. King. 1983. The Cruciferae of the Arabian peninsula: a check-list of species and a key to the genera. (Studies in the flora of Arabia IV). Arab Gulf J. Scientific Research 1: 41–66.

——. 1984. *Dauphinea brevilabra*. Flowering Plants of Africa 48: plate 1889.

——, [ed.]. 1986. Plant life of South-West Asia. Proc. Roy. Soc. Edinburgh Sect. B (Biol. Sci.) vol. 89. 322 p.

——. 1986. Labiatae of South-West Asia: diversity, distribution and endemism, pp. 23–35. *In* I.C. Hedge [ed.], Plant Life of South-West Asia. Proc. Roy. Soc. Edinburgh Sect. B (Biol. Sci.). vol. 89.

——, and D. Podlech. 1987. A first survey of *Astracantha* and *Astragalus* in the Arabian peninsula. Bot. Jahrb. Syst. 108: 259–270.

——, and Kit Tan. 1987. Two remarkable new Cruciferae from Saudi Arabia. Pl. Syst. Evol. 156: 197–206.

——, and J.M. Lamond. 1988. Edinburgh's Indian botanical connections and collections. Commemorative volume celebrating the bicentenary of the Indian Botanic Garden, Calcutta (founded 1787).

ACCOUNTS FOR FLORAS

Flora of Turkey

Hedge, I.C. 1965. Cruciferae: key to genera, pp. 249–262; Tribe *Brassiceae*, pp. 263–278; Tribe *Lepidieae* (except *Isatis*), pp. 279–287 & 307–346; Tribe *Euclidieae*, pp. 346–352; Tribe *Sisymbrieae*, pp. 479–495. *In* P.H. Davis [ed.], Flora of Turkey and the East Aegean Islands, vol. 1. Edinburgh University Press, Edinburgh.

——. 1967. *Saponaria*, pp. 138–147. *In* P.H. Davis [ed.], op. cit. vol. 2.

——. 1970. *Hedysarum*, pp. 549–560; *Onobrychis*, pp. 560–589; *Sartoria*, pp. 589–590. *In* P.H. Davis [ed.], op. cit. vol. 3.

——, and J.M. Lamond. 1972. Umbelliferae: multi-access key to Turkish genera, pp. 280–288; accounts of 39 genera, pp. 289–292, 304–333, 342–351, 367–378, 380–381, 393, 418–421, 425, 429–431, 513–518 & 537–538. *In* P.H. Davis [ed.], op. cit. vol. 4.

——. 1975. *Echinops*, pp. 609–622. *In* P.H. Davis [ed.], op. cit. vol. 5.

——. 1979. Scrophulariaceae: *Rhamphicarpa*, pp. 687–688; *Lagotis*, pp. 753–754; *Wulfenia*, pp. 754–755; *Melampyrum*, pp. 755–756; *Odontites*, pp. 763–766; *Parentucellia*, pp. 766–767; *Bellardia*, p. 768; *Pedicularis*, pp. 768–777; *Rhinanthus*, pp. 777–778; *Rhynchocorys*, pp. 778–781; *Lesquereuxia*, pp. 781–782; *Bungea*, pp. 782–783; *Lathraea*, pp. 783–784. *In* P.H. Davis [ed.], op. cit. vol. 6.

——. 1982. Key to genera of Labiatae, pp. 37–42; *Salvia*, pp. 400–461; *Dorystoechas*, pp. 461–462. *In* P.H. Davis [ed.], op. cit. vol. 7.

——, and J.M. Lamond. 1982. *Nepeta*, pp. 264–288. *In* P.H. Davis [ed.], op. cit. vol. 7.

——, and F. Yaltirik. 1982. *Quercus*, pp. 659–683. *In* P.H. Davis [ed.], op. cit. vol. 7.

Flora Iranica

Hedge, I.C. 1968. Cruciferae: key to genera, pp. 1–31; Tribe *Brassiceae*, pp. 33–61; Tribe *Lepidieae*, pp. 63–122; Tribe *Euclidieae*, pp. 123–131; Tribe *Drabeae*, pp. 175–191; Tribe *Arabideae*, pp. 193–218; Tribe *Sisymbrieae*, pp. 309–342. *In* K.H. Rechinger [ed.], Flora Iranica Lieferung 57. Akademische Druck- u. Verlagsanstalt, Graz.
——, and J.M. Lamond. 1970. Capparidaceae. *In* K.H. Rechinger [ed.], op. Lieferung 68. 32 p.
——, and J.M. Lamond. 1975a. Aizoaceae. *In* Rechinger, [ed.], op. cit. Lieferung 113. 8 p.
——, and ——. 1975b. Molluginaceae. *In* K.H. Rechinger [ed.], op. cit. Lieferung 114. 8 p.
——. 1982. Labiatae: key to genera, pp. 3–9; *Salvia*, pp. 403–476. *In* K.H. Rechinger [ed.], op. cit. Lieferung 150.
——, and K.H. Rechinger. 1982. *Sulaimania* (gen. nov.), pp. 345–347. *In* K.H. Rechniger [ed.], op. cit. Lieferung 150.
——, and J.M. Lamond. 1987. Umbelliferae: general remarks, pp. 5–6; characters, pp. 6–7; a synoptic guide to characters and genera, pp. 8–13; dichotomous key to artificial groups, pp. 14–36; Tribus *Echinophoreae*, pp. 65–82; Tribus *Scandicineae*, pp. 83–110; *Neocryptodiscus*, pp. 207–209; *Stewartiella*, pp. 213–214; *Hyalolaena*, pp. 233–236; *Berula*, pp. 305–306; *Diplotaenia*, pp. 372–374; *Azilia*, gen. nov., pp. 386–387; *Korovinia*, pp. 436–437; *Oedibasis*, pp. 437–438; *Hohenackeria*, p. 526; *Pinacantha*, pp. 527–528; economic Umbelliferae of the Flora Iranica area, pp. 528–531. *In* K.H. Rechinger [ed.], op. cit. Lieferung 162.
——, ——, and K.H. Rechinger. 1987. *Trachyspermum*, pp. 336–343. *In* K.H. Rechinger [ed.], op. cit. Lieferung 162.
——, and K.H. Rechinger. 1987. *Sium*, pp. 303–304; *Cortia*, pp. 362–364. *In* K.H. Rechinger [ed.], op. cit. Lieferung 162.

Flora Europaea

Hedge, I.C. 1972. *Salvia*, pp. 188–192. *In* T.G. Tutin, V.H. Heywood, N.A. Burges, D.M. Moore, D.H. Valentine, S.M. Walters and D.A. Webb [eds.], Flora Europaea vol. 3. Cambridge University Press, Cambridge.

Flora of Iraq

Hedge, I.C., and J.M. Lamond. 1981. Crucifera: Tribe *Brassiceae*, pp. 845–885; Tribe *Lepidieae*, pp. 885–932; Tribe *Euclidieae*, pp. 933–943; Tribe *Arabideae*, pp. 997–1014; Tribe *Sisymbrieae*, pp. 1067–1085. *In* C.C. Townsend and E. Guest [eds.], Flora of Iraq, vol. 4 part 2 (Bignoniaceae to Resedaceae). Ministry of Agriculture and Agararian Reform, Baghdad.

Flora of Cyprus

Hedge, I.C. 1985. *Salvia*, pp. 1287–1299. *In* R.D. Meikle, Flora of Cyprus vol. 2. Bentham-Moxon Trust, Kew.

BOOK REVIEWS

Hedge, I.C. 1959a. A guide to British woody plants. Notes Roy. Bot. Gard. Edinb. 23: 46.
——. 1959b. A new German colour chart. Notes Roy. Bot. Gard. Edinb. 23: 86.
——. 1966a. A family key. Notes Roy. Bot. Gard. Edinb. 26: 405.
——. 1966b. Two of the earliest works on the British Flora. Notes Roy. Bot. Gard. Edinb. 27: 100.

——. 1967. The Introduction to the Flora of Iraq. Notes Roy. Bot. Gard. Edinb. 27: 210.
——. 1968a. American wild flowers. Notes Roy. Bot. Gard. Edinb. 28: 162.
——. 1968b. A Linnaean antagonist. Notes Roy. Bot. Gard. Edinb. 28: 258.
——. 1928c. Floristic work in South-West Asia: collective reviews. Notes Roy. Bot. Gard. Edinb. 28: 329–338.
——. 1969a. European taxonomic literature 1966. Notes Roy. Bot. Gard. Edinb. 29: 14.
——. 1969b. Nomenclatural terms. Notes Roy. Bot. Gard. Edinb. 29: 128.
——. 1969c. Plant taxonomy for Everyman. Notes Roy. Bot. Gard. Edinb. 29: 156.
——. 1969d. Flora Europaea. Notes Roy. Bot. Gard. Edinb. 29: 283. [Review of vol. 2].
——. 1975a. Flora of Egypt. Notes Roy. Bot. Gard. Edinb. 34: 106.
——. 1975b. Quercus in Britain. Notes Roy. Bot. Gard. Edinb. 34: 112.
——. 1975c. Flora of Iraq. Notes Roy. Bot. Gard. Edinb. 34: 127–128. [Review of vol. 3].
——. 1976. Botanical publications in Iran. Notes Roy. Bot. Gard. Edinb. 35: 234.
——. 1977. South Australian botany. Notes Roy. Bot. Gard. Edinb. 35: 390.
——. 1978a. Plants for Mediterranean gardens. Notes Roy. Bot. Gard. Edinb. 36: 104.
——. 1978b. The Flora of Cyprus vol. 1. Notes Roy. Bot. Gard. Edinb. 37: 73–74.
——. 1979a. Chorology of European plants. Notes Roy. Bot. Gard. Edinb. 37: 326.
——. 1979b. The tangle of tamarisks. Notes Roy. Bot. Gard. Edinb. 37: 474.
——, L.A. Lauener, and H.K. Tan. 1979. The flora of China. Notes Roy. Bot. Gard. Edinb. 37: 467–468.
——. 1980. What's in a name? Notes Roy. Bot. Gard. Edinb. 38: 286.
Davis, P.H., and I.C. Hedge. 1980. A synopsis of the Middle East flora. Notes Roy. Bot. Gard. Edinb. 38: 486.
Hedge, I.C. 1983a. Hegi re-vitalised. Notes Roy. Bot. Gard. Edinb. 40: 459–460.
——. 1983b. A history of plant collecting in Pakistan. Notes Roy. Bot. Gard. Edinb. 40: 519.
——. 1984a. Plants and medicine in North Africa. Notes Roy. Bot. Gard. Edinb. 41: 472.
——. 1984b. Plant life of Sinai. Notes Roy. Bot. Gard. Edinb. 41: 490.
——. 1985a. A memorial to Prof. Michael Zohary. Notes Roy. Bot. Gard. Edinb. 42: 351–352.
——. 1985b. A world monograph of *Trifolium*. Notes Roy. Bot. Gard. Edinb. 42: 599–600.
——. 1986. Flora Iranica Papilionaceae. Notes Roy. Bot. Gard. Edinb. 43: 337–339.

1. *FILAGO PETRO-IANII*, A NEW SPECIES FROM MALLORCA

MANFRED DITTRICH

Conservatoire et Jardin botaniques, Genève, Switzerland

AND

JUAN RITA

Dep. de Biologia i C. de la S. Univers. de les Illes Balears, 07071 Palma de Mallorca, España

ABSTRACT

Filago petro-ianii Rita & Dittrich (Compositae-Inuleae), a new species from Mallorca, is described and illustrated. Floral and carpological investigations show that the new species is to be placed in subgenus *Evax* (Gaertn.) Wagenitz. Accounts of the habitat, phenology and ecology of the new species are also provided, together with a distribution map.

Key words: Compositae-Inuleae, *Filago*, *F. petro-ianii*, Mallorca (Balearic Islands), plant sociology, phenology, ecology.

INTRODUCTION

About three years ago, one of us (J.R.) collected in southern Mallorca (Islas Baleares) an *Evax*, locally known as 'Marina de Llucmajor'. All attempts to determine this plant by means of the existing floras always led to *Evax rotundata* Moris (= *Filago tyrrhenica* Chrtek & Holub), a species of Sardinia, Corsica and N.W. Italy. This species, however, differs from our plant in a number of characters, of which the most important are listed in Table 1 (see also Fig. 1). Comparative studies of other "*Evax*" species in the Geneva herbarium caused us to conclude that our plant has to be a new species of the *Filago-Evax* group.

The morphological investigations and description (by M.D.) are supplemented by a detailed presentation of the habitat and geographical distribution as well as phenological and ecological observations (by J.R.). We would like to thank Prof. Dr. K.H. Rechinger (Vienna) for the Latin diagnosis.

TAXONOMY

Filago petro-ianii Rita & Dittrich, **sp. nov.** Figs. 1–3.

Herba annua acaulis capitulo singulo terminali vel saepius ramis usque ad 8 brevibus appressis. Folia ramealia alterna oblonga leviter araneosa. Capitula in ramis aut singula terminalia aut plura depresso-congesta. Folia capitula rosulato-fulcrantia usque ad 6 mm longa, apice subrhombica, supra capitula conniventia, in apiculum contracta, margine revoluta, subtus tomentosa. Capitula globosa, transsectione rotundata, ad 4 mm diametro; involucrum 2–3-seriale; involucri phylla ± lanceolata, acuminata, hyalina, apice brunn-

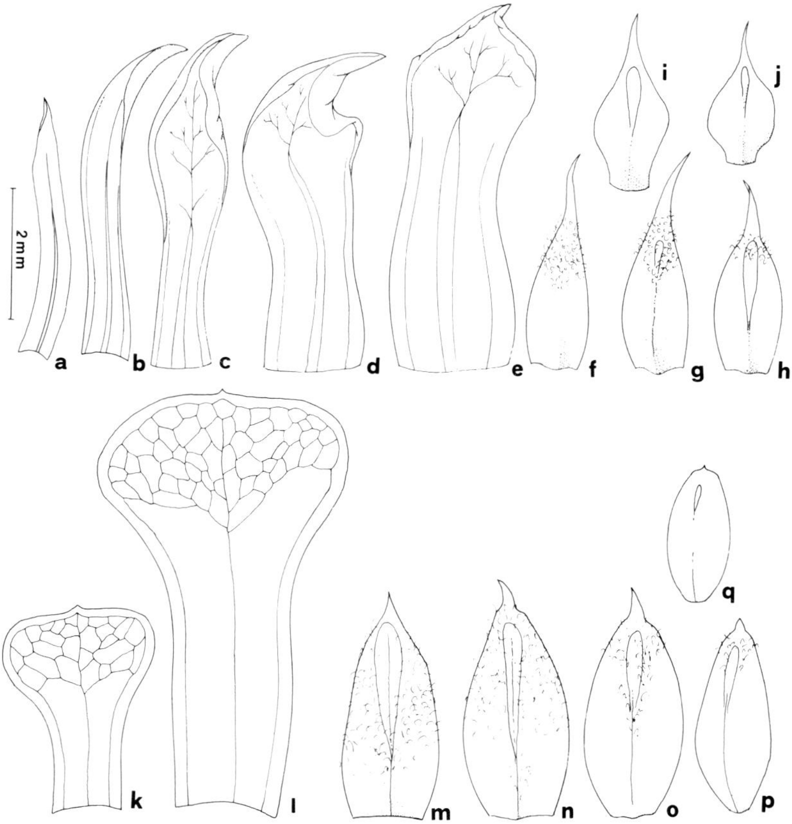

Figure 1. Leaf sequences of *Filago petro-ianii* (a–j) and *F. tyrrhenica* (k–q). — a–b. Cauline leaves. — c–e. Cauline leaves surrounding capitulum. — f–h. Phyllaries. — i–j. Paleae. — k–l. Cauline leaves (l, one of the upper cauline leaves near the capitulum). m–o. Phyllaries. — p–q. Paleae. The indumentum has been omitted to show the different types of venation. — (With 2 mm scale.)

escentia, crispule pilosa. Bracteae sequentes involucri phyllis similes, florem singulum functione femineum fulcrantes. Receptaculum conico-cylindricum, superne glabrum, flores 5–8 functione masculos proferens. Achaenia brunnescentia, 0.7 mm longa, obovoidea, latere ventrali subapplanata, apice rotundata non coronata. Hilum annulato-torosum, collo obliquo basin versus directo. Pericarpii superficies laevis, pilis numerosis papillosis clavatis obsitae. Pappus nullus.

Typus: Mallorca, Llucmajor, Acantilados del Pas de Sa Senyora (DD 85), 80 m, 12 iv 1987, Rita (holotypus G; isotypus UMP).

Specimens seen (G: Conservatoire botanique de Genève; UMP: University of Palma de Mallorca, Laboratory of Botany; grid references, e.g. DD 88, refer to Fig. 4): SPAIN. MALLORCA (ISLAS BALEARES). Palma de Mallorca, Es Puntiró, cerca de la carretera de Sineu (DD 88), 40 m, iv 1986, Bibiloni (G, UMP); *ibid.*, 3

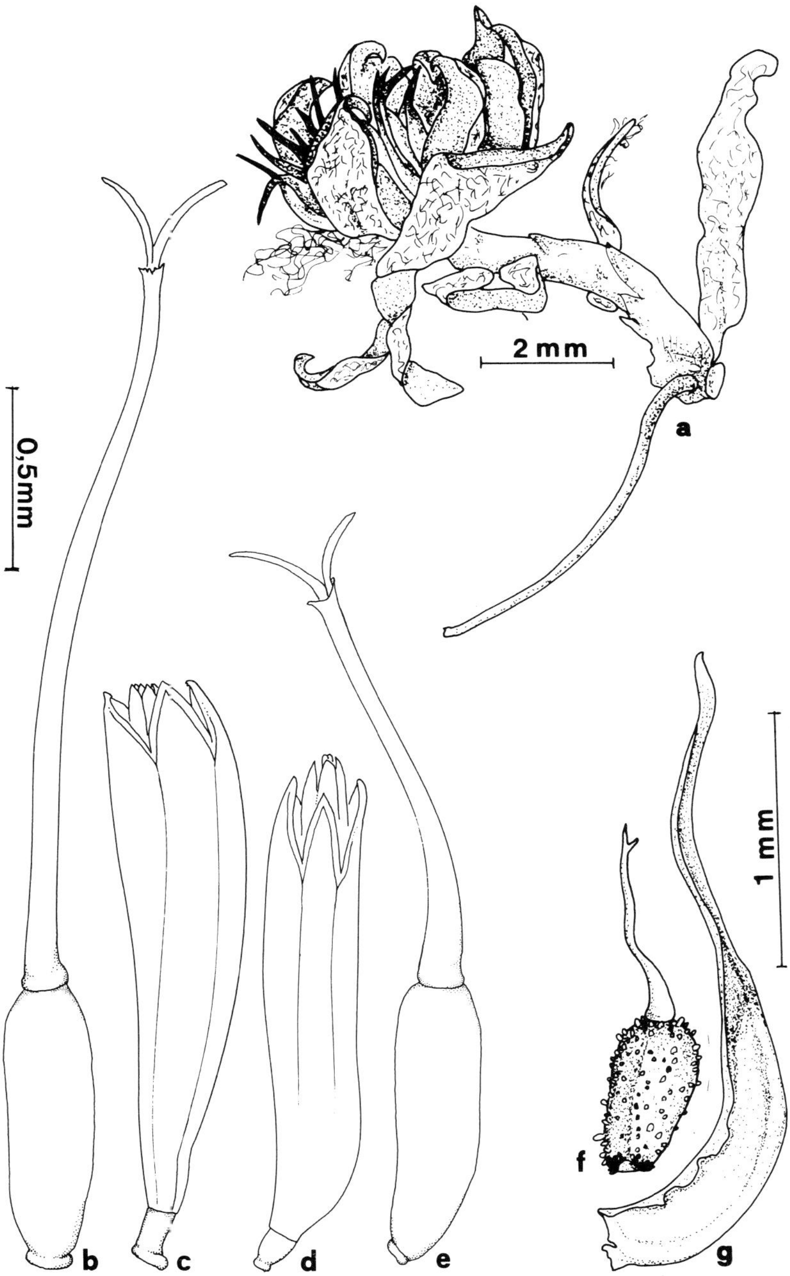

Figure 2. a. *Filago petro-ianii*, axillary branch. — b–c. *F. tyrrhenica*. — b. Female flower. — c. Functionally male flower. — d–g. *F. petro-ianii*. d. Male flower. — e. Female flower. — f. Young achene showing papillae. — g. Palea (side view). — (a, with 2 mm scale; b–e, with 0.5 mm scale; f–g, with 1 mm scale.)

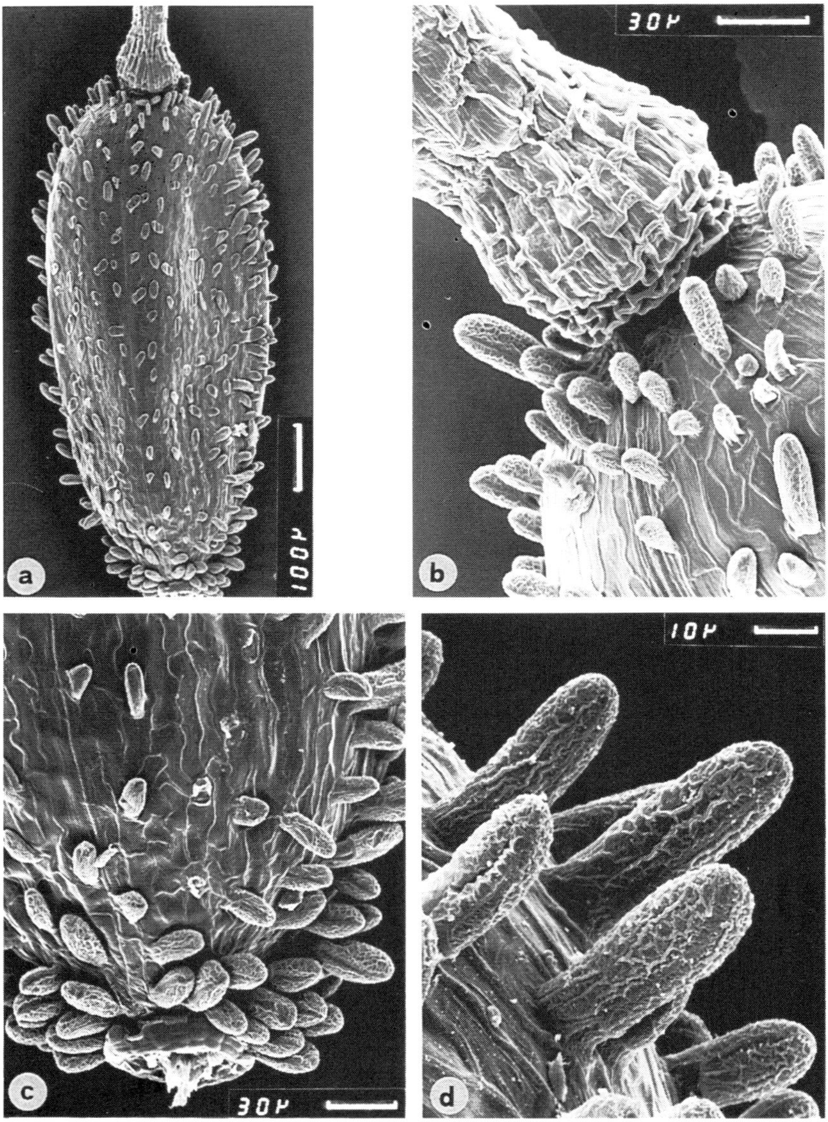

Figure 3. Achene details of *Filago petro-ianii.* — a. Ventral surface of achene. — b. Apical part of fruit with corolla base. — c. Detachment area with its central vessel bundle. — d. Papillae of pericarp. — (Each with scale.)

v 1987, Rita (G, UMP); Llucmajor, S'Aguila de'n Cuart (DD 85), 100 m, 16 iii 1984, Rita (G, UMP); Llucmajor, Pinares próximos a Cala Pí (DD 85), 40 m, 12 iii 1985, Rita (UMP); *ibid.*, 26 iv 1985, Rita & Bibiloni (G, UMP); Llucmajor, garrigas proximas a la urbanización Sa Torre (DD 76), 120 m, 6 iv 1985, Rita (G,

UMP); Llucmajor, Cap Blanc, 80 m, 19 iv 1985, Rita (UMP); Llucmajor, garrigas próximas a la urbanización Bahia Azul (DD 76), 140 m, 26 iv 1985, Rita (UMP); Llucmajor, Acantilados del Pas de Sa Senyora (DD 85), 80 m, 28 iii 1986, Rita (G, UMP — type); *ibid.*, 12 iv 1987, Rita (G, UMP).

MORPHOLOGICAL DESCRIPTION

Procumbent annuals (to biennials?), ± stemless. *Stem* very short (c. 5 mm), unbranched or with up to 8 procumbent axillary branches c. 10 mm long. *Cauline leaves* linear, 4 × 0.5 mm, sparsely arachnoid. *Branches* each terminating in a simple capitulum or a group of capitula, surrounded by a rosette-like arrangement of cauline leaves which are c. 6 mm long and enlarged into a rhombic apical part with an acuminate tip, revolute margins and densely tomentose inner surface (Figs. 1a–e, 2a).

Capitula (2–)4 mm diam., ± spherical, circular in cross-section, often forming flat cushion-shaped clusters together with 3–5 peripherally inserted smaller capitula.

Involucre 2–3-merous; phyllaries numerous, inserted at base of ± cylindrical receptacle, 3–3.5 × 0.8–0.9 mm, hyaline and glabrous except for a greenish oblong spot on midrib on the scarcely tomentose apical part (Fig. 1f–h). Inner paleae 2.3–2.9 × 0.9–1.2 mm, similar to phyllaries but differing in being ovate to broadly ovate, concave (Fig. 1i–j); each palea bears one of the numerous filiform female flowers (Fig. 2e–f).

Female flowers up to 100 in larger (central) heads; corolla yellowish-white, 0.95–1 mm long, 0.15 mm wide at the slightly club-shaped base and c. 0.05 mm wide in the middle of the filiform tube (Fig. 2e); apical margin irregularly laciniate; tube and limb indistinguishable due to the regular tubular structure of the flowers and the absence of staminodes. *Style* completely unstructured, smooth, circular in cross-section, towards apex with two 0.3 mm long divaricately divergent branches bearing two marginal stigmatic rows on their inside surface. *Ovary* similar in shape to mature achene (Fig. 2e–f), flattened ventrally, densely covered with club-shaped papillae.

Functional male flowers 5–8, borne at apex of receptacle where paleae were absent, tubular, terminating in 4 corolla lobes 0.1 mm wide at base and 0.2 mm long; longitudinal veins 4, diverging at each sinus and following margins to lobe tips (Fig. 2d). *Stamens* c. 1 mm long; filaments 0.4–0.5 mm long, band-like, glabrous; anthers 0.6 mm incl. apical and basal appendages each c. 0.15 mm long; apical appendages hyaline, oblong-triangular, basal ones sagittate, equalling apical ones. *Ovary* rudimentary, only 0.1 mm long, with completely unstructured, glabrous surface (Fig. 2d). The style of the 'hermaphrodite' flowers clearly serves only to brush out the pollen grains; whereas the base and scape of the style are totally smooth, the elongated epidermis cells of its tip provide a rough surface for pollen transport.

Achenes 0.7–0.75 × 0.3 mm, obovate, slightly flattened on ventral side; apical region evenly rounded, bearing a small central cone-shaped obtuse nectary, surrounded by circular scar of corolla tube (Fig. 3a–b). Base of achene with a basal-laterally orientated detachment area having the shape of a collar-like volva

(Fig. 3c); from here 2 median longitudinal vessels, a ventral and a dorsal one, go up to the apical plate. Whole surface of pericarp smooth, covered with numerous club-shaped papillae (Fig. 3d). Pappus not developed.

RELATIONSHIPS

Systematic position within Filago

As Wagenitz (1969) has pointed out, every detailed study of a taxon of the *Filago-Evax* group inevitably leads to the question of the delimitation of the genera. Once again it must be emphasised that the two genera, on the basis of their type specimens, can be easily delimited by several characters. Unfortunately, however, there are some species in which these features are combined in such a way that a clear assignment to one or other of the two genera becomes doubtful. Furthermore, there are characters, for example the presence or absence of a pappus, which are considered important in the delimitation of both genera (cf. Boissier 1875), yet which in some species can vary even on the same plant. We, therefore, adopt the opinion of Wagenitz (1969), who conbines the three groups of species of *Filago sensu stricto* (= *Gifola* Cass.), *Evax* Gaertn. *sensu stricto* and *Oglifa* Cass. (incl. *Logfia* Cass.) in the genus *Filago* L. *sensu lato* and who regards these three groups as subgenera.

The following characters of *Filago petro-ianii* show that it has to be regarded as a taxon of *F.* subgenus *Evax* (Gaertn.) Wagenitz:

a. The central capitulum and the smaller marginal ones form hemispherical cushion-shaped synflorescences.
b. Capitula circular in cross-section.
c. Capitula with cylindrical/obtusely conical receptacle.
d. Phyllaries numerous, also appressed at fruiting time.
e. All flowers without a pappus.
f. Inner flowers sterile and functionally male.
g. Achenes flattened from adaxial (ventral) side.

Relationships of F. petro-ianii *with other Mediterranean species of* Filago

On the basis of achene indumentum, one can distinguish two groups of species in *Filago sensu lato*, one with glabrous or hairy achenes (but not papillate), the other with ± densely papillose achenes. The species of the first group, to which primarily *F. carpetana* (Lange) Chrtek & Holub and *F. asterisciflora* (Lam.) Sweet belong, differ from *F. petro-ianii* not only in carpological characters but also in general appearance; both of the above-named species have a clearly developed stem, the synflorescences attain at least twice the diameter (10–40 mm) and are surrounded by spathulate-lanceolate rosette-like leaves, tomentose on both sides, which overtop the clusters of capitula.

Of similar importance are the differences between the new species and those belonging to the group with papillose achenes. To this group belongs the highly variable *F. pygmaea* L., which occurs throughout the Mediterranean region; its large synflorescences (up to 35 mm diam., comprising numerous heads ± equal in size, are also enveloped in a rosette-like arrangement of leaves; the latter often

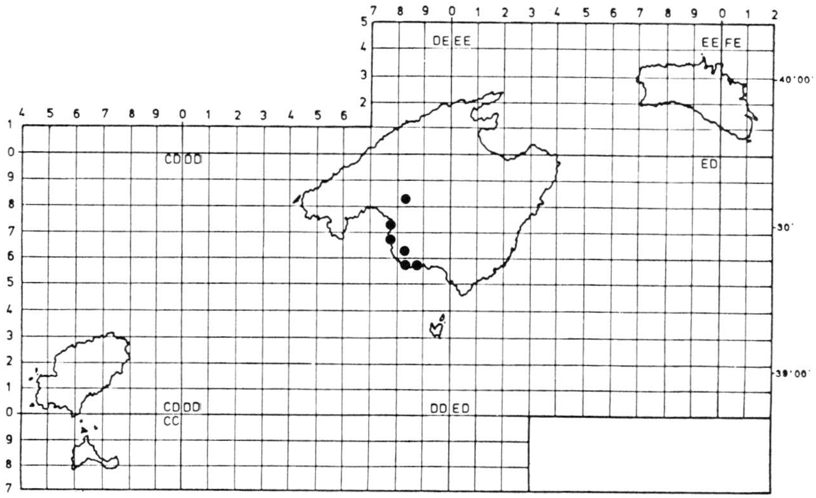

Figure 4. Distribution of *Filago petro-ianii.*

overtop the capitula and may even be several times longer (up to 30 mm), as in *F. contracta* (Boiss.) Chrtek & Holub.

Evax nevadensis Boiss., closely related to *Filago perpusilla* (Boiss. & Heldr.) Chrtek & Holub, can be allocated to the same group. However, *E. nevadensis* has an often branched stem up to 6 cm tall, which terminates in a synflorescence up to 12 mm wide, whose obovate 'rosette leaves' are whitish-pilose on both sides and whose achenes are twice as long as those of *F. petro-ianii* (c. 1.5 mm).

Judging by its appearance, *F. tyrrhenica* Chrtek & Holub (= *Evax rotundata* Moris 1841, non *Filago rotundata* Moench 1794) is most closely related to our new species. Apart from size aspects there are, however, a large number of differential characters, of which the most important are summarised in Table 1.

GEOGRAPHICAL DISTRIBUTION AND HABITAT

So far, the new species has only been found in the south of Mallorca, primarily between Arenal and Punta Plana (for distribution map see Fig. 4). This region has a semi-arid climate with annual precipitation below 400 mm, an arid period of at least 5 months, a mean annual temperature of 17° C (10° C in January, 26° C in August) and a mean relative humidity of 70 % (Llorens & Guijarro 1982).

The geological substratum consists of reef limestone and Pleistocene solidified dunes. The soil derived from these materials is seldom more than 30 cm deep. Often there is a high content of loam and clay, which is accompanied by a ± intensive rubefaction process. Such alkaline soils with a pH of 7.5–8.5 can show a great variety of identifiable carbonates.

The vegetation of these habitats consists of a 2–4 m tall macchie which is to be classified as Al. *Oleo-Ceratonion* Br.-Bl. (Bolòs & Molinier 1958). The macchie is not homogeneous in structure; on the contrary, it is intermixed with open areas,

Table 1. Differences between *Filago petro-ianii* and *F. tyrrhenica*.

Character	*F. petro-ianii*	*F. tyrrhenica*
Cauline leaves	oblong, sparsely arachnoid	spathulate, with densely woolly indumentum
Cauline leaves surrounding capitulum	oblong, ending in a rhombic, acuminate terminal part with revolute margins, tomentose beneath	apical part broadly spathulate, not acuminate at tip, without revolute margins, pilose on both sides
Phyllaries and paleae	acuminate tip 0.6 – 1 mm long	acuminate tip at most 0.5 mm long
Diameter of capitulum	max. 4 mm	max. 3 mm
Receptacle	± cylindrical	cone-shaped
Length of filiform corolla	1 mm	2 mm
Achene surface	pericarp smooth with clavate papillae	pericarp with longitudinal grooves, smooth, rarely with small papillae

which are occupied by different therophytes and geophytes; the latter can be classified in several associations of the Cl. *Thero-Brachypodieta* Br.–Bl. (marked with an asterisk in the list that follows). The most frequently associated species of *Filago petro-ianii* are the following: *Gymnostyles stolonifera* (Brot.) Tutin, *Crepis pusilla* (Sommier) Merxm., *Filago pyramidata* L., **Sagina apetala* Ard., **Desmazeria rigida* (L.) Tutin, **Galium murale* (L.) All., **Euphorbia exigua* L., **Campanula erinus* L., **Astericus aquaticus* (L.) Less., **Neatostema apulum* (L.) I.M. Johnston, *Lophochloa cristata* (L.) Hyl., *Anagallis arvensis* L. *s.l.*, *Blackstonia perfoliata* (L.) Hudson, *Centaurium pulchellum* (Swartz) Druce, *Teucrium polium* L., *Collema tenax* (Sw.) Ach. and *Cladonia foliacea* (Hudson) Willd.

PHENOLOGY AND ECOLOGY

Filago petro-ianii is a therophyte with a relatively short life cycle. The achenes do not germinate until the first half of January; in this fact, *F. petro-ianii* differs from other therophytes, whose seedlings can already be seen after the first autumn rains in October/November. Our plant flowers in March, the achenes are mature in April and after that the plants dry up.

F. petro-ianii apparently prefers open areas, with plant societies whose degree of coverage lies below 40 %. It is a pioneer plant that colonises little-used paths or abandoned agricultural areas, which are exposed to the influence of the sea. Despite its pioneer character it avoids markedly ruderal areas, which are preferred for example by *F. pygmaea* L. *F. petro-ianii* belongs to a group of about 70 species whose classification in the plant sociological system is extremely difficult. Due to the existence of numerous representatives of the Cl. *Thero-Brachypodietea* Br.-Bl. (marked by an asterisk in the list above) together with our *Filago*, this plant association is the most likely to which our plant can be assigned.

LITERATURE CITED

Boissier, P.E. 1875. Flora Orientalis, vol. 3. Geneva & Basel. 1033 p.

Bolòs, O. de, and R. Molinier. 1958. Recherches phytosociologiques dans l'ile de Majorque. Coll. Bot. (Barcelona) 5: 699–865.

Llorens, L., and J.A. Guijarro. 1982. La vegetatión de los taludes de la costa sur de la isla Mallorca. Estudio de Salsolo-Lycietum (ass. nov.). Biol. Bot. Médit. 9: 95–106.

Wagenitz, G. 1969. Abgrenzung und Gliederung der Gattung *Filago* L. s.l. (Compositae-Inuleae). Willdenowia 5: 395–444.

2. TAXONOMIC NOTES ON SOME SPECIES OF THE GENUS *CICHORIUM* (COMPOSITAE-*LACTUCEAE*)

G. WAGENITZ AND UTE BEDARFF

Lehrstuhl für Pflanzensystematik, Systematisch-Geobotanisches Institut der Universität Göttingen, D-3400 Göttingen, Untere Karspüle 2, West Germany

ABSTRACT

Cichorium intybus L. is a widespread and variable species. Several variants which have been described are of no taxonomic value. However in the Mediterranean region (and extending into SW Asia) there occurs a recognizable subspecies: subsp. *glabratum* (C. Presl) Wagenitz & Bedarff. This has some superficial similarity to the wild form of *C. endivia* L. and has been the source of some confusion. Revision of herbarium material shows that the wild form of *C. endivia* (subsp. *divaricatum* (Schousb.) P.D. Sell) is almost entirely restricted to the coastal regions of the Mediterranean. The differences between the two taxa mentioned are tabulated. *Cichorium calvum* Schultz-Bip. ex Aschers. is a recently recorded adventive plant found in fields of *Trifolium alexandrinum* L. and *T. resupinatum* L. in Central Europe. Few collections from other countries exist but it may have been confused with *C. endivia* and should be looked for.

Key words: *Cichorium*, endive, chicory, Compositae (Asteraceae), Mediterranean region, adventive plants.

INTRODUCTION

Cichorium L. (c. 7 species) is one of rather few genera in the Compositae which can be used as vegetables (endive and chicory; Schultze-Motel 1986). Moreover the roots of *Cichorium intybus* L., when dried and roasted, have been used as a substitute for coffee. The same species is widespread in Europe and parts of Asia and occurs as an adventive in many countries. In spite of this, taxonomists have paid very little attention to the genus. The first motive for this study was the occurrence of a rare, long-forgotten species as an adventive plant in Central Europe: *Cichorium calvum* Schultz-Bip. ex Aschers. During the investigation the problem of the delimitation of *C. intybus* and *C. pumilum* Jacq. became prominent and led to a subspecific division of *C. intybus*.

TAXONOMY

1. A Mediterranean Southwest-Asian subspecies of Cichorium intybus *L.*

Cichorium intybus, the Common Chicory*, is very variable in height, type of branching, indumentum and form of leaves. There are also some variants in flower

*The Common Chicory, also known in America as 'blue dandelion' or 'blue sailors' (van Wijk 1911) and in Germany as 'Wegwarte' (Hegi 1928) is associated in folklore with medicinal or magical properties (Marzell 1938–1941).

colour. This variation has led to the listing and description of some varieties or forms, e.g. by Hegi (1928) and Soó (1970). Most of these differences seem to have no taxonomic significance and have accordingly not been taken into consideration in modern floras. The investigation of material from different parts of the species range has shown that a certain combination of characters can be used to define a subspecies which prevails in the Mediterranean and South-West Asia.

Cichorium intybus L. subsp. **glabratum** (C. Presl) Wagenitz & Bedarff, **stat. nov.** Fig. 1

Basionym: *Cichorium glabratum* C. Presl, Fl. Sicul. 1: 32 (1826).
Syn.: *C. glaucum* Hoffmanns. & Link, Fl. Portug. 2: 178 (1820) e descr.
C. intybus L. var. *glabratum* (C. Presl) Gren. in Gren. & Godron, Fl. France 2: 286 (1850).
C. byzantinum Clementi in Mem. Accad. Sci. Torino, Cl. Sci. Fis. Mat. Ser. 2. 16: 285 (1855) e descr.
C. intybus L. var. *eglandulosa* Freyn & Sint. in Österr. Bot. Z. 44: 259 (1894) (Type: Sintenis 4919, B!)

No type material of *C. glabratum* Presl has been available to us, but the description together with the provenance of the material leave no doubt that it belongs here. Soó (1970) mentions a "subsp. *illyricum* (Borb. 1877 pro sp.) Jáv. 1925". We have not been able to trace the name *Cichorium illyricum* Borb.; it is not listed in 'Index Kewensis'. The name is also not in the list of combinations of subspecies validly published in Jávorka's Mágyar Flóra (Holub 1984). Probably the name "subsp. *spicatum* Ricci (1964)" could be added to the synonymy, but this name is not validly published as there is no indication of the type. According to Ricci's photograph it represents a rare form with an elongated spike-like arrangement of the capitula.

Subsp. *glabratum* can be distinguished from subsp. *intybus* by the following characters: stem usually less than 60 cm tall, often branched from near the base; stem leaves rapidly decreasing in size from the base of the plant upwards; primary capitula often on long and slightly thickened peduncles; involucre glabrous or with few simple or glandular hairs.

Distribution (Fig. 1): Specimens have been seen from the following countries or regions: Morocco, Spain, southern France, Italy, Yugoslavia, Romania, Albania, Bulgaria, Greece, USSR (Caucasus, Transcaspia), Turkey, Iraq, Iran, Afghanistan, Pakistan (in the Karakorum up to c. 2600 m!).

Not much attention has been paid to this taxon although it was first described by Presl in 1826. Grenier (in Grenier & Godron, op. cit. 286–287, 1850) is the first to mention explicitly the distribution: "bords de la Méditerranée" and the possibility of a confusion with *C. pumilum* (Syn: *C. divaricatum*): "Nous avons souvent reçu, des bords de la Méditerranée, la var. ß. de l'espèce précedente [i.e. var. *glabratum*] sous le nom de *C. divaricatum*". The confusion in the herbaria exists even today. It has been expressed by Meikle (1985: 989; without mention of var. *glabratum*): "The differences between *C. intybus* and *C. endivia* have been so poorly indicated that the two, distinct, species have been repeatedly and hopelessly confused. No record for either species, from the eastern Mediterranean region, should be accepted without examination of supporting specimens." We will return to this problem after the treatment of '*C. pumilum*'. The 'Flore de France' by Rouy

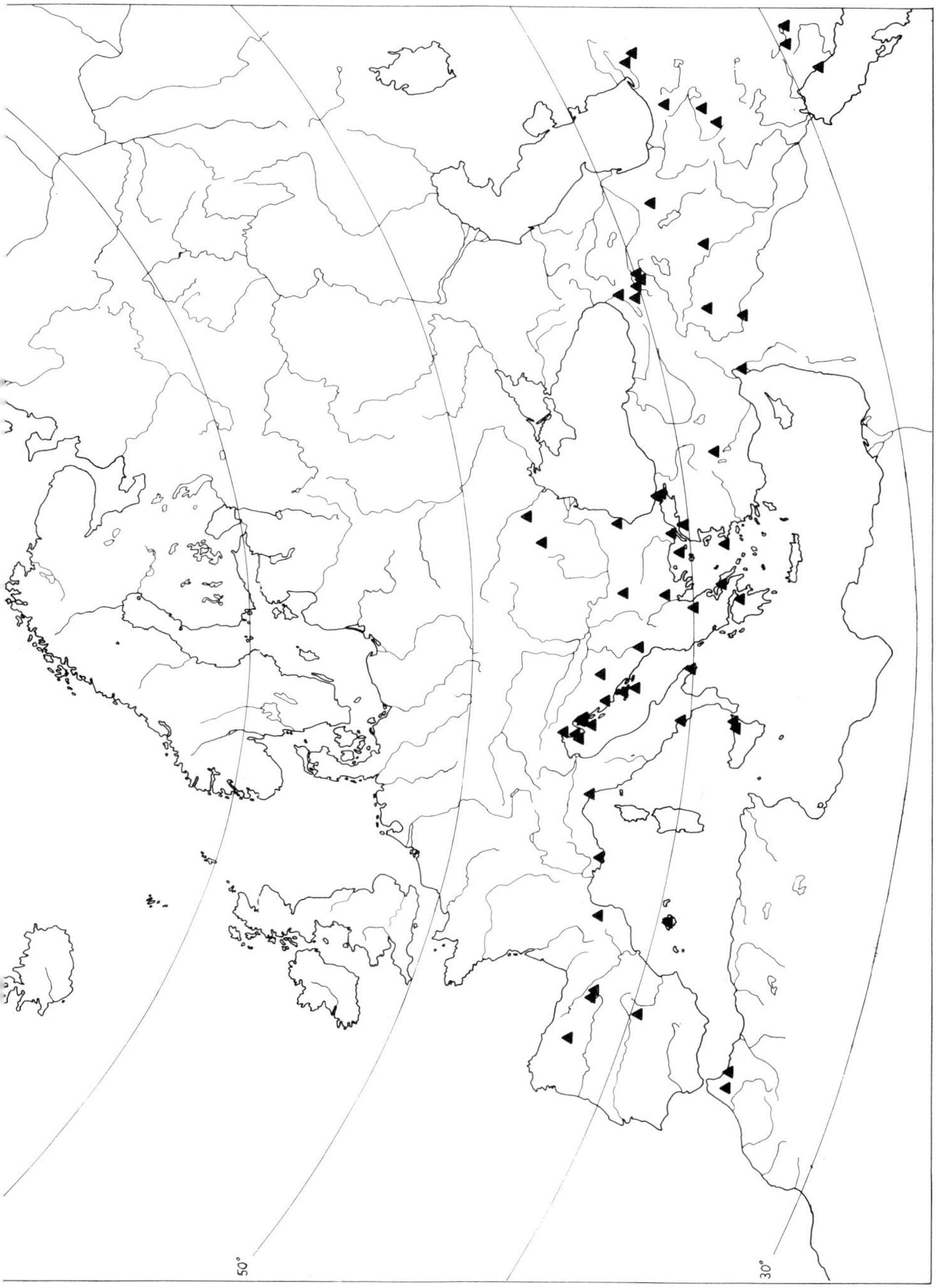

Figure 1. Distribution of *Cichorium intybus* subsp. *glabratum* in southern Europe and southwest Asia based on herbarium specimens examined. It is also known from Iran, Afghanistan and Pakistan outside the map area.

(1908) also treats var. *glabratum*, but there is an error — it is not this variety which is regarded as typical for the Mediterranean region, but another variety, var. *leucophaeum* Grenier, with hirsute involucres! Other floras of Mediterranean countries list var. *glabratum* but do not give details of the distribution (e.g. Hayek 1928–31, Rechinger 1943). As is often the case with varieties, the taxon has been neglected in recent treatments; it is not mentioned in 'Flora Europaea' and is treated as a plain synonym of *C. intybus* in 'Flora of Turkey' (Matthews in Davis 1975). Lack (1977) includes *C. glabratum* in the synonymy of *C. pumilum*, but we see no reason for this.

Typical plants of subsp. *glabratum* are quite distinct from subsp. *intybus*, but transitional specimens occur. North of the Alps scattered populations with glabrous involucres occur which, however, are different in habit from subsp. *glabratum*. On the other hand one can rarely find plants with distinctly hairy involucres in the Mediterranean (e.g. in Northern Italy), which according to the type of branching and the leaves are better placed in subsp. *glabratum*.

2. *Taxonomic position, nomenclature and distribution of* 'Cichorium pumilum *Jacq.*'

Cichorium pumilum Jacq. is without doubt very closely allied to *C. endivia* L. and represents the wild plant which gave rise to the cultigens known as endive. The first botanist to unite these taxa under *C. endivia* (the earlier name) seems to have been Visiani (1847); this has been confirmed by Boissier (1875), A. de Candolle (1884:12) and several others. Recently Meikle (1985) took up a combination made by Webb in 1847–1848, reducing the wild type to a form. We prefer to treat it as a subspecies; the different cultigens may thus be united as different varieties of subsp. *endivia*.

There is another tradition going back to A.P. de Candolle (1838), which unites *C. pumilum* (Syn: *C. divaricatum*) with *C. intybus*. Recently this has been done by Quézel & Santa (1963). This view is apparently influenced by the fact that the Mediterranean plants of *C. intybus*, which we place in subsp. *glabratum*, show some superficial similarities with *C. pumilum*. This is especially noteworthy in smaller plants (Fig. 3).

The nomenclature of this taxon at the level of the subspecies has been controversial. Sell in 1975 used subsp. *divaricatum*, but later Jeffrey (1979) has claimed that subsp. *pumilum* is the older name, being published by Ball (1878). But in Ball's paper only the following note can be found under *C. pumilum*: "Hoc mihi videtur prioris [i.e. *C. intybus*] subspecies facta annua vel interdum biennis". Ball did not publish a new combination. The pertinent synonymy is as follows:

Cichorium endivia L. subsp. **divaricatum** (Schousb.) P.D. Sell in J. Linn. Soc. Bot. 71:240 (1975). Figs. 2 & 3

Basionym: *C. divaricatum* Schousb., Iagtagg. Vextr. Marokko, 197 (1800).
Typus: [Morocco] Marokko, 1800, Schousboe (C!)

Syn.: *C. pumilum* Jacq., Observ. Bot. 4: 3 (1771).
C. minimum Portenschl., Enum. Pl. Dalmat. 16 (1824).
C. intybus L. var. *divaricatum* (Schousb.) DC., Prodr. 7: 84 (1838).
C. divaricatum Schousb. var. *pumilum* (Jacq.) Guss., Fl. Sic. Syn. 2: 427 (1843).
C. endivia L. var. *pumilum* (Jacq.) Vis., Fl. Dalm. 2: 97 (1847).

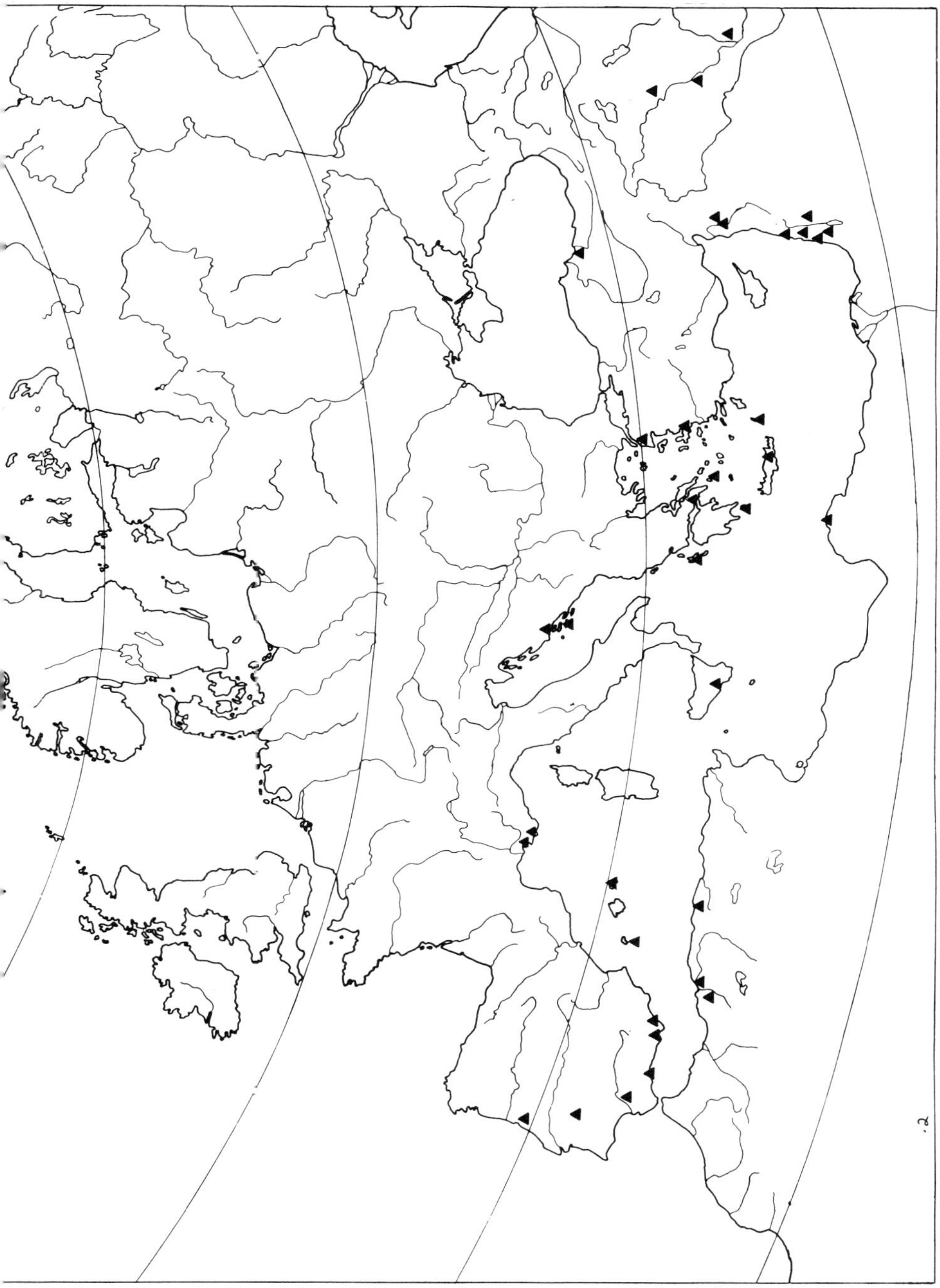

Figure 2. Distribution of *Cichorium endivia* subsp. *divaricatum* in southern Europe and southwest Asia based on herbarium specimens examined. It also occurs in the Canary Islands and south Iran outside the map area.

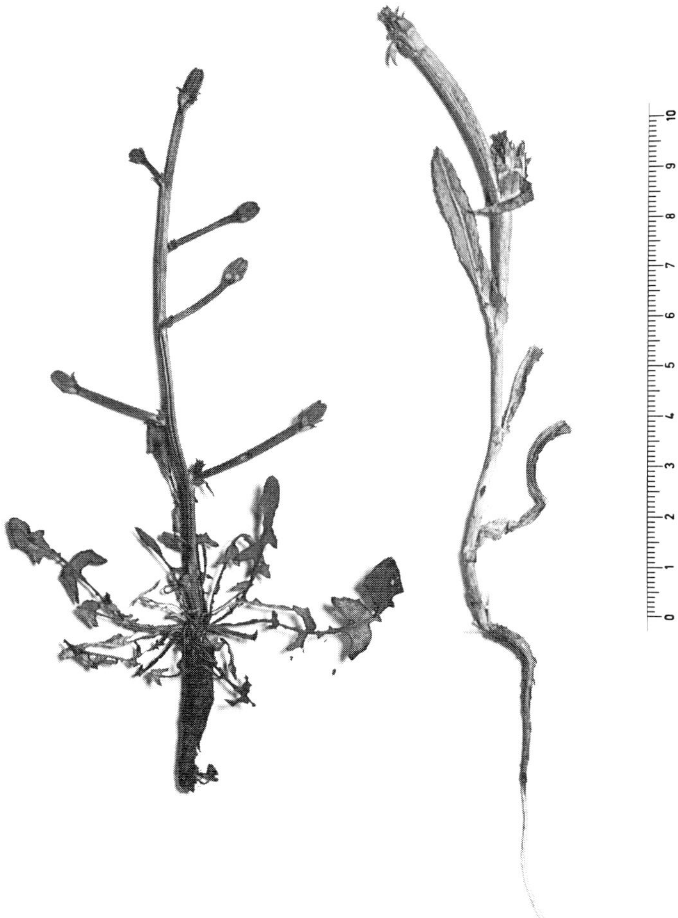

Figure 3. Left: *Cichorium intybus* subsp. *glabratum* (Yugoslavia: Dalmatia, Velika Solina, 11 vi 1927, leg. *E. Korb*, W). — Right: *C. endivia* subsp. *divaricatum* (Greece: Cephalonia, *Bornmüller 1926: 962*, B).

C. endivia L. f. *divaricatum* (Schousb.) Webb in Webb & Berth., Phytogr. Canar. 2: 391 (1847–1848).
C. intybus L. var. *pumilum* (Jacq.) Fiori, Fl. Anal. d'Italia 3: 386 (1904).
C. intybus L. subsp. *divaricatum* (Schousb.) Bonnier & Layens, Tabl. Synopt. Fl. France 183 (1894). — We have seen only one of the numerous reprints with the title "Flore portative de la France".
C. endivia L. subsp. *pumilum* (Jacq.) Hegi, Ill. Fl. Mitteleur. 6 (2): 998 (1928).

Distribution (Fig. 2): Specimens have been seen from the following countries or regions: Canary Islands, Algeria, Libya, Portugal, Spain, southern France, Italy, Yugoslavia, Greece, Turkey, Lebanon, Syria, Israel, Jordan, Iraq, Iran (only in the south). A single collection from Pakistan is from the campus of the university at Peshawar and probably adventive.

Table 1. Differences between *C. intybus* subsp. *glabratum* and *C. endivia* subsp. *divaricatum*.

C. intybus subsp. *glabratum*	*C. endivia* subsp. *divaricatum*
perennial (rarely annual ?)	annual
taproot stout	root slender
basal leaves numerous, in a distinct rosette	basal leaves few, not in a distinct rosette
leaves usually runcinate-pinnatifid	leaves often undivided, dentate
bracts in middle and upper part of stem small, lanceolate, nearly entire, exauriculate	bracts in middle and upper part of stem large, triangular, dentate with semi-amplexicaul base
phyllaries of involucre glabrous or nearly so	involucral phyllaries long-ciliate, hairs with firm basal part and slender tip
outer phyllaries only 1/3–1/2 × length of inner ones	outer phyllaries usually longer
pappus-scales 0.2–0.3 mm long	pappus-scales 0.4–0.6 mm long

Fig. 3 shows small specimens of both taxa which had been determined as *C. pumilum*, but show quite convincingly the different habit. The following table gives the differences between the two taxa.

C. endivia subsp. *divaricatum* (Fig. 2) is almost entirely restricted to an area close to the Mediterranean coasts. Only in Iraq and Iran (Prov. Kerman) does it penetrate into the interior. In the 'Flora Palaestina' area this is the only indigenous taxon of *Cichorium* according to the material we have seen and the Flora (Feinbrun-Dothan 1978). The same seems to be the case for Tunisia (Bonnet & Barratte 1896) and Libya (Jeffrey 1979).

3. Cichorium calvum *Schultz-Bip. ex Aschers.* Fig. 4.

This species was described by Ascherson in Schweinfurth (1867), the name having been previously given by Schultz-Bipontinus (first mentioned without description in the 'Index Seminum Horto Botanico Berolinensi' for 1858). The seeds came from Ethiopia and were probably collected by Schimper. Fig. 4 shows a specimen cultivated by Schultz-Bipontinus himself; very probably this came from the same seed-collection which was used in the Berlin Botanical Garden. Ascherson did not mention any locality; apparently he did not know the species from wild material. The treatment of the family in the 'Flora of Tropical Africa' (Oliver & Hiern 1877) refers to the publication by Schweinfurth, but no native locality is indicated. It came as a surprise when Merxmüller in 1957 published the rediscovery of this lost species from achenes mixed with the seeds of *Trifolium alexandrinum* imported from Egypt.

Since 1977 the species has been found in stands of *Trifolium resupinatum* (T.r.) and *T. alexandrinum* (T.a.) in several different regions of Germany and Austria:

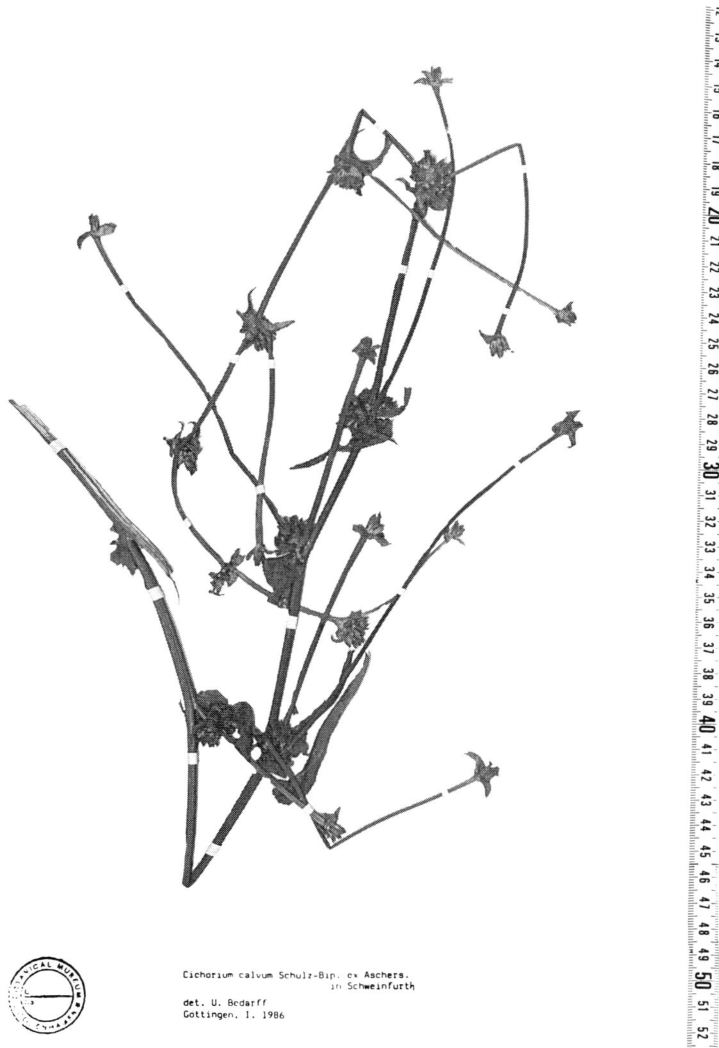

Figure 4. Cichorium calvum. Specimen cultivated by Schultz-Bipontinus (C).

Rhineland (T.r.)	Gerstberger 1977*
Hassia (T.a.)	Schnedler 1977*/1978
Austria: Burgenland (T.a.)	Traxler 1979
Franconia (T.a., T.r.)	Walter 1979
Westphalia (T.r.)	Lienenbecker & Raabe 1979
Mecklenburg (T.r.)	Funk & Henker 1981, Danckert et al. 1984
Westphalia (T.r.)	Raabe 1984

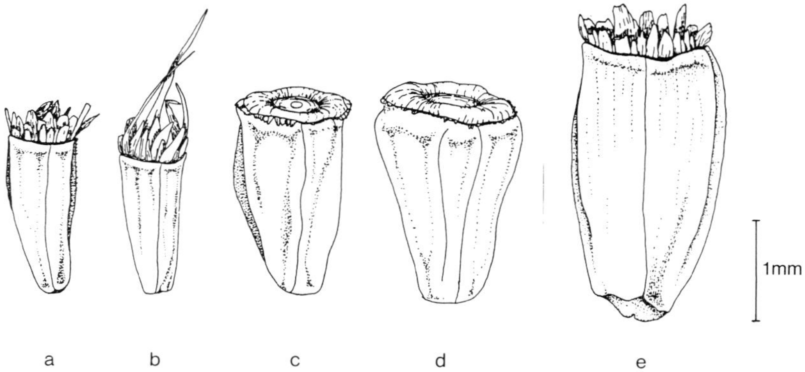

Figure 5. Cichorium achenes. — a & b. *C. endivia* subsp. *divaricatum*. — c & d. *C. calvum*. — e. *C. intybus* subsp. *intybus*.

In the references marked with an asterisk (*) the plants had been identified as *C. endivia*.

In Hassia the species is widespread as shown by the material collected by W. Ludwig and given to us for revision. It is very probable that the species is present with these two species of *Trifolium* almost everywhere in fields. We are not aware of any records in the literature from Europe outside Germany and Austria. This may be due to the fact that the journals from Central Europe in which the floristic records have been published are not very well known outside this area! The only recent flora where *C. calvum* is mentioned is 'Flora Iranica' (Lack 1977), in which it is reported from Afghanistan and Pakistan.

Revision of herbarium material of the genus has brought to light several findings of *C. calvum* outside Central Europe:

FRANCE: Port Juvénal, Touchy (collected before 1900, M).

PORTUGAL: Cascais, S. Domingos de Rana, Rebelva, Cova de Raposa, 25 m, 14 vi 1964, *Vasconcellos 64143* (M).

EGYPT: Kairo, ad Gise, ad agrorum margines, 21 iv 1908, *Bornmüller 10812* (B); Zeitoun, au N du Caire, lieux incultes, *Burdet 918* (G).

PAKISTAN: Peshawar University Campus, 5 vi 1971, *Salim* (M); Karachi, University Campus, 15 iv 1974, *Kazmi 5421* (M).

Cichorium calvum is very close to *C. endivia* and has often been confused with the latter species. The main difference is in the achenes. In *C. calvum* these are shorter (c. 2 mm long) and relatively thicker and are epappose or with some minute scales (c. 0.1 mm) only visible under at least × 10 magnification (see Fig. 5; the pappus is somewhat more distinct in the two collections cited from Egypt). In *C. endivia* the achenes are c. 3 mm long and the pappus 0.4–0.6 mm. In both species some of the capitula are situated at the end of long peduncles, but these are slender in *C. calvum*, while in *C. endivia* they are usually distinctly thickened towards the apex. The large triangular and acuminate upper leaves also seem to be characteristic for *C. calvum*.

ACKNOWLEDGMENTS

Material has been examined from the following herbaria (abbreviations according to Index Herbariorum, ed. 7): B, C, G, GOET, HBG, M, W, WU. We are grateful to the directors and curators of these herbaria. Dr. W. Ludwig (Marburg) sent us a considerable number of his own collections of *C. calvum* from Hassia. We thank Dr Kit Tan for improving the English text and the careful editing of the manuscript.

LITERATURE CITED

Ball, J. 1877–1878. Spicilegium Florae Maroccanae. J. Linn. Soc., Bot. 16: 281–772.

Boissier, P.E. 1875. Flora Orientalis. Vol. 3. H. Georg, Genève & Basel. 1033 p.

Bonnet, E., and G. Barratte. 1896. Catalogue raisonné des plantes vasculaires de la Tunisie. Imprimerie Nationale, Paris. 519 p.

Candolle, A.P. de. 1838. Prodromus systematis naturalis regni vegetabili. Pars 7, sect. prior. Treuttel & Würtz, Paris. 330 p.

Candolle, A. de. 1884. Der Ursprung der Culturpflanzen. Leipzig. F.A. Brockhaus, Leipzig. x + 590 p. [German translation by E. Goeze, of A. de Candolle's 'Origine des Plantes cultivées'].

Danckert, H., W. Kintzel, and H. Koch. 1984. III. Nachtrag zur Flora des Kreises Lübz. Arch. Freunde Naturgesch. Mecklenburg 23: 9–27.

Feinbrun-Dothan, N. 1978. Flora Palaestina. Part 3. Israel Academy of Sciences and Humanities, Jerusalem. 481 p.

Funk, B., and H. Henker. 1981. Die Segetalflora von Perserklee-Ansaaten. Bot. Rundbr. Bez. Neubrandenburg 12: 32–36.

Gerstberger, P. 1977. *Silene conoidea* L. im der Begleitflora von *Trifolium resupinatum* L. im Rheinland. Götting. Florist. Rundbr. 10: 91–94.

Hayek, A. von. 1928–1931. Prodromus Florae peninsulae Balcanicae. Vol. 2. Repert. Spec. Nov. Regni Veg. Beih. 30(2). Verlag des Repertoriums, Dahlem, Berlin. 1152 p.

Hegi, G. 1928. Illustrierte Flora von Mitteleuropa. Vol. 6(2). J.F. Lehmanns Verlag, München. 838 p.

Holub, J. 1984. Subspecies names in Jávorka, Mágyar Flóra (1924–1925). Preslia 56: 303–318.

Jeffrey, C. 1979. A check-list of the Libyan flora 3. Compositae. Candollea 34: 307–332.

Lack, H.W. 1977. *Cichorium*, pp. 6–9. *In* K.H. Rechinger and H.W. Lack, Flora Iranica Lfg. 122 (Compositae II). Akademische Druck-u. Verlagsanstalt, Graz. 352 p.

Lienenbecker, H., and U. Raabe. 1979. Adventivarten in *Trifolium resupinatum* - Äckern in Ostwestfalen. Götting. Florist. Rundbr. 13: 22–23.

Marzell, H. 1938–41. Wegwarte, pp. 227–235. *In* H. Bachtold-Staubli [ed.}, Handwörterbuch des deutschen Aberglaubens. Band 9.

Matthews, V.A. 1975. *Cichorium*, pp. 626–629. *In* P.H. Davis [ed.], Flora of Turkey and the East Aegean Islands 5. Edinburgh University Press. Edinburgh.

Meikle, R.D. 1985. Flora of Cyprus 2: 988–991. The Bentham-Moxon Trust, Kew.

Merxmüller, H. 1957. Compositen-Studien VI. *Cichorium calvum* Schultz-Bip. ex Aschers. Mitt. Bot. Staatssamml. München 2: 317–318.

Oliver, D., and P. Hiern. 1877. Compositae, *Cichorium*, pp. 446–447. *In* P. Oliver [ed.], Flora of Tropical Africa 3. L. Reeve & Co., London.

Quézel, P., and S. Santa. 1963. Nouvelle Flore de l'Algérie et des régions désertiques méridionales. Tome II. Éditions du Centre National de la Recherche Scientifique, Paris. 571–1170 p.

Raabe, U. 1984. *Anthemis hyalina* DC. – auch in westfälischen Kleefeldern. Natur und Heimat, Münster 44: 19–22.

Rechinger, K.H. 1943. Flora aegaea. Denkschr. Akad. Wiss. Wien, Math.-Naturw. Kl. 105(1): i–xx, 1–924.

Ricci, I. 1964. *Cichorium intybus* L. subsp. *spicatum* subsp. nova. Studio cito-tassonomico. Ann. Bot. (Roma) 28: 219–223.

Rouy, G. 1908. Flore de France. Vol. 10. G. Rouy, Asnières & Les Fils d'Émile Deyrolle, Paris. 404 p.

Schnedler, W. 1977. Pflanzen von denen in der mitteleuropäischen Literatur selten oder gar keine Abbildungen zu finden sind. IV. Göttinger Florist. Rundbr. 11: 88–97.

——. 1978. Pflanzen von denen in der mitteleuropäischen Literatur selten oder gar keine Abbildungen zu finden sind. VIII. Göttinger Florist. Rundbr. 12: 97–102.

Schultze-Motel, J. [ed.]. 1986. Rudolf Mansfelds Verzeichnis landwirtschaftlicher und gärtnerischer Kulturpflanzen (ohne Zierpflanzen). Band 3. Springer-Verlag, Berlin. 1998 p.

Schweinfurth, G. 1867. Beitrag zur Flora Aethiopiens. Erste Abtheilung. Druck und Verlag von G. Reimer, Berlin. 311 p.

Soó, R. 1970. Synopsis systematico-geobotanica florae vegetationisque Hungariae IV. Akadémiai Kiadó, Budapest. 614 p.

Traxler, G. 1979. *Cichorium calvum* Schultz-Bip. ex Asch., Kahlfrüchtige Wegwarte (Zichorie) – neu für Österreich. Burgenländ. Heimatbl. 41: 91–92.

Visiani, R. de. 1847. Flora dalmatica. Vol. 2. F. Hofmeister, Lipsiae. x + 268 p.

Walter, E. 1979. Bemerkenswerte Adventivarten in fränkischen Kleeäckern. Ber. Naturf. Ges. Bamberg 54: 69–117.

Wijk, H.G.G. van. 1911. A dictionary of plant names. Vol. 1. The Dutch Society of Sciences, Haarlem. 1444 p.

3. A NEW ANNUAL *BUPLEURUM* FROM SOUTHERN PELOPONNISOS

SVEN SNOGERUP

The Botanical Museum, University of Lund, Ö. Vallgatan 18, S-22361 Lund, Sweden.

ABSTRACT

Bupleurum greuteri Snogerup is described from Greece (the province of Lakonia and the island of Kithira). The characters separating it from related species are listed and illustrated.

Key words: *Bupleurum* (Apiaceae), annual, taxonomy, Greece.

Bupleurum greuteri Snogerup, **sp. nov.** Fig. 1–6. Map: Fig. 7.
Typus: *Runemark & Snogerup 20833* (Holotype LD).

Bupleurum annuum, 6–20 cm altum, sparse pseudodichotome ramosum. Folia inferiora juvenilia petiolata, lanceolata, superiora anguste linearia. Umbellae pedunculis 1–3 cm longis suffultae, radii 3–5, pedunculo multo breviores. Involucri phylla 5 vel raro 4, 10–30 mm, radiis longiora. Umbellulae plerumque 4–6-florae, pedicelli brevi. Involucelli phylla 5 vel 4, 5–12 mm longa, lanceolata ad anguste ovata, cuspidata, centro herbacea, margine late scariosa, serrulata. Petala 1.2 mm, apice angusta, cucullata, lobulo inflexo angusto obtuso instructa. Antherae 0.3–0.4 mm, stylopodium 0.8–1 mm latum, styli 0.3–0.4 mm longi. Fructus (immaturus) 1.2 mm vel longior, prismaticus, juga filiformia.

Annual, 6–20 cm, small individuals only sparingly branched apically, the main stem terminating in a large central umbel, umbels 1–10, large specimens often pseudodichotomously branched from the base, with up to c. 25 umbels, last-formed umbels often depauperate. Stem in basal part 0.5–1.5 mm thick, with 3–4 ridges or basally terete, smooth. Cotyledons only seen when withered, probably linear. All leaves entire, serrulate with teeth 0.03 mm or shorter. First leaves soon withering, long-petiolate, their lamina 5–15 mm, broadly to narrowly lanceolate with a prominent midrib, marginal veins and inconspicuous reticulate venation. Upper leaves successively narrower, the largest often 5–8 cm long and 1.5–3 mm broad, basal part, corresponding to a petiole, with parallel veins and a scarious margin, distal part with venation as in the lamina of basal leaves. Umbels borne on peduncles 10–25(–30) mm long, rays usually 3–5, unequal, in central umbels the longest usually 5–10 mm, all much shorter than peduncle. Bracts 5 or rarely 4, 10–25(–30) mm, sometimes unequal, usually enclosing the umbellules or almost so, similar to bractlets in shape and texture though often narrower at base. Bractlets usually 5, rarely 4, 5–12 × 1–4 mm, lanceolate to narrowly ovate, acuminate-cuspidate, serrulate, teeth less than 0.05 mm, with 3 or sometimes 5 prominent longitudinal veins and inconspicuous veinlets, herbaceous between veins, with a usually 0.5–1 mm wide whitish scarious margin. Well-developed umbellules 4–6-flowered, with subequal pedicels c. 0.5 mm long. Flowers protandrous; petals c.

Figures 1–6. Bupleurum greuteri. — 1–3. Habit, small medium and one of the largest individuals. — 4. Bract. — 5. Bractlets. — 6. Petals, larger one seen from inside. Drawings from *Runemark & Snogerup 20724* and *20833*.

1.2 mm, often with a few teeth on margin, apically narrow, cucullate; inflexed lobe 0.5–0.6 mm, narrow, blunt or slightly emarginate; anthers 0.3–0.4 mm, filaments c. 0.5 mm; stylopodium 0.8–1 mm wide, styles 0.3–0.4 mm, basally erect and apically curved outwards. Mericarps not seen when ripe, 1.2 mm long or probably more, rounded prismatic, smooth, ridges filiform.

Material seen. — GREECE: LAKONIA: Monemvasia, 18 May 1970, *Stamatiadou 8958* (ATH); Spilea, W of Pyrgos, 20 m, 20 May 1964, *Runemark & Snogerup 20560* (E;

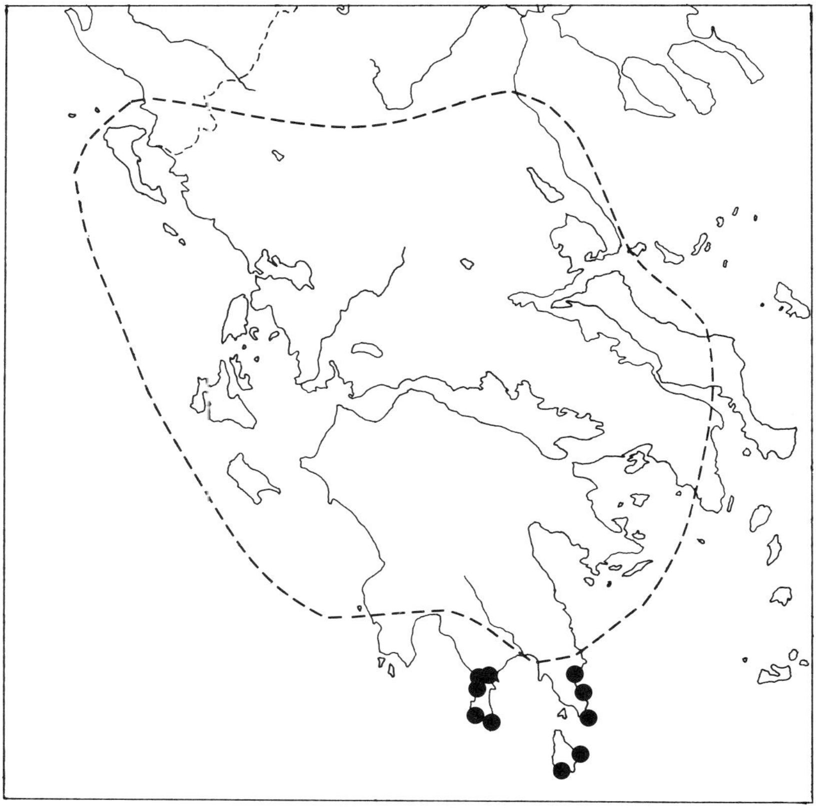

Figure 7. The distribution of *Bupleurum greuteri* (dots) and *B. glumaceum* (broken line) in SW Greece.

LD); 2 km N of Areopolis, 50 m, 22 May 1964, *Runemark & Snogerup 20724* (LD); 2 km ENE of Areopolis, 650 m, 21 May 1964, *Runemark & Snogerup 20547* (LD); 1 km W of Karnoupclis SW of Gythion, 15 May 1977, *Snogerup 1552* (LD); ravine 1 km SSE of Korogenia Nika, c. 6 km N of Cape Teneron, 240–330 m, 16 May 1984, *Strid 23058* (C); 3 km NNW of the village of Kampos N of Neapolis, 24 May 1964, *Runemark & Snogerup 20779* (LD); SW of Gerolimin, 0–50 m, 22 May 1964, *Runemark & Snogerup 20708* (LD); 4 km S of Velanidia, 50–100 m, 1 May 1982, *Runemark & Svensson 48074* (LD); Kithira, 26–31 May 1937, *Werner* (W); KITHIRA: W of Diakostis, 50 m, 26 May 1964, *Runemark & Snogerup 20833* (LD Holotype); old castle of Chora Kithira, 25 May 1964, *Runemark & Snogerup 20818* (LD); maritime cliffs W of Chora Kithira, 25 May 1964, *Runemark & Snogerup 20821* (LD); Chora, O-Abfall des Burghügels, 100 m, 11 May 1964, *Greuter 6514* (B, LD).

B. greuteri grows in communities of annuals of varying density on bare patches in phrygana. Several of the localities are below cliffs and on rocky slopes, all on limestone, from sea level to 650 m. The known distribution is shown in Fig. 7. The map in Snogerup (1964) p. 360 for *B. glumaceum* Sm. contains dots for misidentified *B. greuteri.*

B. greuteri belongs to subsect. *Aristata* (Godr.) Briq. (= Group B in Flora of Turkey). It is most similar to *B. glumaceum* Sm. in bractlets and petals, and this is no doubt the most closely related species. In the habit, with few umbels, and the large bracts *B. greuteri* more closely resembles *B. baldense* Turra, a species of the C and W Mediterranean and W Europe. It differs from *B. glumaceum* in the length of the bracts, in the venation of the bractlets and the slightly larger stylopodium and longer style. It is also less tall than *B. glumaceum* and the umbels are generally much fewer. *B. greuteri* and *B. glumaceum* differ from all other species of subsect. *Aristata* in the venation and texture of the bractlets and the unusual shape of the petals.

The species is named in honour of one of the first botanists who collected it and who also considered it a new species.

LITERATURE CITED

Snogerup, S. 1964. Studies in the Aegean Flora. IV. *Bupleurum flavum* Forsk. and related species. Bot. Notiser 115: 357–375.

4. ENDEMISM AND SPECIATION IN THE GREEK FLORA*

ARNE STRID

Botanical Laboratory, University of Copenhagen, 140 Gothersgade, DK-1123 Copenhagen K, Denmark

ABSTRACT

Greece has c. 4,900 species of vascular plants, including c. 742 endemics. The age of the flora, the dissected topography and the variety of rock substrates are the main reasons for this high diversity. Phytogeographical elements are discussed, in particular the Anatolian connections. Examples of geographical, ecological and seasonal differentiation patterns are given, and conservational aspects are discussed.

Key words: Greek flora, endemism, speciation, phytogeography.

HISTORICAL INTRODUCTION

The richness and diversity of the Greek flora has long been recognised by taxonomists, and indeed the foundation of botany as a science goes back to Theophrastos (4th century B.C.). Modern botanical exploration of Greece may be said to have started 200 years ago with John Sibthorp (1758–1796) who, on his 'grand tour' during 1786–87 with the artist Ferdinand Bauer, visited many places in present-day Greece as well as western Turkey and Cyprus. Their specimens and drawings formed the basis of the magnificent *Flora Graeca* by J. Sibthorp and J.E. Smith, one of the greatest illustrated Floras of all time. This was published from 1806–1840 in 10 leather-bound folio volumes and contained 966 hand-coloured illustrations engraved on copper from the original drawings by Ferdinand Bauer. Only 25 complete sets of the original issue were published; they are amongst some of the costliest botanical works ever produced. The illustrations are unsurpassed in their combination of botanical accuracy and artistic finish. However, many errors regarding distribution of species have arisen and persisted even to this day because 'Dr. Sibthorp ... trusted to memory, and dreamed not of dying'.

One of the localities visited by Sibthorp was the Athos peninsula in northern Greece. This is a semi-independent territory inhabited only by monks, and a true remnant of the Byzantine world, with some of the monasteries dating back to the 10th century. Mt. Athos or Agion Oros, an isolated, cone-shaped limestone mountain at the tip of the peninsula, rises to 2033 m and is of considerable botanical interest with several local endemics. One of the species discovered by Sibthorp and named after him is *Helichrysum sibthorpii* Rouy (Asteraceae). This is particularly interesting in the context of endemism, since it represents a section

* Based on a lecture given at the symposium 'Advances in Plant Systematics and Evolution', Rancho Santa Ana Botanic Garden, Claremont, California, May 23, 1987.

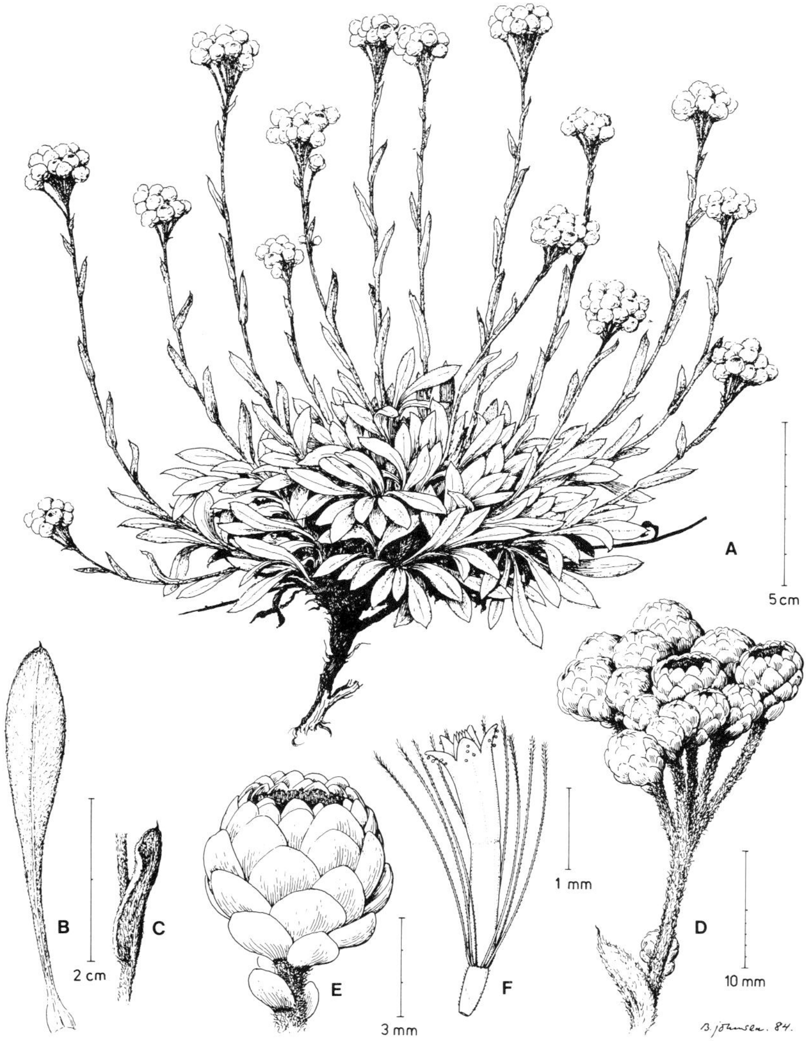

Figure 1. Helichrysum taenari (Asteraceae), a rare local endemic occurring in a single locality near the southern tip of Peloponnisos. It was first collected by Rothmaler in 1942 (material destroyed in Berlin in 1943) and not rediscovered until 1984, presumably in the same spot where it is restricted to a rock face of a few hundred square metres. Other members of sect. *Virginea* (characterised by white involucral bracts) are local endemics of other parts of the Mediterranean region (see text). — A. Habit. — B. Basal leaf. — C. Cauline leaf. — D. Inflorescence. — E. Flowering head. — F. Floret.

of the genus, sect. *Virginea* (DC.) Fiori, in which all members are regional or local endemics in various parts of the Mediterranean region: *H. sibthorpii* on Mt. Athos, *H. amorginum* Boiss. & Orph. on the Aegean island of Amorgos and two adjacent smaller islands, *H. doerfleri* Rech. fil. in eastern Crete, *H. billardieri* Boiss. & Blanche from Lebanon, *H. frigidum* (Labill.) Willd. from Corsica and Sardinia, *H. lacteum* Coss. & Dur. from Algeria, and the recently rediscovered *H. taenari* Rothm. emend. Strid from the southern tip of Peloponnisos (see Strid 1986b). This pattern undoubtedly indicates an old group in which fragments of a widespread species have survived and differentiated in isolated parts of what was once a continuous distribution area.

Dominant figures in Greek floristics during the 19th century were August Heinrich Rudolph von Grisebach (1814–1879), Theodor von Heldreich (1822–1902) and Theodoros Georgios Orphanides (1817–1886). By the end of the century the basic characteristics of the Greek flora were known, and the period culminated with the publication of Halácsy's *Conspectus Florae Graecae* (3 vols., 1900–1904 with supplements 1908 and 1912), an accurate and critical work which is still a major source of information pending the production of a modern *Flora Hellenica*.

The 20th century has been characterised by 2 trends: (1) continued and intensified floristic exploration, particularly in the less accessible areas, and (2) biosystematic studies of selected groups, especially in the last 20 years. A leading collector and taxonomist from the late nineteen-twenties up to the present is Karl Heinz Rechinger (born 1906). A Scandinavian group, started and led by Hans Runemark, has made a series of experimental taxonomic studies, particularly in the Aegean region. Greek botanists now play an increasingly important part in the exploration of their country, and a particularly active group has been built up in Patras by Dimitrios Phitos.

We can now form a reasonably clear picture of species richness and endemism in the Greek flora. On the basis of the relatively conservative taxonomy of *Flora Europaea*, Webb (1978) calculated the total number of species of vascular plants in Greece as being 4,100–4,250 (excluding apomictic taxa). Crete, which was treated as a separate entity in *Flora Europaea*, was estimated to have 1,700–1,850 species. There are no exact figures as to how many of these that are not found in the rest of Greece, but a reasonable guess would be about 250. The East Aegean Islands, which were not included in *Flora Europaea*, probably add a further c. 200 species. Finally, there are perhaps 250 species now known to occur in Greece although not listed in *Flora Europaea*; many of these have been discovered in recent years in the still poorly explored mountains along the northern borders (cf. Strid & Papanicolaou 1981, Strid & Franzén 1982, etc.). This brings the total number of species in Greece to c. 4,900, a very high number for a comparatively small country in the temperate zone ('splitters' might recognise up to c. 6,000). In a recent printout from the database of the Threatened Plants Unit at Kew (December 1986, courtesy Hugh Synge and Christine Leon) 1,008 species were listed as nationally rare or threatened. About 742 species are endemic to Greece, the highest figure for comparable countries or territories in the Mediterranean area (Morocco has 536, the Canary Islands 514, mainland Spain 501, etc.; cf. Leon, Lucas and Synge 1985).

There are two major reasons for the high incidence of endemism in Greece: (1)

the age of the flora which has remained *in situ* relatively undisturbed by glaciations since the major upheavals of the Miocene, and (2) the dissected topography and the variety of rock substrates (limestones, schists, granite, ultramafic rocks, etc.) which have favoured speciation in small-population systems. Although relatively poor in species in absolute terms it is the islands and the high mountains that are particularly rich in endemics; these are the great natural laboratories in which we can study the process of evolution and speciation at work.

ENDEMISM IN THE GREEK FLORA

Endemic species are usually referred to one of two broad categories, paleo-endemics and neo-endemics, although more intricate systems of classification have been proposed. Generally speaking paleo-endemics are taxonomically isolated species representing an evolutionary standstill, often with primitive characteristics (woody perennials) and usually found in rocky habitats. Typical examples are *Jankaea heldreichii* (Boiss.) Boiss. (Gesneriaceae) from Mt. Olympus and *Anchusa cespitosa* Lam. (Boraginaceae) from the White Mountains (Levka Ori) of western Crete. Neo-endemics on the other hand have close relatives in adjacent geographical areas or environments, and occur in groups in which active speciation is taking place. Examples are the *Nigella arvensis* L. complex (Ranunculaceae), annuals of the Aegean islands, and the *Nepeta sibthorpii* Bentham complex (Lamiaceae), short-lived perennials at moderate to high altitudes in the mountains. Both complexes have been subject to extensive biosystematic studies (Strid 1970 and Baden 1987, respectively).

Crete

Crete is the only territory in Greece combining isolation with high mountains, and predictably it is, within Greece, the region where the concentration of endemics is highest (Greuter 1980). Of the 742 Greek endemics some 244 occur in the Cretan Area (Crete + Karpathos and adjacent small islands). Of these, 157 are Cretan Area endemics in a restricted sense, while the remaining 87 are more widespread Greek endemics. Some of the Cretan endemics are far from rare or threatened; indeed some of them are among the commonest and most characteristic species of the island, e.g. *Phlomis lanata* Willd. (Lamiaceae) and *Ebenus cretica* L. (Fabaceae). At the other end of the scale are perhaps 10 species whose known occurrence in the world is restricted to less than 100 individuals; examples are *Onobrychis sphaciotica* W. Greuter (Fabaceae) and *Nepeta sphaciotica* P.H. Davis (Lamiaceae). Such species are in imminent danger of extinction owing to breeding collapse, even if no further disturbance of the habitat takes place.

The Mountain Flora

The plants of the high mountains of Greece are being treated in a current flora project, the *Mountain Flora of Greece* (vol. 1 in 1986, second and final volume expected in 1990). This Flora has been planned to include all vascular plants commonly occurring above an altitude of 1800 m within the borders of Greece ('full members'). In addition, taxa which only occasionally reach this altitude or are commonly found in open, treeless habitats above 1500 m have also been included, but are less fully treated ('associate members'). The rules have been interpreted with some flexibility, the aim being to exclude typical forest taxa and to include all taxa of open, alpine and subalpine habitats. The timberline can scarcely be used to delimit the mountain flora; it often lies around 1800 m (up to c. 2300 m on Olympus), but varies greatly as a consequence of logging, fires etc. Inevitably, there will be a number of borderline cases where there are doubts as to whether to include a taxon or not. Phytogeographically and ecologically, however, we believe the delimitation of the project is reasonable and practicable.

Vol. 1 of the *Mountain Flora of Greece* (Strid 1986a) includes 926 numbered taxa (species and subspecies), 768 of these being 'full members' and 158 'associate members'. A comparison with *Flora Europaea* (Tutin et al. 1964–1980) shows that of our 926 taxa no less than 304, or c. 33 %, were either not recorded from Greece in *Flora Europaea* or were recorded under a different name and/or with a different taxonomic rank. If 'full members' only are considered, the number is 262 out of 768, or about 37 %. Finally, 30 taxa listed for Greece in *Flora Europaea* have been excluded, either because the records have been shown to be erroneous or because the names have been reduced to synonymy. Since we have seldom used narrower concepts of species and subspecies, these figures serve to demonstrate the considerable increase in floristic and taxonomic research that has taken place in Greece since, and largely inspired by, the publication of the first volumes of *Flora Europaea*.

The Greek mountain flora is composed of a number of phytogeographical elements, reflecting different affinities and migration routes. A relatively large group consists of widespread boreal or Euro-Siberian species which often occur at low altitudes in N Europe and ascend to alpine levels in the south (e.g., *Dianthus deltoides* L., *Geranium sylvaticum* L., *Vaccinium myrtillus* L., *V. vitis-idaea* L., *Nardus stricta* L.). Several arctic-alpine species reach their southernmost localities on granitic or schistose mountains in N Greece (e.g., *Saxifraga stellaris* L., *Gnaphalium supinum* L., *Juncus trifidus* L., *Eriophorum vaginatum* L.), more rarely on limestone (e.g. *Saxifraga oppositifolia* L., *Dryas octopetala* L.). There are significant connections with the Alps and the Carpathians as well as with the mountain floras of Italy and Anatolia.

Many Greek mountain taxa have undoubtedly evolved from more widespread, often omni-Mediterranean lowland taxa. Particularly in the dry mountains of S Greece (Peloponnisos and even more on Crete), common lowland species may ascend to high altitudes, often forming special mountain ecotypes (e.g. *Euphorbia acanthothamnos* Heldr. & Sart. ex Boiss., *Trifolium uniflorum* L., *Hypericum empetrifolium* Willd.) It is a significant feature of the vegetation of Crete that the evergreen Mediterranean lowland scrub ascends to at least 1500 m where it merges into subalpine and alpine vegetation types. Deciduous forest is virtually lacking, as are the forest-floor herbs and grasses, a fact that would seem to indicate that the absence of deciduous forest is not a result of man's activity but a natural feature. Aridity and isolation have contributed to the evolution and preservation of a number of local endemics, especially in the White Mountains of western Crete, which in this respect is the richest area in Greece.

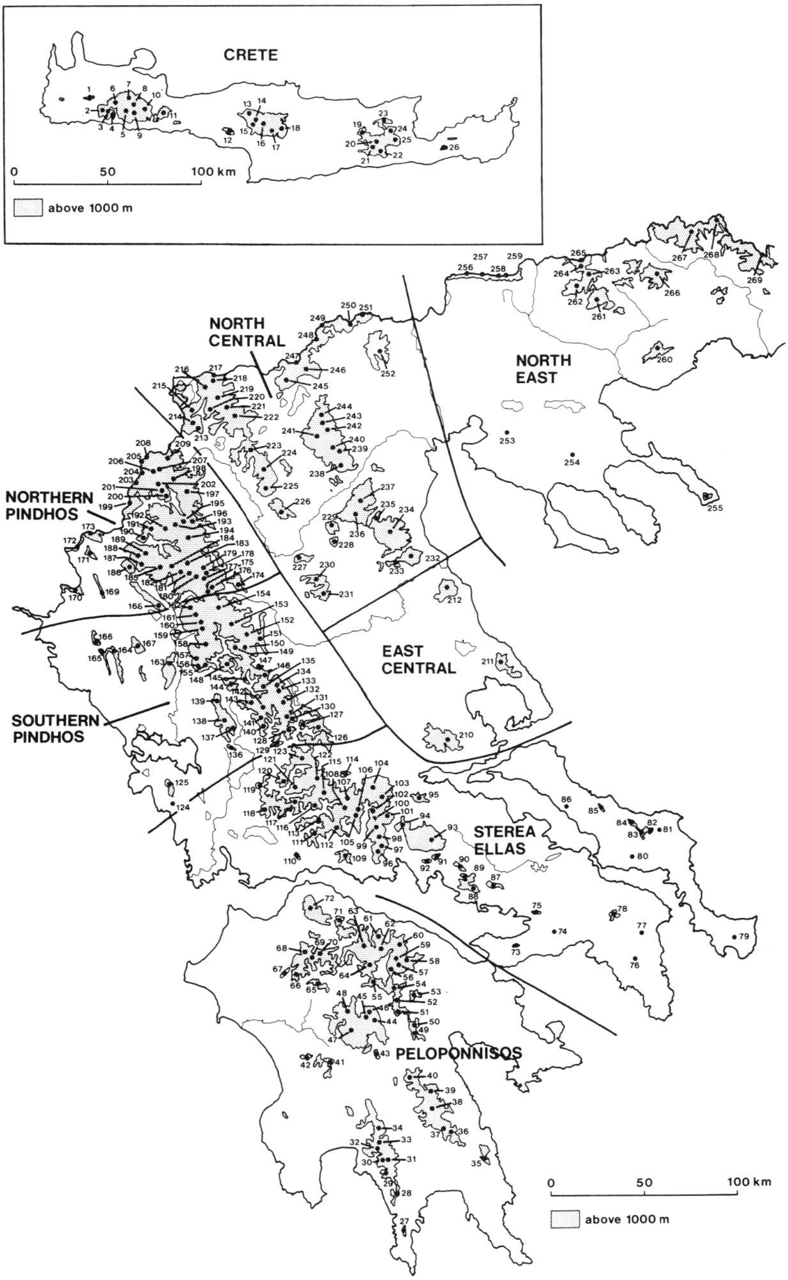

Figure 2. Map of the Greek mainland and Crete, showing the eight regions used in the Mountain Flora of Greece project. The small numerals refer to a list of names of mountains (not included here).

Phytogeographical Analysis

For the purpose of the *Mountain Flora of Greece* the country was divided into eight more or less natural regions (see Fig. 2). A phytogeographical analysis of 816 taxa included in Vol. 1 demonstrated that 398 of these, or 49 %, are 'widespread' (defined as taxa extending beyond Balkan Peninsula + Italy or Balkan Peninsula + Anatolia). The remaining 418 taxa, or 51 %, are thus 'narrow'. The latter were classified (Table 1) with respect to (a) their total distribution (six progressively narrower categories), and (b) their distribution within the eight regions of Greece. The table should be read as follows: the figures 14 and (21; 9) in the upper left-hand corner indicate that on Crete 14 taxa belong to the phytogeographical element 'Balkan Peninsula + Anatolia' which constitutes 21 % of the 'narrowly distributed' taxa found on Crete (68 in all) or 9 % of the total mountain flora of Crete (149 taxa). Similarly, the figures 53 and (13; 6) in the lower left-hand corner indicate that 53 out of 418 (816) taxa, or 13 % (6 %), belong to this element when the total mountain flora of Greece is taken into account.

With respect to endemism, some significant features can be observed from the table, for instance:

(1) Balkan endemics constitute the largest phytogeographical element, comprising 155 taxa, or 37 % (19 %). In the central and northern mainland regions 50–55 % (17–20 %) of the taxa belong to this element. It decreases in importance in Sterea Ellas and Peloponnisos, and is only poorly represented in Crete.

(2) Greek endemics are particularly abundant in Peloponnisos and Sterea Ellas, extending across the Gulf of Corinth which forms only a weak phytogeographical barrier. Taxa with similarly small distribution areas in N Greece (e.g. 'serpentine' species in the NW and limestone species in the NE) usually extend across the borders to Albania and Bulgaria respectively, and have thus been classified as Balkan endemics.

(3) Of the regions considered here, Crete has by far the highest percentage of single-area endemics (if lowland taxa were to be included the difference would certainly be even greater). Both paleo- and neo-endemics are well represented.

(4) Single-mountain endemics are chiefly found in some of the large and/or isolated limestone massifs, for example, Levka Ori in Crete (nos. 2–11 in fig. 2), Taygetos on Peloponnisos (nos. 28–34), Olimbos in the North Central region (no. 234) and Athos in the North East (no. 255), and less frequently on 'serpentine' (e.g. Smolikas in N Pindhos, nos. 191–193; Vourinos in North Central, no. 226), granite (Varnous in North Central, nos. 216–219), or micaceous schist (Kajmakčalan in North Central, no. 247).

The Anatolian Element

An analysis of the Anatolian element in the Greek mountain flora (Strid 1986c) has shown that the percentage of 'Turkish species' in Greece is c. 3 times as high as the percentage of 'Greek species' in Turkey. The migratory pressure from east to west has thus been considerably higher than from west to east. The northern route (via Thrace and Turkey-in-Europe) and the southern one (along the South Aegean Island Arc) are of approximately equal importance in terms of number of species. However, there are significant differences in habitat between species representing

Table 1. *Endemism in the Greek mountain flora. Explanation in the text.*

	Phytogeogr. elem.						
Area in Greece	Balkan Penins. + Anatolia[a]	Balkan Penins. + Italy[b]	Balkan endemics[c]	Greek endemics	Single-area endemics	Single-mt endemics[d]	Total
Crete	14 (21; 9)	6 (9; 4)	4 (6; 3)	7 (10; 5)	26 (38; 17)	11 (16; 7)	68 – 149
Peloponnisos	23 (17; 7)	17 (12; 5)	35 (26; 11)	41 (30; 13)	10 (7; 3)	11 (8; 4)	137 – 311
Sterea Ellas	24 (14; 6)	26 (15; 6)	61 (35; 15)	46 (27; 11)	9 (5; 2)	7 (4; 2)	173 – 418
S Pindhos	17 (14; 5)	28 (23; 8)	60 (50; 17)	16 (13; 5)	0 (0; 0)	0 (0; 0)	121 – 348
N Pindhos	19 (12; 4)	30 (19; 7)	84 (53; 19)	17 (11; 4)	5 (3; 1)	4 (3; 1)	159 – 439
East Central	10 (20; 5)	8 (16; 4)	25 (51; 13)	6 (12; 3)	0 (0; 0)	0 (0; 0)	49 – 190
North Central	23 (13; 5)	25 (14; 5)	98 (54; 20)	13 (7; 3)	1 (1; 0)	21 (12; 4)	181 – 495
North East	26 (22; 7)	12 (10; 3)	54 (55; 17)	7 (6; 2)	1 (1; 0)	8 (7; 2)	119 – 388
Total	53 (13; 6)	37 (9; 5)	155 (37; 19)	59 (14; 7)	52 (12; 6)	62 (15; 8)	418 – 816

[a] Including taxa extending slightly beyond the borders of E or S Anatolia.
[b] As defined here Italy includes Sicily, Sardinia and Corsica.
[c] Including taxa extending locally to SW Romania or the Carpathians.
[d] Large massifs such as Levka Ori and Taygetos have been defined as a "single mountain". Taxa endemic to border mountains such as Kajmakčalan have been classified as "single-mt endemics" even if they occur on both sides of the border.

the northern and southern connections. The former tend to be widespread Euro-Siberian forest species (e.g. *Vaccinium myrtillus* L., *Daphne mezereum* L.) or more regional species of wet, non-calcareous habitats (e.g. *Bruckenthalia spiculifolia* (Salisb.) Reichenb., *Geum coccineum* Sm.). The southern species tend to prefer arid conditions and to grow on limestone. Some are found on the mountains of Crete (e.g. *Ranunculus brevifolius* Ten., *Potentilla speciosa* Willd.), but many have now disappeared although they presumably occurred there earlier (e.g. *Minuartia juniperina* (L.) Maire & Petitmengin, *Omphalodes luciliae* Boiss.). A few reach Crete, but are absent from the Greek mainland (e.g. the wild form of *Cupressus sempervirens* L.).

Within the Anatolian element is a group of perhaps 15–20 species that are widely disjunct, with a main distribution area in C and E Anatolia and even further east, with outposts in the mountains of Peloponnisos and Sterea Ellas. A classical example is *Biebersteinia orphanidis* Boiss. (Geraniaceae) which was collected by Orphanides on Mt. Killini (N Peloponnisos) in 1851 and has never been seen again in its original and only European locality. It occurs, however, in a few places in SC Anatolia (cf. Davis in *Fl. Turkey* 2: 451, 1967). Two members of Boraginaceae, *Arnebia densiflora* (Ledeb. ex Nordm.) Ledeb. and *Solenanthus stamineus* (Desf.) Wettst. in Stapf, have similar patterns of distribution; both occur locally in rocky habitats on Mt. Chelmos (N Peloponnisos) and Giona (Sterea Ellas), and are widely distributed in C and E Anatolia. A herbaceous member of Berberidaceae, *Gymnospermium odessanum* (DC.) Takht., related to *Leontice* L. and *Bongardia* C.A. Meyer (cf. p. 42), has an isolated locality on Mt. Panachaikon (NW Peloponnisos) far from its Asiatic centre. *Juniperus drupacea* Labill., a small tree with conspicuously large, ovoid fruits, has its main area of distribution in SC Anatolia and W Syria, but forms a small forest on the N slopes of Mt. Parnon (E Peloponnisos) around 1000 m. Most of the species showing these wide disjunctions appear to be perennials of rocky habitats in the mountains, with no obvious adaptations for long-range dispersal and often with supposedly primitive morphological characteristics. Their isolated occurrences in Greece are probably the result of vicariance events, although long-distance dispersal cannot be entirely ruled out with species such as *Thlaspi kotschyanum* Boiss. & Hohen. (Brassicaceae) and *Geum heterocarpum* Boiss. (Rosaceae).

Speciation Mechanisms

The mechanisms of speciation are as diverse as the flora itself, but the dominant principle seems to have been differentiation at diploid level in geographically isolated populations, an example of which is the *Nigella arvensis* L. complex (Ranunculaceae) in the Aegean region that has been studied experimentally (Strid 1970). On practically all the larger islands there are distinct local variants of this complex, recognisable by unique combinations of features particularly with respect to the shape and colour of the floral parts. The plants are diploid, outbreeding annuals suitable for biosystematic study. A large number of crossing experiments were carried out, and the pollen fertility of the F_1 hybrids was recorded as an indication of genetic similarity between the parent populations. The resulting pattern of geographical differentiation (Fig. 4) coincides remarkably well with the phytogeographical subdivision of the Aegean region based on floristic data (Rech-

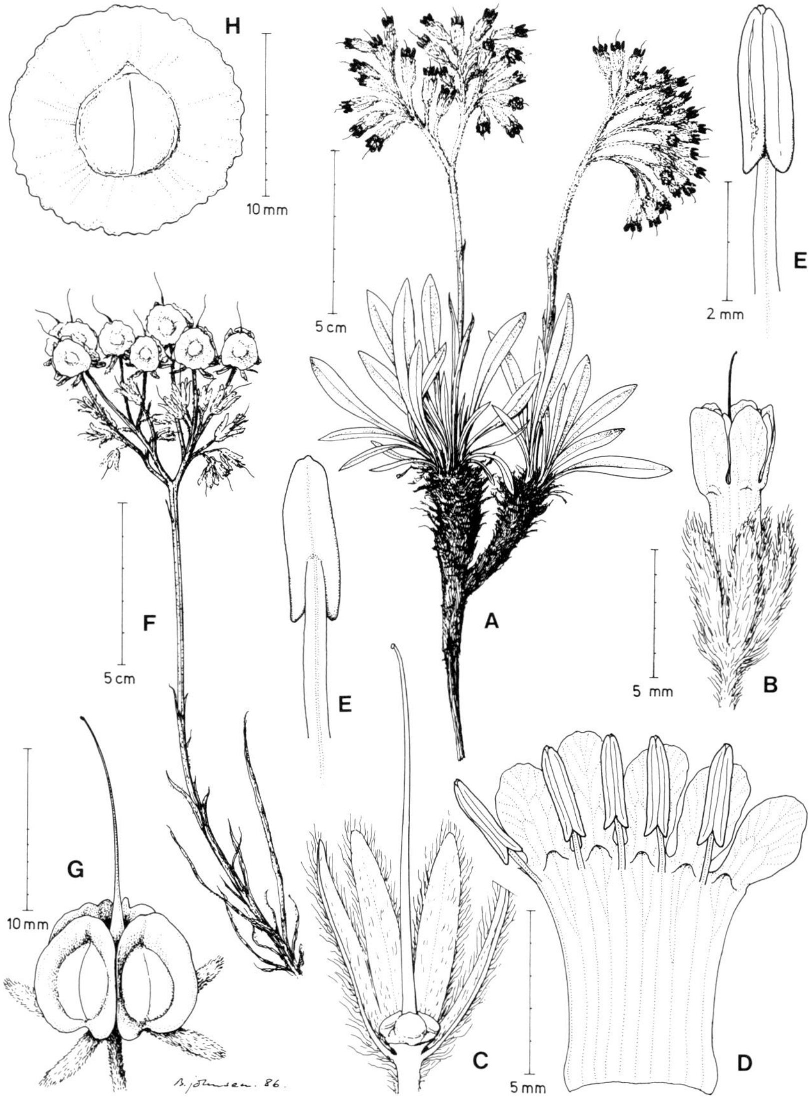

Figure 3. Rindera graeca (Boraginaceae), a Greek endemic occurring on limestone mountains from Taygetos to Timfi; one of the westernmost representatives of a chiefly Asiatic genus with c. 27 species. — A. Habit. — B. Flower. — C. Calyx opened (corolla removed). — D. Corolla opened. — E. Stamen. — F. Infructescence. — G. Fruit. — H. Single nutlet.

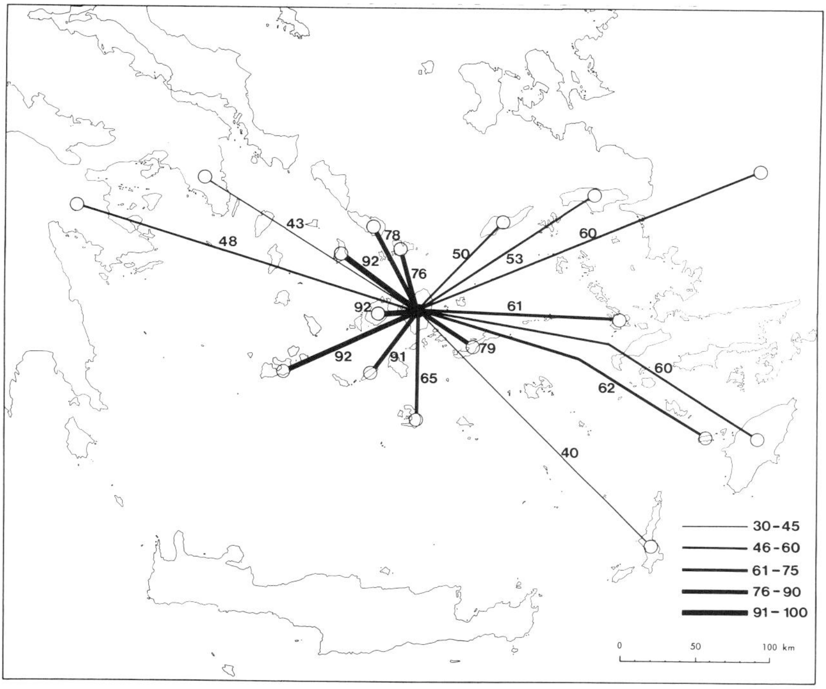

Figure 4. Diagrammatic representation of crossing experiments between populations of the *Nigella arvensis* complex (Ranunculaceae), a group of outbreeding, diploid annuals of the Aegean region. Individuals from the central island (Naxos) have been crossed with representatives of 18 other island and mainland populations. The numerals and the relative thickness of the connecting lines indicate the percentage of stainable pollen in F_1 hybrids. Note that populations from the central group of islands (the Cyclades) are largely interfertile, whereas partial sterility barriers are encountered vis-à-vis the Greek mainland, Anatolia and the East Aegean Islands. Plants with pollen stainability values below c. 50% produce few or no seeds. The differentiation pattern reflects paleogeographical events in the area. — Redrawn from Strid (1970).

inger 1943) and is undoubtedly a reflection of paleogeographical developments in the region.

There are now a number of similar studies on the Aegean region as well as the mainland mountains, some of the most recent being on *Scutellaria* L. (Lamiaceae; Bothmer 1981, 1985, 1987), the *Nepeta sibthorpii* Bentham group (Lamiaceae; Baden 1987) and the *Achillea clavennae* L. and *A. ageratifolia* (Sm.) Boiss. groups (Asteraceae; Franzén 1986, and unpublished thesis). In particular, Franzén has studied variation in chemical compounds (flavonoids) in addition to carrying out traditional biosystematic studies. All three groups comprise short-lived perennials chiefly found in the mountains; all are outbreeding, sometimes self-incompatible diploids (2n = 34 in *Scutellaria*, 2n = 18 in *Nepeta*, 2n = 18 in *Achillea*), and there is a general similarity in the differentiation patterns which reflect the age and degree of geographical isolation.

Speciation as a consequence of *habitat specialisation* has occurred particularly in the 'serpentine' endemics, most of which are found in the mountains of NW Greece. The rocks are dark, with a greenish or blackish tinge when freshly broken; they weather easily developing a rust to purplish patina, and produce extensive screes. Ultramafic rocks (traditionally known as serpentine in botanical literature) occur in many parts of the world; the unusual chemical composition of the rocks and soils (large amounts of Mg and heavy metals) makes them inhospitable to many species, others being strictly adapted to this substrate, constituting an element of serpentine endemics. Examples are *Campanula hawkinsiana* Hausskn. & Heldr. (Campanulaceae), *Bornmuellera baldaccii* (Degen) Heywood (Brassicaceae), *Orobanche rechingeri* Gilli (Orobanchaceae) and taxa of *Alyssum* sect. *Odontarrhena* (Meyer) Hooker (Brassicaceae). The area of ultramafic rocks continues through Albania to Serbia, and most of the species (possibly 50 in all) are restricted to this area, although a few have disjunct occurrences on similar substrates in SW Turkey. Further ecological and evolutionary studies of this interesting element are needed.

Speciation as a consequence of *seasonal differentiation* is an interesting phenomenon found particularly among geophytes of the Mediterranean lowlands. The commonest annual cycle of the geophytes includes flowering and rapid seed maturation during spring and early summer (March to June); above-ground parts of the plants have usually completely died down by midsummer, new leaves are produced in response to the autumn rains around October, vegetative growth occurs throughout the usually frost-free winter, and flowers are produced again in the spring. However, in a number of taxonomic groups, mostly among the monocotyledons (*Crocus, Sternbergia, Colchicum, Scilla, Urginea, Narcissus, Galanthus,* Araceae, etc.), but including also a few dicotyledon genera (*Cyclamen, Ranunculus*) certain species have reversed this pattern. They flower in the autumn (September to December), often without leaves, vegetative growth takes place in the winter and spring, seed maturation is slow and often not complete until May or June, the plants are dormant throughout the dry hot summer, and flowers are produced again following the autumn rains. In some genera like *Galanthus* and *Ranunculus* autumn-flowering is rather an exception (*G. reginae-olgae* Orph., *R. bullatus* L.); in others such as *Colchicum* it has become the dominant pattern. *Crocus* is of particular interest in this respect, with approximately equal numbers of spring-flowering and autumn-flowering taxa (Fig. 5). Examples of the former are *C. chrysanthus* (Herbert) Herbert and *C. sieberi* Gay, of the latter *C. boryi* Gay and *C. goulimyi* Turrill; a few taxa such as *C. laevigatus* Bory & Chaub. and certain taxa of the *C. biflorus* Miller complex can be characterised as winter-flowering.

It is reasonable to assume that many of the geophytes flowered in the winter in periods when the climate was slightly milder. With winters becoming colder or when migrating to higher latitudes or altitudes these plants were faced with selective forces driving them either to flower earlier (in the autumn) or later (in the spring). This is a strong mechanism for inducing reproductive isolation; as soon as there is no overlap in the flowering time of two parts of a formerly continuous population, they have reached the point of no return and represent two biological

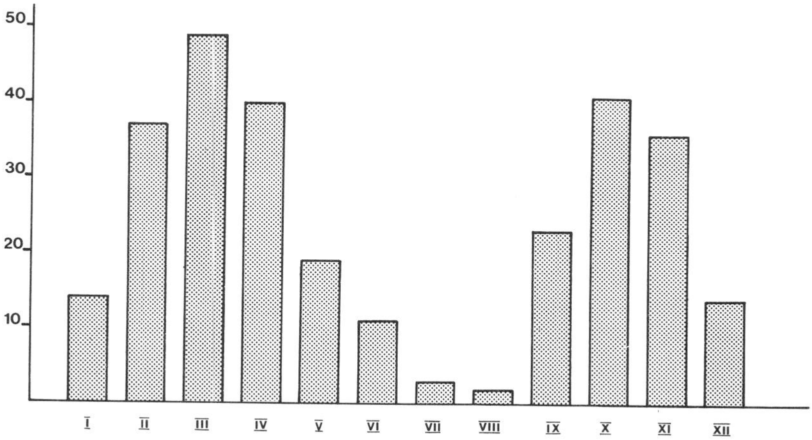

Figure 5. Flowering times for species and subspecies in *Crocus* (Iridaceae), a genus with its centre of diversity in Greece and Turkey (horizontal axis: months; vertical axis: number of taxa). Note the peaks in spring and autumn, with some taxa flowering in winter and very few (from high mountains) in summer. — Data from Mathew (1982), with minor additions.

species. In some of these the morphological differences are still somewhat obscure, e.g. between autumn- and spring-flowering subspecies of the *Crocus biflorus* complex, and between *Galanthus reginae-olgae* and *G. nivalis* L., the former sometimes being regarded as merely an autumn-flowering variant of the latter.

Specialised pollination mechanisms and co-evolution with pollinating insects have been vehicles for speciation in certain groups such as *Aristolochia* L. and even more so among the orchids. The sexual mimicry of the genus *Ophrys* L. is well known in this respect. The shape, tactile movements and pheromone scents of the flowers attract males of certain species of solitary bees and wasps which effect pollination. Pollinator specialisation is an important factor in developing and maintaining species integrity, although it sometimes breaks down as witnessed by the frequent occurrence of hybrids in *Ophrys* which is notoriously difficult taxonomically. Anything from c. 12 to 30 species are recognised in Greece; cf. a series of papers (Vöth 1984, 1987; Hölzinger et al. 1985; Baumann & Künkele 1986) in recent issues of Mitt.-Bl. Arbeitskr. Heim. Orch. Baden-Württ.

Agricultural Weeds — Origin and Speciation

The Mediterranean region is the centre of origin of most of the weeds now found in temperate areas throughout the world, and especially in other regions with a mediterranean climate. They present many fascinating problems in the field of evolutionary biology.

In the *Nigella arvensis* complex (see above) most of the taxa occur in semi-natural habitats in the Aegean region. A single taxon, *N. arvensis* subsp. *arvensis*, has become a weed of cereal fields in C Europe where it probably evolved together with primitive agriculture over thousands of years. With modern agricultural methods (seed purification and use of herbicides) it is now being pushed back to its original area in SE Europe (Strid 1971). *Centaurea cyanus* L. (Asteraceae), once a common weed of cereal fields throughout Europe, has a similar story to tell: a presumably primitive form of it, differing only in quantitative characters, is found in rocky habitats in the Aegean region. In both cases differentiation has taken place at the diploid level and has not resulted in reproductive isolation.

There are other cases where polyploidy has been involved. *Agrostemma githago* L. (Caryophyllaceae), until recently a common weed of ryefields throughout Europe, is a tetraploid with 2n = 48, probably derived from the diploid *A. gracilis* Boiss. which is scattered in Turkey. The weed variant of *Buglossoides arvensis* (L.) I.M. Johnston (Boraginaceae) is either tetraploid (2n = 28) or hexaploid (2n = 42). A closely related plant, variously referred to as *B. arvensis* subsp. *gasparrinii* (Heldr. ex Guss.) R. Fernandes or *B. incrassata* (Guss.) I.M. Johnston, occurs on rocky mountainsides in the Mediterranean region and is, at least sometimes, diploid (2n = 14). The *Capsella bursa-pastoris* L. complex (Brassicaceae), annual, inbreeding weeds of cultivated fields and ruderal habitats now occurring throughout almost the whole world, comprises diploids (chiefly in the Mediterranean region) and tetraploids (widespread). A local taxon from W Greece and Albania, *C. grandiflora* (Fauché & Chaub.) Boiss., is of particular interest. It occurs in semi-natural habitats and meadows, has larger flowers as implied by the name, produces nectar and is presumably insect-pollinated. The chromosome number is not known, but can be expected to be diploid.

It can be asked to what extent speciation has actually occurred in response to the development of agriculture which in the eastern parts of the Mediterranean region goes back perhaps 10,000 years. If we look at the areas of the Southern Hemisphere that have a mediterranean climate it is significant that practically all weeds in the SW Cape Province and SW Australia (where European-style agriculture has a history of some 330 and 160 years, respectively) are of Mediterranean origin. The same is true, although presumably not to such an extent, for the floras of California and C Chile. It is a characteristic feature of the native floras of the Cape and SW Australia that the role of annuals is insignificant compared with the flora of the Mediterranean. Virtually no species were pre-adapted to become weeds, and when the land was opened up to agriculture the new ecological niches were rapidly taken over by the aggressive specialists from the Mediterranean region. The abundance of more or less weedy annuals in the flora of the Mediterranean, as compared with the floras of other lands with a similar climate that have only recently been settled by agrarian man, would seem to indicate that many of the weeds are of more recent origin than is generally assumed. When selection pressure is strong, speciation may be a relatively rapid process requiring only thousands or even hundreds rather than millions of years.

Endemism and Conservation

The large numbers of rare endemics in the flora of Greece make the question of conservation particularly urgent. For the mountain species grazing is often believed to be a major threat, but as pointed out by Greuter (1980) 'grazing (not overgrazing!) has been a natural environmental factor even before man came onto the scene, and has continued through historical times, for several millennia, in its

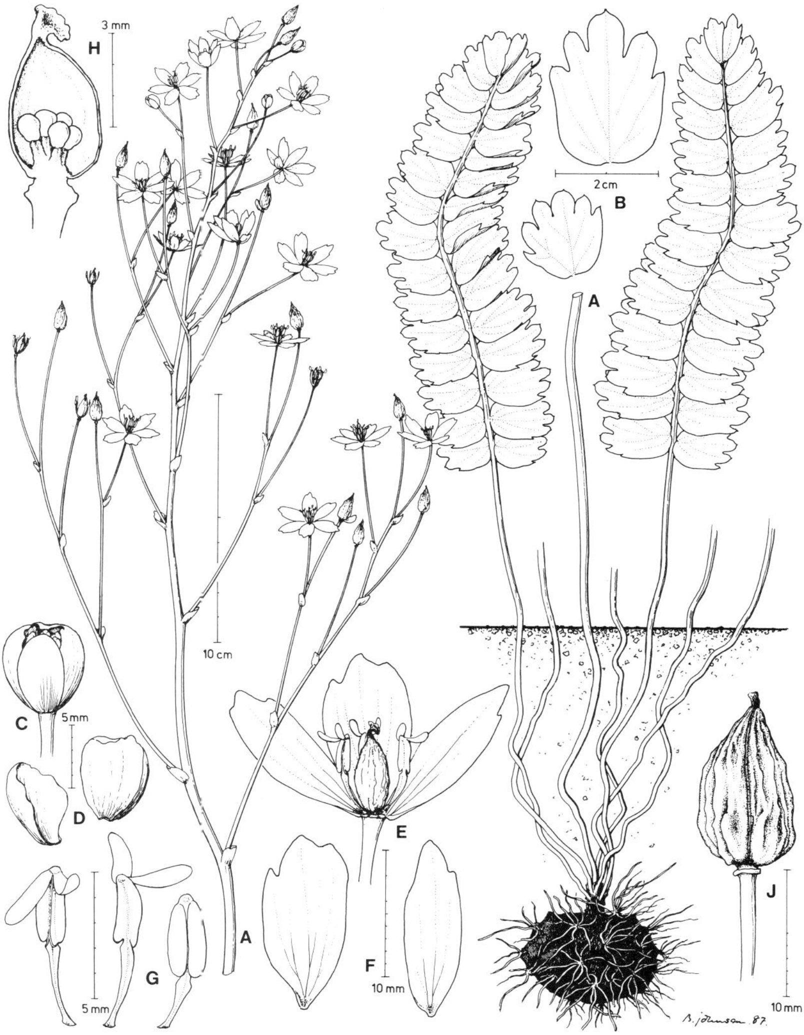

Figure 6. Bongardia chrysogonum (Berberidaceae), rediscovered in Greece after 168 years (see text). — A. Habit. — B. Leaflets. — C. Flower bud. — D. Deciduous perianth segments. — E. Flower with 3 honey-leaves removed. — F. Honey-leaves. — G. Stamens, 2 of them with dehisced anthers. — H. Longitudinal section of young fruit showing ovules. — J. Young fruit. (All from *Strid et al. 25491*).

present form. The plants now in existence are fit either to withstand or to escape it, and may even depend on it in some cases... There is absolutely no need to antagonise the shepherds and peasants — whose sympathy for and understanding of our conservational goals would be essential — by threatening the very base of their existence in trying to abolish grazing in their traditional pasture areas'. In fact, most of the mountain endemics will be safe as long as the present form of management continues. Exceptions are some rare species consisting of only one or a few very small populations, which thus face the danger of extinction as a result of breeding collapse. Overgrazing in the dry mountains of southern Greece (in particular Crete) may pose a threat to some rare species, but grazing pressure is probably lower now than it was 100 years ago and may well decrease even more in the coming years for purely economic reasons; it gives a meagre outcome at best, and the younger generation in the mountain villages tend to seek work in the cities (or even in Canada or Australia) rather than shepherd goats and sheep in their native mountains.

Some of the rare species in lowland habitats are in more immediate danger from urbanisation, tourism and changes in agricultural practices. In this context I should like to mention a small group of perennial, bulbous or tuberous weeds of ploughed fields, which survive traditional tilling by having sufficiently deep-seated underground organs, but which now face the double threat of herbicides and of uprooting by modern steel ploughs. Some of these such as *Gladiolus italicus* Miller (Iridaceae), *Bellevalia ciliata* (Cyr.) Nees (Liliaceae) and *Ornithogalum nutans* L. (Liliaceae) are able to survive at the edges of fields or in semi-natural habitats, whereas others such as *Tulipa boeotica* Boiss. & Heldr. (Liliaceae) and *Leontice leontopetalum* L. (Berberidaceae) are decreasing rapidly. Of particular interest is a relative of the latter, *Bongardia chrysogonum* (L.) Griseb. (Berberidaceae), which was last seen in Greece by Dumont d'Urville in 1819 (!) on the remote Aegean island of Astipalea, and was unexpectedly rediscovered in 1987 when a small population was found in an agriculturally marginal area in the northern Peloponnese. It is undoubtedly on the verge of extinction in Greece (although it occurs further east, in Turkey and Soviet Central Asia). It is apparently dependent on annual tilling by oldfashioned methods (it is uprooted by modern ploughing and unable to withstand competition if the field is abandoned), and thus raises an intriguing evolutionary question: Where did such a species grow before man turned to agriculture — or is it the outcome of evolution in response to agriculture?

ACKNOWLEDGMENTS

My studies of the Greek flora have been supported from a number of sources, in particular the Danish Natural Science Research Council and the Carlsberg Foundation. I wish to thank the director and staff of the Rancho Santa Ana Botanic Garden for the invitation to speak at their annual symposium and for hospitality extended on this occasion. Thanks are also due to Mrs Margaret Greenwood Petersson who corrected the English text of the manuscript and to Dr Kit Tan who edited the paper, both with their customary accuracy.

LITERATURE CITED

Baden, C. 1987. Biosystematic studies in the *Nepeta sibthorpii* group (Lamiaceae) in Greece. Opera Bot. 93: 1–54.

Baumann, H., and S. Künkele. 1986. Die Gattung *Ophrys* L. — eine taxonomische Übersicht. Mitt.-Bl. Arbeitskr. Heim. Orch. Baden-Württ. 18: 305–688.

Bothmer, R. von. 1981. Differentiation patterns in the genus *Scutellaria* from the Balkan peninsula. Bot. Jahrb. Syst. 102: 271–283.

——. 1985. Differentiation patterns in the *Scutellaria albida* group (Lamiaceae) in the Aegean area. Nordic J. Bot. 5(5): 421–439.

——. 1987. Differentiation patterns in the E. Mediterranean *Scutellaria rubicunda* group (Lamiaceae). Pl. Syst. Evol. 155: 219–249.

Davis, P.H. 1967. *Biebersteinia*, p. 451. *In* P.H. Davis [ed.], Flora of Turkey and the East Aegean Islands. Vol. 2. Edinburgh University Press, Edinburgh.

Franzén, R. 1986. Taxonomy of the *Achillea clavennae* group and the *A. ageratifolia* group (Asteraceae, Anthemideae) on the Balkan Peninsula. Willdenowia 16(1): 13–33.

——. 1987. Biosystematics of the *Achillea clavennae* and *A. ageratifolia* groups (Asteraceae). Unpublished Thesis, Department of Systematic Botany, University of Lund. 43 p.

Greuter, W. 1980 [as 1979]. The endemic flora of Crete and the significance of its protection, pp. 91–97. *In* Proceedings, Conference on the Protection of the Flora, Fauna and Biotopes in Greece, Athens 11–13 October 1979. Athens.

Halácsy, E. von. 1900–1908. Conspectus florae graecae. Vol. 1, pp. 1–576, 1900; Vol. 1, pp. 577–825, 1901; Vol. 2 (612 p.), 1902; Vol. 3 (519 p.), 1904; Supplementum (132 p.), 1908. G. Engelmann, Leipzig. Second supplement in Magyar Bot. Lapok 11: 114–202 (1912). Whole set reprinted 1968 by Verlag J. Cramer, Lehre.

Hölzinger, J., and A. and S. Künkele. 1985. Die Verbreitung der Gattung *Ophrys* L. auf dem griechischen Festland. Mitt.-Bl. Arbeitskr. Heim. Orch. Baden-Württ. 17: 1–101.

Leon, C., G. Lucas, and H. Synge. 1985. The value of information in saving threatened Mediterranean plants, pp. 177–196. *In* C. Gómez-Campo [ed.], Plant conservation in the Mediterranean area. W. Junk, Dordrecht (no. 7 in Geobotany series, ed. M.J.A. Werger).

Mathew, B. 1982. The Crocus. B.T. Batsford Ltd, London. 224 p.

Rechinger, K.H. fil. 1943. Flora Aegaea. Flora der Inseln und Halbinseln des ägäischen Meeres. Denkschr. Akad. Wiss. Wien, Math.-Nat. Kl. 105(1): xx + 924 p. (Reprinted 1973 by Otto Koeltz Antiquariat, Koenigstein).

Sibthorp, J., and J.E. Smith. 1806–40. Flora Graeca, vols. 1–10. J. White, London (cf. Stearn in Taxon 16: 168–178, 1967).

Strid, A. 1970. Studies in the Aegean flora. XVI. Biosystematics of the *Nigella arvensis* complex, with special reference to the problem of non-adaptive radiation. Opera Bot. 28: 1–168.

——. 1971. Past and present distribution of *Nigella arvensis* L. ssp. *arvensis* in Europe. Bot. Not. 124(2): 231–236.

——. [ed.]. 1986a. Mountain Flora of Greece, vol. 1 (vascular cryptogams to Plumbaginaceae). Cambridge University Press, Cambridge. xxx + 822 p.

——. 1986b. *Adonis cyllenea* (Ranunculaceae) and *Helichrysum taenari* (Asteraceae) rediscovered in Peloponnisos. Ann. Musei Goulandris 7: 221–231.

——. 1986c. The mountain flora of Greece with special reference to the Anatolian element. Proc. Roy. Soc. Edinburgh 89B: 59–68.

——, and R. Franzén. 1982. New floristic records from the mountains of northern Greece. (Materials for the Mountain Flora of Greece, 12). Willdenowia 12(1): 9–28.

——, and K. Papanicolaou. 1981. Floristic notes from the mountains of northern Greece. (Materials for the Mountain Flora of Greece, 7). Nordic J. Bot. 1(1): 66–82.

Tutin, T.G., V.H. Heywood, N.A. Burges, D.M. Moore, D.H. Valentine, S.M. Walters, and D.A. Webb. 1964–1980. Flora Europaea. 5 vols. Cambridge University Press, Cambridge.

Vöth, W. 1984. Bestäubungsbiologische Beobachtungen an griechischen

Ophrys-Arten. Mitt.-Bl. Arbeitskr. Heim. Orch. Baden-Württ. 16: 1–20.
——. 1987. Neue bestäubungsbiologische Beobachtungen an griechischen *Ophrys*-Arten. Mitt.-Bl. Arbeitskr. Heim. Orch. Baden-Württ. 19: 112–118.
Webb, D.A. 1978. Flora Europaea — a retrospect. Taxon 27(1): 3–14.

5. *GRIMMIA* (MUSCI): A VARIETY NEW TO THE LEBANON AND A NEW SPECIES FROM GREECE

C.C. TOWNSEND

392, Staines Road, Twickenham, Middlesex TW2 5JA, England

ABSTRACT

Grimmia crassifolia Lindb. ex Broth. var. *cucullata* Baumg. new to the Lebanon is described and figured, including the fruit, hitherto apparently unknown in this species. *G. meteorae* sp. nov. is described from Greece (Thessaly).

Key words: Bryophyta (mosses), *Grimmia*, new species, new variety, Greece, Lebanon.

Recent retirement has enabled me to make further inroads into a large backlog of random collections of mosses from the Mediterranean region, almost entirely gathered during many years' travelling as a botanical guest lecturer with Swans' Hellenic Cruises. Although the greater proportion of these are of little interest, two gatherings of *Grimmia* seem worthy of notice.

GRIMMIA CRASSIFOLIA Lindb. ex Broth. var. CUCULLATA Baumg. in Ann. Naturhistor. Mus. Wien 50: 534 (1934). Figs. 1 & 2

LEBANON: on rocks on the hillside above Baalbek, very dry situation, 5 May 1974, *C.C. Townsend 74/224* (E & Herb. Townsend), *74/233* (Herb. Townsend).

When I finally came to examine this moss in earnest I was considering only the first gathering, which is pure but sterile, as no doubt most further collections will prove to be. Its size and appearance led me to believe that I had gathered non-fruiting material of *G. pitardii* Corb. (with which I find *G. gibbosa* S. Agnew to be identical, from a fragment of a paratype in my herbarium kindly given to me by Mrs Agnew — as already suggested in Long, Crundwell & Townsend, 1981). The areolation of the two species is, however, quite different, the basal marginal cells of *G. pitardii* being equally thickened and the inner basal cells longer, larger and thin-walled. The areolation of the Lebanese moss is much more evocative of *G. laevigata*. Another species considered was *Grimmia singarensis* Schiffn., but this is a *Schistidium*, without the thick transverse walls of the basal marginal cells of the Lebanese moss, which are those of a *Grimmia*. *G. singarensis* was placed by Agnew & Vondráček (1975) in the synonymy of *Schistidium apocarpum* (Hedw.) B.S.G. var. *atrofuscum* (Schp.) Husn., and as a mere synonym of *S. apocarpum* by Bremer (1980, p. 95). *Grimmia cucullata* Henderson was also placed under *S. apocarpum* by Bremer (ibid. p. 96).

I was considerably misled for a while by the description of *G. crassifolia* Lindb. ex Broth. given by Savicz-Ljubitskaya & Smirnova (1970), as well as the original

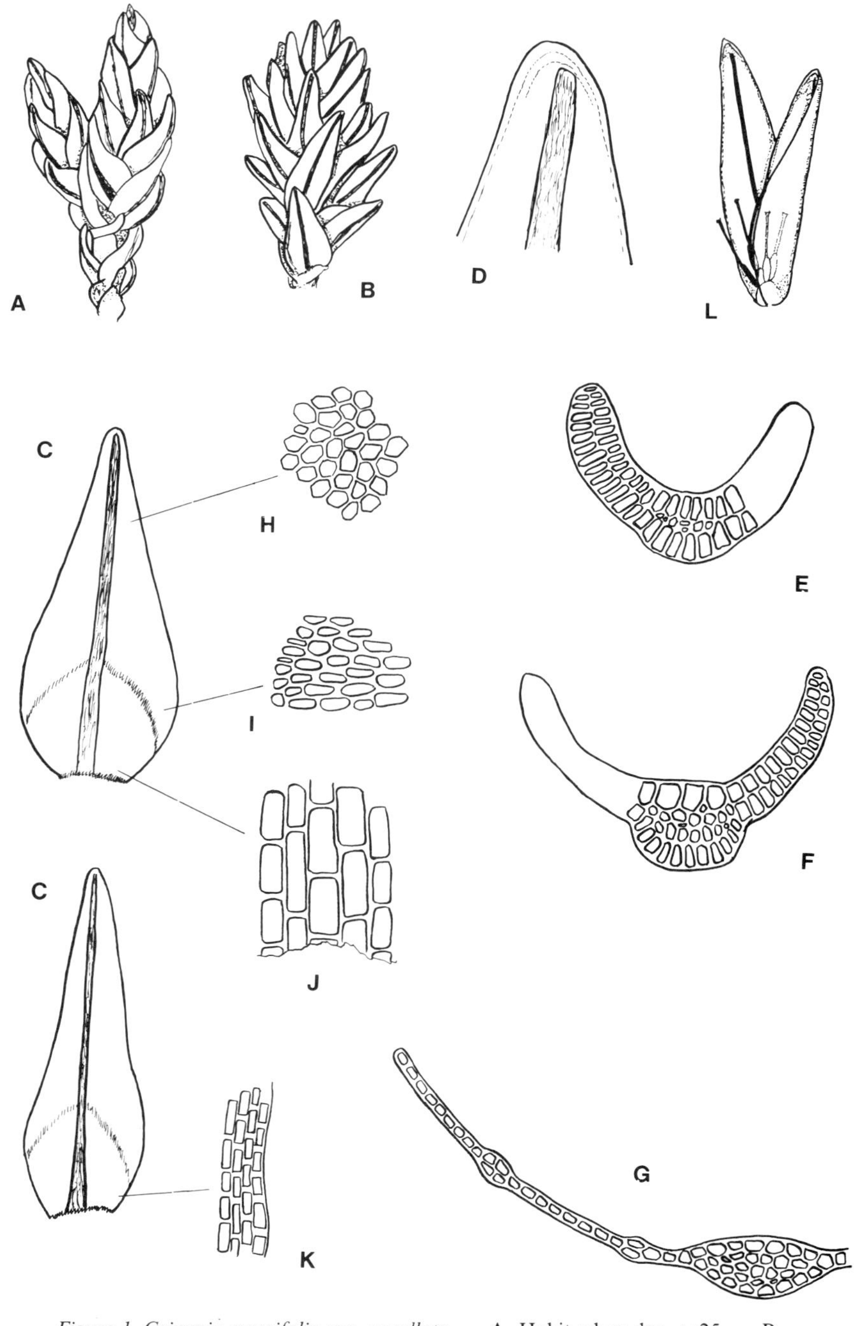

Figure 1. Grimmia crassifolia var. *cucullata*. — A. Habit when dry, × 25. — B. Habit when moist, × 25. — C. Leaves, × 60. — D. Leaf tip, × 110. — E. Transverse section of leaf towards apex, × 400. — F. Transverse section of leaf above middle, × 400. — G. Transverse section of leaf near base, × 550. — H. Cells of lamina above mid-leaf, × 700. — I. Cells of lamina within shoulder, × 700. — J. Cells of lamina at mid-base, × 700. — K. Cells at basal margin, × 250. — L. Female inflorescence, × 60. — All from *Townsend 74/224*.

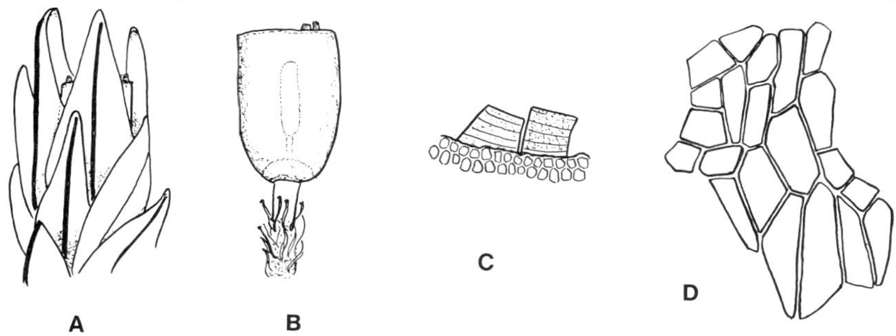

Figure 2. Grimmia crassifolia var. *cucullata.* — A. Perichaetium enclosing capsule, × 20. — B. Capsule, × 30. — C. Fragments of peristome, × 230. — D. Cells of exothecium, × 700. — All from *Townsend 74/233.*

description of this species, which make no allowance for its having a (then unknown) muticous variety. It was not until later, when seriously considering describing the Lebanese plant as new, that I came across *G. crassifolia* var. *cucullata* Baumg. whilst searching the moss literature of the Middle and Near East known to me. This could easily have been overlooked since it was not picked up for the 'Index Muscorum'. The variety appears to be more widespread than *G. crassifolia* var. *crassifolia*, in which the upper leaves are piliferous. It has been found in N & NE Afghanistan, N Iran and now the Lebanon. As far as I have ascertained, var. *crassifolia* has only been found in the Caucasus and extreme NC Afghanistan.

It was not until this paper had actually been submitted for publication that scattered taller stems of the same moss (*T.74/233*) were found in a mixed gathering of other mosses collected in the same locality on the same day — *Grimmia crinita* Brid., *G. laevigata* (Brid.) Brid., *G. ovalis* (Hedw.) Lindb., *Tortula caninervis* (Mitt.) Broth., and two fruiting stems of *Funaria pulchella* Philib. Much of the *G. crassifolia* var. *cucullata*, however, grew from among a collemaceous lichen. To my great surprise and delight two capsules were found; as far as I am aware, the capsule of *G. crassifolia* has not previously been encountered. It was more surprising that the capsules were immersed, In view of the similarity of the leaf structure to that of *G. laevigata*, one might have expected a similar long-exserted capsule. In spite of the short seta, however, *G. crassifolia* must surely be placed in Section *Guembelia*; the capsule base is quite regular with the seta centrally inserted.

The bryophyte flora of the arid regions of the Near and Middle East being so poorly known, a figure and full description of *G. crassifolia* var. *cucullata* is given; it is thought expedient to provide a Latin description of the fruit (see below). The original description of *G. crassifolia* (Brotherus 1892) is good and in Latin.

Dioicous. Stems simple or slightly branched, densely caespitose, c. 3–5 mm tall, saxicolous. Leaves closely appressed when dry, slightly spreading when moist, lanceolate, c. 2 mm long with the apex obtuse and cucullate, margins plane throughout; lamina for the most part bistratose with obscure, smooth, rounded-hexagonal cells 6–9 μm wide, $\pm$ equally incrassate or slightly collenchymatous, the lower part unistratose or with a few bistratose striae, in the widest part and slightly below near the margin with transversely elongate, very incrassate cells; interior basal cells quadrate or shortly (to c. 2.5 × 1) rectangular; basal marginal cells quadrate or shortly rectangular, hyaline or pale, in c. 5 rows, the transverse walls much thicker than the longitudinal. Nerve dorsally obtuse, the widest part up to c. 60 μm across, the median ventral cells in the lower one-third rectangular and often ± sinuose, quadrate above, in section with the cells homogeneous or the central smaller. Capsule described below. Female inflorescence with the outer perichaetial leaves similar to those of the stem, the inner shorter, more sheathing and with the basal cells laxer.

Capsula breviter oblonga, 1.25 mm longa, immersa; seta brevissima, recta, 0.4 mm longa; peristomium verosimiliter perfectum, in specimine meo vetustum fractum, c. 56 μm longum dentibus rubellis leviter papillosis non perforatis; cellulae oris subquadratae, rubellae, c. 12 μm latae; cellulae exothecii valde irregulares, c. 21–52 μm, haud collenchymatosae; sporae 12–14 μm, leviter papillosi, numerosae vacuae. Columella persistentia. Calyptra et operculum ignoti.

Var. *crassifolia* not only has the upper leaves with a hair-point, but the leaves are marginally broader with a broader apex.

It is difficult to say how frequent this moss is at Baalbek. I had only about an hour to search, and the attendance for most of this time of a polite but adhesive Palestinian bearing an unpleasant looking machine pistol tended to reduce concentration! It was certainly not common.

Grimmia meteorae C.C. Townsend **sp. nov.** Fig. 3

G. trichophyllae Grev. proxima sed foliis subsecundis superne bistratosis marginibus usque 4-stratosis inferne striis numerosis bistratosis denique prope basin unistratosis cellulis haud sinuosis atqua capsula matura laevi differt; a *G. elatiore* Bruch ex Bals. & De Not. cellulis laevibus atque striis bistratosis foliorum differt.

Planta dioica dense caespitosa caulibus ± ramosis, c. 15–25 mm alta, saxicola, Folia in siccitate sursum subappressa pilis hyalinis rectis vel flexuosis, humectata distincte subsecunda, lanceolata, usque 2.5 mm longa pilis hyalinis exclusis; pili hyalini usque c. 0.75 mm, leviter denticulati; margines plerumque utrinque saltem prope basin anguste reflexi; lamina parte superiore bistratosa margine usque 4-stratosa, parte maxima media striis numerosis bistratosis eis *Coscinodonti cribrosi* simillimis praedita, parte basali (c. 1/6) unistratosi; cellulae superiorae laevissimae hexagonae vel subquadratae 11–14 μm parietibus haud incrassatis; cellulae basales quadratae vel breviter (usque 2.5, rarissime 3, × 1) rectangulares; cellulae basales marginales quadratae vel breviter rectangulares parietibus transversalibus quam eis longitudinalibus multo magis incrassatis, in seriebus c. quinque dispositae; cellulae omnes haud sinuosae. Costa usque c. 90 μm lata cellulis ventralibus centralibus ad apicem elongatis, cellulis dorsalibus quadratis, in sectione transversali cellulis ± homogeneis internis minoribus. Folia perichaetialia eis caulis similes sed magis abrupte angustata, exteriora longe pilosa, interiora breviter pilosa vel mutica non bistratosa, magis vaginantes, cellulis inferioribus laxioribus. Seta usque c. 8 mm longa in siccitate recta humectata cygnea; capsula (vetusta) breviter oblongo-elliptica, c. 1.5 mm longa, laevis, cellulae oris atque exothecii eis *G. trichophyllae* simillimi; sporae c. 12 μm, tenuiter papillosae; operculum longe rostratum; dentes exteriores peristomii vetusti, fracti, truncati, usque c. 90 μm longi, sanguinei, leviter papillosi.

GREECE. THESSALY: on damp stony bank near the Varlaam Monastery, the Meteora, near Kalabaka, apparently calcareous, 10 May 1987, *C.C. Townsend 87/125* (E, holotype; Herb. Townsend, isotype).

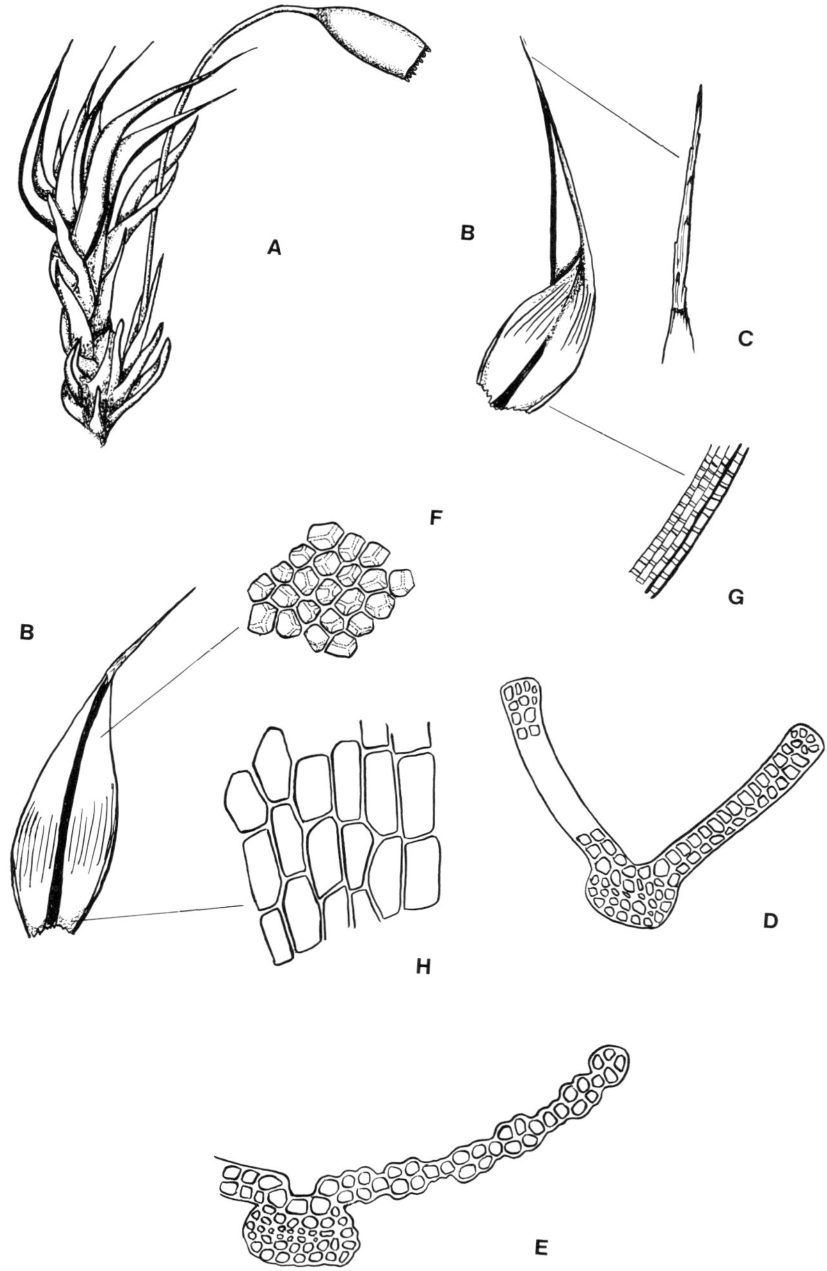

Figure 3. Grimmia meteorae. — A. Habit when moist, × 18. — B. Leaves, × 40. — C. Hairpoint of leaf, × 120. — D. Transverse section of leaf above middle, × 380. — E. Transverse section of leaf towards base, × 380. — F. Cells of lamina above mid-leaf, × 640. — G. Cells at basal margin, × 280. — H. Cells at mid-base, × 720. — All from the type.

The extraordinary conglomerate boulders and kopje-like extrusions of and around the Meteora are not rich in mosses, bearing great quantities of *Grimmia ovalis* (Hedw.) Lindb., together with other species such as *Hedwigia ciliata* (Hedw.) P. Beauv. *Grimmia trichophylla* Grev. also occurred as the usual southern European var. *meridionalis* Schp., with short basal cells and very divergent leaves. The new species was found on rather damp rocky ground in fairly close proximity to such mosses as *Funaria pulchella* Philib., *Gymnostomum luisieri* (Serg.) Serg. ex Crundw. and *Tortula cuneifolia* (With.) Turn. — the first two calcicoles and the last in my experience more indifferent as to substratum. Nevertheless, I would hesitate to cite *Grimmia meteorae* as a calcicole in such a complex area.

G. meteorae was certainly collected on its general appearance as something differing from *G. ovalis* and *G. trichophylla*, but it was a considerable surprise on viewing the leaves under the microscope to find the *Coscinodon*-like pattern of bistratose ridges. As far as I was then aware, the only European *Grimmia* species displaying such a pattern are *G. caespiticia* (Brid.) Jur. and *G. alpestris* (Web. & Mohr) Schleich. ex Hornsch., from which last moss hours of endeavour have not enabled me to separate *G. sessitana* De Not. The cygneous seta totally eliminates the Greek moss with respect to either of these. *G. elatior*, with leaves entirely bistratose above without bistratose striae, is ruled out by its papillose cells (always clear at least here and there in section) and coarser habit with the leaves not subsecund. The species is certainly closest to *G. trichophylla* but cannot I think fall within the range of variation even of that protean species, for the following reasons:

1. The *Coscinodon* – like striae coupled with the uniformly bistratose upper part of the lamina and up to 4-stratose margins. Such characters are normally very reliable in *Grimmia* and I have seen no approach to them in any material of *G. trichophylla* which I have examined.
2. The subsecund posture and more rigid appearance of the leaves of *G. meteorae*.
3. Not a trace of sinuosity has been seen in any of the cells of the lamina of the many leaves of the new species examined. In my experience at least some cells, particularly in the supra-basal region, are found to be obviously if not strongly sinuose in *G. trichophylla*.
4. The capsules are rather old, but appear to have none of the ridges normal in those of *G. trichophylla*. Admittedly these are sometimes less distinct in var. *meridionalis* but nevertheless usually evident.
5. The material of *G. trichophylla* var. *meridionalis* found nearly at the Meteora was quite typical of this moss, with no deviation at all to the new species; as with other material of the variety, it is hard to get an upper surface view of the leaves, as they are strongly reflexed and lie sideways on the slide. This is not the case with the new species.

I am grateful to Mr M.F.V. Corley for kindly examining this moss in view of his recent investigations of European *Grimmia* and for his observation '... I cannot get any nearer to naming it than you have', as well as for other suggestions which were in accord with my own line of thought, viz. the consideration of some more obscure European species. Two of these clearly needed closer investigation, viz.:

GRIMMIA PILOSISSIMA Herzog in Rev. Bryol. 38: 12 (1911). This Sardinian species has not been seen, but the original description and figure are enough to exclude it from consideration as the Greek plant. Certainly, Herzog describes the leaves as 'lamina superne bistratosa... striisque viridibus bistratosis notata...', and this is confirmed by Loeske (1930), who states 'Durch Doppelschichtigkeit ist der schmale Blatteil verdunkelt, und stärker verdickte Langsstreifen sind oft zu bemerken'. But characters separating *G. pilosissima* from *G. meteorae* are the leaves abruptly narrowed from a broad base; leaf margins plane below, incurved above, seta abruptly curved at the apex so that the capsule is cernuous (rather than a genuinely cygneous seta); the striate capsule; and the deep, yellow tufts with long hairpoints.

GRIMMIA FRAGILIS Schp. in Syn. Musc. Europ. ed. 2: 257 (1876). This was described from 'ad rupes graniticas in alpestribus provinciae lusitanicae Beira', and collected by Welwitsch in 1848. It was placed by Schimper between *G. trichophylla* and *G. hartmani*, and described as having the leaves '... e duplici strato cellularum quadratarum formatis'; Loeske (1930), however, also observed: 'Doppel – bis dreischichtige Striefen siehen sich hier und da neben der Rippe gegen den Halbscheidenteil herab'. I located the type among Schimper's specimens (BM-K), in the *G. donniana* folders in BM. The gathering number, not cited by Schimper, is *Welwitsch* 53. It is certainly not the same as the Greek moss. The leaf margins are plane below and incurved above; the seta, as with *G. pilosissima*, is abruptly curved at the apex, not cygneous; the bistratose ridges are much less distinct — as is the entire upper areolation since the cells are only c. 6–8 μm, about half the size of those of *G. meteorae*; and the leaves, as Schimper described them, are very fragile and broken across (though this is not the case with any other material in BM determined as *G. fragilis* — *Levier* 81 & 81 bis, with five specimens from Herb. Dixon). Whatever *G. fragilis* is, and most authors have associated it with *G. montana* (though whether from a study of the type I am unsure), it is certainly not the same plant as *G. meteorae*. If it is recognised as a species it needs a new name, since *G. fragilis* Schp. (1876) is a later homonym of *G. fragilis* F. Web. (1804). *G. fragilis* has usually been recognised as a species by Portuguese authors — e.g. Guimaraes (1929–30), though the binomial does not occur in Corley et al. (1981).

Since there is no other European *Grimmia* described which can seriously be considered, it will be seen that *G. meteorae* is the only European species of the genus which combines bistratose ridges with at least one basally reflexed leaf margin and a cygneous seta.

I have attempted quite unsuccessfully to identify the plant with N American and Asian *Grimmia* species — the latter rendered much easier by such excellent works as Savicz-Ljubitskaya & Smirnova (1970), Deguchi (1948) and Cao & Vitt (1986). Outside Europe its closest relative seems to be *G. percarinata* (Dix. et Sak.) Nog. ex Deguchi, which is an apparently rare Japanese endemic with distinctly sinuose and mammillose upper cells, only known in sterile condition.

LITERATURE CITED

Agnew, S., and M. Vondráček. A moss flora of Iraq. Feddes Repert. 86: 341–489.

Bremer, B. 1980. A taxonomic revision of *Schistidium* (Grimmiaceae, Bryophyta) 2. Lindbergia 6: 89–117.

Brotherus, V.H. 1892. Enumeratio Muscorum Caucasi. Act. Soc. Sc. Fenn. 19(12): 1–170.

Cao, Tong, and D.H. Vitt. 1986. A taxonomic revision and phylogenetic analysis of *Grimmia* and *Schistidium* (Bryopsida; Grimmiaceae) in China. J. Hattori Bot. Lab. 61: 123–247.

Corley, M.F.V., A.C. Crundwell, R. Düll, M.O. Hill, and A.J.E. Smith. 1981. Mosses of Europe and the Azores; an annotated list of species, with synonyms from the recent literature. J. Bryol. 11: 609–689.

Deguchi, H. 1978. A revision of the genera *Grimmia*, *Schistidium* and *Coscinodon* (Musci) of Japan. J. Sc. Hiroshima Univ. Ser. B, Div. 2 (Bot.) 16: 121–256.

Guimaraes, A.L. Machado. 1929–30. Sinopse das Briofitas de Portugal, Segunda Parte, Musgos. Bol. Soc. Brot. Ser. 2, 6: 180–265.

Loeske, L. 1930. Monographie der europaîschen Grimmiaceen. Bibl. Bot. 101: 1–236.

Long, D.G., A.C. Crundwell, and C.C. Townsend. 1981. New records of bryophytes from the Canary Islands. J. Bryol. 11: 521–536.

Savicz-Ljubitskaya, L.I., and S.N. Smirnova. 1970. Opredelitel' listostebel'nych mchov C.C.C.P. Verchoplodnye mchu. Komarov Botanical Institute of the Academy of Sciences of the U.S.S.R., Leningrad.

6. THREE NEW ENDEMIC TAXA OF *VERONICA* (SCROPHULARIACEAE) FROM TURKEY

MANFRED A. FISCHER

Institute of Botany, University of Vienna, A-1030 Wien, Austria

AND

AVNİ ÖZTÜRK

Biology Department, Faculty of Sciences and Arts, Atatürk University, Erzurum, Turkey

ABSTRACT

Two new, endemic species of *Veronica* L. (Scrophulariaceae) are described from Turkey: *V. kopgecidiensis* A. Öztürk & M.A. Fischer (Sect. *Veronicastrum*, in the *V. gentianoides* Vahl complex) and *V. cetikiana* A. Öztürk (Sect. *Veronica*, allied to *V. orientalis* Miller and *V. leiocarpa* Boiss.). Morphological and karyological variation in the *V. gentianoides* polymorphic complex is discussed; a new variety of *V. gentianoides*, var. *alpina* [Hausskn. ex] A. Öztürk & M.A. Fischer, is described to include most high-altitude ecotypes, while the rather distinct taxon var. *glacialis* Nábělek is raised to subspecific rank.

Key words: Scrophulariaceae, *Veronica*, new species, Turkey, taxonomy, variation.

INTRODUCTION

Recent research by the authors has further clarified the taxonomy of the difficult polymorphic species complex centred on *Veronica gentianoides* Vahl (Sect. *Veronicastrum*) and *V. orientalis* Miller (Sect. *Veronica*). In the former group, one new species and a new variety are described, and another variety raised to the rank of subspecies; in the latter, a second new species is described.

All specimens cited have been seen except where indicated otherwise by 'n.v.'. The citation of Turkish specimens follows the grid system used in *Flora of Turkey* (Edinburgh University Press, Edinburgh, 1965–1985) and explained therein by Davis (Vol. 1 pp. 1–3 & 9–11 and Maps 1–2, 1965).

TAXONOMY

Sect. Veronicastrum

The unpublished binomial '*Veronica pontica*' nom. nud. was used by Bornmüller (in sched.) to designate one of his collections of *V. gentianoides sensu lato* from Yildiz Dağ in Sivas, Turkey. Given the large amount of variation in the *V. gentianoides* complex, it is easy to understand that he later changed his opinion and reduced it to a variety, *V. gentianoides* var. *pontica* Bornm. Fischer (1979) did not recognise this taxon at all. However, Öztürk (1983, and unpublished doctoral and habilitation theses: '*V. gentianoides* subsp. *kopgecidiensis*, *V. gentianoides* subsp. *kopgetschidiensis*, *V. gentianoides* var. *latifolia*', all nomina nuda), as a result of extensive investigations of c. 60 populations, found sufficient evidence to distinguish it at subspecific level. Further research shows that this taxon can be recognised by nine correlated differential characters, and shows constancy of characters over a 1000 km range from Ak Dağ in central W Anatolia (Kütahya) east to Kop Dağ near Erzurum; moreover, it also has distinct ecological preferences. We therefore consider this evidence, presented below, adequate justification for the specific status awarded here.

Veronica kopgecidiensis A. Öztürk & M.A. Fischer, **sp. nov.** (Sect. *Veronicastrum*).
Icon.: Figs. 1a, 2a. Map: Fig. 4a

V. gentianoides Vahl var. *pontica* Bornm., Mitth. Thüring. Bot. Vereins, N.F., 20: 41 (1905).

V. pontica Bornm. in sched., nom. nud.

V. gentianoidi Vahl affinis sed praecipue foliorum rosula basali absente et foliorum basalium vaginis subnullis differt, clare perspicueque etiam foliorum caulinorum (3–) 4–5 (–6)–paribus distincta, praeterea rhizomate tenuiore (1–2 mm diam.), foliorum marginibus non vel brevissime tantum cartilagineis, corolla atrocaerulea, stylo longiore (6–) 7–9 (–10) mm longo, foliis plerumque obtusiusculis, pedicellis paulo brevioribus itaque racema densiore recedit. *Habitat* in pratis humidis montanis et subalpinis, alt. 1300–2400 m. *Fl.* Junio (— ad initium Julii).

Hemicryptophytic, herbaceous perennial with creeping stolons ending in rosettes (during vegetative stage), forming rather loose turfs (not dense mats). *Rhizome* thin, naked, at most 1–2(–3) mm diam. *Basal leaves* absent or withered by flowering time, with 1–4 thin stolons (3–7 cm × c. 1 mm) arising from their axils; sheath of basal leaves very short (at most 1 mm long) or ± absent; petioles rather broad, ± winged, c. 5–20 mm long; lamina oblong to elliptic, 20–45 × (3–) 4–9 (–11) mm. *Cauline leaves* (3–) 4–6 pairs, internodes subequal and leaves thus ± evenly spaced along stem (not mainly near base), oblong to elliptic, upper ones sessile, usually linear-oblong to oblanceolate or subspathulate, apex usually obtuse; margins not or only very slightly cartilaginous, indistinctly and remotely crenulate-serrulate to subentire. *Stem and leaves* glabrous to ± densely glandular-pubescent; glandular hairs 0.2–0.6 mm long, 4–6-celled. *Transitional sterile bracts* only 0–3 (–4), usually alternate. *Flowering shoots* suberect to erect, 30–50 cm tall, unbranched, ending in a single terminal raceme. *Raceme* (15–) 20–40 (–50)-flowered; lowest bracts slightly larger than upper; pedicels 4–10 mm long in flower, 7–13 mm long in fruit, always (0.7–) 1–2 × bracts; raceme densely glandular-pubescent with 0.2–0.6 mm long glandular hairs and few thin very short scattered eglandular hairs. *Calyx* 3–5 mm long, lobes 4, subequal or 2 twice as long as other 2. *Corolla* very dark, deep blue, 14–17 mm diam., usually slightly more zygomorphic than in *V. gentianoides*; tube conical, 1–1.5 mm long, 1–1.5 mm diam., glandular-puberulent. *Style* (6–) 7–9 (–10) mm long, distinctly longer than in most specimens of *V. gentianoides*. *Capsule* suborbicular or broadly ovate, (4–) 5–6 × (4–) 5–5.5 mm, emarginate, glandular-pubescent (hairs 5-celled, 0.2–0.4 mm long). *Seeds* 30–40. *Fl. June (– July). Moist to wet meadows, 1300–2400 m.*
Typus: TURCIA BOREO-ORIENTALIS: A8 Gümüşhane/Erzurum: jugi transitum

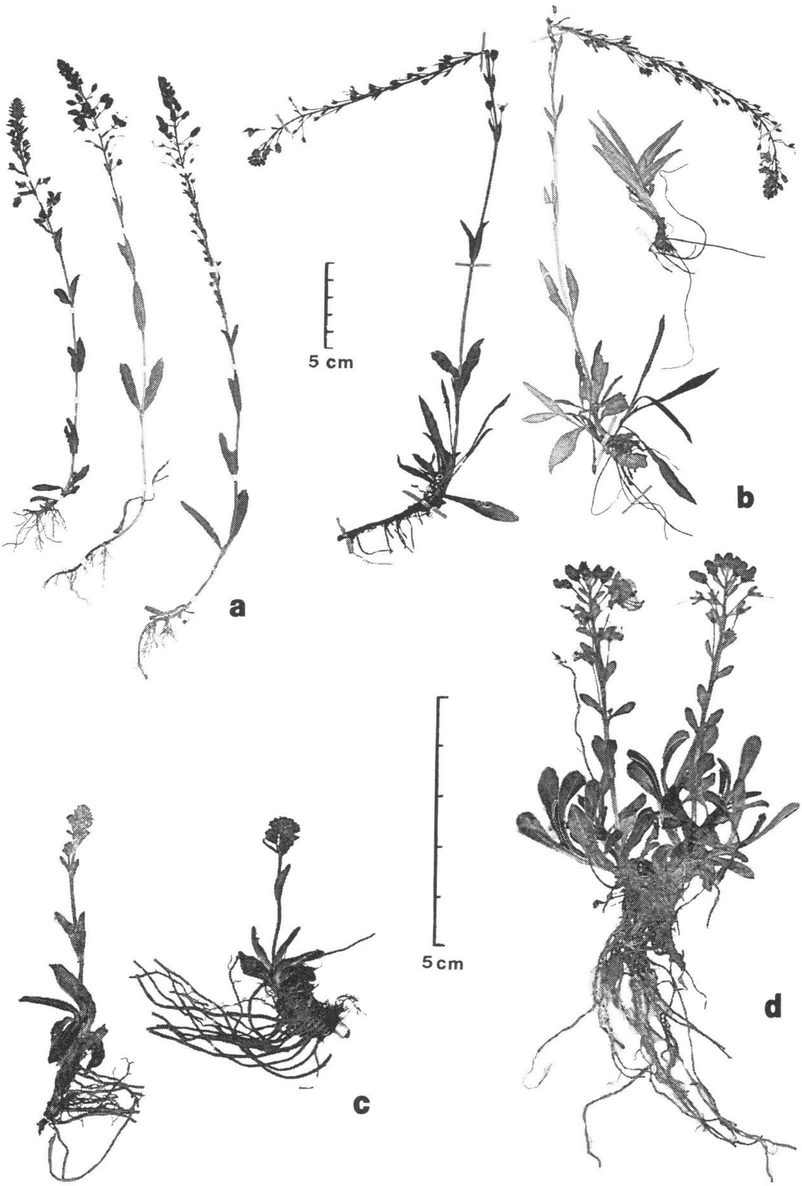

Figure 1. Photographs of herbarium specimens of *Veronica* taxa to show habit. — *a. Veronica kopgecidiensis* (*I. Bozakman & K. Fitz 813*). — *b. V. gentianoides* subsp. *gentianoides* var. *gentianoides* (*Davis & Hedge D. 29597*). — *c. V. gentianoides* subsp. *gentianoides* var. *alpina* (*Siehe 1898: 201*, isotype). — *d. V. gentianoides* subsp. *glacialis* (*Mitchell, Cheese & Watson 3586*). — (Scale for *a*, *b* different from that for *c*, *d*; both indicated.)

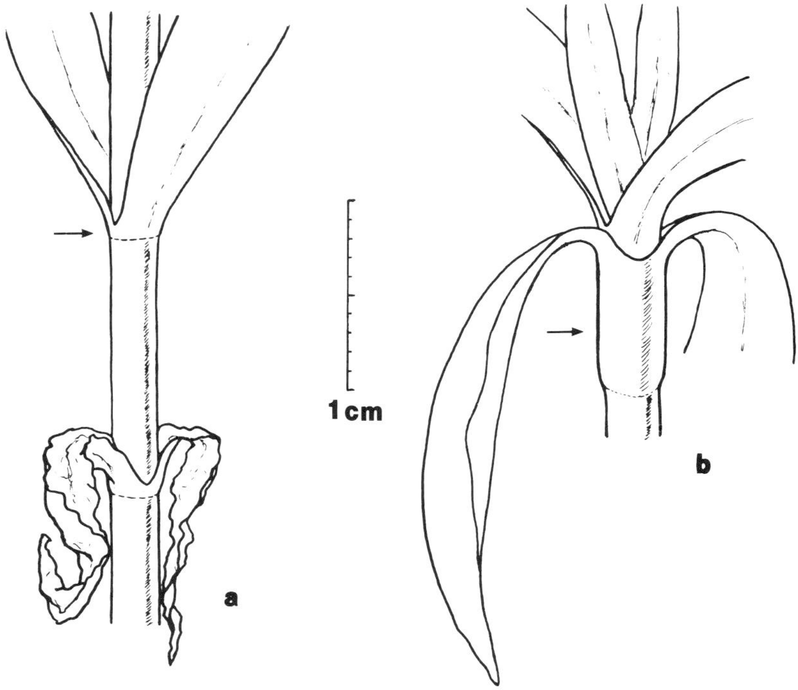

Figure 2. Base of flowering stem of *Veronica kopgecidiensis* (*a*) and *V. gentianoides* (*b*) to show differences in the basal sheaths (arrowed) of the lower leaves.

Kopdaği geçidi (= Kopgeçidi dictum) inter pagos Aşkale et Bayburt, 82 km viarii ab urbe Erzurum, in prato humido subalpino, unacum *V. hispidula* Boiss. & Huet, *V. pusilla* Kotschy, *V. reuteriana* Boiss. et *V. serpyllifolia* L., 2360 m, 28 Jun. 1976, *Avni Öztürk 46* (holo. Hb. A. Öztürk, iso. ATA, ANK, EGE, HUB, ISTE, ISTF, WU, Hb. Selçuk Üniv. Konya).

Specimens seen: TURKEY. A3 Bolu: Akçakavak yaylasi, 1500 m, 9 Jun. 1980, *Y. Akman & E. Yurdakulol 10816* (ANK). A4 Ankara: 26 km N of Kizilcahamam, moist meadow, 1300 m, 24 Jun. 1970, *F. Sorger 70-5-93* (Hb. Sorger); *ibid.*, pond margin, 1300 m, 14 Jun. 1971, *F. Sorger 71-3-7* (Hb. Sorger, WU), Çamkoru, NW of Kizilcahamam, 1350 m, 4 Jun. 1970, *I. Bozakman & K. Fitz 813* (W). Karagöl, 40 mi (64 km) N of Ankara, long grass clearing in dense *Carpinus-Quercus* woodland on N-facing slope, 1600 m, 10 Jun. 1965, *M.J.E. Coode & B.M.G. Jones 2164* (E). A4 Kastamonu: Ilgaz Dağlari, 35 km S of Kastamonu, 1 km E of top of pass, grassy slopes, 1760 m, 28 Jun. 1971, *J.R. Edmondson 491* (E, W). A4/5 Kastamonu: d. Tosya, 'Küsentschair ad Giaurdagh' (= Gâvur Daği), 23 Jul. 1892, *P. Sintenis 1892: 3870 p.p.* (with *V. gentianoides*: WU). A6 Tokat/Sivas: [Yildiz Dağ] in pratis alpinis, 2100–2300 m, 6–7 Jun. 1890, *J. Bornmüller 1659* (type of var. *pontica*; G, JE, W, WU, Z). A8 Gümüşhane: Kop Dağ pass summit, 2400 m, 3 Jul. 1960, *Furse & Synge 860* (E); *ibid.*, 11 Jul. 1981, *A. Öztürk 662* (ATA, Hb. A. Öztürk). B2 Kütahya: d. Simav, between Kiçir and Ak Dağ, clearing in *Pinus nigra* forest, 1700 m, 19 Jun. 1965, *M.J.E. Coode & B.M.G. Jones 2672* (E). B3 Eskişehir: Sündiken Daği, 22 May 1971, *T. Ekim* (*ANK 818*) (E); *ibid.*, 1700 m, 30 Jun. 1970. *T. Ekim* (*ANK 808*) (E).

V. kopgecidiensis is endemic to Turkey and closely allied to the very polymorphic species *V. gentianoides* Vahl. The latter is distributed almost throughout the mountains of Asia Minor (Donner 1985: 97; Fischer unpublished), Caucasia, NE Iraq and W Iran (Kurdistan) (Fischer 1981: 140–141).

The following taxa, together with *V. kopgecidiensis*, are included in the *V. gentianoides* group and are discussed below:

V. gentianoides sensu lato (incl. *V. charadzae*, *V. imerethica*, *V. kemulariae*, *V. sumilensis*):
- subsp. *gentianoides*
  - var. *gentianoides* var. *alpina*;
- subsp. *glacialis*

V. kopgecidiensis
V. schistosa
V. davisii

V. kemulariae Kuth. (Great and Lesser Caucasus), *V. charadzae* Kem.-Nath. and *V. imerethica* Kem.-Nath. (both Central Caucasus) were all treated as distinct species by Borisova (1955). Elenevskij (1978) reduced the first two to synonymy but accepted – with some reserve – *V. imerethica* as a separate species. The latter is said by both the above authors to differ from *V. gentianoides* by having only 1–2 pairs of cauline leaves (not 3 or more – evidently including the transitional sterile bracts), which are softer in texture, and green (not glaucous). Other taxa in the complex are *V. sumilensis* Freyn & Sint. (N Anatolia: Fischer 1979 in obs., p. 706) which resembles dwarf forms of *V. gentianoides* (see below) and which may be a good taxon although it is only known from two syntypes (only one seen by us!) and further material is needed to assess its status; *V. schistosa* E. Busch (C & W Great Caucasus), an ally of *V. gentianoides* which is generally accepted as a good species; and *V. davisii* M.A. Fischer (SE Turkey, NE Iraq: Fischer 1979, 1981), which may be connected to the group through *V. gentianoides* subsp. *glacialis* (see below).

Variation within V. gentianoides. – This concerns size (to 100 cm high); leaf colour (green to ± glaucous), shape (suborbicular or rhombic-elliptic to narrowly linear-lanceolate), margin (entire to coarsely crenate-serrate) and size; leaf and stem indumentum (glabrous to densely pubescent; glandular or eglandular); raceme indumentum (glandular or eglandular); corolla colour and size (8–14 mm diam.); style length (3–7 mm); and shape and size of the capsule.

The number and arrangement of the transitional sterile bracts (between the cauline leaves and flower-bearing bracts of the raceme) are also very variable. Sterile bracts may be 2 to several, the lowest often opposite but at least the rest alternate; quite commonly, all cauline leaves are alternate and ± bract-like. Frequently, these characters display considerable variation even on a single herbarium sheet – an indication of high intra-populational variation. The sheath of the lowermost (rosette) leaves is, however, always present.

Alpine ecotypes. – Large plants (more than 50 cm tall) are more frequent in the western parts of the species' range (Caucasus to easternmost Turkey). Dwarf ecotypes occur at high altitudes; *V. imerethica*, *V. kemulariae* and perhaps even *V. sumilensis* may represent such alpine variants. Another such form, collected by Siehe on Erciyas Dağ (B5 Kayseri) was named var. *alpina* in sched. (nom. nud.)

by Haussknecht, although apart from its reduced size it differs from *V. gentianoides* in no essential characters. We consider that these dwarf alpine ecotypes, though probably of polytopic origin (occurring as they do on at least six different, geographically scattered high mountains) should be united taxonomically and given varietal status within *V. gentianoides*.

Veronica gentianoides Vahl var. **alpina** [Haussknecht ex] A. Öztürk & M.A. Fischer, **var. nov.** Icon.: Fig. 1c. Map: Fig. 4c.

A varietate typica statura minore (4–10 cm alta), foliis minoribus, ovato-lanceolatis usque lineari-lanceolatis, in apicem attenuatis, 5–25 mm longis, 2–6 mm latis, breviter petiolatis vagina basali 2–5 mm longa, 1–2 foliorum caulinorum paribus, racema 5–15(–18)-flora, stylo 3–5 mm longo differt.

Typus: TURCIA. B5 Kayseri: Dömbelek Daği, 'feuchte Alpentriften' [moist alpine pastures], Jul. 1895, *W. Siehe 1895:201* (holo. WU, iso. E, JE).

Specimens seen: TURKEY. A2(A) Bursa: Ulu Dağ, Jun.–Jul. 1968, *J. Contandriopoulos, A. Pamukçuoğlu & P. Quézel 409/19* (MARS). A8 Rize: Kaçkar Dağ, 30 km SE of Çamlihemşin, Kavron valley, c. 2100 m, 10 Sept. 1980, *F. Sorger 80-62-69* (Hb. Sorger). B5 Kayseri: Erciyas Dağ, N crater, c. 2500–3000 m, 29 May 1902, *E. Zederbauer* (W, WU); *ibid.*, 3000 m, 22 Jul. 1898, *W. Siehe 1898: 221* (JE, WU); *ibid.*, 3200 m, Jul., Aug. 1856, *B. Balansa 1028* (W); *ibid.*, 2400 m, 5 Aug. 1976, *F. Sorger 76-28-31* (Hb. Sorger, Hb. M.A. Fischer); *ibid.*, steep scree, 2700 m, 29 Aug. 1977, *F. Sorger 77-76-50* (Hb. Sorger, WU); *ibid.*, E slope, 2500–2750 m, 4 Aug. 1978, *F. Ehrendorfer et al. 787-176-22* (WU). B7 Tunceli: Munzur Dağ, above Ovacik, 2650 m, 19 Jul. 1957, *Davis & Hedge, D. 31422* (E). B8 Erzurum: mts. S of Erzurum towards Çat, stony alpine slopes, c. 3000 m, 3 Aug. 1968, *E.M. Rix et al. 1056* (E).

Other dwarf high-altitude populations of *V. gentianoides* occur in the mountains of SE Turkey and adjacent NE Iraq and W Iran. These have been named var. *glacialis* Nábělek, a taxon mentioned but not fully accepted by Fischer (1979, 1981). Similar plants from Süphan Dağ were recognised by Öztürk in 1977 (unpublished doctoral thesis) under the provisional and ineffectively published name 'subsp. *süphandaghensis*'. Examination of all available material reveals this taxon to be rather uniform, and distinct from the other alpine ecotypes lumped above under var. *alpina*. They have subspathulate laminas and slightly longer pedicels than typical *V. gentianoides*. Although this taxon of Nábělek is restricted to the area indicated, it does not completely replace subsp. *gentianoides* as there exists one collection of the latter from within the range of var. *glacialis*: C9/10 Hakkari: Cilo Dağ, 10 km W of Cilo Tepe, 3600 m, 9 Aug. 1954, *P.H. Davis & O. Polunin, D. 24190* (E, W). This indicates that Nábělek's taxon deserves a higher taxonomic rank than a mere polytopic ecotype and it is here accordingly raised to subspecific rank.

Veronica gentianoides Vahl subsp. **glacialis** (Nábělek) A. Öztürk & M.A. Fischer, **stat. nov.** Icon.: Fig. 1d. Map: Fig. 4d.

V. gentianoides Vahl var. *glacialis* Nábělek, Spisy Přír. Fak. Masarykovy Univ. 70:34 (1926).

Descriptio emendata: A subsp. *gentianoidi*, imprimis a var. *alpina*, foliis oblanceolato-subspathulatis, apice obtuso-rotundatis differt.

Plant 5–10(–13) cm tall. *Leaves* (incl. petiole) 7–22 mm long, 2–5 mm broad, oblanceolate-subspathulate, lower ones cuneate into a long petiole, with distinct 2–5 mm long basal sheath, forming a rosette. *Cauline leaves* sessile, subspathulate (not linear-lanceolate as in var. *alpina*), in 1–2 pairs; leaf margins subentire to serrulate, cartilaginous; transitional sterile bracts 0–10. *Raceme* (5–)10–20-flowered. *Pedicels* 7–10(–12) mm long, 1.5–2.5 × subtending bract. *Style* 3–5 mm long. *Fl. Jun.–Aug. Montane steppe, scree, earthy slopes, often nr snow, 2500–3800 m.*

Syntypes: TURKEY [C9 Siirt] distr. Ramoran: Halakur-Dar (Herakol Dağ) ad septentr. ab oppido Ğezâret-ibn-Omar (Cizre), ad nives tabescentes, c. 2500 m, 25 Jul. 1910, *Nábělek 2007* (BRNU, n.v.); C9 Hakkari/Van: mons Kela Mame (Mamemusa Dağ) supra pagum Hoz inter Hašitha (Aşutka) et Săttak (Çatak), ad nives alt. ca. 2600 m, 28 Jul. 1910, *Nábělek* 2010 (BRNU, n.v.).

Specimens seen: TURKEY. B9 Bitlis: Süphan Dağ, moist alpine steppe on igneous substrate, 2900 m, 9 Jul. 1975, *A. Öztürk 90* (ATA, WU). N cliffs above crater of Nemrut Dağ, rock ledges, 2700 m, 12 Aug. 1956, *McNeill 177* (E). Nemrut Dağ, moist mountain steppe, igneous substrate, 2500 m, 17 Jun. 1975, *A. Öztürk 28* (ATA). Pelli Dağ above Pelli, damp earth, 3000 m, 7 Jul. 1954, *P.H. Davis & O. Polunin, D. 22508* (E). B9 Van/Siirt: Kavuşşahap Dağlari S of Pişvanik, N side of Arpit, 2850–2950 m, 13 Aug. 1967, *Mitchell, Cheese & Watson 3586* (K). B9 Van: d. Şatak, Kavuşşahap Dağ, earthy scree, 3400 m, 23 Jul. 1954, *P.H. Davis & O. Polunin, D. 23090* (E). d. Gevaş, Artos Dağ, near snow, 3000 m, 15 Jul. 1954, *P.H. Davis & O. Polunin, D. 22860* (E, W). C9 Hakkari: Kara Dağ, earthy slope near snow, 3500 m, 16 Aug. 1954, *P.H. Davis & O. Polunin, D. 24458* (E). C10 Hakkari: Sat Dağ, NW of Sat Gölü, by stream from melting snow, 2850 m, 28 Jun. 1966, *P.H. Davis* 45509 (E). IRAQ. Prov. Arbil: Mt. Helgurd (Algurd Dagh) ad confines Persiae, in decl. occid. summi montis, substr. schist. metamorph. et serpent., 3000–3800 m, 10–14 Aug. 1957, *K.H. Rechinger fil. 11438-b* (W). IRAN: Prov. Azarbaijan-e Khavari: mt. Sabalan, 'Meshkim-Shahr', 3600 m, 2 Aug. 1970, *Izadyar 14102-E* (W).

Within the range of variation of *V. gentianoides*, as described above, we found no specimens with sheathless basal leaves – the main diagnostic character of *V. kopgecidiensis*. All specimens, both large and dwarf, had clearly developed sheaths, the lower leaves being aggregated in a rosette and with not more than two pairs of cauline leaves. Only rarely were specimens found showing a tendency towards early withering of the basal leaves, but this was not associated with any reduction in the sheath.

Karyological Variation. – The *V. gentianoides* group also seems to show considerable karyological variation (Tumadžanov et al. 1972, 1977). Many fundamental problems, including the establishment of the true chromosome base number (8 or 12?) remain unsolved and further studies are essential. The most common chromosome number reported for *V. gentianoides* is 2n = 48 (Zhukova 1967, Tumadžanov & Beridze 1969, Tumadžanov et al. 1974, Charadze et al. 1973, Gvinianidze & Avazneli 1980, 1982). Several other numbers are known: 2n = 16 (Tumadžanov & Beridze 1969), 24 (Tumadžanov et al. 1974), 32 (Tumadžanov & Beridze 1969, Pogosjan et al. 1971), 64 (Kliphuis & Wieffering 1979) and 80 (Tumadžanov & Beridze 1969). 2n = 48 is reported from *V. kemulariae* (Meškova 1965), while *V. schistosa* has 2n = 24 (e.g. Gvinianidze & Avazneli 1982).

Sect. Veronica

V. orientalis Miller is an enormously variable species (Fischer 1979, 1980, 1981). Further study of the vast amount of material in herbaria may result in the recognition of additional taxa. One very distinct taxon is very conspicuous within the complex on account of its erect habit, vivid green vegetative parts, and subentire or only minutely serrulate leaves with obtuse base. In these respects it is thus rather similar (by convergence only!) to the European *V. austriaca* L. subsp. *dentata* (Schmidt) Watzl. The distinctiveness of this taxon was first recognised by Öztürk (unpublished doctoral thesis, misidentified as '*V. stenobotrys*'; unpublished habilitation thesis, as '*V. çetikii*'; 1983, as '*V. tchetikii*', nom. nud.; '*V. akdoganensis*' in sched.). Fischer (1979), on account of the scanty material available at that time, did not separate it taxonomically at all. It is here described below as the new species *V. cetikiana*.

Veronica cetikiana A. Öztürk, **sp. nov.** (Sect. *Veronica*).

Icon.: Fig. 3. Map: Fig. 4f.

V. orientali Miller affinis, a qua differt habitu suberecto (nec prostrato nec ascendente), statura 30–50 cm alta, foliorum sub racemis paribus 5–7, foliis omnibus prasinoviridibus (nec glaucoviridibus), sessilibus, ovato-lanceolatis basi obtusa usque truncata vel subcordata (nec cuneata), margine aequaliter serrulatis usque subintegerrimis (nec basi incisis), supremis sicut in *V. orientali* lanceolato-linearibus integerrimis.

Suffruticose perennial, (20–)30–40(–50) cm tall, strongly branched at base. *Stems* suberect, unbranched, with (2–)4(–8) opposite lateral racemes, indistinctly bifariously pilose; hairs subadpressed, crisp, curved upwards, 0.1–0.6 mm long. *Internodes* 10–20 mm long near base, 20–45(–65) mm long in upper part. Leaves in 5–7 pairs below lowest inflorescences, usually with short vegetative shoots in their axils, grass green (drying yellowish-green), not at all glaucous, sessile; lamina ovate-lanceolate to oblong-lanceolate, (17–)22–32(–35) × (4–)7–10(–14) mm, base rounded, obtuse or subcordate, margin subentire or evenly and ± equally serrulate with up to 14 teeth, slightly but distinctly revolute; margin and lamina surfaces ± densely puberulent with upward-curved eglandular hairs 0.05–0.1(–0.2) mm long, upper surface often ± glabrescent. *Vegetative shoot apex* (above racemes) rosette-like to elongate, 1–8 cm long, with (2–)4–8 pairs of linear to oblong-linear, entire, revolute leaves (8–)15–20(–30) × 1.5–3(–6) mm. *Racemes* (20–)40–90-flowered, 4–10 cm long in flower and then ($\frac{1}{2}$–)$\frac{1}{5}$–$\frac{1}{10}$(–$\frac{1}{13}$) × total plant height, elongating to (7–)25–40(–70) cm in fruit; peduncle (2–)3–5(–7) cm long; inflorescence eglandular-puberulent with 0.05–0.15 mm long upwardly curved subadpressed hairs. *Bracts* narrowly linear-lanceolate, acute, entire, 2–5 mm long. *Pedicels* suberect, (3–)4–6(–8) mm long, (0.8–)1–1.5(–2) × bracts. *Calyx lobes* (4–)5, unequal, narrowly linear-lanceolate, acute, 2–4 mm long in flower, 4–5 mm long in fruit. *Corolla* sky- to violet-blue, 8–10 mm diam., lobes usually elliptic, acute, glabrous; tube whitish, 0.3–0.8 × 0.3–0.8 mm. *Capsule* obcordate, (2.5–)3–3.5(–5) × (2.5–)3–3.5(–5) mm, with shortly cuneate to rounded base, eglandular-puberulent; style (2.5–)4(–5) mm long. *Seeds* 5–10, broadly elliptic, flat, c. 1 × 1.7 mm. *Fl. June–July. Calcareous mountain steppes and meadows, field margins, (1300–)1800–2400 m.*

Typus: TURCIA: B8 Muş/Erzurum: in districto oppidi Hinis, in monte Hamorit Daği (= Akdoğan Daği), in declivium borealium steppa montana, 2200 m, 4 Aug. 1976, *A. Öztürk 304* (holo. Hb. A. Öztürk, iso. ANK, ATA, EGE, HUB, ISTE, ISTF, WU, Hb. Selçuk. Üniv. Konya).

Specimens seen: TURKEY. B8 Muş: Kizilağaç köyü, steppe, N-facing slope, c. 1340 m, 13 Jun. 1981, *Y. Altan 3424* (ATA). Buğlan geçidi, *F. Sorger 81-6-23* (Hb. Sorger). B8 Muş/Erzurum: Akdoğan (Hamorit) Dağ, N slope, 1950–2000 m, 4 Jul. 1976, *A. Öztürk 288*

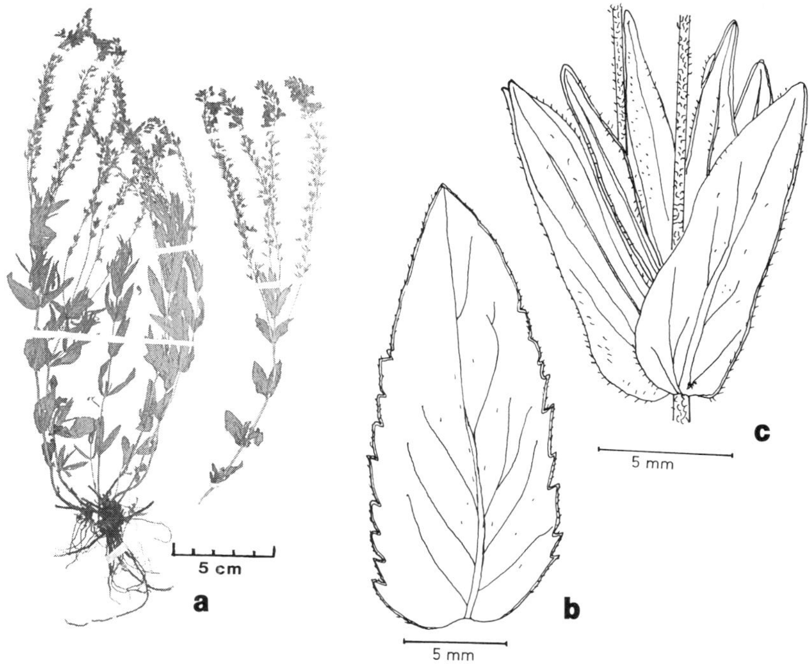

Figure 3. Veronica cetikiana (type collection). *a*. Photograph of two specimens to show habit. — *b*. Stem leaf. — *c*. Pair of stem leaves with vegetative axillary shoot initials. (Different scales as indicated.)

(ATA); *ibid.*, 2300 m, 4 Jul. 1976, *A. Öztürk 306* (ATA). B8 Erzurum: Hasankale (Pasinler) to Hinis, c. 1800 m, 10 Jul. 1975, *A. Öztürk 97* (ATA).

V. cetikiana, a stenoendemic species, also somewhat resembles *V. leiocarpa* Boiss. from S Anatolia (Içel, Adana, Hatay) and W Syria (sensu Davis, *Flora of Turkey* 1:13, 1965). The latter species is treated in Feinbrun-Dothan (1978) and Fischer (1980). *V. leiocarpa* differs from *V. cetikiana* by its usually larger, deeply serrate-crenate leaves which turn blackish after some time when dried; longer racemes with longer indumentum (hairs 0.3–0.7 mm long); fruiting pedicels strongly adpressed-erect, and larger capsules (4–6 × 3.5 mm) with ± thickened margin and truncate to subcordate base.

ACKNOWLEDGMENTS

We are grateful to Mag. Elvira Hörandl and Dr Ernst Vitek for help with the figures and to the following herbaria for allowing us to study their material: ANK, ATA, E, EGE, HUB, HUJ, ISTE, ISTF, JE, W, WU, Hb. Selçukensis Universitatis, Konya; Hb. Friederike Sorger, Wien.

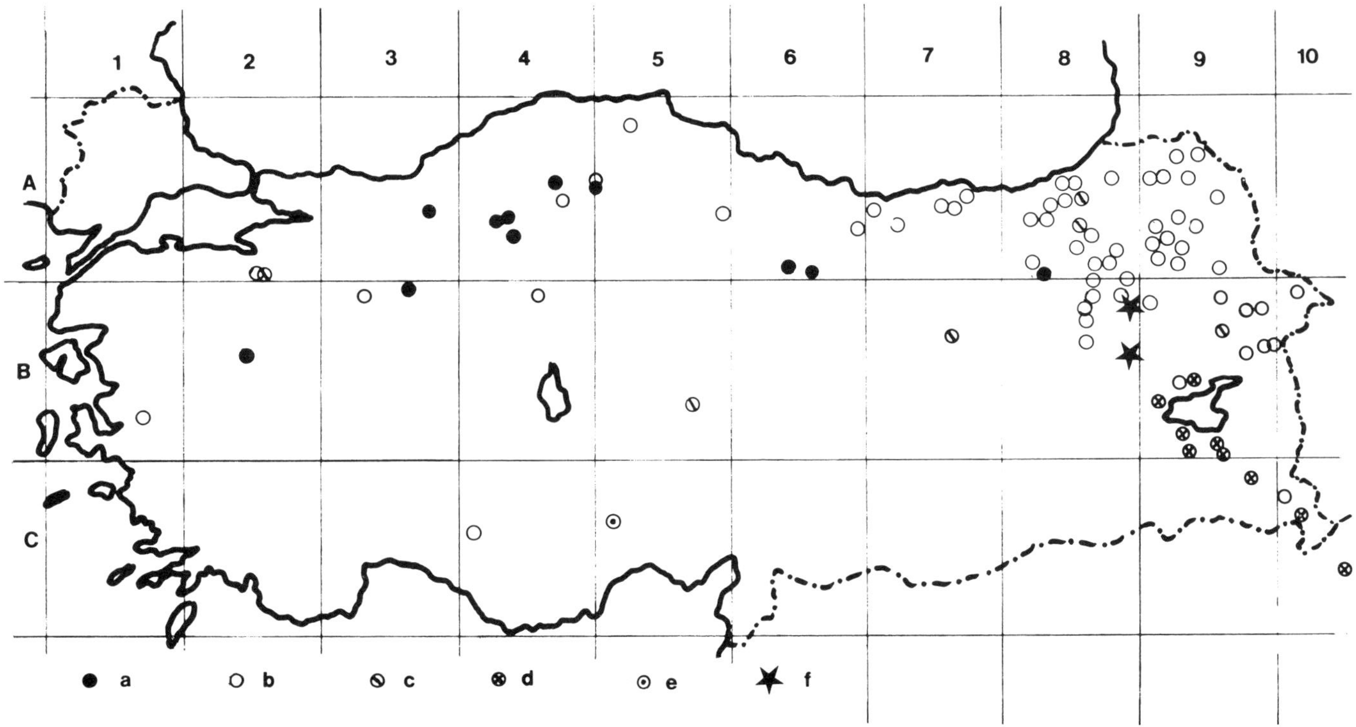

Figure 4. Geographical distributions of *Veronica* taxa in Turkey (and adjacent areas). — *a*. *V. kopgecidiensis* (all known records). — *b*. *V. gentianoides* subsp. *gentianoides* var. *gentianoides* (Turkish records only). — *c*. *V. gentianoides* subsp. *gentianoides* var. *alpina* (all known records). — *d*. *V. gentianoides* subsp. *glacialis* (total range with the exception of Mt. Sabalan in W Iran). — *e*. *V. gentianoides* s.l. (unrevised material, possibly including *V. kopgecidiensis*). — *f*. *V. cetikiana* (all records).

LITERATURE CITED

Borisova, A.G. 1955. *Veronica*, pp. 329–500. *In* B.K. Šiškin and E.G. Bobrov [Eds.], Flora SSSR. Vol. 22. Izdatelstvo Akademii Nauk SSSR, Moskva & Leningrad.

Charadze, A.L., Z.I. Gvinianidze, and M.T. Davlianidze. 1973. K kariologičeskomu izučeniju subnival'nogo florističeskogo kompleksa, I. Zam. Sist. Geogr. Rast. Akad. Nauk Gruzinsk. SSR (Tbilisi) 30: 78–83.

Davis, P.H. 1965. Flora of Turkey and the East Aegean Islands. Vol. 1. Edinburgh University Press, Edinburgh. 567 p.

Donner, J. 1985. Verbreitungskarten zu P.H. Davis: 'Flora of Turkey 1–8'. Linzer Biol. Beitr. 17(1): 1–120.

Elenevskij, A.G. 1978. Sistematika i geografija veronik SSSR i priležaščich stran. Nauka, Moskva. 258 p.

Feinbrun-Dothan, N. 1978 (as '1977'). Flora Palaestina. 3: Ericaceae to Compositae. The Israel Academy of Sciences and Humanities, Jerusalem. Text: 481 p. Plates: tt. 757, i–xxxiii, 21 p. in Hebrew.

Fischer, M.A. 1979 (as '1978'). *Veronica*, pp. 685–753. *In* P.H. Davis [Ed.], Flora of Turkey and the East Aegean Islands. Vol. 6. Edinburgh University Press, Edinburgh.

——. 1980. *Veronica*, pp. 242–265. *In* P. Mouterde [ed. A. Charpin], Nouvelle flore du Liban et de la Syrie. Vol. 3. Dar El-Machreq Éditeurs, Beyrouth.

——. 1981. *Veronica*, pp. 52–165 and tt. 43–110, 247, 248, 255. *In* K.H. Rechinger [Ed.], Flora Iranica 147. Akademische Druck- u. Verlagsanstalt, Graz.

Gvinianidze, Z.I., and A.A. Avazneli. 1980. Kariologičeskoe issledovanie florističeskich kompleksov. Investigatio caryologica complexus floriscíti [sic: *recte*, floristici]. Zam. Sist. Geogr. Rast. Akad. Nauk Gruzinsk. SSR (Tbilisi) 36: 71–74.

——, and ——. 1982. Chromosomnye čisla nekotorych predstavitelei vysokogornych florističeskich kompleksov Kavkaza. Soobšč. Akad. Nauk Gruzinsk. SSR (Tbilisi) 106(3): 577–580.

Kliphuis, E., and J.H. Wieffering. 1979. *In* Á. Löve [ed.], IOPB chromosome number reports, LXIV. Taxon 28: 398–400.

Meškova, L.Z. 1965. Primenenie kariologičeskogo issledovanija k sistematike roda *Veronica* L Avtoref. Kand. Diss. Kazan', pp. 1–16.

Öztürk, A. 1983. Türkiye'nin *Veronica* L. (Scrophulariaceae) taksonlari için yeni durumlar ve yeni yayilişlar. [New records and new distribution data of *Veronica* taxa in Turkey.] Doğa Bilim Derg. ser. A_2, 7(3): 532–537.

Pogosjan, A.I., S.G. Narinjan, and V.E. Voskanjan. 1971. Materialy k kariogeografičeskomu izučeniju flory massiva Aragac. [Data on the caryogeographical study of flora of the Aragatz massif.] Biol. Žurn. Armen. 24(11): 36–43.

Tumadžanov, I.I., and R.K. Beridze. 1969. O vidoobrazovanii v rjade *Gentianoides* Boris. roda *Veronica* L. [On the speciation within the series *Gentianoides* Boris. of the genus *Veronica* L.] Bot. Žurn. (Moscow and Leningrad) 54(11): 1647–1661.

——, ——, and A.I. Pogosjan. 1972. Opyt analiza populjacij *Veronica gentianoides* Vahl s.l. po profilju gor Malogo Kavkaza. Bot. Žurn. (Leningrad) 57(12): 1495–1515.

——, ——, and ——. 1974. O 48- chromosomnom citotipe poliploidnogo kompleksa *Veronica gentianoides* Vahl s Malogo Kavkaza. [On the 48 chromosome cytotype of the polyploid complex *Veronica gentianoides* in the Lesser Caucasus.] Soobšč. Akad. Nauk Gruzinsk. SSR 75(2): 233–236.

——, ——, and ——. 1977. [Population structure, phylogenetic affinities and origin of the polyploid complex *Veronica gentianoides* Vahl agg. (Scrophulariaceae)]. Bot. Žurn. (Leningrad) 62(11): 1548–1559.

Žukova, P.G. 1967 [Karyology of some plants, cultivated in the Arctic-Alpine Botanical Garden.], pp. 139–149. *In* N.A. Avronin [Ed.], Plantarum in zonam polarem transportatio. II. Leningrad (n.v.).

7. THE GENUS *ROEMERIA* IN TURKEY

ASUMAN BAYTOP

Faculty of Pharmacy, Department of Pharmaceutical Botany, University of Istanbul, Istanbul, Turkey

ABSTRACT

Descriptions of the three Turkish species of *Roemeria* (Papaveraceae) are given, together with a new key to the species. Variation in stem and capsule indumentum, and the angle of curvature of the fruiting peduncles is commented on, with observations on their usefulness to taxonomy.

Key words: *Roemeria*, Papaveraceae, taxonomy, Turkey.

INTRODUCTION

The genus *Roemeria* Medikus (Papaveraceae) is a small one of three species, all of which occur in Turkey. It was recently revised by Kadereit (1987) who, however, seemed to be unaware of the existence of the recently published *R. carica*. The species are slender annuals closely resembling annual *Papaver* species in habit, from which they are easily distinguished by their fruits. The latter are linear-cylindrical unilocular capsules, much longer than broad, dehiscing longitudinally from top to bottom by 3 or 4 valves. The three Turkish *Roemeria* species are *R. hybrida* (L.) DC., *R. refracta* DC. and *R. carica* A. Baytop. The first two were included in volume 1 of *Flora of Turkey* (Cullen 1965), while the third was described later (Baytop 1983).

Descriptions of these species based on Turkish material present in ISTE and other Turkish herbaria are given here, with special reference to the variability concerning the indumentum on stems and fruits and the angle of curvature of the fruiting peduncles.

TAXONOMY

1. **R. hybrida** (L.) DC., Syst. 2: 92 (1821). Fig. 1.

Stems 1–7, branched nearly from base, ± villous with fine hairs to 3 mm long to densely hispid with spreading setae. *Leaves* bi- to tripinnatisect; segments linear to oblong, 0.5–3 mm broad, lobes obtuse to acute, setose mainly on lower surface or subglabrous; basal leaves long-petiolate, to 15 cm (incl. petiole), lamina oblong to triangular in outline; upper cauline leaves sessile and smaller. *Buds* oblong, obtuse at apex, 3–4 × as long as broad, to 18 × 6 mm, sparsely villous to pilose all over. *Sepals* oblong-spathulate, truncate at base, cucullate at apex. *Petals* broadly ovate, 1–2.2 cm long, violet. *Stamens* numerous. *Filaments* linear, 8–15 mm, purplish-black, shortly attenuate at apex. *Anthers* oblong, 0.7–1 mm. *Pollen grains* spherical, smooth, mostly 36 μm in diameter. *Capsules* linear-cylindrical, straight or arcuate, rarely twisted, 3–7 cm × 2–2.5 mm,

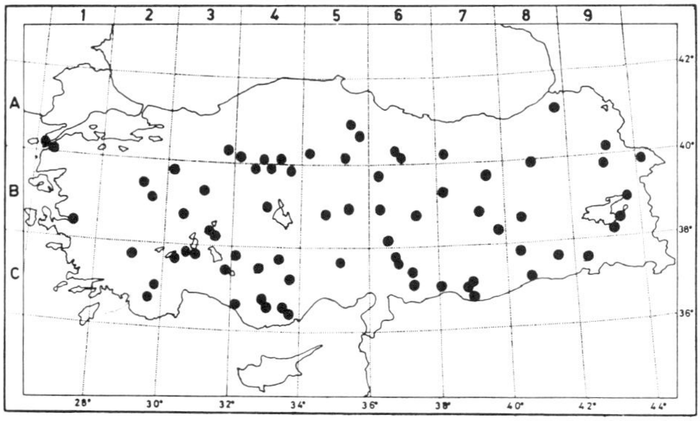

Figure 1. Map showing distribution of *Roemeria hybrida.*

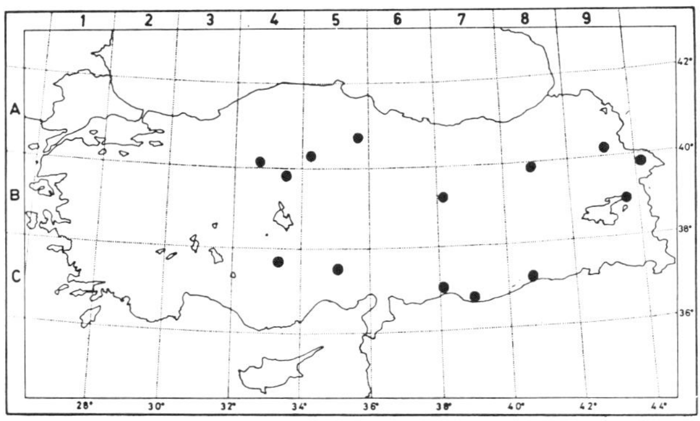

Figure 2. Map showing distribution of *Roemeria hybrida* with hispid-setose stems.

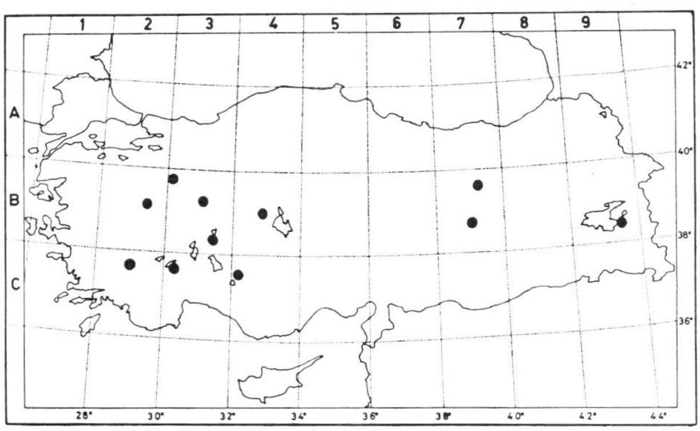

Figure 3. Map showing distribution of *Roemeria hybrida* with fruits setose on the upper half.

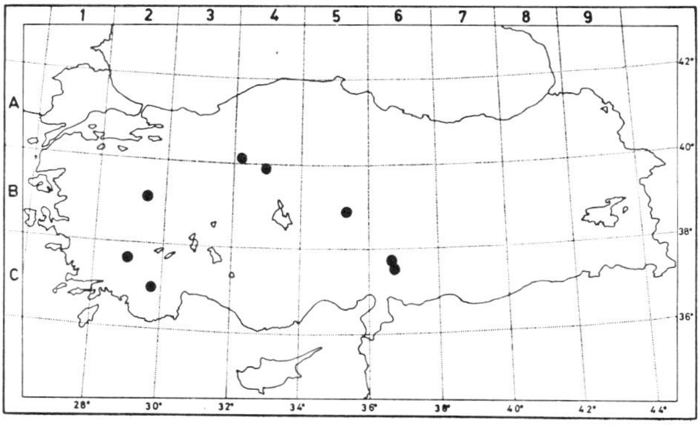

Figure 4. Map showing distribution of *Roemeria hybrida* with glabrous fruits.

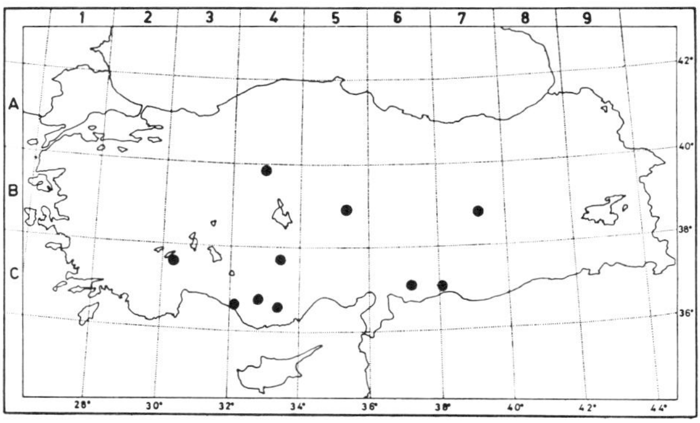

Figure 5. Map showing distribution of *Roemeria hybrida* with scabrous fruits.

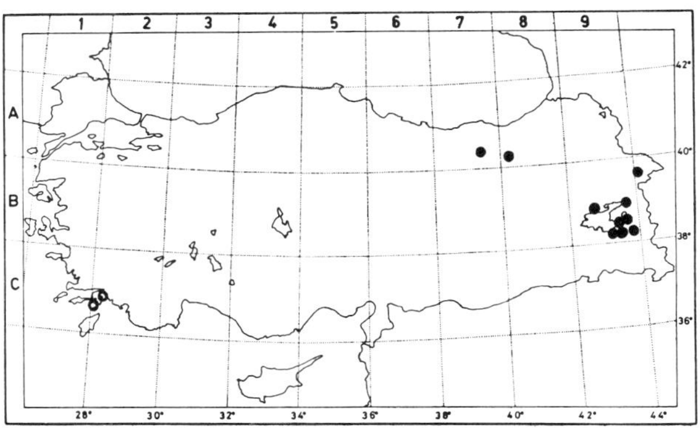

Figure 6. Map showing distribution of ● *Roemeria refracta*, ○ *R. carica*.

dehiscing by 3 or 4 valves, setose with spreading setae present either all over, or on the upper half, or only at the top, sometimes scabrid. *Fruiting peduncles* thickened, villous or pilose, rarely spreading setose, 1–10 cm, erect, spreading, arcuate, recurving to twisted. *Seeds* reniform, grey-brown, foveolate, 0.8–1.2 mm long, foveoles in 4–6 concentric rows on each face of the seed.

This is the commonest species of *Roemeria* in Turkey. The distribution of the species, based on localities cited in *Flora of Turkey* and on examination of 114 herbarium specimens, is plotted in Fig. 1.

A morphological variant characterised by a densely hispid-setose stem exists, which was named var. *hispidissima* Fedde (Fedde 1909). It is adapted to dry and steppic regions, but does not form distinct populations. In many gatherings, it is found together with individuals which commonly possess villous stems. The distribution of this variant is shown in Fig. 2. The specimen *Warburg* 488 (Anatolia, Phrygien, Koniah), cited by Fedde (1909) as *R. dodecandra* (Forsskål) Stapf (= *R. orientalis* Boiss, = *R. hybrida* (L.) DC. subsp. *dodecandra* (Forsskål) Maire) may represent this variant of *R. hybrida*, correlated with arid habitat.

The fruits of *R. hybrida* are commonly spreading-setose all over. Individuals with all capsules setose only on the upper half (Fig. 3) or all capsules glabrous except for a few setae at the apex (Fig. 4) may also occur. It is not rare to find these two forms growing together in a population where most of the individuals have capsules setose all over.

Another variant has ± scabrous fruits, thus possessing short stiff hairs in addition to setae (Fig. 5). This feature is mostly observed on individuals within populations in dry habitats.

Great diversity is shown by the curvature of the fruiting peduncles, which may be erect, spreading, arcuate, recurved or even twisted. It is common to find most of these states together even on one plant, but in general the most recurved peduncles are those of the lowest fruits.

2. **R. refracta** DC., Syst. 2: 93 (1821). Fig. 6.

Stems 1–5, branched from base, subglabrous or pilose to subvillous with thin hairs to 3 mm long, sometimes bearing fine spreading setae as long as hairs. *Leaves* bipinnatisect; segments oblong to linear, 1–4 mm broad, entire or 2–3 lobed with lobes obtuse to acute, subglabrous or sparsely setose mainly on lower surface; basal leaves long-petiolate, to 12 cm long (incl. petiole), lamina triangular to oblong in outline; cauline leaves sessile and smaller. *Buds* oblong, 2–3 × as long as broad, to 20 × 7 mm, sparsely setose all over with spreading setae to 2 mm long, rarely glabrous. *Sepals* broadly oblong, concave, carinate in upper part, shortly cucullate at apex, truncate at base. *Petals* broadly ovate, 1.5–2.5 cm long, bright red with a black blotch at base. *Stamens* numerous. *Filaments* purplish-black, somewhat dilated at the middle, attenuate towards base and apex, to 1.2 cm long. *Anthers* subglobose to oblong, up to 1 mm long. *Pollen grains* spherical, smooth, mostly 36 μm in diam. *Capsules* linear-cylindrical, attenuate at apex, with 3 or 4 setae which project like horns between and beyond the stigmatic lobes, 2–4 × longer than stigma and often curved at their apex, rarely shorter;

fruits 3–8 cm × 3 mm, straight and erect on erect peduncles, pendulous on recurved ones, dehiscing by 3 or 4 valves from top to bottom, each valve bearing an apical horn; valves glabrous, rarely sparsely setose with adpressed or slightly spreading setae, or densely adpressed setose with short setae up to 1 mm long. *Fruiting peduncles* thickened, antrorsely adpressed setose, rarely spreading setose, 2–12 cm long, erect in terminal fruits, recurved downwards and shorter in lower fruits, rarely contorted. *Seeds* reniform, grey brown, foveolate, 1–1.2 mm long, foveoles in 4 or 5 concentric rows on each face of the seed.

In *R. refracta*, the stems are softly and sparsely pilose to subvillous, but it is not uncommon to find in some populations individuals with additional slender setae on their stems, not shorter than the hairs.

The fruits of *R. refracta* are normally glabrous, though two rare variants having setose capsules occur. In the first the capsules are sparsely provided with adpressed or slightly patent setae 1–2 mm long; this was found, pro parte, on two of the herbarium sheets examined (ISTE 42792 p.p. and ISTE 46641 p.p.). In the second variant, the fruits are densely covered with adpressed setae to 1 mm, where the projecting horns may even bear some setae (ISTE 46652 and ISTO 10222 p.p.). Such a variety has been described from a specimen collected in Iran, and named var. *trichocarpa* Fedde & Schlokow (Fedde 1909).

The angle of curvature of the fruiting peduncles does not show any variation. In fully fruiting specimens, the 1–3 lowermost peduncles are short and recurved back, whereas the upper ones are all longer and erect.

3. **R. carica** A. Baytop in Notes Roy. Bot. Gard. Edinb. 41: 281 (1983). Fig. 6.

A full description of this species, which is endemic to SW Anatolia (C2 Mugla), already exists (Baytop, loc. cit., 1983). The herbarium material present in ISTE (6 sheets) has been studied in order to establish the nature of the variation in the indumentum on the stem and fruits.

No variation pertaining to stem indumentum has been detected. All the 89 specimens present on the ISTE sheets had on their stems only antrorsely adpressed setae 1–3 mm long.

Examination of the type collection (ISTE 44137) gives an idea of the approximate percentage of the differently indumented capsules. Of 66 specimens mounted on these sheets, 41 (62.1%) have setose capsules, 21 (31.8%) have glabrous capsules, 3 (4.5%) have capsules non-setose but shortly hispid with short stiff hairs, and 1 (1.5%) has capsules setose and shortly hispid.

No recurved fruiting peduncles are seen in *R. carica*. All are erect or spreading-erect with straight fruits.

DISCUSSION

The present investigation has led to the following results:

1. Stem indumentum, petal colour, filaments and fruit apex offer good key characters for identifying the three species. The number of stigmatic lobes (or fruit valves) has proved to be 3 or 4 in all three species and is therefore not a diagnostic character. The key previously given (Baytop 1983) should be corrected and replaced by the following:

1. Fruit apex with 3 or 4 setiform horns projecting beyond stigmatic lobes; corolla bright red (E. Anatolia) . . . . 2. *refracta*
1. Fruit apex without setiform horns; corolla pale red or violet
   2. Stem with antrorsely adpressed setae; filaments dilated; corolla pale red (SW Anatolia) . . . . 3. *carica*
   2. Stem villous, with or without spreading setae; filaments linear; corolla violet (widespread) . . . . 1. *hybrida*

2. *R. hybrida* and *R. refracta* typically have villous stems, more or less densely so or subglabrous. Variants where the stems are provided with setae in addition to soft hairs have been observed in both species. *R. carica* on the contrary has antrorsely adpressed setae on stems and shows no variation in this respect.

3. *R. refracta* rightly deserves its epithet as it has always strongly bent peduncles at least on the lowermost fruits. *R. carica* has all fruiting peduncles erect or erect-spreading and shows no variation in this character. *R. hybrida*, however, exhibits a great range of variation in the direction of its fruiting peduncles. This diversity may be attributed to its wide distribution and to the climatic conditions.

4. Typical fruits are glabrous in *R. refracta* and setose in both *R. hybrida* and *R. carica*. Yet the three species show a similar variability of capsule indumentum character states, being either setose or glabrous, and with or without supplementary short stiff hairs. These four kinds of fruits are found on separate individuals which do not form distinct populations.

ACKNOWLEDGMENTS

I wish to thank the directors of the following herbaria for giving me permission to examine their collections of *Roemeria*: AEF, ANK, ATA, HUB, ISTO.

LITERATURE CITED

Baytop, A. 1983. A new *Roemeria* from Turkey. Notes Roy. Bot. Gard. Edinb. 41: 281.

Cullen, J. 1965. *Roemeria* Medik., pp. 218–219. *In* P.H. Davis [ed.], Flora of Turkey and the East Aegean Islands, volume 1. Edinburgh University Press, Edinburgh.

Fedde, F. 1909. *Roemeria*, pp. 238–244. *In* A. Engler, Das Pflanzenreich, 40 (IV. 104): (*Papaveraceae-Hypecoideae* et *Papaveraceae-Papaveroideae*).

Kadereit, J.W. 1987. The taxonomy, distribution and variability of the genus *Roemeria* Medik. (Papaveraceae). Flora 179: 135–153.

8. TWO NEW SPECIES OF *ASTRAGALUS* FROM IRAN AND THEIR AFFINITY TO *A. WEIRIANUS* FROM AFGHANISTAN

D. Podlech

Institut für Systematische Botanik der Universität München, Menzingerstrasse 67, D-8000, München 19, Federal Republic of Germany

AND

A.R. Maassoumi

Research Institute of Forests and Rangelands, P.O. Box 13-116, Tehran, Iran

ABSTRACT

The systematic position of *Astragalus weirianus* (Leguminosae–Papilionoideae) is discussed. Together with two new species, *A. cercidophacos* and *A. venustus*, described here, it is considered to belong to a new section, sect. *Cercidophaca*. A key to the three species is provided.

Key words: *Astragalus*, Leguminosae, Iran, Afghanistan, new section, new species.

INTRODUCTION

One hundred years ago, Aitchison & Baker (in Aitchison 1888) described *Astragalus weirianus* from northwest Afghanistan. Because of the occurrence of bifurcate hairs, they put it into Sect. *Cercidothrix*, which in their sense meant the whole of Subgen. *Cercidothrix* Bunge (Bunge 1868). An appendix to the description comments, 'A woody undershrub about eighteen inches high, extremely common in sandy loam in the Badghis, where it forms good pasturage for goats and sheep.' Since that time the species has never been re-collected, in spite of the fact that H. Freitag and D. Podlech spent several days in the Badghis region in 1969 and 1977 respectively without rediscovering it. It is highly probable that this good pasture plant is now extinct or very rare due to overgrazing. *A. weirianus* was once more mentioned and also figured by Rechinger (1957) who considered, probably in error, that it belonged in Sect. *Irinaea* Boriss. The systematic position of this interesting species has remained unsolved up to now.

In 1984, Maassoumi collected two different plants in Gorgan, Iran, which could not be assigned to any of the existing sections of the genus *Astragalus*. Very careful analysis of this material has led to the conclusion that these plants are two new species closely related to *A. weirianus*. All three species show a remarkable combination of characters, so far not known within *Astragalus* Subgen. *Cercidothrix*. All are suffruticose perennials with a short stem; the indumentum of the vegetative parts consists of adpressed white or sometimes also black bifurcate hairs, but on the calyx it consists of bifurcate black and simple white hairs; the stipules are connate-vaginate around the stem and partly adnate to the petiole; the calyx is shortly tubular; the corolla is glabrous, lilac; the standard is distinctly bent back; the legume is linear, compressed, acute at both ends, nearly completely bilocular and with thin valves.

These characters show that these three species have a primitive position within the bifurcate-hairy Astragali (see Podlech & Felux 1974 and Podlech 1982). The fruit characters show an overall similarity with some primitive sections of the former subgen. *Phaca* Bunge. Within Subgen. *Cercidothrix*, the new section described below to accommodate these species occupies the same 'basal taxonomic' position in the hierarchy as sections *Chlorostachys* Bunge, *Macropodium* O. & B. Fedtsch. and *Scheremeteviana* O. & B. Fedtsch. have within Subgen. *Astragalus*.

TAXONOMY

Astragalus L. sect. **Cercidophaca** Podl. & Maassoumi, sectio nova ex Subgen. *Cercidothrice* Bunge.
Typus sectionis: *Astragalus cercidophacos* Podl. & Maassoumi (*vide infra*).

Plantae perennes, suffruticosae, breviter caulescentes, in parte vegetativa sparse pilis albis vel nigris immixtis appressis medifixis obtectae. Stipulae vaginantes, petiolo usque ad medium adnatae in latere opposito caulis inter se distincte connatae. Folia imparipinnata. Inflorescentiae singulae in uno caule in parte superiore caulis exeuntes, pedunculatae, multiflorae, primo densae deinde elongatae laxae. Calyx breviter tubulosus, dense pilis appressis nigris submedifixis et insuper pilis patentibus albis basifixis obtectus. Corolla glabra, lilacina. Vexillum supra unguem rectangulariter recurvum. Fructus pendulus linearis e lateribus compressus basi apiceque angustatus, fere complete bilocularis, valvis tenuibus.

Plants perennial, suffruticose, shortly stemmed, in the vegetative parts sparsely covered with adpressed white or sometimes also black bifurcate hairs. Stipules connate-vaginate, partly adnate to the petiole. Leaves imparipinnate. Inflorescence single in the upper part of the stems, pedunculate, many-flowered, dense but becoming lax and elongated. Calyx shortly tubular, densely covered with subadpressed black bifurcate hairs and patent white simple hairs. Corolla glabrous, lilac. Standard strongly bent back. Legume pendulous, linear, compressed, acute at both ends, almost completely bilocular with thin walls.

KEY TO SPECIES

1. Leaflets 9–13, narrowly linear to filiform, 7–23 mm long and 1–2 mm broad; legume weakly curved with a stalk 9–12 mm long . . . 2. *A. cercidophacos*
1. Leaflets 3–5, much broader; legume distinctly curved, sessile
  2. Leaflets sparsely hairy on both sides; calyx 4–5 mm long; standard c. 8 mm long, broadly elliptic to circular; ovary densely hairy; legume hairy at first but soon becoming glabrate . . . . 3. *A. venustus*
  2. Leaflets glabrous on upper side; calyx 6–7 mm long; standard 12–14 mm long, ovate to trullate; ovary and legume glabrous . . . . . . 1. *A. weirianus*

1. **Astragalus weirianus** Aitch. & Baker in Trans. Linn. Soc. London, Bot. ser. 3, 1:54 (1888). Figs. 1, 4

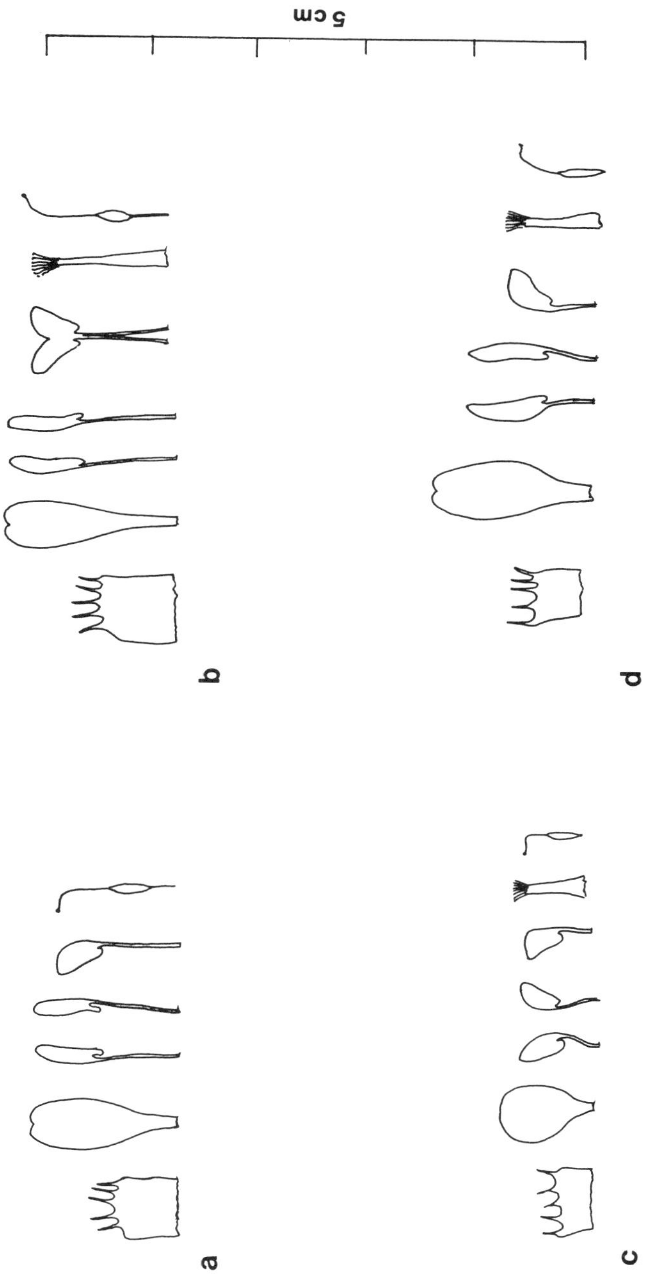

Figure 1. Floral dissections of species of *Astragalus* Sect. *Cercidophaca*. — a. *A. cercidophacos* (*Maassoumi 47554* (Hb. Podlech, isotype).) — b. *A. cercidophacos* (*Maassoumi 47547* (Hb. Podlech).) — c. *A. venustus* (*Maassoumi 47584* (Hb. Podlech, isotype).) — d. *A. weirianus* (*Aitchison 336* (G, isotype).) — (With 5 cm scale.)

Figure 2. Astragalus cercidophacos (isotype). — (With 5 cm scale.)

Icones: Rechinger in Biol. Skr. 9(3):135 (fig. 156), 136 (fig. 157) (1957).

Subshrub, 15–30 cm. Caudex woody, with erect to prostrate branches. Stems 3–15 cm long, erect, at first densely covered with adpressed white bifurcate hairs, above the nodes also with some black hairs, finally glabrescent. Stipules greenish, amplexicaul, 7–9 mm long, adnate to petiole nearly to its middle, connate high up into a contrapetiolar sheath, glabrous or with some black or white hairs on teeth. Leaves erect, 1.5–3.5 cm long; petiole 0.3–0.7 cm long, glabrous like the short rachis. Leaflets 1–2-paired, narrowly elliptic or narrowly obovate to elliptic or obovate, 8–25 × 4–12 mm, shortly acuminate, glabrous above, with scattered white adpressed hairs beneath. Inflorescences solitary in upper part of each stem; peduncle 4–6 cm long, glabrous; raceme laxly many-flowered, finally distinctly elongated. Bracts membranous, 1.5–2 mm long, glabrous or with some white and

Figure 3. Astragalus venustus. — a. *Maassoumi 47584* (Hb. Podlech, isotype). — b. *Wendelbo, Assadi & Shirdelpur 11874* (W). — (With 5 cm scale.)

black hairs. Pedicels 0.5–1 mm long, black-hairy. Flowers spreading, finally pendulous. Calyx 6–7 mm long, shortly tubular, densely covered with unequally bifurcate adpressed black hairs and with lax, simple, subpatent or patent white hairs; teeth subulate, 1.5–2 mm long, predominantly black-hairy. Corolla glabrous, lilac. Standard 12–14 mm long, recurved through nearly 90° above claw, emarginate at apex, ovate to trullate, 6–7 mm broad, narrowed into short, broad claw. Wings 10–12 mm long, blade curved, narrowly oblong, 6–7 × 2 mm, with a minute auricle and a 4–5 mm claw. Keel 8–9 mm long, blade obliquely elliptic-curved, 4–4.5 × 2 mm. Ovary sessile, glabrous, Pod (unripe) sessile, linear,

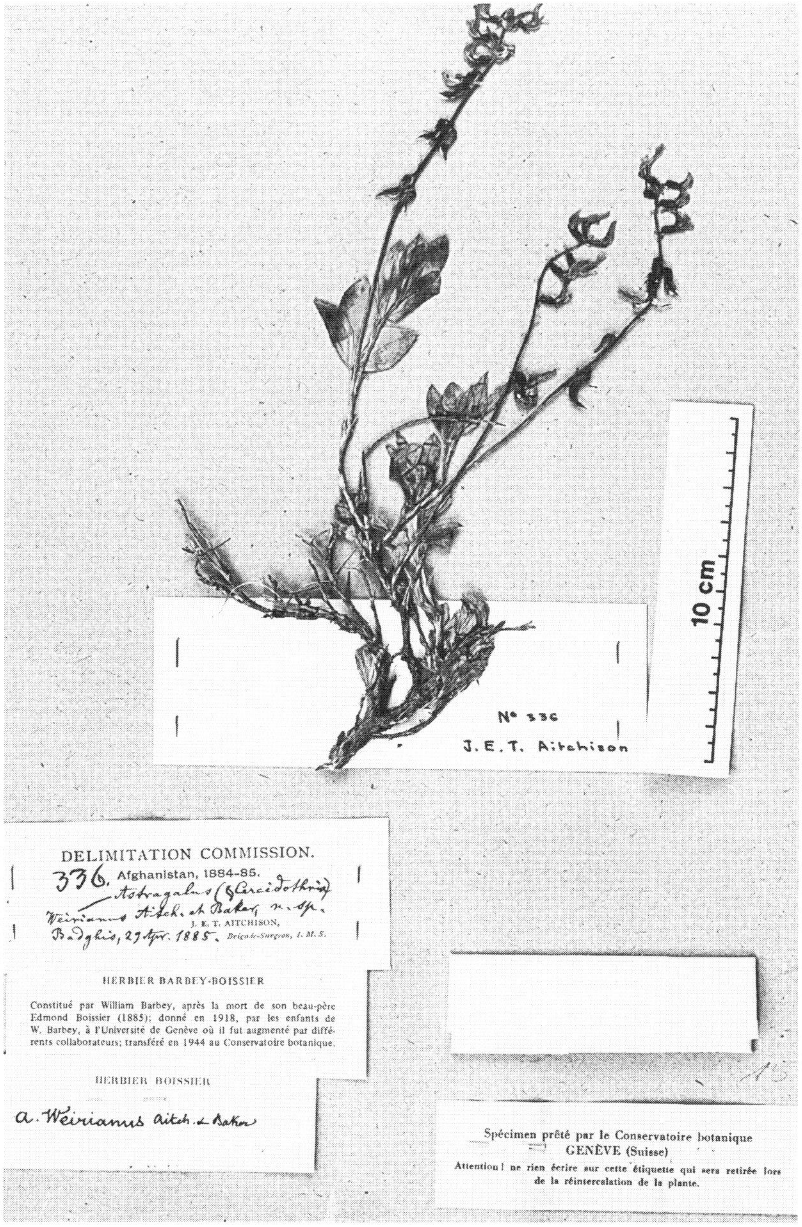

Figure 4. Astragalus weirianus (G, isotype). — (With 10 cm scale.)

curved, acuminate at both ends, glabrous. Flowering time: April–May. Distribution: Afghanistan (Badghis).

Typus: [AFGHANISTAN] Badghis [probably near Gal-icha W. of Gulran (according to the date of collection)], 29 iv 1885, *Aitchison* 336 (holo. K! iso. C, G!).

2. **Astragalus cercidophacos** Podl. & Maassoumi, **sp. nova**. Figs. 1, 2

Differt ab *Astragalo weiriano* Aitch. & Baker foliolis 4–6-jugis, anguste linearibus vel filiformibus, leguminibus stipite 9–12 mm longo suffultis.

Plantae suffruticosae, 10–18 cm altae. Caules ad 10 cm longi, dense appresse pilosi. Stipulae 4–6 mm longae, connati-vaginantes, petiolo adnatae, ciliatae vel pilosae. Folia erecta, 2–3.5 cm longa, petiolo 0.3–0.7 cm longo sparse piloso. Foliola 4–6-juga, 7–23 mm longa et 1–2 mm lata, anguste linearia vel filiformia, supra glabra, subtus appresse albopilosa. Inflorescentiae solitares in parte superiore caulium, laxe multiflorae, pedunculo 3.5–7 cm longo, albopiloso vel in parte superiore nigripiloso suffultae. Bracteae 1.5–2 mm longa, nigripilosae. Pedicelli c. 1 mm longi, refracti, nigripilosi. Calyx 7–9 mm longus, laxe pilis medifixis appressis nigris et insuper dense pilis basifixis subpatentibus albis obtectus, dentibus 1.5–2 mm longis. Corolla verosimiliter lilacina. Vexillum 14–16 mm longum refractum, lamina 4–5 mm lata, elliptica, apice vix emarginata. Alae 13–15 mm longae, lamina recta, anguste oblonga, 6–7 mm longa. Carina 11–12 mm longa, lamina oblique elliptico-triangulari, c. 4 mm longa. Ovarium glabrum, stipite 2–3 mm longo suffultum. Legumen stipite 9–12 mm longo suffultum, lineare, leviter curvatum, 12–20 mm longum (sine stipite et rostro), ventro carinatum dorso anguste rotundatum, valvis tenuibus, fere complete biloculare.

Subshrub, 10–18 cm high. Hairs in the vegetative parts bifurcate, white, on the calyx either bifurcate and black or simple and white. Caudex long, woody, with short branches. Stems to 10 cm long, erect or ascending, densely covered with short white adpressed hairs, at the nodes also with some black hairs. Stipules greenish, amplexicaul, 4–6 mm long, adnate up to middle of petiole, connate up to 2 mm into a contrapetiolar sheath, white-ciliate at the margin, on the surface sparsely or sometimes more densely covered with adpressed white and black hairs. Leaves erect, 2–3.5 cm long; petiole 0.3–0.7 cm long, like the rachis sparsely covered with short white and also black adpressed bifurcate hairs. Leaflets 4–6-paired, narrowly linear to filiform, 7–23 × 1–2 mm, obtuse to acuminate at apex, glabrous on upper surface, sparsely to laxly white-hairy beneath, mostly with involute margins. Inforescences solitary in upper part of each stem; peduncle 3.5–7 cm long, sparsely covered with white hairs, in upper part also with black hairs; raceme lax, many-flowered, 2–4 cm long. Bracts 1.5–2 mm long, linear, blackish-hairy. Pedicels c. 1 mm long, recurved, black-hairy. Flowers finally pendulous. Calyx 7–9 mm long, shortly tubular, covered with bifurcate adpressed black hairs and more densely with subpatent simple white hairs; teeth subulate, 1.5–2 mm long. Corolla glabrous, probably lilac. Standard 14–16 mm long, recurved through nearly 90° above claw, slightly emarginate at apex, elliptic, 4–5 mm broad, narrowed into a broad claw. Wings 13–15 mm long, blade straight, narrowly oblong, 6–7 × 1.5 mm; auricle c. 0.5 mm long, claw 7–9 mm long. Keel 11–12 mm long, blade obliquely elliptic-triangular, 4 × 2–2.5 mm. Ovary with 2–3 mm stipe, glabrous. Pod (unripe) with 9–12 mm stipe, linear, slightly curved, narrowed at both ends, 12–20 mm long (excl. stipe and beak), 3–4 mm high and 2–2.5 mm broad, carinate on ventral side, narrowly rounded on dorsal side, at apex narrowed into a slender beak, almost completely bilocular, valves thin-walled. Flowering time: April. Altitude: c. 1350 m. Distribution: Iran (Gorgan).

Typus: IRAN. Gorgan, Golestan-Park, inter Sharlegh et Cheshme Khan, 1350 m, 7 v 1984, Maassoumi 47554 (holo. TARI! iso. Hb. Podlech!).

Specimens seen: IRAN. Prov. MAZANDERAN: Gorgan, Golestan Park, between Sharlegh and Cheshme Khan, 1350 m, 7 v 1984, Maassoumi 47547 (TARI! Hb. Podlech!); *ibid.*, Maassoumi 47554 (type).

3. **Astragalus venustus** Maassoumi & Podl., **sp. nova.** Figs. 1, 3

Differt ab *Astragalo weiriano* Aitch. & Baker foliolis utrinque sparse pilosis, calyce corollaque multo breviore, lamina vexilli late elliptica vel fere orbiculata, ovario dense hirsuto.

Plantae suffruticosae 7–12 cm altae. Caules 1–3 cm altae. Caules 1–3 cm longi, dense appresse albo-pilosi. Stipulae 2–5 mm longae, connati-vaginantes, petiolo adnatae, ciliatae vel pilosae. Folia erecta, 1–2 cm longa, petiolo 2–5 mm longo, sicut rachis glabro. Foliola bijuga, saepe flabellatim congesta, 5–14 mm longa et 1.5–4 mm lata, anguste obovata, acuminata, utrinque sparse vel laxe pilis albis verrucosis appressis medifixis obtecta, raro fere glabra. Inflorescentiae solitares in parte superiore caulium, pedunculo 3–5 cm longo, albopiloso, racema dense multiflora, deinde elongata. Bracteae c. 2 mm longae. Pedicelli brevissimi, pilosi. Calyx 4–5 mm longus, breviter tubulosus, pilis basifixis vel submedifixis patentibus albis et nigris dense obtectus, dentibus 1.5–2 mm longis. Corolla glabra, sordide lilacina. Vexillum c. 8 mm longum, lamina c. 8 mm lata, late elliptica vel fere orbiculata. Alae c. 7 mm longae, lamina oblique elliptica, curvata, auricula c. 1 mm longa provisa. Carina c. 6 mm longa, lamina fere triangulari. Ovarium subsessile, villosum. Legumina pendula, sessilia, linearia, falcata, 22–25 mm longa, ventro carinata, dorso anguste sulcata, subbilocularia valvis tenuibus.

Subshrub, 7–12 cm high. Hairs in vegetative parts bifurcate, on the calyx also simple, white or black. Caudex slender, with short or elongated erect or creeping branches. Stems 1–3 cm long, densely covered with white adpressed hairs. Stipules greenish, amplexicaul, 2–5 mm long, adnate to the petiole for up to half their length, connate high up into a contrapetiolar sheath, white-ciliate or sometimes pilose on the whole surface. Leaves erect, 1–2 cm long; petiole short, 2–5 mm long, glabrous as in the very short rachis. Leaflets 2-paired, mostly congested, glaucous, sometimes red-margined, 5–14 × 1.5–4 mm, narrowly obovate, acuminate, sparsely or laxly covered on both sides with robust, warty, adpressed white hairs, more rarely almost glabrous. Inflorescences solitary in upper part of each stem; peduncle slender, 3–5 cm long, covered in lower part sparsely, in upper part more densely with white, bifurcate hairs; raceme dense, many-flowered, finally elongated. Bracts membranous, narrowly triangular or ovate, 2 mm long, with white and black hairs. Pedicels very short, 0.5–1 mm long, black-hairy. Calyx 4–5 mm long, campanulate-tubular, densely covered with mixed simple and unequally bifurcate black and white, adpressed or partly patent hairs; teeth subulate, 1.5–2 mm long, predominantly black-hairy. Corolla glabrous, dirty-coloured, standard pale lilac-violet or whitish with blue-violet stripes, wings and keel whitish-greenish-bluish. Standard c. 8 mm long, recurved through 90° above claw, slightly emarginate at apex, broadly elliptic or orbicular in shape, narrowed into a short, broad claw. Wings c. 7 mm, blade obliquely elliptic, curved, 4 × 2 mm; auricle rounded, c. 1 mm, claw broad, 3 mm long. Keel c. 6 mm long, blade nearly triangular, c. 4 × 2 mm, with short auricle and a claw 2–2.5 mm long. Ovary subsessile, densely white-hairy. Pod pendulous, sessile, linear, falcate, narrowed at both ends, 22–25 mm long, 3–4 mm high and 2.5–3 mm broad, keeled on ventral side, narrowly furrowed on dorsal side, at apex with a short beak, almost completely bilocular, valves thin-walled, hairy at first but finally completely glabrescent. Flowering time: May. Altitude: c. 1300–1900 m. Distribution: Iran: lower parts of the Elburz Mountains.

Typus: IRAN. Gorgan, Golestan-Park, Sharlegh Station, 1300 m, 8 v 1984, *Maassoumi* 47584 (holo. TARI! iso. Hb. Podlech!).

Specimens seen: IRAN. PROV. ZANJAN: 15 km from Zanjan on road to Bijar, 1900 m, 30 v 1974, *Wendelbo, Assadi & Shirdelpur* 11874 (W). – PROV. MAZANDERAN: Gorgan: Golestan-Park, Sharlegh Station, 1300 m, 8 v 1984, *Maassoumi* 47584 (type). – PROV. SEMNAN: 40 km NE of Shahrud, 1560 m, 18 v 1966, *Pabot* 8116 (G).

LITERATURE CITED

Aitchison, J.E.T. 1888. The botany of the Afghan Delimitation Commission. Trans. Linn. Soc. London, Bot. ser. 3, 1: 1–139.

Bunge, A. 1868. Generis Astragali species gerontogeae. Pars prior: Claves diagnosticae. Mém. Acad. Imp. Sci. Saint Pétersbourg ser. 7, 11(16): 1–140.

Podlech, D. 1982. Neue Aspekte zur Evolution und Gliederung der Gattung *Astragalus* L. Mitt. Bot. Staatssamml. München 18: 359–377.

——, and F. Felux. 1974. Beiträge zur Gliederung der Gattung *Astragalus* L. I. Zur Wertigkeit der Untergattungen *Phaca* Bge. und *Caprinus* Bge. Mitt. Bot. Staatssamml. München 11: 489–522.

Rechinger, K.H. 1957. Symbolae Afghanicae III. Leguminosae. Biol. Skr. 9(3): 1–208.

9. A TAXONOMIC REVISION OF *IRIS* SUBGENUS *HERMODACTYLOIDES* (IRIDACEAE)

BRIAN MATHEW

with palynological studies by S. Zmarzty

Royal Botanic Gardens, Kew, Richmond, Surrey, TW9 3AB

ABSTRACT

Iris subgenus *Hermodactyloides* constitutes a small group of *Iris* species characterised by having a bulbous rootstock with reticulate-fibrous tunics, leaves which are usually quadrangular in section and flowers which normally possess a long perianth tube with a subterranean ovary, although there are exceptions in each case. Ten species are recognised; these belong to four sections, one of which is described here for the first time. A key to the sections and species is provided and each species is given a full description. This primarily morphological study is based on herbarium specimens, on cultivated wild source material and on fieldwork by B. Mathew in Turkey, Iran and Soviet Central Asia. Photographs of the species are included, and the results of a palynological investigation by S. Zmarzty are presented.

Key words: Iridaceae, *Iris* subgenus *Hermodactyloides*, taxonomy, morphology, palynology.

INTRODUCTION

The history of the small bulbous *Iris* species which constitute subgenus *Hermodactyloides* dates from 1808 with the description of *I. reticulata* M. Bieb. This remained the only known member of the group, as it is recognised today, until 1872 when H.G. Reichenbach added the Lebanese *I. histrio*; several more were discovered and named before 1900. Only three more species have been described during the present century although some subordinate taxa have also been named.

The classification of this rather distinct group has varied greatly since 1846 when Spach first recognised that *I. reticulata* was sufficiently distinguishable from other species to merit placing it in a monotypic subgenus, *Hermodactyloides*; the name is based on the observation that the leaves are quadrangular and similar to those of *Hermodactylus tuberosus* (L.) Salisb.

Ålefeld (1863) took a motley assemblage of *Iris* species, including *I. reticulata*, to form his genus *Neubeckia*; the eight species which he placed in this new genus would, in terms of most of the more recent classifications, fall into several different subgenera, sections and series of *Iris*. Klatt (1865–66) improved on the situation, placing *I. reticulata* in *Xiphion* Miller, a genus made up primarily of the western Mediterranean bulbous *Iris* species which are today recognised as constituting subgenus *Xiphium* (Miller) Spach. Klatt's concept of *Xiphion*, however, also included *X. diversifolium* Steud. ex Klatt from Ethiopia which is now known to be a species of *Moraea* (*M. schimperi* (Hochst.) Pichi-Serm.).

In 1876 J.G. Baker described *Xiphion danfordiae* (now *Iris danfordiae* (Baker) Boiss.) which he placed in a new subgenus, *Micropogon*; the name refers to minute papillae on the claw of the outer perianth segments. A year later Baker's *Systema Iridacearum* was published; in this the genus *Xiphion* was divided into three subgenera encompassing all the various bulbous irises. In Mathew's classification (1981) these are also housed in three subgenera, *Hermodactyloides* Spach, *Scorpiris* Spach and *Xiphium* (Miller) Spach. Baker's groups are however not equivalent to these in their content; his subgen. *Euxiphion* placed *X. reticulatum* and the related *X. histrio* together with the tall unrelated western Mediterranean species which are now recognised as subgen. *Xiphium*; his second group, subgen. *Juno*, does cover the same assemblage as that now recognised as subgen. *Scorpiris*; the third, subgen. *Micropogon*, contained only *X. danfordiae*, a species now included in subgen. *Hermodactyloides*, although in a monotypic section since it does appear to be cytologically and morphologically distinct.

Boissier (1882) accepted the content and name of each of Baker's groups but transferred them to the genus *Iris* with the status of section, under the subgenus *Xiphion*.

In 1892 Baker revised his own views and included all the known bulbous species within a more broadly conceived genus *Iris*, giving their groups subgeneric status. However his subgenus *Xiphion* still combined *I. reticulata* and its relatives with the tall western Mediterranean species, while *I. danfordiae*, because of its small inner segments, was absorbed into subgen. *Juno*, containing *I. persica* L., *I. caucasica* Hoffm. and their allies.

Also in 1892, M. Foster gave a lecture to the Royal Horticultural Society (later that year published as '*Bulbous Irises*'), in which he described an assemblage of species called 'The Reticulate Group', the constitution of which conforms with that recognised today as subgen. *Hermodactyloides*. His system, although informal, must be seen as the forerunner of the presently recognised classification of bulbous irises into three distinct subgenera. Following this, W.R. Dykes (1913) proposed a formal classification which brought together these 'reticulate' species under the mantle of section *Reticulata* Dykes; consequently this is, at sectional level, the correct name for the group.

G.H.M. Lawrence (1953), in his reclassification of the genus *Iris*, recognised a broad subgen. *Xiphium*, divided into two sections, *Xiphion* and *Reticulata*, so that, in this system, the 'reticulate' species were still regarded as constituting a distinct unit. G.I. Rodionenko (1961) narrowed the concept of *Iris* and elevated the 'Reticulata' irises to generic level, as *Iridodictyum*, whereas in the classification of Mathew (1981) they were restored to their original subgeneric status in the genus *Iris* under the name of subgen. *Hermodactyloides* Spach.

The distinctness of this group of species has therefore not been in contention since the time of Foster, the only variation being in the level of taxonomic hierarchy applied to it. Although it might at first appear that *Hermodactyloides* is an advanced and isolated group, and should accordingly be given generic status, it must be noted that one of its members, *I. pamphylica* Hedge, has certain features which suggest a relationship with those *Iris* species assigned to subgen. *Limniris* series *Syriacae* (Diels) Lawrence, and notably *I. masia* Stapf ex Foster. This group, in turn, can be associated with other series in subgen. *Limniris*. Further comments are made below under *I. pamphylica*. In view of this apparent relationship it is considered unwise to remove subgen. *Hermodactyloides* from the genus *Iris*.

TAXONOMY

Before presenting the taxonomic treatment of subgenus *Hermodactyloides* it is necessary to indicate in the form of a key its relationship with the two other 'bulbous' subgenera which inhabit the Mediterranean region and western Asia.

KEY TO SUBGENERA OF *IRIS* WITH BULBOUS ROOTSTOCKS

1. Leaves bifacial, channelled; bulb with membranous to coriaceous tunics, sometimes splitting lengthways but not netted; stem produced above ground or, less commonly, subterranean; capsule usually carried well above ground level
  2. Bulbs with fibrous non-persistent roots; inner perianth segments erect, well-developed (except in *I. serotina*); perianth tube at most 5 cm long — Subgen. *Xiphium* (Miller) Spach
  2. Bulbs with swollen fleshy roots always present; inner perianth segments horizontal or deflexed, much-reduced (except in *I. cycloglossa*); perianth tube 3.5–11 cm long — Subgen. *Scorpiris* Spach
1. Leaves quadrangular in section or ± terete, or rarely bifacial and channelled (*I. kolpakowskiana*); bulb with netted tunics (?rarely membranous – see *I. winkleri* below); stem subterranean (except *I. pamphylica*); capsule usually carried at ground level — Subgen. *Hermodactyloides* Spach

Iris subgenus **Hermodactyloides** Spach in Ann. Sci. Nat. sér. 3, 5: 91 (1846). Type of subgenus: *I. reticulata* M. Bieb.

Syn: *Iris* section *Reticulata* Dykes, Genus Iris: 220 (1913).
Iris section *Tetragonae* Arber in Ann. Bot. 35: 310 (1921).
Iridodictyum Rodionenko, Rod Iris: 201 (1961).

Rootstock a bulb; tunics reticulate-fibrous (or possibly rarely membranous, see *I. winkleri* below); roots fibrous, not usually thickened or fleshy (but see *I. pamphylica*). Leaves basal, quadrangular, ± terete (*I. bakeriana*) or channelled (*I. kolpakowskiana* & *I. winkleri*). Stem subterranean or produced above ground (*I. pamphylica*). Bract and bracteole membranous, usually sheathing the perianth tube. Flowers with 3 large outer perianth segments; inner 3 segments erect and well-developed or reduced and bristle-like (*I. danfordiae*); perianth tube well-developed; ovary usually subterranean (except *I. pamphylica*); stigma usually bilobed. Capsule usually erect, borne at or just above ground level or on an aerial peduncle and pendent (*I. pamphylica*); seeds with or without an appendage.

KEY TO THE SPECIES OF SUBGENUS *HERMODACTYLOIDES*

1. Leaves channelled, enclosed by a single sheath; stigma entire (B. Sect. *Monolepis*)
  2. Bulb tunics reticulate-fibrous — 2. *I. kolpakowskiana*
  2. Bulb tunics membranous — 3. *I. winkleri*
1. Leaves quadrangular to ± terete with several membranous sheaths; stigma bilobed
  3. Perianth tube c. 2 cm long; capsule pendent at maturity, carried on an aerial peduncle to 20 cm long (A. Sect. *Brevituba*) — 1. *I. pamphylica*
  3. Perianth tube 3 cm or more long; capsule erect, borne at ground level at maturity

4. Inner perianth segments subulate, 3–5 mm long; flowers yellow (D. Sect. *Micropogon*) 10. *I. danfordiae*
4. Inner perianth segments oblanceolate, 35–50 mm long; flowers violet, purple, blue or yellow (C. Sect. *Hermodactyloides*)
5. Flowers yellow, spotted green on outer segments 9. *I. winogradowii*
5. Flowers violet, purple or blue (rarely albino)
6. Lobes of style branches 2–2.5 cm long, 3–4 mm wide, acuminate; flowers pale or slate-blue, not conspicuously blotched on outer segments 6. *I. vartanii*
6. Lobes of style branches 0.6–2(– 2.2) cm long, 2–10 mm wide, obtuse to acute, rarely shortly acuminate; flowers various shades of blue, purple and violet, often conspicuously blotched on outer segments
7. Bract and bracteole green, rather rigid and closely sheathing the perianth tube
8. Leaves ± terete in cross-section with 8 veins or ribs 5. *I. bakeriana*
8. Leaves quadrangular in cross-section with 4 veins or ribs 4. *I. reticulata*
7. Bract and bracteole thin and papery, whitish, sometimes faintly tinged or veined green
9. Falls with a clearly defined sinus (notch) between lamina and claw; lamina spotted darker blue in the central area 8. *I. histrioides*
9. Falls with no clearly defined sinus, merging gradually from lamina to claw; lamina with large dark blue blotches over most of its area 7. *I. histrio*

A. Section **Brevituba** B. Mathew, **sect. nov.**

Bulbus tunicis reticulatis compositus. Folia in sectione quadrangularia. Pedunculus bene evolutus. Tubus perianthii brevis; stigma bilobatum. Capsula maturitate pendens; semina appendiculata. Typus sectionis: *I. pamphylica* Hedge.

Bulb with reticulate tunics and small needle-like fibres at its base. Leaves quadrangular, with several sheathing leaves at the base. Peduncle well-developed above ground. Perianth tube short; inner segments well-developed; stigma bilobed. Capsule fusiform, pendent at maturity; seeds with an appendage.
Type of section: *I. pamphylica* Hedge.

1. **I. pamphylica** Hedge in Notes Roy. Bot. Gard. Edinb. 23: 557 (1961); Bot. Mag. n.s. t. 648 (1973); Mathew, Dwarf Bulbs: 147, t. 50 (1973); Mathew, The Iris: 177 (1981); Mathew in Davis, Flora of Turkey 8: 404 (1984); Baytop & Mathew, The Bulbous Plants of Turkey: 47 (1984). Fig. 1.

Type: TURKEY, Antalya Prov., Manavgat to Akseki, above Fersinuluni, 10 April 1956, *Davis & O. Polunin* D. 25845 (holo. E!; iso. K!, fruiting specimens), cult. Sir Frederick Stern, Highdown, Sussex, fl. 2 March 1958 (E!, flowering specimen).

Plant 15–25 cm tall at anthesis. Bulb with a finely netted tunic, producing thick, rather fleshy roots annually and furnished with short needle-like hairs around the base; bulblets few. Leaves 1–2, 17–24 cm long at anthesis, elongating to 55 cm, 4-angled in section. Peduncle produced above ground for up to 20 cm, concealed by several green ventricose sheathing leaves. Bract and bracteole ± equal, ventricose and closely sheathing the perianth tube. Perianth tube c. 2 cm long. Flowers bicoloured; outer segments narrowly obovate, 3.5–4 cm long, claw suberect, 2.5–2.7 cm long, 0.4–0.7 cm wide, green with darker spots, lamina reflexed, elliptic, obtuse, c. 1.5 cm long, 0.7–1 cm wide, brownish-purple with a yellow, purple-spotted median ridge; inner segments erect, narrowly oblanceolate, acute, c. 4 cm long, c. 0.6 cm wide, pale to deep blue veined darker and shading to green at the base. Stamens with filaments c. 1.5 cm long, anthers 1 cm long. Style branches narrowly

Figure 1. I. pamphylica Hedge.

oblanceolate, c. 3.5 cm long, 0.3–0.4 cm wide at base of lobes; lobes narrowly lanceolate, acute, 0.8 cm long, 0.15–0.2 cm wide; stigma bilobed. Capsule erect at first, becoming pendulous and fusiform with a long-tapering beak at maturity, carried on a peduncle to 20 cm long; seeds with an appendage. 2n = 20 (Gustafsson & Wendelbo, 1975).

Flowering period: March.

Habitat: Edge of *Pinus brutia* forest in *Quercus* scrub and margins of fields, 700–850(–1500) m.

Distribution: S Turkey, Antalya vilayet (province).

Specimens seen: TURKEY. Antalya, Manavgat to Akseki road, *T. Baytop, Brickell & Mathew* 8415; ibid., *Rix* 1340.

I. pamphylica is a very local species known only from this one small area where it is probably severely threatened due to over-collecting. The specific epithet refers to the southern Turkish classical region of Pamphylia, the type locality. Taxonomically speaking it is a somewhat isolated species having several features which distinguish it clearly from all other members of the subgenus. In some ways it appears to form a link with the species of subgenus *Limniris* series *Syriacae* (Diels) Lawrence, notably via *I. masia* Stapf ex Foster. The two are particularly similar in their underground parts, sharing the same type of bulb tunic, rather thickened fleshy roots of annual duration, and sharp needle-like fibres surrounding the base of the bulb; these 'needles' are small and almost hair-like in *I. pamphylica* but are developed to a much greater extent in *I. masia*.

Figure 2. I. kolpakowskiana Regel.

In the early stages of seedling development, *I. masia* has a small bulb with little or no rhizome attached; at this stage it very closely resembles the bulb of *I. pamphylica*, but later it tends to develop a compact, almost vertical, rhizome and the bulb is somewhat suppressed. The structure and colouring of the flowers of *I. masia* and *I. pamphylica* are also somewhat similar but the leaves are, in cross-sectional shape at least, quite different, those of *I. pamphylica* being quadrangular while those of *I. masia* are narrowly ensiform with unifacial blades. However, Arber (1921), in an anatomical study of leaf structure of the Iridaceae, considers that there are no fundamental differences between these two types of leaf and indicates that the 'radial' leaves of section *Tetragonae* (i.e. subgen. *Hermodactyloides*), and of the genus *Hermodactylus*, 'are merely variants upon the ensiform type'. Rodionenko (1961, Engl. transl. Blanco-White 1987: 176) commented upon a possible relationship between subgenus *Hermodactyloides* and series *Syriacae*, noting that there was 'reason to suppose that the ancestral form of *Iridodictyum* (i.e. subgen. *Hermodactyloides*) may have been a species close to the Asia Minor species *I. grant-duffii* Baker' (i.e. a relative of *I. masia*, in series *Syriacae*). Rodionenko was at the time unaware of the existence of *I. pamphylica* which actually has more features in common with series *Syriacae* than any other species of subgenus *Hermodactyloides*; his views have therefore been strengthened by the discovery of this somewhat outlying species from southern Turkey.

The beaked, pendent capsule, produced on a stem, gives the fruiting stage of *I. pamphylica* a likeness to that of *Hermodactylus tuberosus* (L.) Salisb. However, in the latter the ovary is 1-locular, not 3-locular as in *Iris*. The pollen morphology of

I. pamphylica (see below) adds weight to the view that there is a link between this species and *I. masia* and that *I. pamphylica* is best placed in a monotypic section within subgen. *Hermodactyloides*.

Clearly some further research into the relationships and origins of *I. pamphylica* is desirable since this species may hold at least some of the clues as to the origin of the whole subgenus.

B. Section **Monolepis** (Rodionenko) B. Mathew, **comb. nov.**

Syn: *Iridodictyum* Rodionenko section *Monolepis* Rodionenko, Rod Iris: 205 (1961).

Bulb with reticulate (? or membranous) tunics, not furnished with needle-like fibres at its base. Leaves channelled, all enclosed at the base within one tubular sheath. Peduncle short, subterranean. Perianth tube long; inner segments well developed; stigma entire (not bilobed). Capsule cylindrical, erect at maturity, carried at or just above ground level; seeds with no appendage.
Type of section: *I. kolpakowskiana* Regel.

2. **I. kolpakowskiana** Regel in Trudy Imp. S.-Peterburgsk. Bot. Sada 5: 263 (1877); Gartenflora 27: 40, t. 939 (1878); Foster, Bulbous Irises: 63 (1892); Dykes, The Genus Iris: 228 (1913); B.A. Fedtschenko in Fl. U.R.S.S. 4: 556 (1935), Engl. ed. 4: 423 (1968); Mathew, The Iris: 177 (1981); Mathew, The Smaller Bulbs: 107, t. 62 (1987). Fig. 2.

Type: U.S.S.R., Kazakh S.S.R., 'prope Wernoje' [Alma-Ata], *Fetisow* (K, LE).
Syn: *Xiphion kolpakowskianum* (Regel) Baker in Bot. Mag.: t. 6489 (1880).
Iridodictyum kolpakowskianum (Regel) Rodionenko, Rod Iris: 202, t. 76 (1961).

Plant 15–20 cm tall at anthesis. Bulb with strongly reticulate-fibrous tunics, producing a few bulblets at the base. Leaves 3–4, 3.5–11 cm long at anthesis, elongating slightly in the fruiting stage, linear, canaliculate, prominently ribbed beneath, c. 2 mm wide; margin glabrous or papillose-scabrid. Peduncle subterranean. Bract and bracteole green, rather rigid. Perianth tube 5–9 cm long. Flowers lilac to violet, the outer segments with a darker violet or purple lamina and yellow median, pubescent, ridge, the claw white veined violet, pubescent; outer segments lanceolate, 3.5–4.5 cm long, claw suberect, 1.5–2.5 cm long, 0.6–0.8 cm wide, lamina reflexed, ovate, acute or obtuse, 2–2.5 cm long, 0.9–1.4 cm wide; inner segments erect, obovate or oblanceolate, acute, obtuse or rounded, 3.5–5 cm long, 0.6–1.3 cm wide. Stamens with filaments 0.5–0.9 cm long, anthers 0.7–1 cm long. Style branches oblanceolate, 2.5–3.5 cm long, c. 0.6 cm wide at the base of the lobes; lobes obliquely lanceolate, acute, 0.9–1.2 cm long, 0.2–0.3 cm wide; stigma entire. Capsule erect, produced above ground level, cylindrical, c. 6 cm long, 0.7 cm wide, with a shortly beaked apex; seeds with no appendage. 2n = 20 (Zakharyeva & Makushenko 1969).
Flowering period: March–April.
Habitat: Open grassy or stony hillsides, flowering near melting snow, 800–1300 m.
Distribution: U.S.S.R. (C Asia, Tien Shan mountains).
Specimens seen: U.S.S.R., W Tien Shan, Ak Tash, *Korovin* 421b; Werny, 6 Mar. 1877, *A. Regel**; Tashkent, Kaplanbek, *Granitov & Vvedensky* 421a; Karasu, 1886, *Krassnov*. *It seems likely that Regel did not visit Werny (Wernoje = Alma-Ata) in 1877 and that this specimen was collected by *Fetisow*, and is therefore type material (see Regel in Petermann, Mitt. 1879: 381).

I. kolpakowskiana is the most distinct of all species in subgenus *Hermodactyloides*, distinguished on account of several major features from all others, except for the dubious *I. winkleri*. The flowers are of rather different form from the other species and have an entire stigma; the 3–4 leaves are channelled with a pale median stripe on the adaxial (upper) surface, and the whole of the aerial shoot is enclosed within one tubular sheath. In the fruiting stage the species is also clearly distinguishable since its capsule is long and cylindrical and contains seeds which are not furnished with an aril-like appendage.

There is perhaps some argument for transferring *I. kolpakowskiana* to subgenus *Scorpiris* on account of the channelled leaves but it has only a few features in common with that group. However it is possible that its origins lie with *Scorpiris* for it does bear some resemblance to certain Central Asiatic species in that subgenus, notably *I. rosenbachiana* Regel and *I. nicolai* Vved. (these two may be conspecific). These species belong to a group within subgenus *Scorpiris* whose long-tubed flowers (similar in length to that of *I. kolpakowskiana*) have the margins of the claw of the outer segments folded downwards; in most species of subgenus *Scorpiris*, and all the others of subgenus *Hermodactyloides*, the claw margins are upturned giving a channelled 'gutter' shape to the claw. In *I. kolpakowskiana* the margins also tend to fold downwards and there is an overall, albeit rather slight, similarity between the flowers of this species and *I. nicolai*. It is of interest also to note the form of the three stigmatic 'flaps' on the undersides of the style branches; in *I. kolpakowskiana* each flap is entire whereas in the other species it is bilobed. Most of the species of subgenus *Scorpiris*, including *I. nicolai*, have almost entire stigmatic flaps with an emarginate apex. It seems likely, in view of these features which *I. kolpakowskiana* shares with some members of subgenus *Scorpiris*, that it has links with that subgenus, possibly through common ancestry. Chromosome numbers for the various taxa involved in this discussion may or may not be significant; $2n = 20$ has been recorded for *I. kolpakowskiana* and *I. rosenbachiana*, and also for *Crocus alatavicus* which is sympatric with these. Rodionenko (1961) (taken from the English translation 1987) noted that *I. kolpakowskiana* was 'markedly different from the rest' (i.e. of subgenus *Hermodactyloides*) and that 'the characters differentiating it . . . make it resemble species of *Crocus*. The historical possibility of a hybrid having arisen between an *Iridodictyum* [= subgenus *Hermodactyloides*] and a *Crocus* is suggested by their coincidental rhythms of growth, genetic closeness and territorial and ecological contact.' The idea of *Crocus* being involved in the ancestry of *I. kolpakowskiana* has certain attractions, particularly on account of the several narrowly linear leaves (other *Hermodactyloides* have only one or two) which possess a pale median stripe on the upper surface. Additionally, the bulb tunics of *I. kolpakowskiana*, although reticulate-fibrous, are of a rather different structure from that of *I. reticulata* and its allies but resemble those of many *Crocus* species. Arber (1921), however, when discussing the leaf anatomy of *I. kolpakowskiana*, noted that 'its leaf is described as *Crocus*-like in external appearance, but the similarity is in reality quite superficial'. More likely, perhaps, is that *I. kolpakowskiana* is the result of hybridisation between the ancestral stocks of subgenus *Scorpiris* and *Crocus*. If this hypothesis were true there would be little justification for maintaining *I. kolpakowskiana* in subgen. *Hermodactyloides* but on the basis of its present observable characteristics the most practical approach would seem to be to retain it in this alliance in a separate section, together with the little-known *I. winkleri*. The palynological survey by S. Zmarzty (see below) demonstrated that from this viewpoint *I. kolpakowskiana* is indeed distinct from other members of the subgenus and that its pollen is not similar to that of *I. rosenbachiana* in subgenus *Scorpiris*.

3. **I. winkleri** Regel in Trudy Imp. S.-Peterburgsk. Bot. Sada 8: 677 (1884); Foster, Bulbous Irises: 64 (1892); Dykes, The Genus Iris: 229 (1913); B.A. Fedtsch. in Fl. U.R.S.S. 4: 554 (1935), Engl. transl. 4: 421 (1968); Mathew, The Iris: 179 (1981).

Type: U.S.S.R., Kirgizia, 'in valle fluvii Naryn inter Urgent et Alabuga', *A. Regel* (LE).

Syn: *Iridodictyum winkleri* (Regel) Rodionenko, Rod Iris: 202 (1961).

Plant 10–20 cm tall at anthesis. Bulb with brown membranous tunics, not reticulate-fibrous. Leaves 3–4, enclosed within one basal sheath, equalling or exceeding the flowers at anthesis, narrowly linear, 3-nerved beneath, the central nerve forming a keel, 1–2 mm wide. Bract and bracteole acuminate, green. Perianth tube equalling the limb. Flowers bluish-violet; outer segments oblanceolate with a patent lamina; inner segments erect, oblong, slightly broader than the outer. Style branches with oblong lobes. Fruiting characters unknown. No chromosome count recorded.

Flowering period: June.

Habitat: High mountain slopes, 3000–4000 m.

Distribution: U.S.S.R., Tien Shan, in the Fergana Range.

A poorly understood species which is known only from the holotype in Leningrad. It is undoubtedly similar in several respects to *I. kolpakowskiana* and it has been speculated that the specimen actually represents a plant of that species which has lost its outer fibrous-reticulate bulb tunics during collecting. However, in the absence of any further material from the area it is impossible to comment upon the authenticity or otherwise of this taxon. Rodionenko (1972) states that 'the existence of the latter [*I. winkleri*] must remain doubtful for the time being'.

C. Section **Hermodactyloides**

Syn: *Iridodictyum* section *Iridodictyum* Rodionenko, Rod Iris: 203 (1961).

Bulb with reticulate-fibrous tunics, not furnished with needle-like fibres at its base. Leaves quadrangular, or nearly terete with 8 ribs, with several sheaths at the base. Peduncle short, subterranean. Perianth tube long; inner segments well-developed; stigma bilobed. Capsule ellipsoid or fusiform, erect at maturity, carried at or just above ground level; seeds with an appendage.

Type of section: *I. reticulata* M. Bieb.

Section *Hermodactyloides* constitutes the largest group in the subgenus, consisting of six species which are apparently inter-fertile. There are however no convincing records of hybrids between these and any of the species in the other three sections. Palynological studies (see below) show that five of the six species have similar pollen morphology (material of the sixth was not available) and that the pollen of these differs from that of the species in sections *Brevituba* and *Monolepis*, but not from that of *I. danfordiae* in sect. *Micropogon*.

Figure 3. I. reticulata M. Bieb.

4. **I. reticulata** M. Bieb., Fl. Taur.-Cauc.: 34 (1808); Regel, Gartenflora 13: 133, 323, t. 452 (1864); Bot. Mag. 92: t. 5577 (1866); Regel, Gartenflora 22: 354, t. 779 (1873); Boissier, Fl. Orient. 5: 120 (1882); Foster, Bulbous Irises: 56 (1892); Dykes, The Genus Iris: 222 (1913); B.A. Fedtsch. in Fl. U.R.S.S. 4: 555 (1935), Engl. transl. 4: 422 (1968); Rech. fil., Fl. Iran. 112: 41, t. 16, fig. 2 (1975); Mathew, The Iris: 178, col. t. 34 (1981); Baytop & Mathew, The Bulbous Plants of Turkey: 47, t. 53 (1984); Mathew in Davis, Flora of Turkey 8: 402 (1984). Fig. 3.

Type: U.S.S.R., Caucasus, 'habitat in Iberia', *D. Adam* (LE, photo E!).

Syn: *Neubeckia reticulata* (M. Bieb.) Ålefeld in Bot. Zeitung (Berlin) 21: 297 (1863).
Xiphion reticulatum (M. Bieb.) Klatt in Linnaea 34: 572 (1866).
Iridodictyum reticulatum (M. Bieb.) Rodionenko, Rod Iris: 202 (1961).
Iris hyrcana Woron. in Grossh., Fl. Kavk. 1: 250 (1928). Type: U.S.S.R. Azerbaijan, Lenkoran (not traced).

Plant 7–15 cm tall at anthesis. Bulb with fine to coarse fibrous-reticulate tunics, with few to many bulblets. Leaves 2, varying from scarcely visible at anthesis to slightly exceeding the flower, elongating later to 30 cm, 4-angled in section. Peduncle subterranean. Bract green, rather rigid and closely sheathing the perianth tube; bracteole membranous. Perianth tube 4–7 cm long. Flowers very variable in colour, blue, violet or purple, the outer segments with a yellow or orange median ridge and often bicoloured with the lamina darker than the rest of the flower, the claw dark-veined; outer segments narrowly panduriform, 3.2–4.8 cm long, claw suberect, 2.1–3 cm long, 0.5–1.2 cm wide, lamina deflexed, ovate, obtuse to subacute, 1.1–1.8 cm long, 0.7–1.3 cm wide; inner segments erect, oblanceolate,

3.5–5 cm long, 0.4–0.6(–1) cm wide. Stamens with filaments 1–2 cm long, anthers 0.6–1.1 cm long. Style branches oblanceolate-cuneate, 3–4 cm long, 0.5–0.9 cm wide at the base of the lobes; lobes obliquely obtuse or acute, occasionally acuminate, often irregularly dentate to varying degrees, 0.6–1.4 cm long, 0.2–0.5 cm wide at the base; stigma bilobed. Capsule erect, carried at or just above ground level, ellipsoid, shortly beaked at the apex, 3.5–5.5 cm long; seeds with an appendage. 2n = 20 (Simonet 1928).
Flowering period: March–June.
Habitat: Mountain grassland, stony slopes, in scrub and rarely in woods, 600–2700 m.
Distribution: NE, E and SE Turkey; NE Iraq; N, NW & W Iran; U.S.S.R. (Caucasus and Transcaucasus).
Selection of specimens seen: U.S.S.R., ARMENIA, Aghveran, near Arzakan, *Gabrielian*; GEORGIA, near Tiflis, *Fischer*; Tiflis, Teleti, *Grossheim*; Tiflis, Kur Valley, *Wittman*. TURKEY. Kars, Sarikamiş pass, *Furse & Synge* 210; Elaziğ/Diyarbakir, Maden to Ergani, *Sauer* 57/141; Erzincan, Kemaliye, Sarikonak, *Güner* 1639; Bingöl, Çukurca, *Güner* 1632; Erzurum, Tahir pass, *Lamond* 2646; Maraş, Akher Dagh (Ahir Dağ), *Balls* 931; Urfa, Karacadağ, *T. Baytop* ISTE 41767; Mardin, Mardin to Savur, Dereyani, *Güner* 1601; Hakkari, Hakkari to Uludere, *Güner* 1669; Hakkari, Şemdinli, Ümit Tepesi, *T. Baytop* ISTE 26121. IRAQ. Sefin Dagh, *Rawi* 10438; Shaqlawa, *O. Polunin* 5042; Sulaimaniya, *Graham* 649; N of Sulaimaniya, *Thesiger* 417; Qopi Qaradagh, *Barkley* 8520. IRAN. Gorgan, Gulestan Forest, *Iranshahr* 8019E, *Furse* 5090; Almeh, *Kukkonen* 5712; Mazanderan, above Zardman, *Wendelbo* 144; Azerbaijan, near Tabriz, *Gilliatt-Smith* 1325; Mianeh, *Mathew et al.* BSBE 1309; Kuh-e-Savalon (Kuhha-ye Sabalan), *Mathew et al.* BSBE 532; Kuh-e-Sahand, *Furse & Synge* 228; N of Sanandaj, *Furse* 2131; E of Sanandaj, *Furse* 5015; Yasuj to Ardakhan, *Davis & Bokhari* 56478; W of Hamadan, *Trott* 1040; Mt Alward, *Cowan & Darlington* 382; E of Khorramabad, *Furse* 1864; W of Shiraz towards Kazerun, *Grey-Wilson & Hewer*; W of Bojnurd, *Furse* 5168; NE of Firuzkuh, *Grant* 17377; near Qazvin, *Schmid* 5058; Karaj to Chalus, *Mathew et al.* BSBE 743; S of Manjil, *Furse* 5072. AFGHANISTAN: There are two records for *I. reticulata*, from Kabul and Paghman; these probably represent introduced material.

This is the most extensively distributed and most variable species in the subgenus exhibiting a wide range of variation in stature, flower colour, size and proportion of floral organs, degree of coarseness of the bulb tunic fibres and the extent to which 'offset' bulblets are produced by the mature bulbs. It is difficult to find any correlation between these characters and their distribution, although there are certain trends which can be detected.

Material from northern Iraq tends to have bulb tunics which consist of very slender fibres, finely reticulated in contrast to the strongly fibrous coarsely netted tunics of most specimens from Iran. Eastern Turkish representatives tend to produce many bulblets like rice grains around the parent bulbs whereas those from Caucasia do not do this so readily. In flower colour it can be said that the material from northeastern Turkey and Caucasia usually has reddish-purple flowers whereas in other more southern parts of its range they are inclined to be bluish or violet, often markedly bicoloured with a darker lamina to each of the three outer perianth segments. In areas of NW Iran and SE Transcaucasia, forms are found with very clear blue flowers; these have been distinguished as *I. hyrcana*, the name alluding to the Hyrcanian region adjacent to the Caspian Sea. A pale sky blue form has been introduced to cultivation in Europe under the name *I. hyrcana* and, through vegetative propagation, gives the impression that it is a uniform and somewhat distinct taxon. However, the wild populations in the region are variable and observations made in Iran by the author show that *I. hyrcana* may have darker blue or even purple flowers and is thus impossible to distinguish from certain variants in other parts of the range. The shape of the bulb has been used as an additional distinguishing feature for *I. hyrcana*, in which it is reputed to be ± globose, whereas *I. reticulata* has a more elongate-ovoid bulb. This is, however, not consistent and certainly some collections of the pale blue form from NW Iran have rather elongated bulbs. Fedtschenko in Flora U.R.S.S. 4: 422 (Engl. transl. 1968) includes the character of bulblet production to distinguish between the two taxa, indicating that *I. reticulata* has no 'adventitious bulbs', while *I. hyrcana* has many. However, throughout the distribution of *I. reticulata* this feature appears to vary considerably and it is clear that it is not helpful taxonomically.

I. reticulata is a popular garden plant and has been much-selected and hybridized to produce a wide range of cultivars. The plant in commerce as *I. reticulata* has uniformly deep violet-blue flowers and it has been said that no similar variant has been found in the wild. However, field notes supplied with *P. Furse* 5179 from Iran state that it was 'the usual commercial violet-coloured reticulata'.

It is this high degree of variability which makes *I. reticulata* such an important plant commercially since there is almost no end to the range of cultivars which can be produced. The natural vegetative method of reproduction by means of bulblets make it an ideal subject for the nurseryman and it is an easily-cultivated plant in northern European countries, unlike some of the other species in the subgenus. Each year millions of bulbs are handled commercially, mainly originating in Holland.

5. **I. bakeriana** Foster in Bot. Mag. 115: t. 7084 (1889); Foster in Gard. Chron. 1890, 2: 293, t. 45a (1890); Foster, Bulbous Irises: 60 (1892); Dykes, The Genus Iris: 227, t. 45, fig. 3 (1913); Mathew, The Iris: 173 (1981), Baytop & Mathew, The Bulbous Plants of Turkey: 46, t. 49 (1984). Fig. 4.

Type: TURKEY. Mardin, 'a native of Armenia', 1887, *Rev. G.F. Gates*, cult. M. Foster, Feb./March 1889 (holo. K!).

Syn: *Iridodictyum bakeranum* (Foster) Rodionenko, Rod Iris: 202 (1961).
Iris reticulata M. Bieb. var. *bakeriana* (Foster) Mathew & Wendelbo in Rech. fil., Fl. *Iranica* 112: 43 (1975); Mathew in Davis, Flora of Turkey 8: 403 (1984).

Plant 8–15 cm tall at anthesis. Bulb with coarsely fibrous-reticulate tunics, with a few bulblets at the base. Leaves usually 2, ranging from scarcely visible at anthesis to equalling the flower, elongating to 25 cm after flowering, ± terete in section with 8 ribs. Peduncle subterranean. Bract green, rather rigid and closely sheathing the perianth tube; bracteole membranous. Perianth tube 4–7 cm long. Flowers usually pale blue, the outer segments with a darker lamina and with or

Figure 4. I. bakeriana Foster.

without a yellow median ridge; claw dark-veined; outer segments narrowly panduriform, 4–4.8 cm long, claw suberect, 2.5–3.3 cm long, 0.6–1 cm wide, lamina horizontal to reflexed, 1–1.2 cm long, 1–1.2 cm wide; inner segments erect, oblanceolate, 3.5–4.5 cm long, 0.6 cm wide. Stamens with filaments 1.5–1.9 cm long, anthers 0.8–1 cm long. Style branches oblanceolate or cuneate, 3.5–4 cm long, 0.4–0.6 cm wide at the base of the lobes; lobes obliquely obtuse, irregularly dentate, 0.6–0.9 cm long, 0.2–0.3 cm wide at the base; stigma bilobed. Capsule erect, carried at or just above ground level, ellipsoid or fusiform, shortly beaked at the apex; seeds with an appendage. $2n = 20$ (Simonet 1952).
Flowering period: (Feb.–)March–April(–early May).
Habitat: Open stony or rocky mountain slopes, clearings in *Quercus* scrub, often flowering near the snowline, 850–2000 m.
Distribution: SE Turkey, W Iran, NE Iraq.
Specimens seen: TURKEY. Mardin, between Mardin and Savur, *Güner* 1600; *T. Baytop* ISTE 41770. IRAN. Near Kermanshah, *Trott* 332; near Rezaiyeh, *Mathew et al.* BSBE 1429; W of Kermanshah, *Furse* 5011. IRAQ. Hagi Umran (Haji Omran), *Al Rawi* 41386; ibid., *O. Polunin* 5061; ibid., *Shahwani* 25252; ibid., *Haines* 1377; ibid., *Poore* 418.

I. bakeriana is very distinct in its nearly terete leaves with 8 veins, and has a discrete area of distribution within that of the much more widely distributed *I. reticulata*. It appears that in areas where the two meet there is some hybridization, with resulting intermediates which have about 6 veins in the leaves (Grey-Wilson, Iris Year Book 1972: 111).

In SE Turkey, in the region of the type locality, *I. bakeriana* has flowers which lack a yellow median ridge on the outer segments. The original introductions into cultivation were made from this region and through cloning, the general impression now is that the flowers of *I. bakeriana* normally lack a yellow ridge. However this is not the case and in other parts of its range the flowers do possess yellow markings on the outer segments and there is some variation in flower colour in the blue and purplish shades.

6. **I. vartanii** Foster in Gard. Chron. 1885: 438 (1885); Baker in Bot. Mag. 113: t. 6942 (1887); Foster, Bulbous Irises: 10, 61 (1892); Dykes, The Genus Iris: 225 (1913); Mathew, The Iris: 179 (1981); Galil in Pl. Syst. Evol. 141: 177 (1983); Feinbrun-Dothan, Fl. Palaestina 4: 125 (1986). Fig. 5.

Type: ISRAEL. Described from living material, cult. M. Foster Oct. 1884, grown from bulbs sent by *Dr Vartan* of the Medical Mission, Nazareth (no specimen traced).

Syn: *Iridodictyum vartanii* (Foster) Rodionenko, Rod Iris: 202 (1961).

Plant 10–15 cm tall at anthesis. Bulb with strongly reticulate-fibrous tunics and several bulblets at the base. Leaves usually 2, shorter than or sometimes exceeding the flower at anthesis, elongating later to 15–30 cm, 4-angled. Peduncle subterranean. Bract and bracteole thin and membranous, usually greenish-veined towards the apex. Perianth tube 5–8 cm long. Flowers pale or slaty blue, or rarely white, the outer segments with a raised yellow median ridge surrounded by a whitish zone with darker blue veins radiating outwards on to the wholly bluish lamina, the abaxial surfaces of the segments greenish; outer segments oblanceolate, (3.5–)4–4.5(–5) cm long; claw suberect, 2–2.7 cm long, c. 0.4–0.5 cm wide; lamina horizontal or slightly deflexed, oblong, obtuse or subacute, (1.5–)1.8–2.2(–2.3) cm long, c. 1.5 cm wide; inner segments erect, narrowly oblanceolate, obtuse or subacute, 3–3.8 cm long, c. 0.5 cm wide. Stamens with filaments 0.5–0.6 cm long, anthers 0.6–1.2 cm long. Style branches narrowly wedge-shaped, obscurely and irregularly dentate, 3.2–4.8 cm long, c. 0.5 cm wide at the base of the lobes, lobes gradually tapering to an acuminate apex, 2–2.5 cm long, 0.3–0.4 cm wide; stigma bilobed. Capsule erect, carried at ground level, ellipsoid, without a prominently beaked apical portion; seeds with a small appendage. $2n = 20$ (Simonet 1934).

Flowering period: December–February.

Habitat: Open rocky places and clearings in scrub, 300–1000 m (approx.).

Distribution: Israel, Lebanon, S Syria, NW Jordan.

Specimens seen: ISRAEL. S Canaan, *Davis* 5143; above Nazareth, *Davis* 3845; Nazareth-El Reina road, *Davis* 3868; Jerusalem, *Meyers & Dinsmore* 2312. JORDAN. N of Es Salt, 20 Dec. 1920, *Meyers & Dinsmore* 3706; near Salt, *Al-Eisawi* 9004. SYRIA. Damascus, Mouhajje, *Norris* s.n.

I. vartanii is easily distinguished by its long narrow acuminate style-branch lobes and unique colouring, a slate blue shade with darker veining but without any of the dark spotting on the outer segments which is such a feature of *I. histrio*, a species with a related but not identical distribution.

Feinbrun-Dothan, loc. cit., records *I. vartanii* as being ‘fairly rare’ in Israel and it would appear that throughout its range it is not a plentiful plant. Although it is only in the last few decades that any concern has been expressed about the

Figure 5. I. vartanii Foster.

over-collection of bulbous plants in the Middle East, it is of interest to note that Foster in 1885 admitted that 'I feel a pang of conscience, as if I had done a cruel deed, when I worry my far-off friends to send me bulbs and roots, and having received something of beauty from a happy sunny clime, kill it by exposing it to the muddle of frost, rain, darkness, thaw, wind'. From these comments it is clear that *I. vartanii* is not easy to cultivate in the northern European climate although in the mid 20th century an albino variant, known as 'Alba', was propagated and distributed freely by the Dutch bulb industry. Other albino forms have been collected and described as *I. vartanii* var. *alba* Dykes and *I. reticulata* var. *candida* Bornm.

7. **I. histrio** Reichb. fil. in Bot. Zeitung (Berlin) 30: 488 (1872); Boiss., Fl. Orient. 5: 121 (1882); Dykes, The Genus Iris: 223 (1913); Mathew, The Iris: 174, col. t. 36 (1981); Mathew in Davis, Flora of Turkey 8: 403 (1984); Baytop & Mathew, The Bulbous Plants of Turkey: 46 (1984); Feinbrun-Dothan, Flora Palaestina 4: 125 (1986). Fig. 6.

Type: [LEBANON] Sommet du Liban, entre les deux mamelons appelés Taumets et Djerzine, fiès [près] de Saïda, [*Gaillardot*] (iso. K!).

Syn: *Xiphion histrio* (Reichb. fil.) Hooker fil. in Bot. Mag. 99: t. 6033 (1873).
Iris libani Reuter ex Baker in Gard. Chron. 1876, 1: 559 (1876), pro syn.
I. reticulata M. Bieb. var. *histrio* (Reichb. fil.) Foster, Bulbous Irises 57, fig. 6 (1892).
Iridodictyum histrio (Reichb. fil.) Rodionenko, Rod Iris: 202 (1961).

Plant 10–18 cm tall at anthesis. Bulb with finely fibrous-reticulate tunics, with few to many bulblets at the base. Leaves 1–2(–3), varying from rather short at

Figure 6. I. histrio Reichb. fil.

anthesis to much-exceeding the flowers, elongating later to 25–40 cm, 4-angled. Peduncle subterranean. Bract and bracteole thinly membranous, white or faintly tinged green. Perianth tube 4–12 cm long. Flowers predominantly pale blue but with the lamina of the outer segments conspicuously blue-blotched on a creamy or pale blue ground with a yellow median ridge, rarely the whole flower with a darker blue ground colour; outer segments oblanceolate-oblong, (4–)4.3–5(–6) cm long, claw suberect, 2.5–3.5 cm long, 0.8–1.4 cm wide, lamina slightly deflexed, broadly ovate, obtuse or subacute, (1.5–)2–2.2(–2.5) cm long, 0.8–1.6 cm wide; inner segments erect, oblanceolate, 3.5–4.5 cm long, 0.5–0.6 cm wide. Stamens with filaments 1.1–1.6 cm long, anthers 1–1.4 cm long. Style branches oblanceolate or cuneate, 3.8–5 cm long, 0.8–1.1 cm wide at the base of the lobes; lobes obliquely acute, obtuse or nearly rounded, usually irregularly dentate, 1–2 cm long, 0.3–0.8 cm wide at the base; stigma bilobed. Capsule erect, carried at or just above ground level, ellipsoid with a short beak at the apex; seeds with an appendage. 2n = 20 (Simonet 1934).

Two subspecies may be recognised:

a. subsp. **histrio**

Lamina of outer segments 1.2–1.6 cm wide, sparsely but conspicuously blue-blotched over its whole surface; filaments 1.1–1.4 cm long, rather wide and flattened.

Flowering period: Jan.–March(–April).

Habitat: Open rocky slopes and in scrub, 500–1700 m.

Distribution: Israel, Lebanon, Syria, S Turkey.

Specimens seen: TURKEY. Hatay, between Antakya and Yayladağ, *Güner* 1456, 1486; Amanus Mts, Belen Pass, *Watson et al.* 2451, 2455; Hatay, Akra Dağ, *Thiébaut*; Içel, Silifke, Uzuncaburç, *Güner* 1575; E of Gaziantep, *Haradjian* 1766; between Adana and Gaziantep, *Roper* 162. LEBANON. Sannine, *Aslet*; Jabal el Knaisse, *O. Polunin* 5240; Dahr-Ul-Baidar, *Trott* 2091; Yahshoush, *Mooney* 4632; Barouk, *Maitland*; Meyranba, *Birkinshaw* 29; Jebel Jermak, *Davis* 5148; above Djouni, 1901, *Siehe* s.n.; Eden, *Trench* s.n.

b. subsp. **aintabensis** (G.P. Baker) B. Mathew, **stat. nov.**

Syn: *I. histrio* var. *aintabensis* G.P. Baker in Gard. Chron. ser. 3, 89: 137 (1931); Güner & Peşmen in Temel Bilm. 4: 33 (1980); Mathew, The Iris: 174 (1981); Mathew, The Smaller Bulbs: 106 (1987). Type: S Turkey: a plant cultivated by G.P. Baker, exhibited at a Royal Horticultural Society Show, 10 Feb. 1931, from bulbs collected near Gaziantep (holo. K!).

Lamina of outer segments 0.8–1.1 cm wide, with the blue blotches usually aggregated towards the centre, in front of and surrounding the median ridge; filaments at least 1.4–1.6 cm long, filiform.

Flowering period: Feb.–April.

Habitat: Stony or rocky areas, often in or near scrub, 610–1100 m.

Distribution: S Turkey, in the neighbourhood of Gaziantep.

Specimens seen: Turkey, Gaziantep, *Haradjian* 823; Gaziantep, 'Kara Tash', *Balls* 2157; between Gaziantep and Bahçe, *T. Baytop* ISTE 36728, ibid., *Güner* 1581.

I. histrio can usually be fairly readily recognised by the flower colour, pale blue with conspicuous blotching over most of the surface of the lamina of the outer segments, although in subsp. *aintabensis* this tends to be confined to a smaller zone in the centre. Although *I. histrioides* also has prominently spotted outer segments the two species can be distinguished by the shape of these segments as indicated in the key to species: the flower colour and markings are also distinct when seen in the living state. *I. histrio* has a wider and more southerly distribution in drier habitats than the very restricted *I. histrioides* which occurs at higher altitudes in more mesophytic conditions in northern Turkey. It is therefore not surprising that the former is little-known horticulturally in the more northern climates of Europe and North America whereas *I. histrioides* is easily cultivated and has become commercially valuable with thousands of bulbs being produced by the Dutch bulb industry each year.

I. histrio subsp. *aintabensis* is recognisable by its generally smaller flowers, less heavily blotched on the outer segments; it is more tolerant of cultivation in Europe and is seen more frequently than subsp. *histrio*.

8. **I. histrioides** (G.F. Wilson) Arnott in Journ. Hort. ser. 3, 24: 121, f. 18 (1892); Dykes, The Genus Iris: 224, t. 46, fig. 1 (1913); Bot. Mag. 157: t. 9341 (1934); Mathew, The Iris: 175, col. t. 35 (1981); Mathew in Davis, Flora of Turkey 8: 403 (1984); Baytop & Mathew, The Bulbous Plants of Turkey: 46 (1984). Fig. 7.

Type: TURKEY: Living material, cultivated by G.F. Wilson, probably from bulbs obtained from the German nursery of Max Leichtlin (specimen not traced).

Syn: *I. reticulata* M. Bieb. var. *histrioides* G.F. Wilson in Gard. Chron. ser. 3, 9: 117 (1891); Foster, Bulbous Irises: 9, 59, figs. 7, 37 (1892).

Iridodictyum histrioides (G.F. Wilson) Rodionenko, Rod Iris: 202 (1961).

Figure 7. I. histrioides (G.F. Wilson) Arnott.

Plant 6–15 cm tall at anthesis. Bulb with coarsely fibrous-reticulate tunics, with a few bulblets at the base. Leaves 2, often scarcely visible at anthesis but elongating later to 20–30 cm, 4-angled. Peduncle subterranean. Bract and bracteole membranous, white or slightly greenish with distinct green cross-veins. Perianth tube 4–7 cm long. Flowers blue, the outer segments with a dark-blotched zone on a pale ground around the yellow median ridge; outer segments broadly oblanceolate, 4.2–5 cm long, claw patent or slightly inclined, 2.2–3 cm long, 1–1.2(–1.4) cm wide, lamina reflexed, broadly ovate, obtuse or rounded, 1.7–2.5 cm long, 1.4–2(–2.3) cm wide; inner segments oblanceolate, sometimes narrowly so, obtuse or rounded, 4–4.5(–5.6) cm long, 0.6–0.8(–1) cm wide. Stamens with filaments 1.3–1.5 cm long, anthers 0.9–1 cm long. Style branches strongly cuneate, 3.6–5(–5.7) cm long, 1.3–1.6 cm wide at the base of the lobes, lobes obtuse to acute, 1.1–1.7(–2.2) cm long, 0.6–1.0 cm wide at the base; stigma bilobed. Capsule erect, carried at ground level, ellipsoid, without a prominently beaked apical portion; seeds with an appendage. 2n = 16 (Mitra & Randolph 1959).
Flowering period: March–April.
Habitat: Mountain slopes in turf or in open coniferous woods, 1300–1750 m.
Distribution: N Turkey.
Specimens seen: TURKEY. Amasya, Ak Dağ, *T. Baytop* ISTE 38980; Suluova near Amasya, *Manissadjian* 1090; Amasya, Yenice, *Güner* 1616; Samsun, Ladik, Ak Dağ, *Tobey* 977; Rize, Ikizdere, *Güner & Vural* 5500; 'Tavshan Dagh', *Ioannes Papazoghlou* s.n.

I. histrioides is rather more closely related to the yellow-flowered *I. winogradowii* than to any of the other blue- or purplish-flowered species. It has been a popular garden plant in Europe since its first introduction into cultivation in the late 19th century and is now important commercially. It is cytologically similar to *I. winogradowii* and has been hybridized with the latter to produce intermediates with flowers having yellow and blue mixtures; two of these hybrids have been named 'Katharine Hodgkin' and 'Frank Elder'. Some of the many cultivars ascribed to *I. reticulata* are almost certainly hybrids of *I. histrioides* and *I. reticulata* but these have not been investigated in detail. *I. histrioides* 'George' is a cultivar which has become commercially available in recent years; it has deep purple flowers, a colour quite unlike that of any of the wild forms, and although morphologically very similar to *I. histrioides* it is presumed to be a hybrid between this and one of the purple-flowered cultivars of *I. reticulata.*

I. histrioides is known from a fairly limited area of central-northern Turkey and is in consequence probably somewhat vulnerable to the attentions of collectors and to overgrazing. However, although it was previously thought to occur only in the region of Amasya it has now been found further to the east in Rize province.

9. **I. winogradowii** Fomin in Woron. & Schelk., Sched. Herb. Fl. Cauc. no. 166 (1914); Medwedew in Trudy Tiflissk. Bot. Sada 18: 482 (1918); Stapf in Bot. Mag. 154: t. 9220 (1930); B.A. Fedtsch. in Fl. U.R.S.S. 4: 554 (1935), Engl. transl. 4: 422 (1968); Mathew, The Iris: 179 (1981); Mathew, The Smaller Bulbs: 108 (1987). Fig. 8.

Type: U.S.S.R., TRANSCAUCASIA, Tiflis, Gori, 'in regione subalpina montis Lomis-Mta jugi Adžaro-Imeretici', 28 Apr. 1913, *Mlokosiewicz* (holo. LE, photo. K!).

Syn: *Iridodictyum winogradowii* (Fomin) Rodionenko, Rod Iris: 202 (1961); Rodionenko in Iris Year Book 1972: 84 (1972).

Plant 8–15 cm tall at anthesis. Bulb with coarsely fibrous-reticulate tunics, with a few bulblets at the base. Leaves 2–3(–4?), shorter than the flower at anthesis, elongating later to 15–30 cm, 4-angled. Peduncle subterranean. Bract and bracteole thin and membranous, often suffused green towards the apices. Perianth tube 5–7 cm long. Flowers pale yellow, the outer segments with a deeper yellow–orange median ridge and sparsely spotted green on the lamina; outer segments oblanceolate, 4.3–5.2 cm long, claw suberect, 2.5–2.6 cm long, 0.8–1.2 cm wide, lamina reflexed, ovate, obtuse, 1.7–2.6 cm long, 1.2–2 cm wide; inner segments erect, narrowly oblanceolate, acute, 3.7–5 cm long, 0.6–1 cm wide. Stamens with filaments 1.1–1.2 cm long, anthers 0.9–1 cm long. Style branches oblanceolate, 4–4.3 cm long, 0.6–1.1 cm wide at the base of the lobes; lobes obliquely ovate, obtuse or subacute, 1.2–1.6 cm long, 0.6 cm wide at the base; stigma bilobed. Capsule erect, carried at ground level, ellipsoid without a prominently beaked apical portion; seeds with an appendage. 2n = 16 (Mitra & Randolph 1959).

Flowering period: April–May.

Habitat: Subalpine meadows.

Distribution: Known only from the type locality in U.S.S.R., W Georgian SSR.

Specimens seen: U.S.S.R., W GEORGIAN SSR, Gori, Mt Lomis-Mta, *Kozlowsky* 153.

I. winogradowii is named after a colleague of Fomin, P.Z. Winogradow-Nikitin, who first noted its existence. It is related to *I. histrioides* and will hybridize with it

Figure 8. I. winogradowii Fomin.

to produce variable intermediates (see above, under the latter). Apart from the obvious difference in flower colour the two are distinguishable by the shape and dimensions of the style branches, oblanceolate and 0.6–1.1 cm wide (at the base of the lobes) in *I. winogradowii*, strongly cuneate (wedge-shaped) and 1.3–1.6 cm wide in *I. histrioides*.

I. winogradowii occurs in only one locality and is noted as being reduced to only a few hundred plants (Rodionenko 1972). Fortunately it is relatively easy to cultivate in northern Europe and, although by no means common, it is available commercially to some extent.

D. Section **Micropogon** (Baker) Boiss., Fl. Orient. 5: 124 (1882).

Syn: *Xiphion* subgenus *Micropogon* Baker in Journ. Bot. 14: 265 (1876).

Bulb with reticulate tunics, not furnished with needle-like fibres at its base. Leaves quadrangular, with several sheaths at the base. Peduncle short, subterranean. Perianth tube long; inner perianth segments much-reduced, subulate; stigma bilobed. Capsule ellipsoid, erect at maturity, carried at or just above ground level; seeds with an appendage. Type of section: *I. danfordiae* (Baker) Boiss.

10. **I. danfordiae** (Baker) Boiss., Fl. Orient. 5: 124 (1882); Bot. Mag. 116: t. 7140 (1890); Foster, Bulbous Irises: 61 (1892); Mathew, The Iris: 173, col. t. 38 (1981); Baytop & Mathew, The Bulbous Plants of Turkey: 46 (1984); Mathew in Davis, Flora of Turkey 8: 404 (1984). Fig. 9.

Figure 9. I. danfordiae (Baker) Boiss.

Type: TURKEY. Adana Prov., Cilician Taurus, near the village of Anascha, 1220 m, 24 March 1876, *Mrs Danford* (holo. K!).

Syn: *Xiphion danfordiae* Baker in J. Bot. (London) 14: 265 (1876).

Iris bornmuelleri Hausskn. in Flora 72: 141 (1889); Gard. Chron. 1890, 1: 293, t. 49b (1890). Type: Turkey, 'supra Amasiam Anatoliae', *Bornmüller* (iso. K!).

I. amasiana Bornm. in Möller's Deutsche Gärtn. Zeitung 4: 129 (1889). Type: not specified but probably as for *I. bornmuelleri.*

I. crociformis Freyn in Oesterr. Bot. Z. 44: 326 (1894), in syn. sub *I. bornmuelleri.*

Iridodictyum danfordiae (Baker) Rodionenko, Rod Iris: 202 (1961).

Plant 6–7 cm tall at anthesis. Bulb with coarsely fibrous-reticulate tunics, with many small bulblets at the base. Leaves (1–)2(–3), shorter than the flower at anthesis, elongating later to about 20 cm, 4-angled. Peduncle subterranean. Bract and bracteole dissimilar, the bract green, the bracteole narrower and semi-transparent. Perianth tube 3–4(–6) cm long. Flowers yellow with sparse green spots and a deep yellow or orange median ridge in the centre of the lamina of the outer segments; outer segments oblanceolate, 2.5–4 cm long, claw suberect, 1.5–2.5 cm long, 0.5–0.7 cm wide, lamina patent or reflexed, ovate, obtuse, 0.7–1.8 cm long, 0.6–1.3 cm wide; inner segments erect, subulate, 0.3–0.5 cm long. Stamens with filaments 0.4–0.5 cm long, anthers 0.5–0.6 cm long. Style branches cuneate, 2.2–3.3 cm long, 0.6–1 cm wide at the base of the lobes; lobes ovate, usually obtuse, irregularly dentate, 1–1.3 cm long, 0.4–0.7 cm wide at the base. Capsule erect, carried at ground level at maturity, ellipsoid without a conspicuously beaked apical portion, c. 2.5 cm long, 1 cm wide; seeds with an appendage. $2n = 18$ (Johnson and Mathew 1989 in press).

Flowering period: March–April(–May).
Habitat: Stony slopes in exposed places or in sparse coniferous woods, usually flowering near the snowline, 1000–2000 m.
Distribution: Central-north, central and central-south Turkey.
Specimens seen: TURKEY. Amasya, *Manissadjian* 577; Ordu, between Mesudiye and Gölköy, *Güner* 1474; Gümüşhane, between Bayburt and Gümüşhane, *Stainton* 8215; Sivas, near Kangal, *T. Baytop* ISTE 12783; Erzincan, Sipikor Dağ, *Sintenis* 1889: 117; near Erzurum, *Holmboe*; Niğde, Ulukişla, *Güner* 1499; SE of Ulukişla, *McMurtrie* 2325; Adana, Burücek, *Balls* 1201; Anascha, *Siehe*; Hacin Dağ, *Siehe* 1896: 681.

A distinctive species, easily recognised by its smallish yellow flowers with reduced bristle-like inner perianth segments. It has long been cultivated in European gardens and is an important species in the Dutch bulb trade. The clone which is propagated by the thousand in Holland is a sterile triploid ($3n = 27$, Mitra & Randolph 1959) which is a more vigorous larger-flowered plant than its wild counterparts.

Apart from the unique floral morphology, *I. danfordiae* is cytologically distinct from other species of subgenus *Hermodactyloides* and appears not to hybridize with any other. Although there have been reports of man-made hybrids these are unsubstantiated. In view of the apparent isolation of *I. danfordiae* it has been decided to maintain Baker's group *Micropogon* at sectional level. In pollen morphology, however, it is similar to the species of sect. *Hermodactyloides* (see below).

POLLEN MORPHOLOGY

The descriptions of *Iris* pollen presented here are based on material taken from herbarium sheets deposited in the Herbarium of the Royal Botanic Gardens, Kew.

In most species of *Iris* subgenus *Hermodactyloides*, the outer layer of the pollen grain wall consists of discrete pieces of exine, the relative arrangement of which is maintained by the non-exinous layer directly below. Acetolysis of the pollen destroys this lower layer, causing the arrangement of the exine pieces to disintegrate. Immersing the pollen grains in a standard acetolysis mixture at room temperature has a similar destructive effect.

An acetone-series method is more appropriate for the preparation of *Iris* pollen. This process dissolves the 'pollenkitt' but leaves the rest of the pollen grain intact. Grains prepared in this way do, however, tend to expand significantly, and this is especially true of any non-exinous areas of the grain (Rodionenko 1961, Fig. 43; Heslop-Harrison 1979, Figs. 6–8). It is important to note, therefore, that a sample of acetone-series treated pollen grains may have not only a greater mean-size and wider size-range than an acetolysed sample of the same species, but may also be markedly different in shape.

The acetone-series method used to prepare pollen for the present work is described below.

Materials & method

The pollen grains were agitated and centrifuged in a series of acetone concentrations of increasing strength (50%, 70%, 90%, 100% at room temperature). After washing to remove the acetone, 50% of the pollen from each species was centrifuged in 50% glycerol and slide-mounts prepared using glycerol jelly. These were examined and photomicrographed with a Nikon Labophot light-microscope (LM) fitted with a Nikon microflex automatic photomicrograph attachment (model AFM). Measurements of the pollen grains are expressed as means, calculated from a sample of 10 grains unless otherwise stated.

The remaining pollen was reserved for scanning electron microscopy. After washing, 95% ethanol was added to the samples which were then pipetted onto cleaned aluminium stubs and left to air dry at room temperature, protected from contamination by an inverted flask. The prepared stubs were coated with c. 20 nm of platinum evaporated in a Balzer's Union SCD 030 sputter coater before examination with a Hitachi S510 scanning electron microscope (SEM).

Species of *Iris* included in the present survey are listed below. Herbarium specimens from which material was obtained are listed in parentheses.

Subgenus *Hermodactyloides*

Section *Brevituba*: *I. pamphylica* (B. Mathew et al. 8415, S Turkey)

Section *Monolepis*: *I. kolpakowskiana* (collector unknown, coll.: 4.4.1912, C Asiatic USSR)

Section *Hermodactyloides*: *I. reticulata* (Polunin 5042, N Iraq; E. Gabrielian s.n. collected 29.3.1976, Armenian SSR; Mathew & Tomlinson 4315, E Turkey; H. Fouroughi 5301, N Iran; B. Mathew 532, N Iran; P. Furse 5072, N Iran); *I. bakeriana* (syn. *I. reticulata* var. *bakeriana*) (B. Mathew 1529, W Iran; A. Güner 1593, S Turkey; T. Baytop ISTE 41770, S Turkey); *I. histrio* (A. Güner 1456, S Turkey, A. Güner 1575, S Turkey; M. Cheese MCW 2451, S Turkey); *I. histrio* subsp. *aintabensis* (A. Güner 1581, S Turkey; T. Baytop ISTE 36728, S Turkey); *I. histrioides* (A. Güner & M. Vural 5500, N Turkey; A. Güner 1616, N Turkey; T. Baytop ISTE 38980, N Turkey); *I. winogradowii* (B. Mathew, cult. material; Kozlowsky 153, S Caucasus).

Section *Micropogon*: *I danfordiae* (E.K. Balls 1201, S Turkey; A. Güner 1499, C Turkey).

Subgenus *Limniris* Series Syriacae: *I. masia* (Güner 1518, S Turkey)

Subgenus *Scorpiris*: *I. rosenbachiana* (Regel s.n. coll. Feb. 1884, C Asiatic USSR).

Terminology

Many palynological terms, particularly those referring to the positions of germinal apertures, depend on the orientation (polarity) of each grain within the tetrad (Chanda, Ghosh & Nilsson 1979). It is often difficult to apply these terms to a single, mature pollen grain unless its former tetrad position is known.

For 'reticulata-type' pollen described below, the long (LA) and short (SA) axes were taken from a view in which the perimeter of the pollen grain is formed by the ring-like sulcus (as in Fig. 10, D–F). The polar axis (PA) was assumed to be the diameter of the pollen grain in a plane perpendicular to that of the sulcus, and was measured from the type of view shown in Fig. 10, G–I. References to a grain's 'equator' are not made in the strict palynological sense but are used to imply the largest circumference of the grain.

For *I. pamphylica* and *I. masia* LA and SA measurements were taken from a view showing the sulcus in its entirety at mid-focus. Fig. 11, A shows pollen grains (numbered 1 and 2) lying in an equivalent position under SEM. The PA was taken as the diameter of the grain perpendicular to the plane of the sulcus. A grain in a suitable position is shown in Fig. 11, F. Slide-mounted pollen grains of these two species tend to lie in a position from which LA and SA can be measured. PA measurements, therefore, were difficult to obtain.

Results

In the present work, three distinct pollen types are described from *Iris* Subgenus *Hermodactyloides*:

i. reticulata-type pollen (Fig. 10, A–L): large (Erdtman 1971, size classes), ellipsoid, with a ring-like sulcus. Of the six species within Sect. *Hermodactyloides*, five, including *I. reticulata*, have this type of pollen. The sixth species of the section, *I. vartanii*, was not examined due to lack of material. *I. danfordiae* (Sect. Micropogon) also has reticulata-type pollen.

Schulze (1971) refers to this pollen type as 'syncolpate' in his discussion of the pollen morphology of *Iridodictyum* (syn. *Iris* Subgenus *Hermodactyloides*).

ii. pamphylica-type pollen (Fig. 11, A, D–F): large, ellipsoid, monosulcate. Within Subgenus *Hermodactyloides*, found only in *I. pamphylica* (Section *Brevituba*, Fig. 11, A), but *I. masia* of Subgenus *Limniris* has very similar pollen (Fig. 11, D–F). It should be noted that this comparison is based only on SEM and LM analysis of whole grains. Techniques of sectioning and fine-detail examination were beyond the scope of the present work.

iii. kolpakowskiana-type pollen (Fig. 11, B–C): large, spheroidal, with irregular patches of exine. Aperture not clearly defined. Found only in *I. kolpakowskiana* (Sect. *Monolepis*). Material of *I. winkleri*, the only other species of Sect. *Monolepis*, was unavailable. Rodionenko (1961) describes kolpakowskiana-type pollen as 'non-aperturate'.

The pollen of *I. rosenbachiana* (Fig. 11, G–K) was also examined to explore the possibility of a relationship between *I. rosenbachiana* and *I. kolpakowskiana*. Pollen of the two species was found to be dissimilar, that of *I. rosenbachiana* having discrete plates of exine rather than the irregular patches characteristic of *I. kolpakowskiana* pollen.

Rodionenko (1961) illustrates several species of *Juno* (i.e. *Iris* Subgenus *Scorpiris*, syn. genus *Juno*) with a similar pollen type to *I. rosenbachiana*. He refers to this pollen type as 'pseudo multi-furrowed'. Rodionenko's germination experiments with *Juno* pollen demonstrate the separation of the exine plates during hydration and the emergence of the pollen tube through the spaces which develop between the exinous plates.

Pollen Descriptions

I. reticulata (Fig. 10, A–B, D–I)

Pollen ellipsoid; LA 80–88–94 μm, SA 70–79–86 μm, PA 69–70 μm (sample of 3). Sulcus ring-like, wide or very wide (as much as $\frac{1}{3}$ surface area of grain). The sulcus encircles the grain either at its equator, dividing the exine into more-or-less equal parts, or above its equator so that one exinous area is very much smaller than the other (Fig. 10, B). Numerous small, exinous insulae are irregularly dispersed over the sulcus, their spacing is somewhat influenced by the degree of expansion of the

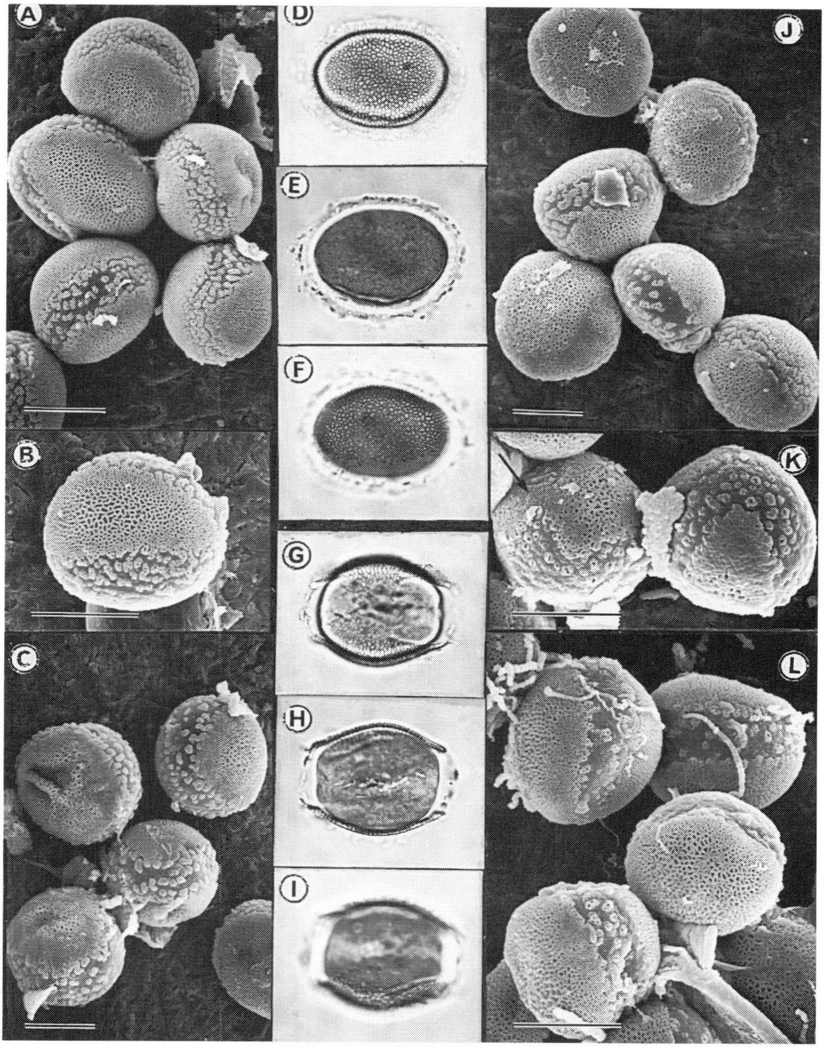

Figure 10. Pollen of *Iris*. — A–B, D–I. *I. reticulata:* A–B, *Gabrielian s.n. coll. 29 Mar. 1976*; D–I, *Mathew & Tomlinson 4315*. — C. *I. histrio* (*A. Güner 1575*). — J, K. *I. bakeriana:* J, *Mathew 1429*; K, *Güner 1593*. — L. *I. danfordiae* (*E.K. Balls 1201*). — (Micron line: A–C, J–L, 50 μm. Magnification: D–I, × 600.)

sulcus membrane. Wall thickness c. 1.5–2 μm. Exine, including that of the sulcus insulae, semitectate reticulate, lumina with exinous granules.

Other taxa with reticulata-type pollen: *I. histrio* (Fig. 10, C), *I. histrio* subsp. *aintabensis*, *I. histrioides*, *I. winogradowii*, *I. bakeriana* (Fig. 10, J–K), *I. danfordiae* (Fig. 10, L).

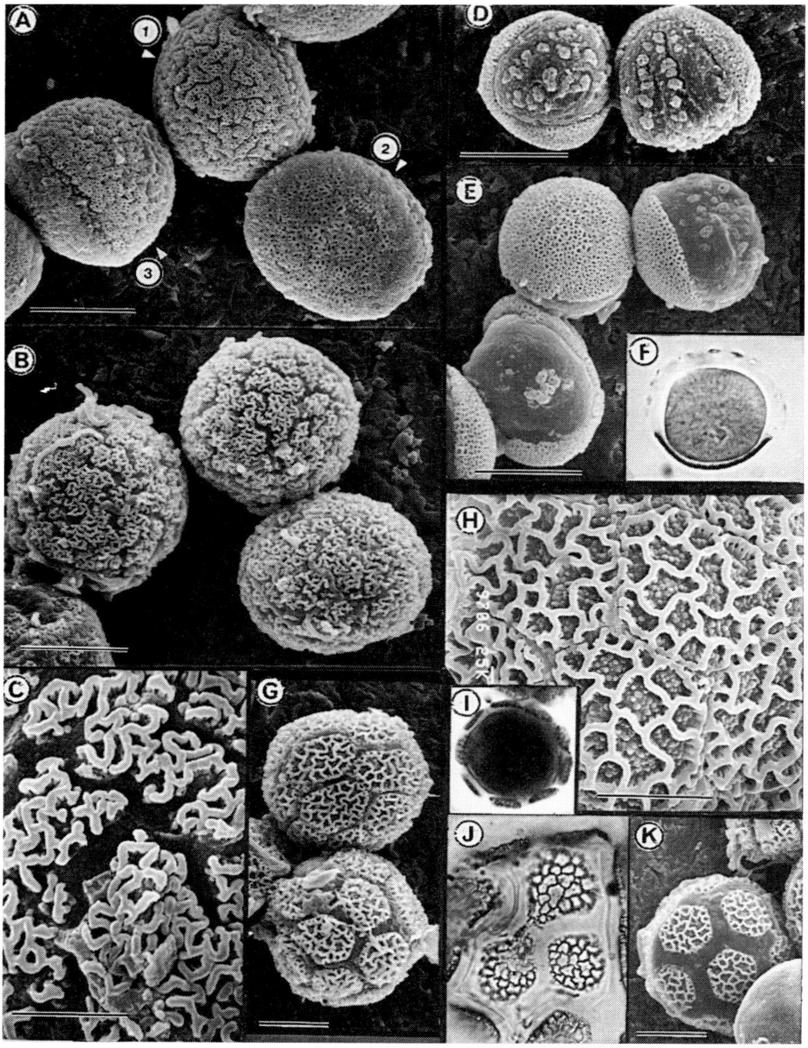

Figure 11. Pollen of *Iris*. — A. *I. pamphylica* (*Mathew et al. 8415*). — B–C. *I. kolpakowskiana* (*coll. unknown, 4 Apr. 1912*). — D–F. *I. masia* (*Güner 1518*). — G–K. *I. rosenbachiana* (*Regel s.n. coll. Feb. 1884*). — (Micron line: A, B, D, E, G, K, 50 μm; C, 10 μm; H, 20 μm. Magnification: F. J. × 600; I. × 300.)

Variation within group: *I. bakeriana* has relatively large pollen (LA 90–99–105 μm, SA 86–93–100 μm). Very occasionally, pollen of this species has a sulcus which does not form a complete ring (Fig. 10, K, arrowed).

I. pamphylica (Fig. 11, A)

Pollen ellipsoid; LA 88–91–95 μm, SA 75–81–86 μm, PA not available. Monosulcate, sulcus very wide, $\frac{1}{3}$–$\frac{1}{2}$ surface area of grain, covering one face completely and extending into the other. Sulcus covered with anastomosing ridges of exine and bordered by a narrow zone of exinous insulae. Wall thickness c. 2–3 μm. Exine, including that partly covering the sulcus, semitectate reticulate. Lumina with exinous granules.

I. masia (Fig. 11, D–F)

Pollen ellipsoid; LA 60–69–75 μm, SA 58–63–66 μm (sample of 5), PA 66–68–70 μm (sample of 6). Monosulcate, sulcus very wide, $\frac{1}{3}$–$\frac{1}{2}$ surface area of grain, covering one face completely and extending into the other. Sulcus with several to many exinous insulae located mostly towards centre of sulcus. Wall thickness c. 2–3 μm. Exine, including that of sulcus insulae, semitectate reticulate. It is not clear whether the granules in the lumina of the non-apertural exine are also present in the lumina of the exine of the sulcus insulae.

I. kolpakowskiana (Fig. 11, B–C)

Pollen spheroidal; diameter 83–87–92 μm. Structure of aperture not clearly defined. Exine in irregular patches (Fig. 11, C). Spacing between patches influenced by degree of expansion of underlying membrane. Wall thickness c. 2–3 μm. Exine semitectate reticulate with sparse exinous granules in, or associated with, the lumina.

I. rosenbachiana (Fig. 11, G–K)

Pollen spheroidal; diameter 114–119–125 μm. Structure of aperture not clearly defined. Exine forms more-or-less six-sided plates which separate from one another with expansion of the underlying membrane (Fig. 11, K). Plates 31–38–45 μm × 25–32–37 μm. Wall thickness c. 4–5 μm. Exine semitectate reticulate with exinous granules in the lumina.

ACKNOWLEDGMENTS

We would like to thank the staff of the Palynology Unit at Kew for their helpful advice and the Keepers of the Herbaria at Kew (K), Edinburgh (E), British Museum (BM), Geneva (G), Vienna (W) and Istanbul (ISTE) for the valuable information gleaned from their specimens. There are many people who have assisted by sending living material of these bulbous irises; in particular our thanks go to Dr A. Güner (Hacettepe University, Ankara), A. McMurtrie (Toronto), T. Baytop (Istanbul University) and Mrs K.N. Dryden (Sawbridgeworth, England). We are also indebted to the staff of the Alpine Department at the Royal Botanic Gardens, Kew, for their expertise in cultivating *Iris* species and making them available for research. Several of the photographs were taken by Erich Pasche (Figs. 1, 2, 3, 5, 8, 9) and Adil Güner (Fig. 4); to them we are grateful. The remaining two (Figs. 6 and 7) were taken by B. Mathew. S. Zmarzty prepared Figs. 10 and 11.

LITERATURE CITED

Ålefeld, F. 1863. Über die Gattung *Iris*. Bot. Zeitung (Berlin) 1863: 297–298.

Arber, A. 1921. The leaf structure of the Iridaceae. Ann. Bot. (London) 35: 301–336.

Baker, J.G. 1877. Systema Iridacearum. J. Linn. Soc. Bot. 11: 61–180.
——. 1892. Handbook of the Irideae. G. Bell & Sons, London. 247 p.
Chanda, S., K. Ghosh, and S. Nilsson. 1979. On the polarity and tetrad arrangements in some mono- and diaperturate angiosperm pollen grains. Grana 18: 21–31.
Dykes, W.R. 1913. The genus Iris. Cambridge University Press, Cambridge. 245 p.
Erdtman, G. 1952. Pollen morphology and plant taxonomy. Angiosperms. (An introduction to palynology. I). Almqvist & Wiksell, Stockholm. 539 p. Reprinted 1971. Hafner Publishing Co., New York. 553 p.
Fedtschenko, B.A. 1935. *Iris*, pp. 511–576. *In* V.I. Komarov [ed.], Flora U.R.S.S. Vol. 4. Akademii Nauk S.S.S.R., Leningrad. Engl. transl. pp. 421–423. Israel Program for Scientific Translations, Jerusalem.
Feinbrun-Dothan, N. 1986. *Iris*, pp. 114–127 & tt. 156–171. *In* N. Feinbrun-Dothan, Flora Palaestina Vol. 4. Israel Academy of Sciences and Humanities, Jerusalem.
Foster, M. 1892. Bulbous irises. Royal Horticultural Society, London. 86 p.
Furse, P. 1972. Iris Reticulata Section. Iris Year Book 1972: 103–111.
Galil, J. 1983. Morpho-ecological studies on geophilic plants: vegetative reproduction in *Iris vartani* (Iridaceae). Pl. Syst. Evol. 141: 177–185.
Güner, A., and H. Peşmen. 1980. A taxonomical studies on some Turkish irises. Temel Bilm. 4: 32–34.
Gustaffson, M., and P. Wendelbo. 1975. Karyotype analysis and taxonomic comments on irises from SW and C Asia. Bot. Not. 128: 208–226.
Heslop-Harrison, J. 1979. Pollen walls as adaptive systems. Ann. Missouri Bot. Gard. 66: 813–829.
Klatt, F.W. 1866. *Xiphion*, pp. 568–573. *In* F.W. Klatt, Revisio Iridearum (conclusio). Linnaea 34: 537–739.
Johnson, M.A.T., and B. Mathew. 1989. The identity of *Iris* 'Katharine Hodgkin', a cytological and morphological approach. Kew Bull. 44, 3 (in press).
TheLawrence, G.H.M. 1953. A reclassification of the genus *Iris*. Gentes Herb. 8: 346–371.
Mathew, B. 1981. The Iris. Batsford, London. 202 p.
——. 1984. *Iris*, pp. 382–410. *In* P.H. Davis [ed.], Flora of Turkey and the East Aegean Islands, Vol. 8. Edinburgh University Press, Edinburgh.
Mitra, J., and L.F. Randolph. 1959. Karyotype analysis of bulbous *Iris*. Bot. Gaz. 120(3): 125–131.
Randolph, L.F., and J. Mitra. 1959a. Chromosome numbers of *Iris* species, pp. 537–542. *In* L.F. Randolph [ed.], Garden Irises. American Iris Society, St. Louis.
——, and ——. 1959b. Iris chromosome numbers. Bull. Amer. Iris Soc. 152: 2–6.
Regel, A. 1879. Reisen in Central-Asien 1876–79. Petermanns Mitt. 25: 376–384.
Rodionenko, G.I. 1961. Rod *Iris* – *Iris* L. (voprosy morfologii evolyutsii i sistematiki). Akademii Nauk S.S.S.R., Moskva & Leningrad. 216 p. English translation by T. Blanco White. 1987. British Iris Society, London. 222 p.
——. 1972. Irises. Iris Year Book 1972: 84–97.
Schulze, W. 1971. Beiträge zur Pollenmorphologie der Gattungen um *Iris* L. Feddes Repert. 81(8–9): 507–517.
Simonet, M. 1928. Le nombre des chromosomes dans les genres des *Iris*. Compt. Rend. Soc. Biol. 99(30): 1314–1316.
——. 1932. Recherches cytologiques et génétiques chez les *Iris*. Bull. Biol. France et Belgique 105: 255–444.
——. 1934. Nouvelles recherches cytologiques et génétiques chez les *Iris*. Ann. Sci. Nat. Bot., sér. 10, 16: 229–388.

——. 1952. Nouveaux denombrements chromosomiques chez les *Iris*. Compt. Rend. Acad. Sci. (Paris) 235: 1244–1246.

Spach, E. 1846. Revisio generis *Iris*. Ann. Sci. Nat. sér. 3, 5: 89–111.

Wendelbo, P., and B. Mathew. 1975. Iridaceae. *In* K.H. Rechinger [ed.], Flora Iranica, Lieferung 112. Akademische Druck- und Verlagsanstalt, Graz, 79 p. + 24 p. of plates.

Wu, Q.-G., and D.F. Cutler. 1985. Taxonomic, evolutionary and ecological implications of the leaf anatomy of rhizomatous *Iris* species. Bot. J. Linn. Soc. 90: 253–303.

Zakharyeva, O.I., and L.M. Makushenko. 1969. Chromosome numbers of monocotyledons belonging to the families Liliaceae. Iridaceae, Amaryllidaceae and Araceae. Bot. Zhurn. 54: 1213–1227.

10. TWO NEW SUBSPECIES OF *GALIUM PARADOXUM* (RUBIACEAE – *RUBIEAE*) FROM CENTRAL AND EASTERN ASIA

E. SCHÖNBECK-TEMESY AND F. EHRENDORFER

Institut für Botanik und Botanischer Garten der Universität, Rennweg 14, A-1030 Wien, Austria

ABSTRACT

Galium paradoxum Maxim. from the mountain forests of Asia is apparently differentiated into eco-geographical races. Subsp. *duthiei* from the upper montane zone (with birches, etc.) of the Himalaya in N.W. India, Nepal and Sikkim and of W. China (Yunnan, Sichuan) and subsp. *franchetianum* from the montane conifer zone of Japan are described as new. Subsp. *paradoxum* occurs in mixed montane forests of the Urals, Altai, Sayan, Ussuri and N.E. China.

Key words: *Galium paradoxum* (Rubiaceae – *Rubieae*), new subspecies, India, Himalaya, China, Japan.

INTRODUCTION AND TAXONOMY

Galium paradoxum Maxim. occurs in temperate, lower to upper montane forests from the Far East (Ussuri), Korea and Japan through W. and N.E. China to the Himalaya, the Altai, Sayan and a disjunct area in the Ural Mountains. Its diagnostic characters are as follows: Weak slender perennial, with thin rhizomes and runners; stems angular, simple or branched, glabrous and smooth except for retrorsely hispid nodes; lowermost leaves opposite, the upper (together with smaller stipules) in whorls of 4, lanceolate-ovate to suborbicular, with 1 median and ± distinctly pinnate lateral veins, slightly antrorsely hairy above and along margins; inflorescence cymose, few-flowered; corolla white or whitish, rotate; mericarps covered with hooked hairs.

G. paradoxum can be assigned to Sect. *Platygalium* W. Koch *sensu lato* (incl. Sect. *Cymogalia* Pobed.). However, because of its pinnately veined leaves, smaller stipules and other plesiomorphic characters, it occupies a rather isolated and basic position. This justifies its placement in a monotypic series Ser. *Paradoxa* Pobed. Recent herbarium studies by us have demonstrated an allopatric differentiation of this interesting species and suggest the recognition of three subspecies, one for the type and two described here as new.

KEY TO SUBSPECIES

1. Leaves broadly ovate to suborbicular, (5–)6–10(–15) × (3.5–)4–7(–10) mm, usually rounded to truncate at base; inflorescence reduced, with a terminal 2–3-flowered cyme and 1–2 additional flowers in the axils of the upper leaves; corolla lobes acute to acuminate . . . . .subsp. *duthiei*
1. Leaves larger, up to 40 × 20 mm; inflorescence less reduced; corolla lobes obtuse to subapiculate 2

Figure 1. Galium paradoxum. — A. subsp. *paradoxum* (*Licent* 11991, W). — B. subsp. *franchetianum* (holotype, K). — C. subsp. *duthiei* (*Delavay* 3102, K). (With 1 cm scale.)

2. Leaves lanceolate-ovate, ± attenuate at base, in 2–3 whorls with stipules; hairs of leaves 0.3–0.5 mm long; corolla 2–3 mm diam. . . . . subsp. *paradoxum*
2. Leaves broadly ovate to suborbicular, ± truncate at base, in 3–4 whorls with stipules; hairs of leaves 0.12–0.25 mm long; corolla (2.5–)3–4 mm diam. subsp. *franchetianum*

Galium paradoxum Maxim., Bull. Acad. Imp. Sci. Saint-Pétersbourg 19: 281–282, 1874, subsp. **paradoxum** Fig. 1A.
TYPE: [China] in Mandshuria austro-orientali: ad fontes fluvii Wai-Fudin S-tae Olgae sinus affluentis, in silva mixta, truncis putridis innascens, sat frequens . . . , [*Maximowicz*] (LE).

Syn: *Galium syreitschikowii* Lipsch., Repert. Spec. Nov. Regni Veg. 26: 286, t. 97 (1929). TYPE: [U.S.S.R.] in regione Uralensi australi: Simskij Zavod, distr. Zlatoust, m.

'Shukowa-Schischka' circa stagnum armentali, in piceeto abietino loco saxoso, 10 Jul. 1927, *S. Lipschitz* (Herb. Univ. Mosq. Nr. 240) (holo. MW).

Habitat: Mixed montane forest, c. 1000–2300 m.

Distribution: Urals, Altai, Sayan (Baikal), Ussuri, N.E. China.

Galium paradoxum Maxim. subsp. **franchetianum** Ehrend. & Schönb.-Tem., **subsp. nov.** Fig. 1B.

Syn: *Galium stellariifolium* Franch. & Sav., Enum. Pl. Jap. 1: 213 (1875) & 2: 393 (1877) nom. nud.

A subsp. *paradoxo* et a subsp. *duthiei* indumento breviore, foliorum verticillis quaternis 1–4 et corolla paulo majore distinctum.

Planta 10–20 cm alta. Folia 10–30(–40) × (10–)15–20 mm, cum stipulis subaequalibus vel leviter inaequalibus in verticillis 3–4 quaterna, infima tantum opposita (bina), late ovata usque suborbicularia, apice apiculata, in petiolum (3–)5–12 mm longum cito attenuata, pilis antrorsis 0.15–0.25 mm longis supra secus nervos et marginem praedita, subtus glabra. Inflorescentia subcorymbosa, cymis 1–3, terminalibus 2–3(–4)-floris et lateralibus 3–5-floris. Corolla (2.5–)3–4 mm diam., lobis obtusiusculis usque subapiculatis. *Fl.* Maio-Augusto.

TYPE: Japan, Nikko, *H. Takeda* (holo. K).

Habitat: Damp coniferous woods in mountains (*fide* Ohwi, 1965: 831–832).

Distribution: Probably endemic to Japan (recorded, as *G. paradoxum*, from all four islands by Ohwi, loc. cit., 1965).

Specimens seen: Japan: Honshu: Suruga, Mt. Fuji-san, Jul. 1887, *Sci. Coll. Imp. Univ. Japan* (K); Musashi, Misumine-sun, 29 May 1895, *Watanabe* (WU).

Differs from the other two subspecies by its shorter indumentum, more similar leaves and stipules in 3–4 whorls, and by usually somewhat larger flowers.

As we have not yet studied plants from Korea, it remains to be seen which subspecies occurs there.

Galium paradoxum Maxim. subsp. **duthiei** Ehrend. & Schönb.-Tem., **subsp. nov.** Fig. 1C.

A subsp. *paradoxo* imprimis foliis minoribus latioribus et inflorescentia reducta recedit.

Planta 5–25 cm alta. Folia (5–)6–10(–17) × (3.5–)4–7(–10) mm, late triangulari-ovata usque suborbicularia, apice cuspidata usque minutissime apiculata, in petiolum 2–5 mm longum cito contracta, superiora cum stipulis minoribus in verticillis 2–3 quaterna, inferiora in verticillis 1–2 opposita, pilis 0.3–0.5 mm longis supra sparsim, subtus secus nervum tantum vestita. Inflorescentia valde reducta, cyma terminali bracteosa 2–3-flora, insuperque 1–3 flores in axillis foliorum superiorum. Corolla 2–2.5 mm diam., lobis acutis usque acuminatis. *Fl.* Maio–Julio.

TYPE: Nepal: Nampa Gadhik, c. 3700–4000 m, 27 Jul. 1886, *Duthie* 5646 (holo. K).

Habitat: Shady rocks, c. 2700–4000 m.

Distribution: N.W. India (Terhi Garhwal, Kumaun), Nepal, Sikkim, W. China (Yunnan, westernmost Sichuan).

Specimens seen: N.W. India: Terhi Garhwal: Jángla in Ganges Valley, c. 3400–3700 m, 27 Jul. 1883, *Duthie* 1112 (BM); Kumaun: birch forest above Nábi in Kutti Valley, 2700–4000 m, 12 Sep. 1884, *Duthie* 3004 (BM, K). Nepal: Chilime Kharka, c. 4000 m, Jul. 1949, *O. Polunin* 1185 (BM). Sikkim: Chvangtum, 23 May 1849, *Thomson* (K); Lachen, c. 3700 m, 16 Jul. 1849, *Thomson* (K). W. China: Yunnan: Yen-tze Hai, 3200 m, 18 Jul. 1887, *Delavay* 3102 (K). Sichuan: Mt. Wu, c. 2900 m, Jul. 1903, *E.H. Wilson* 3772 (BM).

Hooker (1881: 207) has already commented on *G. paradoxum*, that 'the Sikkim specimens are smaller than the Manchurian, and resemble a *Nertera*'. Cufodontis

(1940: 234) also calls attention to few-flowered and small-leaved plants gathered by Duthie in Nepal and suggests that they may be specifically different from typical *G. paradoxum*. The more ample material available now substantiates the recognition of a new subspecies, subsp. *duthiei*. It differs from the typical subspecies in its smaller, broadly ovate to suborbicular leaves, only (1.1–)1.3–1.7(–1.9) × as long as broad. In contrast, subsp. *paradoxum* usually has lanceolate-ovate to ovate leaves, 2–2.5 × as long as broad. Furthermore, the inflorescence of the latter appears less reduced and has cymes on longer stalks and with more flowers. The following two specimens are intermediate between subsp. *paradoxum* and subsp. *duthiei*: China, W. Kansu: Sinn-ling-Chan, 12 Jul. 1918, *Licent* 4292 (K): Sichuan: Tienchuan-hsien, 2590 m, 4 Jun. 1936, *Chu* 2516 (W).

Subsp. *franchetianum* is easily distinguished from subsp. *duthiei* by its larger leaves, longer stems with more numerous whorls of 4 leaves, larger flowers, and shorter indumentum.

DISCUSSION

G. paradoxum belongs to an element of relict taxa from the cool-temperate mountain forests of C. and E. Asia, which today do not occur in N. America, S.W. Asia or Europe. Subsp. *paradoxum* has a more northerly distribution with altitudinal range of 1000–2300 m. This appears comparable to the easternmost subspecies, subsp. *franchetianum*, whereas subsp. *duthiei* from the southwest part of the species' area occurs at altitudes of 2700–4000 m.

Preliminary pollen measurements have shown diameters of c. 17.5 μm for subsp. *franchetianum*, whereas subsp. *paradoxum* and subsp. *duthiei* have larger pollen grains c. 21.5 μm diam. This might suggest diploidy and polyploidy (an assumption that will have to be verified by karyological studies), and an eastern centre of origin for *G. paradoxum*.

ACKNOWLEDGMENTS

We gratefully acknowledge financial support by the Fonds zur Förderung der wissenschaftlichen Forschung in Österreich, project P 6189B. The illustration (Fig. 1) was kindly prepared by Mr P. Lassnig.

LITERATURE CITED

Cufodontis, G. 1940. Revision der chinesischen *Galium*-Arten, nebst Bemerkungen über einige ihrer Formenkreise. Oesterr. Bot. Z. 89: 211–251.

Hooker, J.D. 1881. The Flora of British India. Vol. III. Caprifoliaceae to Apocynaceae. L. Reeve & Co., London. 712 p.

Ohwi, J. 1965. Flora of Japan (in English) . . . ed. F.G. Meyer and E.H. Walker. Smithsonian Inst., Washington, D.C. 1067 p.

11. *PIPTATHERUM* AND *STIPA* (GRAMINEAE) IN THE ARABIAN PENINSULA AND TROPICAL EAST AFRICA

HELMUT FREITAG

Arbeitsgruppe Morphologie u. Systematik der Pflanzen, Universität Kassel, Heinrich-Plett-Str. 40, D-3500 Kassel, Federal Republic of Germany

ABSTRACT

As an extension of earlier revisions of *Stipa* L. and *Piptatherum* P. Beauv. (Gramineae – tribe Stipeae) in SW Asia, an account of these genera is given for the Arabian Peninsula and Tropical East Africa. Full descriptions are given only for the 3 new taxa described here and for those species which have not been previously dealt with. *Piptatherum* (Syn: *Oryzopsis* auct. eur. non Michaux) is represented by 2 taxa: the Mediterranean *P. miliaceum* (W Saudi Arabia) and *P. holciforme* subsp. *abyssinicum* Freitag subsp. nov. (SW Saudi Arabia, Yemen to Ethiopia). *Stipa* (*sensu lato*) has 6 species: *S. mandavillei* Freitag sp. nov. (Sect. *Lasiagrostis*; endemic to Oman); *S. keniensis* (Pilger) Freitag comb. nov. (Sect. *Aristella*), with two subspecies, subsp. *keniensis* (Afro-Montane – Ethiopia to Tanzania; Syn: *S. dregeana* var. *elongata* (Nees) Stapf) and subsp. *somaliensis* Freitag subsp. nov. (N Somalia); 3 species of Sect. *Stipella*: the S Mediterranean *S. parviflora* (NW Saudi Arabia), *S. capensis* (northern part of Arabian Peninsula), and the Afro-Montane *S. tigrensis* (SW Saudi Arabia, Yemen Ethiopia to NE Uganda); and the Irano-Turanian *S. hohenackeriana* (Sect. *Barbatae*; N Saudi Arabia). An appraisal of species which might be introduced is included, as well as an assessment of the species occurring in S Africa.

Key words: Gramineae, Arabian Peninsula, Tropical East Africa, *Piptatherum*, *Stipa*, *Oryzopsis*.

INTRODUCTION

The genera *Piptatherum* P. Beauv. (Syn: *Oryzopsis* auct. eur. non Michaux) and *Stipa* L. were revised by the author for SW Asia (Freitag 1975, 1985), where they have centres of distribution. Towards the tropics they fade out and retreat to higher mountain areas. Only three species of *Stipa* occur at and S of the equator, and *Piptatherum* peters out further north. Although both genera have recently been dealt with by competent authors for Tropical East Africa (Clayton 1970) and Saudi Arabia (Cope 1985), the author became interested in extending his study to that area. Through the Directors and Keepers of the herbaria of Bayreüth (UBT), Berlin (B, BSB), British Museum (Natural History), London (BM), Copenhagen (C), Edinburgh (E), Firenze (FT), Genève (G), Kew (K), Nairobi (EA), Paris (P), Pretoria (PRE), Riyadh (RIY), Wageningen (WAG) and Hb. Miehe, most but certainly not all of the specimens available from the area could be used for the study.

As the paper has to be considered an addition to and extension of the two earlier revisions, the presentation follows the same pattern. Detailed descriptions, synonyms and types are only given for taxa not previously dealt with.

The northern limit of the study area results from the fact that Sinai, Jordan and Iraq were included in the previous publications. It agrees with the line drawn by Wickens (1982). In Africa, all countries S of Egypt with indigenous *Stipa* and *Piptatherum* species are taken into consideration, with the exception of S Africa. In the latter country there occur only four taxa: *Piptatherum miliaceum* (L.) Coss. – not mentioned by de Winter (1965) but recorded by Stapf (1889) and Dyer (1976) as occasionally introduced from the Mediterranean; *Stipa capensis* Thunb. – introduced from the Mediterranean and very common; *S. dregeana* Steud. var. *dregeana* – endemic; and *S. dregeana* var. *elongata* (Nees) Stapf, in the present paper raised to species rank under the name *S. keniensis* (Pilger) Freitag – most common in Tropical E Africa. The types of all taxa have been seen, but the comprehensive treatment of de Winter (1965) still remains modern. Also the transfer of two more SW and S African *Stipa* species to *Stipagrostis* Nees and *Aristida* L. respectively by de Winter (1963) has been checked and approved: *S. parvula* Nees = *Aristida parvula* (Nees) de Winter and *S. namaquensis* Pilger = *Stipagrostis anomala* de Winter.

For the present author it remains a surprise that according to the available specimens and data a wide gap remains in the distribution of *Stipa* between Tanzania and S Africa. The higher mountains of Zimbabwe, Malawi and even of Mozambique should likewise offer suitable habitat conditions, at least for *S. keniensis*. Probably further collections will bridge that gap.

For convenience, an abbreviated key for the separation of the two closely related genera *Piptatherum* and *Stipa* is given below.

1\. Spikelet and anthecium dorsally compressed; callus compressed; callus very short, scarcely distinct from lemma, with rounded base; glabrous or with 2 tufts of hairs only; palea exposed, only marginally covered by the clasping lemma; awn usually less than 10 mm . . . . *Piptatherum*

1\. Spikelet and anthecium terete or laterally compressed; callus distinct, with ± pointed base, densely bearded; palea completely or almost completely covered by the clasping lemma; awn usually longer than 10 mm . . . . *Stipa*

TAXONOMY

PIPTATHERUM

KEY TO THE SPECIES IN ARABIA AND TROPICAL E AFRICA

1\. Lemma glabrous, dumbbell-shaped, 1.5–2 mm long; awn 3–4 mm long 1. *P. miliaceum*

1\. Lemma hairy, lanceolate, 4–5 mm long; awn 6–10 mm long. . . . . 2. *P. holciforme* subsp. *abyssinicum*

Enumeration of Species

1\. Piptatherum miliaceum (L.) Coss. (Sect. *Miliacea* [Roshev. ex] Freitag)
Syn.: *Oryzopsis miliacea* (L.) Aschers. & Schweinf.

Specimens studied: SAUDI ARABIA. Hijaz: Jabal Dibbagh, 100 km SW of Tabuk, 1700 m, *Collenette 5244* (K); Jabal Warjan, 80 km SW of Al Madinah, 1800 m, *Collenette 5228* (K). Asir: Al Adanah waterfall near Tanumah at road from Abha to Taif, 1980 m, *Collenette 4564* (K, RIY); Al-Mansah, 8 km SW of Abha, *Boulos & Ads 14103* (K).

This species is restricted to mesophilous habitats of the middle and upper montane belts. It has been reported from rocky ravines, by waterfalls and as a weed along roadsides and on abandoned fields in the *Juniperus* belt. The scattered reports from the higher mountains along the escarpment of W Saudi Arabia show that the distributional area of this well-known and morphologically rather uniform Mediterranean species extends considerably further SE than indicated on the map in Freitag (1975, fig. 17).

2. PIPTATHERUM HOLCIFORME (M. Bieb.) Roemer & Schultes (Sect. *Piptatherum*, subsect. *Holciformia* [Roshev. ex] Freitag)

Subsp. **abyssinicum** Freitag, **subsp. nov.**

Differt a subsp. *holciforme* et subsp. *longiglume* spiculis minoribus glumis 6–7(–8) mm non 8–10 mm longis, lemmate 4–4.5(–5) non 5–6 mm longis.

Type: [ETHIOPIA]: Eritrea, Oculé Cusai, gruppo dei monti Soyrà (= Soira), Monte Mamahot verso il torrente Ariget, 2800–3000 m, 23 viii 1902, *Pappi* 1240 (holo. FT!).

From the higher mountains of the Asir in SW Saudi Arabia through Yemen to C Ethiopia. It should be expected also in N Somalia.

A very typical and rather common associate of the upper montane woodland and forest belts from 2400–3000 m. In the Asir Mts. (König 1987) and in Yemen it grows preferably in the shade of *Juniperus*, but it has also been collected from rock ledges, cliffs and seral communities. In Ethiopia most collections are from more mesic forest types, e.g. *Erica arborea-Hagenia* forests.

Specimens studied: SAUDI ARABIA. ASIR: Jabal Sawdah NW of Abha (42°22′E, 18°19′N), 2800 m, *Baierle, Chaudhary & Mahmoud 1363* (BSB); *ibid.*, Chaudhary 7 i 1980 (RIY), 7 iv 1982 (K, RIY); 13 km NW of Abha at road to Jabal Sawdah, 2590 m, *Collenette 4575* (E, K, RIY); waterfall between Abha and Jabal Sawdah, 2400 m, *Collenette 4447* (E, K); 10 km SW of Abha (42°25′E, 18°10′N), 2500 m, *Baierle et al. 1531* (BSB). YEMEN ARAB REPUBLIC. SW side of Jabal Mahdad (Miswar region), c. 3000 m, *Wood 2516* (BM, K). Jabal Taaga, Jibla, c. 3200 m, *Wood 75/256* (BM). Below waterfall at Bau-an at road from San'a to Menakha, c. 2900 m, *Wood 1241* (K). Jabal Sumarah, c. 2900 m, *Wood 75/582* (K). Sumarah pass just above road to W of Jabal Sumarah, 2800 m, *Radcliffe-Smith 4961* (K). ETHIOPIA. ERITREA PROV.: Soira Mts., Mamahot Mt. near Ariget ravine, 2800–3000 m, *Pappi 1298* (FT); *ibid.*, high plain of Golo, 2800 m, *Pappi 1091* (FT); *ibid.*, forest of Carobozzo, 2700 m, *Pappi 2819* (FT). Mt. Metaten, 2500 m, *Pappi 1542* (FT, G, W). Assarè forest near Halai, 2600 m, *Pappi 1954* (FT). TIGRAY PROV.: Tigre v. Begemver (later addition in K: Urahut, 3000 m), *Schimper 1092* (K, W). HARERGE PROV.: summit of Kondudo Mt. (42°20′E, 9°26′N), 2865 m, *Burger 1268* (K); NW face of Gara Muleta Mt. c. 50 km W of Harer (41°43′E, 9°16′N), 2500–3000 m, *Burger 1482* (K); *ibid.*, 2800 m, *Gillett 5362* (K).

The new subspecies differs from subsp. *holciforme* by the smaller size of all generative organs, and also usually of the vegetative parts. However, much more than in subsp. *holciforme*, the overall size of the individuals is dependent on the environmental conditions. Plants growing under humid conditions in full forest shade attain a height of 120 cm (e.g. *Schimper 1092*) and resemble the usually tall-growing subsp. *holciforme*. In contrast, specimens from open habitats with high insolation and suffering from water stress reach only 12–20 cm, as in some specimens from Yemen and Saudi Arabia. They look very much like *P. laterale* from subalpine mountain areas of the Irano-Turanian region. As in the latter and other high-altitude species, the glumes in subsp. *abyssinicum* are purplish tinged.

The new subspecies has a distinct distributional area with a marked gap separating it from that of subsp. *holciforme*, which has its isolated southernmost locality on the Sinai mountains.

Since the first record in Engler (1892), which refers to the specimen *Schimper* 1092, the material of this taxon has been named *Oryzopsis holciformis*. The labels of the *Pappi* specimens carry the annotation, 'var. *abyssinica*' – a name which evidently has never been published. Instead, Chiovenda himself (1908) erroneously cited the Pappi specimens under the name of the W. Mediterranean *O. paradoxa* (L.) Beauv. Bor (in sched.) suspected a new species, but failed to give a description. After a careful reassessment of characters, subspecies rank for the new taxon seemed justified. Even despite the geographical disjunction between it and subsp. *holciforme*, the differential characters are not stronger than in subsp. *longiglume* of the hotter lowlands. Subsp. *abyssinicum* represents a rather unexpected addition to the range of *P. holciforme*. Certainly it has been derived from subsp. *holciforme* or a common ancestor, which might have migrated to the higher mountains in low latitudes during cooler and wetter periods of the Pleistocene, as did other taxa like *Erica arborea* L. and *Juniperus* L., with which it is often associated.

STIPA L.

KEY TO THE SPECIES IN ARABIA AND TROPICAL E AFRICA

1. Annual; panicle spike-like, condensed, and awns often twisted together . 5. *S. capensis*
1. Perennials; panicle more loose and awns not twisted together 2
 2. Awn straight or irregularly bent, up to 15 mm long, scabrous; plant usually loosely tufted . . . . 2. *S. keniensis*
 2. Awn distinctly geniculate, longer than 15 mm, scabrous throughout or pilose in lower part; plants usually densely tufted 3
  3. Awn scabrous throughout, 7–10(–13) cm long; upper glume much shorter ($\frac{1}{2}$–$\frac{2}{3}$) than lower . . . . 3. *S. parviflora*
  3. Awns distinctly pilose in lower part; upper glume at least $\frac{2}{3}$ × length of lower . . 4
   4. Anthecium 10–13 mm long; awn 9–15 cm long . . . . 6. *S. hohenackeriana*
   4. Anthecium and awn much shorter 5
    5. Anthecium 5.5–7.5 mm long; callus 1.2–1.5 mm long; palea less than $\frac{1}{2}$ of lemma in length, glabrous . . . . 4. *S. tigrensis*
    5. Anthecium 4–5 mm long; callus 0.5–0.7 mm long; palea only 1–1.5 mm shorter than lemma, pilose . . . . 1. *S. mandavillei*

Enumeration of species

1. **Stipa mandavillei** Freitag, **sp. nov.** (Sect. *Lasiagrostis* (Link) Hackel). Figs. 1–3. Type: OMAN. Jabal al-Akhdar, vicinity of Birkat Sharaf, (57°20′E, 23°12′N), 2133 m, 13 iv 1975, lower sides of rocky ravine, *Mandaville 6525* (holo. BM!)

Gramen perenne, dense caespitosum, ramificatione intravaginali, culmis et surculis basalibus paucis. *Culmi* (30–)40–55(–65) cm alti, nodulis 4–5, glabri. *Vaginae* glabrae vel retrorse pubescentes, pallide virides, saepe striatae, plerumque margine ciliatae, juxta ligulam vulgo dense barbatae pilis 0.5–1.3 mm longis, in foliis superioribus culmorum invicem lobis membranaceis margine ciliolatis. *Ligulae* in foliis culmorum 0.5–1.5 mm longae, obtusae vel acutiusculae, margine ciliolatae, postice breve setulosae, in foliis surculorum basalium usque ad 0.3 mm longae, truncatae, laceratae, margine ciliatae. *Laminae* planae vel involutae, in foliis culmorum usque ad 18 cm × 1.5–2.5(–3) mm, in foliis surculorum basalium ad 35 cm × 1–2(–2.5) mm, in statu involuto 0.4–0.7(–1) mm diam., supra dense pubescentes pilis 0.1–0.25 mm longis, subtus glabrae, prominente albo-nervosae. *Panicula* (7–)10–15 (–20) × (1–)1.5–2(–2.5) cm, contracta, exserta vel basi vagina folii supremi inclusa; rami

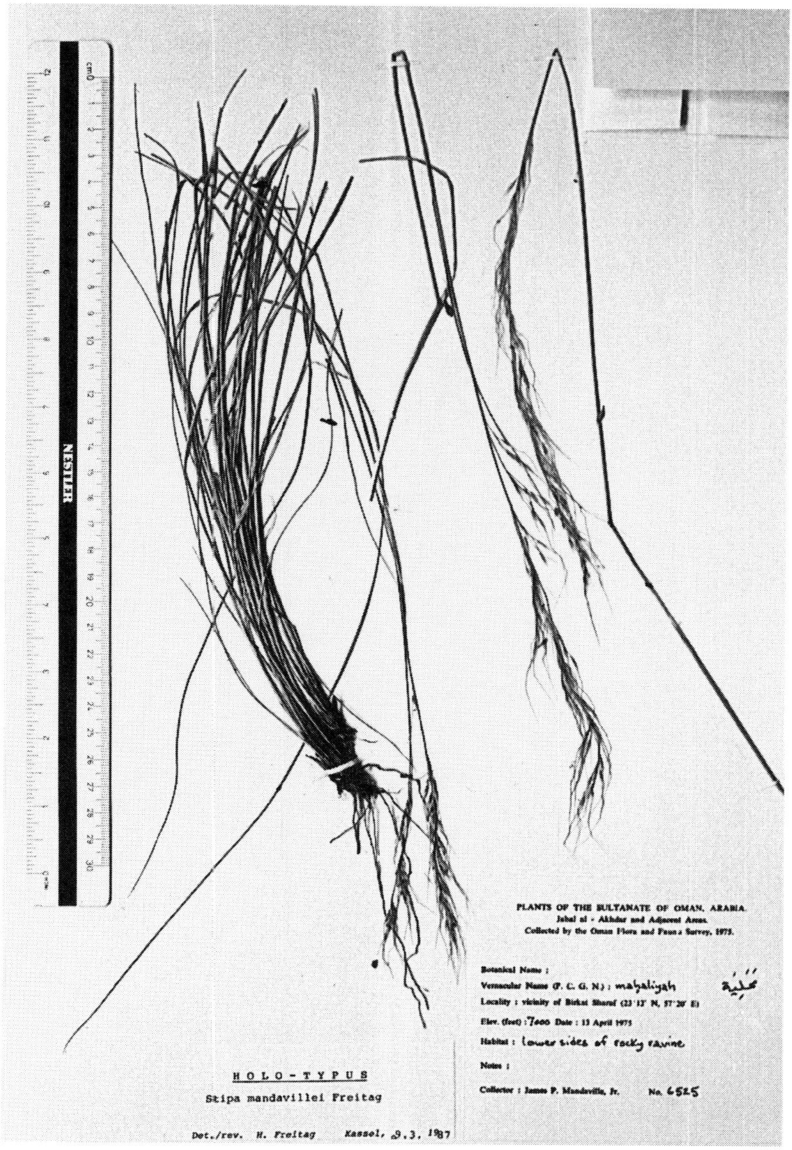

Figure 1. Holotype of *Stipa mandavillei* Freitag (*Mandaville 6525*, BM).

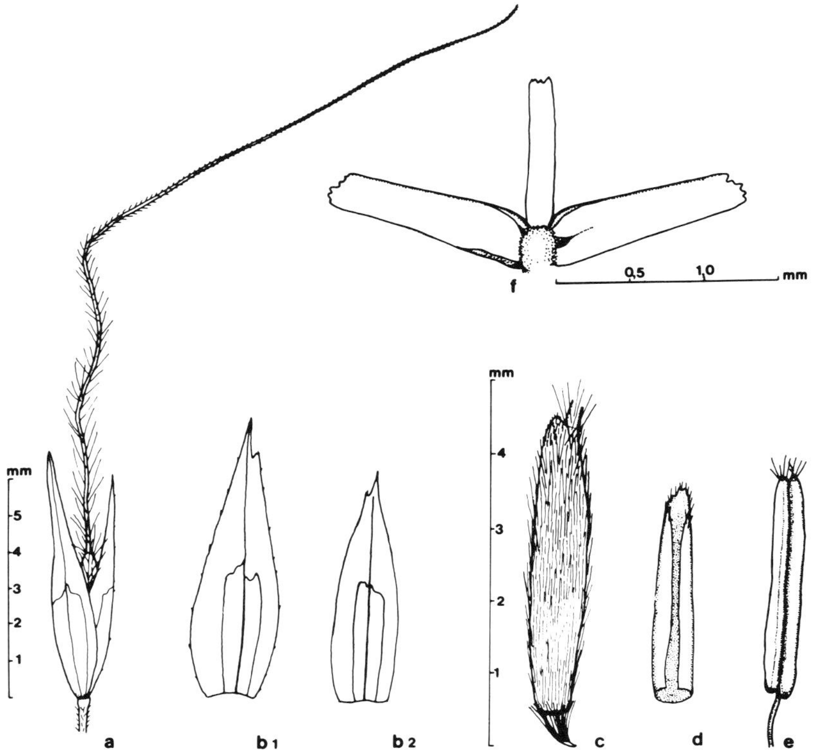

Figure 2. Stipa mandavillei: a. spikelet. — $b_{1,2}$. lower and upper glume. — c. anthecium. — d. palea. — e. stamen. — f. lodicules. — (All from *Mandaville 6487*, BM).

ascendentes vel adpressi, capillares, bini vel singulares, setulosi, usque ad 4.5 cm longi, spiculis numerosis. *Spiculae* lanceolatae, 7–9(–10) mm longae. *Glumae* lanceolatae, acuminatae, hyalinae tantum dorsalis pallide virides, interdum apice purpureo tincto, postice breve setulosae, 3-nervatae, superior 1–1.5 mm brevior. *Anthecium* 4–4.5(–5) mm longum. *Callus* 0.5–0.7 mm longus, dense barbatus pilis usque ad 0.8–1 mm longis, subacutus leviter curvatus, hilo ovali vel fere circulari. *Lemma* tempore florendi pallide viride, membranaceum apice lobos duos minutos 0.2–0.3 mm longos ferente, diffuse pilis ascendentibus (0.6–)0.7–1 mm longis tectum, apicem versus pilis aliquot brevioribus sed sub arista saepe pilis longis coronulatum. *Arista* (1.5–)2.5–3.5(–4) cm longa, unigeniculata vel indistincte bigeniculata, basi articulata; columna (5–)7–15(–17) mm longa, torta, dense pilosa pilis basi 1–1.5 mm longis, sub geniculo tantum (0.3–)0.5 mm; seta recta, in parte inferiore breve pubescens, in parte superiore scabra. *Palea* lemmate 1–1.5 mm breviora, apice obtusa saepe obtuse dentata, inter nervos sparse pilosa pilis setulosis 0.3–0.5 mm longis. *Lodiculae* 3, subaequales, 1.1–1.3 × 0.3 mm, ligulatae vel aliquot spathulatae, apice crenulatae, glabrae, posterior tantum 0.9–1.1 × 0.2–0.25 mm. *Stamina* 3; antherae 2.5–3(–3.2) mm longae, luteae, apicibus dense pilosis pilis setulosis 0.25–0.4 mm longis. *Ovarium* bistylosum stylis fissura profunde separatis. *Caryopsis* immatura 3–3.5 mm longa, fusiformis.

In small to medium-sized tufts, with very short rhizomes, branching extravaginal, with some culms and few vegetative shoots; *culms* (30–)40–55(–65) cm, 4–5-noded, glabrous; *sheaths* glabrous or retrorsely setulose to pubescent, pale green,

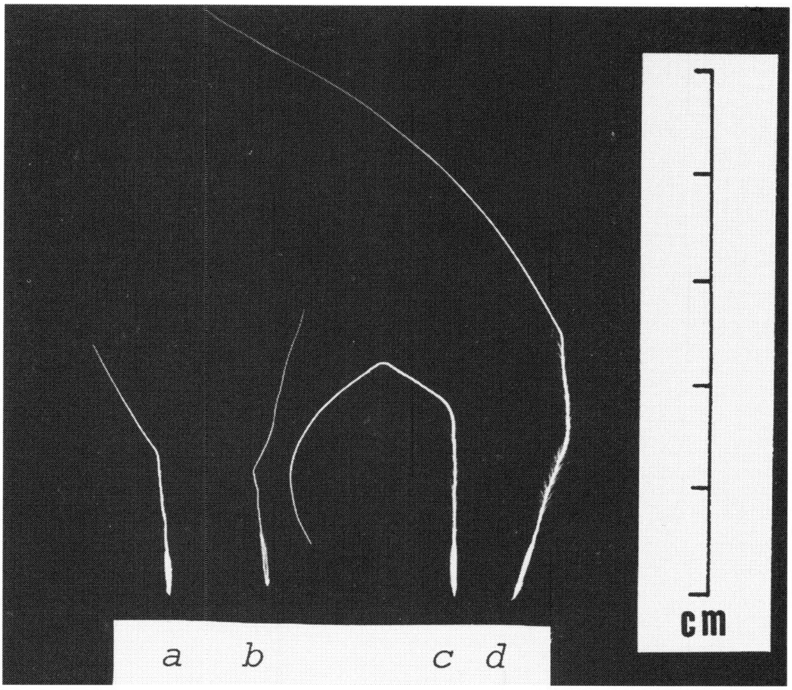

Figure 3. Anthecia with awns of *Stipa* species: a. *S. mandavillei* (*Mandaville 6757*). — b. *S. jacquemontii* (*Freitag 1746*). — c, d. *S. tigrensis* (*Wood Y/75/ 666*, *Wood 3320*).

often striate, usually with ciliate margins, beside the ligule densely bearded by 2 tufts of 0.5–1.3 mm long hairs, the upper culm leaves instead often with 2 membranous lobes with ciliolate margins; *ligules* of culm leaves 0.5–1.5 mm long, obtuse to acutish, with ciliolate margin and shortly setulose back, of vegetative shoots up to 0.3 mm long, truncate, lacerate, with ciliate margin; *blades* flat, those of culm leaves 1.5–2.5(–3) mm wide and up to 18 cm long, those of vegetative shoots 1–2(–2.5) mm wide, up to 35 cm long, involute when dry and then 0.4–0.7 (–1) mm diam., upper surface densely pubescent with 0.1–0.25 mm long hairs, glabrous beneath with prominent whitish veins; *panicle* exserted or ensheathed at base, contracted, (7–)10–15(–20) × (1.0–)1.5–2(–2.5) cm, branches ascending to appressed, capillary, lower and middle ones paired, setulose, up to 4.5 cm long, with numerous spikelets; *spikelets* 7–9(–10) mm long; *glumes* lanceolate, acuminate, with a pale green back and hyaline margins and apex, rarely somewhat purplish tinged, with shortly setulose back and setulose apex, 3-veined, upper 1–1.5 mm shorter; *anthecium* 4–4.5(–5) mm long; *callus* 0.5–0.7 mm long, densely bearded (longest hairs 0.8–1 mm long), curved, acute, scar oval to almost circular, peripheral ring dorsally flattened and protruding; *lemma* at flowering time membranous, pale green, with 2 very short apical lobes 0.2–0.3 mm long, diffusely covered by ascending hairs (0.6–)0.7–1 mm long, near the apex usually shorter, but below the very tip sometimes again with scattered longer hairs forming a coronula; *awn* (1.5–)2.5–3.5(–4) cm long, unigeniculate or indistinctly bigeniculate, with distinct articulation at base, column (5–)7–15(–17) mm long, more densely twisted in lower half than in upper, densely pilose, hairs at base 1–1.5 mm long, becoming shorter to (0.3–)0.5 mm, seta straight, shortly pubescent in lower part, scabrous in upper; *palea* 1–1.5 mm shorter than lemma, with obtuse, often bluntly toothed apex, between the veins and usually up to the apex loosely covered by 0.3–0.5 mm long straight hairs; *lodicules* slightly dissimilar, glabrous, anterior ones ligulate to somewhat spathulate, with crenulate apex, 1.1–1.3 × 0.3 mm, posterior c. 0.9–1.1 × 0.2–0.25 mm; *anthers* 2.5–3(–3.2) mm long, bearded by 2 dense tufts of 0.25–0.4 mm long stiff hairs, yellow, scarcely exserted; *ovary* with 2 styles, separated by a cleft; *caryopsis* 3–3.5 mm long (immature), fusiform.

Endemic to Jabal al-Akhdar and Jabal Aswad in N Oman. According to Mandaville (1977), who collected the four specimens used for the description, the new species belongs to the upper part of the *Reptonia mascatensis-Olea africana* belt and the lower part of the *Juniperus macropoda* (= *J. excelsa*) summit-belt. Its altitudinal range is from 1350 to 2500 m. More extensive field studies certainly will reveal a somewhat wider altitudinal range, but it seems evident that this interesting new endemic is restricted to the comparatively mesic mountains of N Oman, which receive rather regular rains in winter and summer of an estimated total of c. 250–375 mm.

Specimens studied: OMAN. Jabal al-Akhdar: vicinity of Masjid Mu'alla (57°19′E, 23°12′N), 1860 m, rocky slope, *Mandaville 6434* (BM); *ibid.*, 2286 m, *Mandaville 6487* (BM). Jabal al-Aswad: summit (58°37′E, 23°10′N), 1830 m, *Mandaville 6757* (BM).

All relevant characters of the new species point to its placing in Sect. *Lasiagrostis*: extravaginal branching, membranous lemma with apical lobes, short callus, short awn, relative length of palea and of the posterior lodicule. Its closest relative is probably *S. jacquemontii* Jaub. & Spach from N India, N Pakistan, E Afghanistan and N Baluchistan. With that species it shares habit and most characters, except the distinct pubescence of the lower part of the awn, and the somewhat longer callus. *S. jacquemontii* also inhabits temperate evergreen woodlands in areas with overlapping summer and winter rains, but it is a more mesic plant, restricted to higher altitudes with more precipitation and occurring in vegetation belts dominated by evergreen oaks and conifers. *S. mandavillei* very nicely underlines the eastern influence upon the Oman mountain flora stressed already by Mandaville (1977) and Kürschner (1986). Geographically, *S. haussknechtii* Boiss. from the Zagros Mts. in SW Iran comes much closer, but although belonging to the same section it is much less similar morphologically.

Superficially and particularly in awn structure *S. mandavillei* has some resemblance to *S. tigrensis* of Sect. *Stipella* (see below). However, the latter species differs significantly by its much longer callus, very short and glabrous palea, and very short posterior lodicule.

2. **Stipa keniensis** (Pilger) Freitag, **comb. nov.** (Sect. *Aristella*). – For basionym and synonymy see below, sub subsp. *keniensis*.

Usually loosely tufted perennial, with short rhizomes, usually forming large tussocks, branching extravaginal, with about as many culms as vegetative shoots, only in subsp. *somaliensis* densely tufted with mainly intravaginal branching and numerous vegetative shoots; *culms* (50–)80–120(–140) cm, (3–)4–5-noded, 2–3 (–4.5) mm diam. at base, glabrous, nodes usually blackish; *sheaths* usually glabrous, more rarely scaberulous with antrorse prickles or lowermost shortly pubescent between the glabrous veins, striate with prominent whitish veins, margin of lower sheaths ciliate, on one side of the ligule often with a projecting lobe appressed to the culm, these lobes on the middle culm leaves up to 20 mm long, with ciliolate margin, on vegetative shoots replaced by 2 tufts of 0.3–1 mm long hairs; *ligules* of culm leaves 0.3–0.7 mm long, truncate, lacerate, with ciliolate margin, those of vegetative shoots similar but only 0.1–0.3 mm long; *blades* flat, on both shoot systems up to 40(–50) cm long and up to (4–)7–12(–14) mm wide, those of vegetative shoots usually somewhat narrower, involute when dry, upper surface sparsely villous to almost glabrous, with sparse (0.3–)0.5–1(–1.2) mm long hairs and usually short prickles on the very numerous veins, glabrous and with prominent whitish veins beneath; *panicle* exserted, very large, ± effuse, (20–)30–45 (–50) × 4–8(–10) cm, branches ascending, spreading or slightly deflexed, scabrous, in whorls of 3 or 5, more rarely paired, up to 20 cm long, in upper ½ with up to 35(–45) loosely arranged spikelets; *spikelets* (5.5–)6–7.5(–9) mm long; glumes almost equal, lanceolate, acuminate, membranous, 3-veined, with green lines along the veins, otherwise hyaline, shortly scabrous throughout with short prickles and setulose along the veins, the upper usually slightly narrower; *anthecium* 3.5–6 mm long; *callus* 0.5–0.6 mm long, straight, obtuse, densely bearded with 0.3–0.5 mm long hairs; scar ovate to circular, peripheral ring regular; *lemma* becoming coriaceous and brownish during flowering time, chartaceous and dark brown in fruit, margins not covering palea completely, with 2 minute apical lobes of 0.2–0.3 mm, loosely covered throughout with (0.5–)0.6–0.8 mm long ascending to appressed straight hairs, usually without a distinct coronula, lobes ciliate; *awn* (7–)8–14(–16) mm long, straight (except in some specimens of subsp. *somaliensis*), scabrous, with indistinct basal articulation; *palea* almost as long as or 1–1.3 mm shorter than lemma, loosely pilose with 0.3–0.8 mm long straight hairs, more densely so towards the acutish apex; *lodicules* moderately dissimilar, glabrous, anterior broadly ligulate with rounded apex, 0.6–0.9 × 0.25 mm, posterior narrower and shorter, spathulate, with somewhat acuminate and crenulate apex; anthers 2.5–4 mm long, unbearded, yellow, exserted; *ovary* with two horn-like styles; *caryopsis* 3–4.5 mm long, embryo 0.5–0.6 mm long, hilum reaching up to 0.5–0.6 mm below top. *Fl.* usually May–July; fr. July–November.

1. Plant forming loose tufts with comparatively few vegetative shoots; lowermost sheaths light brown or pale; larger blades usually 7–12 mm wide; lower glume usually c. 0.5 mm shorter than upper; anthecium (4.5–)5–6 mm long; awn (11–)12–14(–16) mm long, straight . . . . subsp. *keniensis*
1. Plant forming dense tufts with many vegetative shoots; lowermost sheaths blackish; larger blades up to 3–4 mm wide; lower glume c. 0.5 mm longer than upper; anthecium 3.5–4.5 (–5) mm long; awn 7–10 mm long, often slightly twisted and irregularly bent . . . . subsp. *somaliensis*

a. subsp. **keniensis**

Syn.: *Oryzopsis keniensis* Pilger, Notizbl. Bot. Garten Berlin-Dahlem 9: 509 (1926); *Piptathrum keniense* (Pilger) Roshevitz in Bot. Mater. Gerb. Bot. Inst. Komarova

Akad. Nauk SSSR 14: 112 (1951).

Lasiagrostis elongata Nees, Fl. Afr. Austr. 168 (1841); *L. capensis* var. *elongata* (Nees) Trin. & Rupr., Spec. Gram. Stip. 88 (1842): *Stipa elongata* (Nees) Steud., Syn. Plant. Glum. 1: 132 (1855) non Roth, Ann. Bot. (Usteri) 14: 22 (1795); *S. dregeana* var. *elongata* (Nees) Stapf, Fl. Cap. 7: 573 (1889). – Type: S AFRICA, Cape Prov., in sylvis primaeris Krakakammae, alt. I, Uitenhage, *Drège* sine numero, sine dat. (holo. B, destroyed, fragm. PRE!).

Type: KENYA. Kenia occid. (Mt. Aberdare), Nieri (Nyeri), in silva 'Nieri forest' dicta, *R.E. & Th.C.E. Fries 248*, 21 xii 1922 (holo. UPS n.v., iso. B!).

In most higher mountains from Ethiopia through Kenya and Tanzania down to the Cape Province in S Africa.

In E Africa restricted to the middle and higher montane belts from 1300–2500 (–2800) m with lower temperatures and higher precipitation (more than 800 mm), descending to near sea level only in S Africa; a very common, gregarious tussock grass of evergreen broadleaved forests, particularly of *Olea europaea* L. subsp. *cuspidata* (Wall. ex DC.) Ciferri [syn: *Olea chrysophylla* Lam.], *O. hochstetteri* Baker, *Dombeya* L'Hérit., *Syzygium* Gaertn., and *Juniperus procera–Podocarpus* forests, more rarely in semi-deciduous forests, growing in deep shade as well as along forest edges and in clearings; often the dominant species of the herb layer and not grazed by cattle.

Specimens studied: ETHIOPIA. HARERGE PROV.: Yuka, *Gillett 5422* (K). KENYA. N PROV., TURKANA DISTR.: Moruassigar, *Newbould 7128* (EA). – NE PROV., MARSABIT DISTR.: Mt. Kulal, *Bally B 5637* (K); *ibid.*, 2 km N of Gatab, *Kuchar 9081* (EA). – NYANZA PROV., LONDIANI DISTR.: Tinderet Forest Reserve, *Geesteranus 4915* (G); Mt. Blackett, *Blunt 155* (EA). – RIFT VALL. & CENTRAL PROV., NZOIA DISTR.: Mt. Elgon, E slope above Japata, *Hedberg 13* (EA, K); near Elgon Nat. Park Headquarters, *Mumiukha 229* (EA); Kitum cave, *Taiti 555* (EA). KITALE DISTR.: W slopes of Cherangani Hills, *Bogdan AB 5003* (K). SAMBURU DISTR.: Karisia Hills, *Bono 88* (EA). LAIKIPIA DISTR.: Arabel Forest, *Kuchar 8730* (EA). NAKURU DISTR.: Rongai Forest, *Gachathi 87* (EA); Mt. Londiani Forest, *Perdue & Kibuwa* (EA); Kedowa Forest Reserve, *Magogo 1521* (EA, K); Nakuru Lake Nat. Park, *Kutilek 62* (EA); E Mau Forest Reserve, *Geesteranus 6102* (G). NYANDANIA DISTR.: Thomsons Falls (Nyahururu), *Barneg 1011* (K); *ibid.*, *Natharos 594* (EA); *ibid.*, *Napper 1514* (K); *ibid.*, *Carter 1027* (K). KIAMBU DISTR.: Escarpment SE of lake Naivasha, *Bogdan 424* (K); *ibid.*, *Mean 654* (K); *ibid.*, Lari Forest Reserve, *Hansen 773* (C, EA); Matschallos (?) Kikuyu, *Elliot 616* (K); Muguga, 25 vi 1954, *Clayton* (K); *ibid.*, *Trapvell 2107* (K); *ibid.*, *Brown 2183* (EA); *ibid.*, *Verdcourt 1841* (PRE). NAIROBI DISTR.: Nairobi, *Linton 188* (K); Karen, *van Someren 11* (EA); Langata Forest 6 miles S of Nairobi, *Bogdan 732* (K); Dagoreti Forest, *Battiscombe 1009* (K). Thika, near road junction to Garissa, *Gachathi 743* (B). Nyeri, near forest station, *Fries 510* (B); Nyeri, 26 iv 1938, *Le Pelley* (EA); 7 km from Nyeri to Kiganjo, *Faden & Evans 74/690* (PRE, WAG). W slopes of Mt. Kenya, *Grant 1246* (EA, K). – S PROV., NAROK DISTR.: 5 km NW of Ngorengore, *Davidse 7136* (EA, PRE); Loita Hills 5 km NNE of Morijo, *Kuchar 14426* (EA); *ibid.*, 3 km SW of Morijo, *Kuchar & Msafiri 7060* (EA); 10–15 miles N of Oldurga, *Kratz 52* (EA); Ngong Hills, *van Someren 9548* (EA, K); ibid., *van Someren 2552* (EA). – Further localities: Murna nysigar Peak, *Paulo 1008* (EA, K, PRE); 5 miles E of Nanyuki station, *Moreau 37* (EA, K); Naigane Ngana, *Helan 4c* (EA). TANZANIA. MBULU DISTR.: Neu forest, at Yeada river, *Vesey-FitzGerald 5297* (K). ARUSHA DISTR.: Mt. Meru in Arusha National Park, *Renvoize & Abdallah 2487* (EA, K). Longido Mt., *Peterson 231* (EA). W KILIMANJARO: Lemosho, *Vesey-FitzGerald 5052* (K). Ngare Nairobi river, *Bigger 2156* (EA, K). W USAMBARA MTS.: Shume, *Eichinger 5655* (EA); *ibid.*, *Moreau 40* (EA, K).

b. subsp. **somaliensis** Freitag, **subsp. nov.** Fig. 4.

Differt a subsp. *keniense* caespitibus dense caespitosis, ramificatione intravaginali, vaginis basalibus atris, laminis ad 3(–4) mm latitudine, gluma inferiore c. 0.5 mm longiore quam

Figure 4. Holotype of *Stipa keniensis* (Pilger) Freitag subsp. *somaliensis* Freitag (*Glover & Gilliland 1095*, BM).

superiore, longitudine minore organiorum sequentium: anthecium 3.5–4.5(–5) mm, arista 7–10 mm, antherae 2.5–3.5 mm, caryopsis 3 mm.

Type: SOMALIA. Drainage lines in the shade of Daloh(o) forest, near Erigayo (Ceerigaabo), 4 vii 1945, *Glover & Gilliland 1095* (holo. BM! iso. EA! K!).

Endemic to the high mountains of Northern Somalia, but perhaps also present in NW Somalia and adjacent Ethiopia.

Table 1. Differential characters of *Stipa dregeana* and *S. keniensis*.

	S. dregeana	*S. keniensis*
Panicle	branches short, with numerous spikelets from the very base, panicle narrowly conical, dense; habit of *S. calamagrostis*	branches elongated, with spikelets only in upper ½, panicle very loose, c. as wide as long; habit of *S. brandisii*
Awn	distinctly geniculate, lower segment twisted	straight or only slightly and irregularly bent, lower segment not or only obscurely twisted (subsp. *somaliensis*)
Sheath projection beside ligule	usually absent, if present then short, membranous, unveined	usually present, up to 20 mm long, green, veined
Ecology	in semi-arid open *Acacia* woodland and scrubland	in semi-humid to humid evergreen forest or woodland

Associated with the upper montane *Juniperus procera*-belt and reported growing in the shade of trees only, but altitudinal distribution and ecology very poorly known as yet.

Specimens studied: SOMALIA. SANAAG REGION: Daolo Forest Reserve, *Hansen & Heemstra 6254* (C, K); Surud Rge., Shimba Beris (Shimbiris), 24 xii 1929, *Collenette* (K); *ibid.*, 1930 *Collenette* (K); Golis Rge., *Brockmann 479* (K).

The species was first described by Nees (1841) as *Lasiagrostis elongata* together with *L. capensis*, both on the basis of specimens from S Africa. Nees himself had already questioned its specific rank and had discussed the possibility that it might only be a variety of the latter. Trinius & Ruprecht (1842) included it in that species, and Stapf (1889) formally gave it varietal rank with the remark, 'possibly an extreme shade form.' Because of the earlier *S. capensis* Thunb., its name had to be changed to *S. dregeana* var. *elongata*, which is still the name maintained in modern African floras. A re-investigation of the type of *Oryzopsis keniensis*, described by Pilger (1926), confirmed the earlier statement of de Winter (1965) about its identity with *S. dregeana* var. *elongata*.

After having compared numerous specimens of *Lasiagrostis elongata* and *L. capensis*, including the types, there is still no doubt about the relationship between both taxa. But as they differ in a whole set of independent morphological characters, which contrary to Stapf's view, cannot be sufficiently explained by environmentally induced modifications (see Table 1), they should again be treated as separate species. They also differ significantly in their ecological requirements, and their distributional areas overlap only slightly in S Africa.

S. keniensis is best placed in Sect. *Aristella* (Trin.) Hackel, as already proposed by Freitag (1985). It agrees in all essential characters: the short, obtuse callus, the indurate, darkening lemma which does not completely embrace the palea, and the awn structure (persistent, short, straight or only irregularly bent, scabrous).

However, in ecological respects, in panicle structure and in being not grazed it is very similar to the Himalayan *S. brandisii* Mez of Sect. *Achnatheropsis* Tzvelev. They are the only two Old World *Stipa* species associated with humid warm-temperate or tropical-montane evergreen forests. A very unusual and indeed unique character not mentioned in the hitherto published descriptions is the asymmetrical shape of the upper sheath margin at the culm leaves with a long lobe-like to linear upright projection on one side of the ligule. Usually its position with regard to the ligule alternates with the sequence of leaves. Only in the case of a very long projection does a second lobe develop on the other side of the ligule, but it never exceeds 2 mm.

Certainly the distribution of *S. keniensis* is still not completely known. It will very likely be found on some more of the higher mountains of E Africa, also in Uganda, Zimbabwe and Mozambique. From Uganda it has been already recorded, but erroneously: specimen *Elliot 616* from the Ruwenzori expedition carries 'Uganda' on its label, but from further data it must be concluded that the plant was collected near Nairobi.

The new subsp. *somaliensis* differs in so many characters from typical *S. keniensis* that further collections might justify the establishment of a separate species. More material from NW Somalia and E Ethiopia, where both taxa meet, is needed.

3. STIPA PARVIFLORA Desf. (Sect. *Stipella* Tzvelev)

Specimens studied: SAUDI ARABIA. N HIJAZ: Jabal Dibbagh, c. 100 km SW of Tabuk, *Collenette 491* (K), *4775* (K), *5243* (K); near Shiqri, Tabuk road, *Collenette 4364* (K). S HIJAZ: Jabal Warjan, 80 km SW of Madinah, *Collenette 5227* (K), *4978* (K).

The rich collections of Mrs Sheila Collenette have shown that the distributional area of the S Mediterranean *S. parviflora* extends considerably further south than indicated by my earlier map (Freitag 1985, fig. 12). The species has been found on isolated mountains in N and S Hijaz at altitudes from (1000–)1700–2300 m, usually in granite or sandstone areas and preferably on N-facing slopes. As some other mountains in the central parts of the Hijaz should offer similar ecological conditions, it is expected that the results of further expeditions will bridge the gap between the localities known at present.

One of the specimens from the southernmost locality (*Collenette 4978*) represents an unusual variant. The basal awn segment is distinctly pubescent (hairs 0.5 mm long) instead of being scabrous, and the anthecium carries at its apex a few similar hairs, forming a fragmentary coronula. These characters point to an introgressive influence from *S. tigrensis*, which starts to occur in the neighbouring Asir Mts. On the other hand, in the length of the anthecium and in other characters it still agrees with *S. parviflora*.

4. STIPA TIGRENSIS Chiov., Ann. Bot. 2: 366 (1905). (Sect. *Stipella* Tzvelev).

Type: [ETHIOPIA] ERITREA, OCULÉ CUSAI: Grupo dei monti Soyrà, Monte Mamahot verso il torrente Ariget, 2800–3000 m, 25 viii 1902, *A. Pappi 1180* (holo. FT! iso. K!).

In small, dense tufts, branching intravaginal, with few culms and numerous vegetative shoots; *culms* (15–)35–50(–65) cm, 3–4-noded, glabrous, at and below the nodes sometimes purplish tinged; *sheaths* glabrous or lowermost ones shortly pubescent with retrorse hairs, ± striate, lower ciliate at margins, upper ciliolate only in their uppermost part, beside the ligule with two distinct tufts of hairs up to 1(–2) mm long, or at uppermost culm leaves instead with two membranous lobes; *ligules* short, those of culm leaves up to 1.5(–2) mm long, with rounded to acutish apex and glabrous back, those of vegetative shoots only 0.1–0.2 mm long, truncate, lacerate, with ciliolate margin; *blades* flat, those of culm leaves up to 20 cm long and 2.5 mm wide, involute to convolute when dry and then 0.5–1.1 mm diam., those of vegetative shoots (4–)7–25(–30) cm long, 1–2 mm wide, when dry 0.3–0.7 mm diam., upper surface shortly but densely pubescent, beneath glabrous or scaberulous with short antrorse asperities, very rarely lowermost leaves with scattered hairs up to 1 mm long, usually with prominent veins; *panicle* exserted or ensheathed at base, contracted, (6–)8–12(–17) × 1–2(–3) cm, branches ascending to appressed, capillary, long-setulose, lower ones usually paired, up to 3(–6) cm long, with up to 4(–7) spikelets; *spikelets* (11–)12–15 mm long; *glumes* subequal, lanceolate, acuminate, hyaline, 3-veined, with green or purplish back, glabrous except for scattered long setules along middle vein, upper glume usually 2.5–4 mm shorter; *anthecium* 5.5–6.5(–7.5) mm long; *callus* (1–)1.2–1.5 mm long, densely bearded, longest hairs 0.8–1 mm long, curved, acute, scar oval, peripheral ring dorsally flattened and protruding; *lemma* becoming slightly coriaceous, brownish, distinctly constricted 0.5–0.7 mm below apex, sometimes with 2 short apical lobes up to 0.2 mm long, diffusely covered by ascending to appressed hairs mostly 0.4–0.6 mm long but usually much shorter near apex and just below the tip with a loose coronula of hairs reaching 0.8–1.5(–2) mm; *awn* (2.5–)3–6(–7) cm long, with distinct basal articulation, indistinctly bigeniculate, the prominent upper geniculation at (1–)1.4–2 cm from base, the often indistinct lower one c. 3/5 way up column, column twisted, pilose throughout, with hairs decreasing in length from 2–2.5 mm at base to 0.5–1 mm at top, seta straight or slightly flexuous, scabrous; *palea* less than 1/2 of lemma, 1.5–2 mm long, with obtuse, crenulate apex, glabrous; *lodicules* moderately dissimilar, glabrous, anterior ones ligulate to somewhat spathulate, (0.8–)1.1–1.4(–1.5) mm long, up to 0.3(–0.4) mm wide, posterior ligulate, 0.8–1.3 mm × 0.15 mm, apex rounded, not crenulate; *anthers* 3–3.5 mm long, bearded by 2 thin tufts of 0.25–0.3 mm long stiff hairs, yellow, more rarely purplish, exserted; *ovary* with 2 styles, divided by a deep cleft; *caryopsis* c. 4 mm long, embryo 1.1–1.3 mm long. *Fl.* March.–April; *fr.* April–May.

A species of open habitats from the upper montane to the subalpine zones. It grows preferably on dry, windswept rocky slopes. More rarely it occurs along wadi beds (Asir), in the shelter of shrubs and in *Festuca* – *Pentaschistis* – *Poa* grassland with *Hyparrhenia hirta* (L.) Stapf and *Lavandula pubescens* Decaisne (Jabal Marra: Miehe 1988), and even in degraded *Juniperus* woodlands (Asir, *fide* König 1987). It is probably commoner in the respective areas than the known collections suggest, and might occur also in South Yemen (People's Democratic Republic of Yemen).

Specimens studied: SUDAN. Jabal Marra, near summit, 3070 m, *Miehe 746, 835* (Hb. Miehe). UGANDA. KARAMOJA DISTR.: Moroto Mts., base of Sogolime Cliff, 3100 m, viii 1963, *Wilson* (EA, K). SAUDI ARABIA. ASIR: 50 km NW of Khamis Mushayt (42°22′E, 18°34′N), 2350 m, *Baierle, König & Mahmoud 1658* (BSB). YEMEN ARAB REPUBLIC. Dobson Falls, 3 vi 1977, *Chaudhary* (RIY). Jabal Miswar, near Sheruama, c. 2900 m, *Wood 3320* (BM, K). Jabal Mahdad, Miswar region, c. 3000 m, *Wood 2520* (BM, K). Jabal Mada',

c. 3200 m, *Wood 3073* (BM, K). Near Thulla, San'a prov., 2300 m, *Müller-Höhenstein & Deil 1581* (UBT). Jabal Nabi Shuaib, San'a prov., 3500 m, *Müller-Hohenstein & Deil 1541* (UBT). Menakha, just above town, c. 240 m, *Wood Y/75/666* (BM). Jabal Sumarah, c. 2900 m, *Wood 1663* (BM, K). Jabal Mudhbur (Jabal Lanz), near summit, c. 3300 m, *Wood 2348* (BM, K).

This species exhibits a rather broad range of variation, particularly in size of plant and of generative organs. The smallest individuals, collected from exposed cliffs at 3000 m (*Wood 2520*) have culms only 15–30 cm tall and unusually delicate vegetative tufts 4–6 cm in height, with awns 2.5–3.5 cm. A few other specimens with shorter awns but normal in plant size have been collected from peaks higher than 3000 m in Yemen (*Wood 3073*, *Müller-Hohenstein & Deil 1541*), whereas specimens from lower altitudes normally have longer awns. However, all intermediates exist, and the type itself holds an intermediate position with awns of 4.5–5.5 cm instead of the 3–4 cm mentioned in the diagnosis. As in some SW Asiatic species, the small overall size of the plants is due to modifying influences of harsh environmental conditions. In contrast, the correlation between awn length and altitude seems to be genetically fixed, but the variation is ecoclinal.

Müller-Hohenstein & Deil 1581 from Yemen differs in its abnormally long branches, giving the inflorescence a very lax aspect.

On account of its very short palea and the resulting constriction in the upper part of the lemma in combination with the rather long, pungent callus, 2-styled ovary and membranous lemma, *S. tigrensis* belongs to Sect. *Stipella.* Compared to the two widely distributed Mediterranean species of that section, *S. parviflora* and *S. capensis*, it has retained more primitive characters. The only slightly dissimilar lodicules and the moderately different glumes point to Sect. *Lasiagrostis* as the possible ancestor of Sect. *Stipella.*

S. tigrensis differs from the related SW Mediterranean *S. nitens* Ball not only 'in little more than the hairier lemma and awn' (Clayton 1970, p. 115), but also in usually much longer hairs on those organs, shorter glumes, awn and lemma, and a much looser arrangement of the spikelets which do not become bent together by the awns. In the latter character, as well as in the extremely short, scarcely discernible posterior lodicule (noted as absent by Maire 1953), *S. nitens* is certainly the closest relative of *S. capensis.*

5. Stipa capensis Thunb. (Sect. *Stipella*)

Specimens studied: SAUDI ARABIA. N AND C: 5 km N of Tamriat, 18 iv 1982, *Heemstra* (RIY). Wadi Um Rimth, 10 iii 1983, *Heemstra* (RIY). Hail, 22/23 iii 1981, *Chaudhary* (RIY). Raudha-al Biladlyya near Shamlul, iv 1980, *Chaudhary* (RIY). Near Rumah, 14 iv 1980, *Chaudhary* (RIY). Majmaa, spring 1981, *Iskander* (RIY). Aargah, 22 v 1984, *Chaudhary* (RIY). Hafarat-al-Nisah, 5 ii 1983, *Chaudhary* (RIY). Riyad, 30 iii 1980, *Chaudhary* (RIY); *ibid.*, 13 viii 1984, *Al-Noori* (RIY). Dharan, *Mandaville 101* (RIY). QATAR. Umm Slai Ali, c. 25 km N of Doha, 29 iii 1977, *Boulos* (G).

This unmistakable species is widely distributed and very common in northern, central and eastern Saudi Arabia and in all countries along the southern shores of the Persian Gulf up to E Muscat (Migahid 1978, Batanouny 1981). As *S. capensis* definitely depends upon the occurrence of regular winter rains, it gradually peters out and finally disappears towards the south. At present it would be rather conjectural to draw the southern boundary of its distributional area on the Arabian Peninsula, even if all specimens are taken into consideration.

The habitat of the annual *S. capensis*, which germinates immediately after the first winter rains, is the same as in adjacent Iraq, Jordan, Sinai and Egypt: semidesert communities rich in therophytes on sandy ground, preferably in dry river beds, non-saline depressions and in and around settlements. It withstands even high grazing pressure. Morphological variation is mainly due to the very variable moisture conditions and to trampling effects. Decumbent specimens with culms of only 5 cm have been seen as well as stout plants of 55–60 cm tall.

6. STIPA HOHENACKERIANA Trin. & Rupr. (Sect. *Barbatae* Junge emend. Freitag)

Specimens studied: SAUDI ARABIA. N: Turaif camp, *Collenette 5238* (K); top of Wadi Um Rimth, below Hammad escarpment, *Heemstra 7346* (RIY).

The rare occurrence of this typical Irano-Turanian species in northernmost Saudi Arabia has already been mentioned by Freitag (1985, fig. 26). The second record (*Heemstra 7346*) represents a small southward extension of the distributional area. The species has been recorded growing at about 900 m on shallow sand overlying limestone.

Doubtful and excluded species

STIPA TENACISSIMA L.

Two specimens of this W Mediterranean species have been seen during this study. The labels are 'Central Africa, *W. Oudney* 1822–1824' (BM), and 'Ethiopia: Jebel near Tarkuna, c. 400 m, rainfall 250–300 m; growing gregariously over a wide area on denuded soils over limestone in montane steppe, 12 iii 1960, *Mooney 8541*' (K). The first was certainly collected somewhere in NW Africa, as according to Lanjouw & Stafleu (1954) Oudney collected in N Africa instead of in C Africa. The locality of the second specimen cannot be traced to Ethiopia either. A later addition on the label 'probably Tarkaina in Djibouti' is as conjectural as the interpretation 'Kofele' (Bale prov. in Ethiopia) by Cufodontis (1968). It is even highly improbable that the specimen has been collected on an experimental farm, because climatic conditions in the dry tropical lowlands of E Africa would not allow the species to 'grow gregariously over a wide area'. It is probable that this specimen also was collected somewhere in NW Africa.

STIPA PROLIFERA Steudel, Syn. Plant. Glum. 1: 131 (1855).

This species has been described from a *Kotschy* gathering made in Nubia (Sudan). It was later mentioned only by Schweinfurth & Ascherson (1867) and Clayton (1970), who put it into the synonymy of *Stipagrostis uniplumis* (Licht.) de Winter. The very poor type in P has no generative organs except one detached glume of 6.2 mm. Therefore it is difficult to confirm the judgement, which is in conflict with several statements of the diagnosis (one simple awn, glabrous anthecium etc.). But at least it can be excluded from *Stipa* by the Kranz anatomy of the leaves and the fasciculate branching at the upper nodes.

ORYZOPSIS HYMENOIDES (Roemer & Schultes) Ricker

This true *Oryzopsis* species is native in N America. No specimen from Arabia or Africa has been seen, but Cufodontis (1968) cited a collection from Ethiopia, again made by Mooney in 1963, without locality. He already gives the only plausible interpretation for its occurrence: that it might have come from an experimental plot sown with introduced seed (if the record is correct).

PIPTATHERUM PARADOXUM (L.) P. Beauv. (Syn: *Oryzopsis paradoxa* (L.) Nutt.)
This W Mediterranean species has been erroneously mentioned by Chiovenda (1905). In fact the citations refer to *P. holciforme* subsp. *abyssinicum* (see p. 117).

ACKNOWLEDGMENTS

My thanks are due to the Keepers of the herbaria mentioned in the Introduction, who have so generously sent their material on loan. Moreover, I am very grateful to Mrs. Chr. Döhler for the careful execution of the drawings of *Stipa mandavillei*.

LITERATURE CITED

Batanouny, K.H. 1981. Ecology and flora of Qatar. Alden Press, Oxford. 245 p.

Chiovenda, E. 1905. Diagnosi di Graminacee nueve della Colonia Eritrea. Ann. Bot. (Rome) 2: 365–367.

——. 1908. Graminaceae. *In* R. Pirotta, Flora della Colonia Eritrea I. Annu. Reale Ist. Bot. Roma 8: 275–464.

Clayton, W.D. 1970. Gramineae, Pt. 1, 176 p. *In* R.M. Polhill [ed.], Flora of Tropical East Africa. Whitefriars Press, London.

Cope, T.A. 1985. A key to the grasses of the Arabian Peninsula. (Studies in the Flora of Arabia XV). Arab Gulf Scient. Res., Spec. Publ. 1: 1–82.

Cufodontis, G. 1953–1972. Enumeratio plantarum Aethiopiae Spermatophyta. 2 vols. Jard. Bot. Nat. Belg. Bruxelles. 1657 p.

Dyer, R.A. 1976. Flora of Southern Africa. The genera of South African flowering plants. 2 vols., Dept. Agric. Techn. Serv. Pretoria. 1040 p.

Engler, A. 1892. Über die Hochgebirgsflora des tropischen Afrika. Abh. Königl. Akad. Wiss. Berlin. 461 p.

Freitag, H. 1975. The genus *Piptatherum* (Gramineae) in Southwest Asia. Notes Roy. Bot. Gard. Edinb. 33: 314–408.

——. 1985. The genus *Stipa* (Gramineae) in Southwest and South Asia. Notes Roy. Bot. Gard. Edinb. 42: 355–489.

König, P. 1987. Vegetation und Flora im südwestlichen Saudi-Arabien (Asir, Tihama). Diss. Bot. 101. J. Cramer, Berlin. 257 p.

Kürschner, H. 1986. Osmanisch-makranische Disjunktionen. Ein Beitrag zur pflanzengeographischen Stellung und zu den florengenetischen Beziehungen Omans. Bot. Jahrb. Syst. 106: 541–562.

Lanjouw, J., and F.A. Stafleu. 1954. Index Herbariorum, Pt. 2(1) – Collectors A–D. Regnum Veg. 2, Utrecht. 174 p.

Maire, R. 1953. Flore de l'Afrique du Nord, vol. 2. P. Lechevalier, Paris. 374 p.

Mandaville, J.P. 1977. Plants. J. Oman Stud., Spec. Rep. 1: 229–267.

Miehe, S. 1988. Vegetation ecology of the Jebel Marra Massif in the semiarid Sudan. Diss. Bot. 113. J. Cramer, Berlin. 171 p.

Migahid, A.H. 1978. Flora of Saudi Arabia, ed. 2, 2 vols. Riyadh University Publ., Riyadh. 439 p.

Nees von Esenbeck, Ch.G.D. 1841. Florae Africae australioris illustrationes monographicae. I. Gramineae. Acta Phys.-Med. Acad. Caes. Leop.-Carol. Nat. Cur. 19, Suppl. 135–208.

Pilger, R. 1926. Gramineae. I. Notizbl. Bot. Gart. Berlin-Dahlem 9: 507–522.

Roshevitz, R.Ju. 1951. De genere *Piptatherum* P.B. notae criticae. Bot. Mater. Gerb. Bot. Inst. Komarova Akad. Nauk SSSR 14: 78–129. [In Russian.]

Roth, A.W. 1795. Observationes botanicae. Ann. Bot. (Usteri) 14: 22.

Schweinfurth, G., and P. Ascherson. 1867. Aufzählung sämmtlicher zur Zeit bekannten Phanerogamen- und Gefäßkryptogamen-Arten aus dem Gesammtgebiete der Nil-Länder, pp. 253–311. *In* G. Schweinfurth, Beitrag zur Flora Aethiopiens. G. Reimer, Berlin.

Stapf, O. 1889. Gramineae, pp. 310–750. *In* W.T. Thiselton-Dyer [ed.], Flora Capensis, vol. 7. L. Reeve & Co. Ltd., Ashford.

Steudel, E.G. 1855. Synopsis plantarum glumacearum. I. Gramineae. J.B. Metzler, Stuttgart. 475 p.

Trinius, C.B., and F.J. Ruprecht. 1842. Species graminum stipaceorum. Acad. Imp. Sci., Petropoli. 189 p.

Wickens, G.E. 1982. A biographical index of plant collectors in the Arabian Peninsula (including Socotra). Studies in the flora of Arabia III. Notes Roy. Bot. Gard. Edinb. 40: 301–330.

Winter, B. de. 1963. Notes on the genus *Aristida* L. (Gramineae). Kirkia 3: 132–137.

——. 1965. The South African *Stipeae* and *Aristideae* (Gramineae). Bothalia 8: 201–404.

12. A FAR-ROLLING STONE

Rupert C. Barneby

New York Botanical Garden, Bronx, NY 10458, USA

ABSTRACT

A massive seed dug up in a Leicestershire garden ca 1969 is identified as an endocarp of *Caryomene* (Menispermaceae), a genus of Amazonian lianas not formally described until 1971. Its introduction as a contaminant of Brazilnuts appears plausible.

Key words: *Caryomene*, Menispermaceae, endocarp, Brazilnuts.

In the year 1969 or thereabouts Mrs P. Taylor, who had acquired and restored an old house with adjoining barn and paddock on the main street (No. 51) of Sutton Bonington near Loughborough in Leicestershire, dug up in the topspit of a fallow plot the enigmatic object depicted in Fig. 1. Of woody texture and presumably of vegetable origin, measuring about 4 × 3 × 2.5 cm, it was something exotic and mysterious to emerge from the soil of the English Midlands. In attempts to identify it Mr W.H. Clark, a friend of the Taylors, passed this dimpled egg from hand to

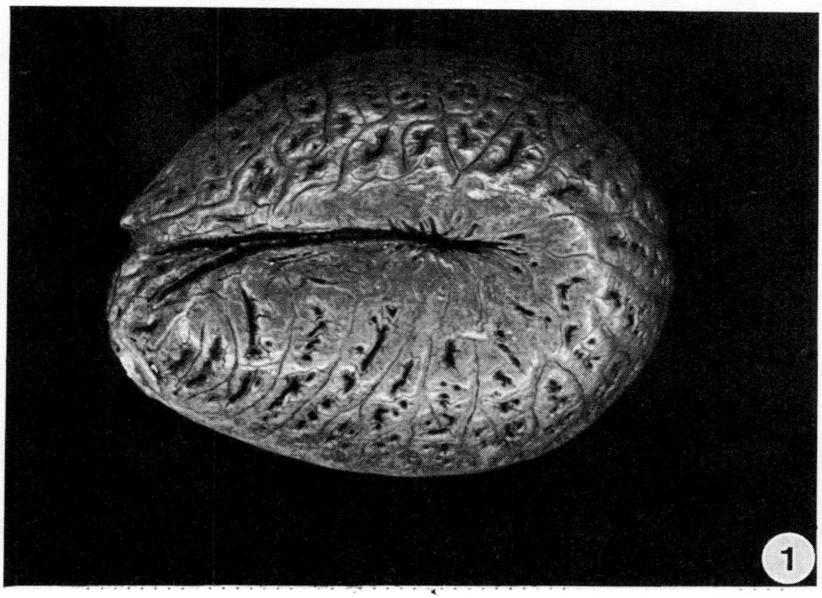

Figure 1. Endocarp of *Caryomene* sp. dug up in Leicestershire. Reproduced by permission of the Director, Royal Botanic Gardens, Kew.

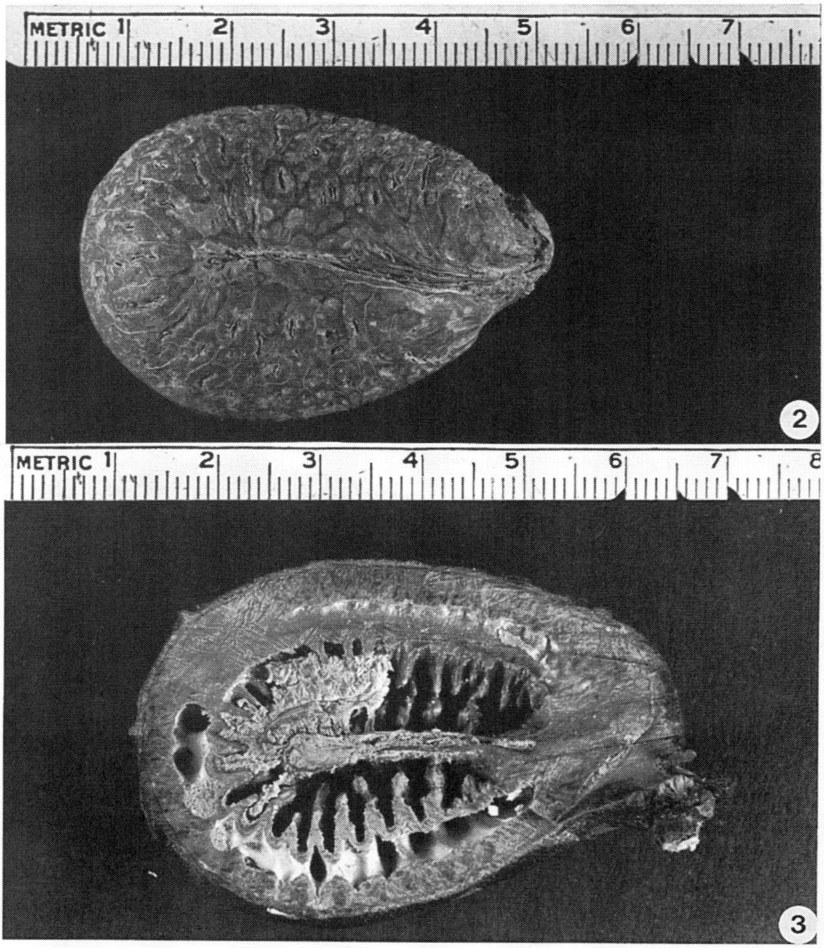

Figures 2 and 3. Endocarp (external view and longitudinal section) of *Caryomene olivascens* Barneby & Krukoff, from the type-specimen *N. T. Silva 1756* (NY). Photographs by Allen Rokach, New York Botanical Garden.

hand and from museum to museum, but no one recognized it. At Kew, where it was received in November 1969 and is now preserved, it was tentatively identified by Mr L.L. Forman as the endocarp of a menispermaceous drupe; but nothing could be found in the herbarium or in the literature to match it. In due course Dr Raymond Harley sent from Kew to the New York Botanical Garden a photograph similar to that reproduced here, and this was passed around the herbarium staff by Dr H.S. Irwin in May 1976.

By coincidence I had recently completed, in collaboration with the late Dr B.A. Krukoff, a taxonomic revision of Neotropical Menispermaceae tribe

Anomospermeae, and had published in December 1971 the description of a new genus *Caryomene*. Because I had so lately spent laborious hours sectioning, studying and illustrating endocarps of this genus, the Sutton Bonington stone was instantly recognizable, indeed as familiar to my eye as the rose in spring. Inexplicably, as it then seemed, the fruit of an Amazonian liana had turned up in an English garden a year or two before the genus to which it belongs had entered the botanical record. The caryomenes, at present four in number – we have tantalizing hints of others in the shape of sterile specimens – are high-climbing woody vines of the Amazonian Hylaea that bear their flowers in the canopy and are consequently seldom observed. Their fruits are notable in the family for their relatively great size, for the dense ligneous texture of the endocarp, and for laminated endosperm complexly folded around several series of intruded lamellae that correspond with shallow pits on the exterior wall. Presented as Figs. 2 and 3 are photographs of the endocarp of the type-specimen of *Caryomene olivascens* Barneby & Krukoff, collected on the Jarí river in Pará in 1969. The generic likeness to the Sutton Bonington stone is unmistakable. It would be imprudent, however, to assume that they represent the same species. So few endocarps of *Caryomene* are as yet available for comparison that we cannot guess how much, if at all, they may vary intraspecifically in size or external sculpture, and associated leaves are required for discrimination of species.

The questions that spring to mind are by what means, and at what time, did this singular seed make its way from the Brazilian rainforest to a Leicestershire village. The time seems unknowable, though the sample's perfect state of preservation suggests that it cannot have lain many years in the ground. We do have a clue to a mechanism of transport. Number 51 Main Street in Sutton Bonington is also known as the Old Bakehouse, and is the site of a former village bakery. Until recent years village bakeries commonly offered cakes and confectionery as well as bread, and Brazilnuts are likely to have been used there as an ingredient. It has been suggested that the *Caryomene* nut was brought to England in a sack of Brazilnuts and, when recognized as foreign matter, was discarded as rubbish. This seems the more plausible because the capsules and seeds of *Bertholletia* are normally harvested from the forest floor, where a *Caryomene* nut might easily be picked up by mistake (or as a makeweight). *Caryomene* is related to *Abuta* Aubl., a genus rich in alkaloids that furnishes some active ingredients of curare, and its fruits may well be highly toxic. Who knows what rural tragedy may have been averted when that peculiar stone was tossed over the wall?

ACKNOWLEDGMENTS

I am indebted to Dr Leonard Forman (K) for encouraging me to put this odd little story into words, and to the Director, Royal Botanic Gardens, Kew, for permission to reproduce the photograph in Fig. 1. Figs. 2 and 3 were photographed at NY by Allen Rokach.

LITERATURE CITED

Barneby, R.C., and B.A. Krukoff. 1971. A generic survey of the American Tricliseae and Anomospermeae. Mem. New York Bot. Gard. 22(2): 1–89.

13. REFLECTIONS ON TYPIFICATION AND APPLICATION OF THE NAMES *MIMOSA PIGRA* L. AND *M. ASPERATA* L. (MIMOSACEAE)

RUPERT C. BARNEBY

New York Botanical Garden, Bronx, NY 10458, U.S.A.

ABSTRACT

Taxonomic revision, to be reported elsewhere, of the pantropical aggregate-species widely known as *Mimosa pigra* Linn. has shown that six major taxa lie concealed within it. When these are segregated, typification and application of numerous older synonyms become of critical concern. The nomenclatural consequences of a new taxonomy for the group, here merely sketched in by means of a key to taxa, are discussed and summarized. *Mimosa pigra* sens. str. is identified as a relatively rare species endemic to the lower Paraná basin in extratropical South America. *Mimosa berlandieri* A. Gray, a North American species often treated as a variety of *M. pigra*, is equated with the much earlier *M. asperata* Linn. The very common and widespread mimosa currently known, in the Palaeotropics as well as in the Americas, as *M. pigra* is to be known as *M. pellita* Humb. & Bonpl. ex Willd. *Mimosa elliptica* Benth. is re-established for a species of Atlantic southeastern Brazil. The existence of two undescribed Neotropical species is implied in the taxonomic key.

Key words: Mimosaceae, *Mimosa*, typification, taxonomy, Linnaeus.

INTRODUCTION

Mimosa pigra Linn. and *M. asperata* Linn., dating back to 1755 and 1759 respectively, have been universally accepted, since Bentham's revision of the genus in 1875, as taxonomic synonyms applicable to the most widespread of all shrubby mimosas, a species common over much of tropical lowland America and Africa, and weedy or cultivated further afield. It would be expected, at this late date and following generations of scrutiny, that the typification of these names had been established and the identity of their types was common knowledge. It was an unwelcome surprise to find, while verifying nomenclature in connection with current studies of Neotropical *Mimosa*, that this is not the case. Two factors, independent but complementary, are to be taken into account, the selection of the type itself, and its taxonomic identity. In the first place the comprehensive concept of *M. pigra* adopted by Bentham (under the now incorrect title of *M. asperata*) is found on analysis to include several distinct specific taxa. When *M. pigra* sens. lat. is subdivided, the precise identity of its nomenclatural type and that of each supposed synonym must be ascertained, and the names distributed according to laws governing legitimacy and priority. Because *M. pigra* and *M. asperata* have been perceived as mere forms of one species, the identity of their types has been nomenclaturally immaterial, and the lectotypifications that have been proposed are careless and perfunctory. This is not the place to present in detail my taxonomic reappraisal of *M. pigra* and its immediate relatives, which will form part of the revision in preparation. In order to place these nomenclatural notes in taxonomic perspective, nevertheless, I append a key to the segregates from *M. pigra*. My prime purpose in this paper is only to correct misconceptions that have grown up around the two names *M. pigra* and *M. asperata* and thereby justify herbarium annotations that I am now making, annotations that would be misleading or even incomprehensible if left unexplained.

1. *MIMOSA PIGRA* L.

The concise protologue of *M. pigra* L. (Cent. Pl. I: 13, 1755, reissued in Amoen. Acad. 4: 274, 1759) was derived exclusively from two antecedent descriptions: *Aeschynomene spinosa quinta* Breyne (Exot. Pl. Cent. 44, t. 20. ± 1678) and *Ae. spinosa quinta . . . seu foliis acaciae latioribus* J. Commelijn (Horti Med. Amstel. 2: 59, t. 30, 1697), each of which is based on plants raised in Europe from American seeds. Breyne's plate depicts leaves only, that of Commelijn a branch bearing flowers and young fruits, and a detached ripe fruit at its foot. The leaf-formula displayed in these two illustrations is essentially the same: about seven pinna-pairs and 16–22 pairs of well spaced (not closely imbricated) leaflets; and both show a characteristic long spicule ascending from the leaf-stalk between the pulvini of each pair of pinnae. In respect to foliage, the two figures corroborate each other. Commelijn's figure is, however, partly compiled and fanciful. The author himself declares that he had never seen flowers ('flores nunquam offendi') of *Aeschynomene spinosa quinta*, yet his plate shows a plant in flower and young fruit, one flowering head (crudely drawn) shown issuing from the same leaf-axil as a cluster of well-formed pods, something unheard of in any living *Mimosa*. Moreover these pods appear to have only half as many articles as the ripe one shown detached at the foot of the plate. Consequently there is uncertainty as to which fruit, if either, belongs with the foliage. It seems likely that the incompleteness of Breyne's description and the obvious flaws in Commelijn's account induced DeCandolle (Prod. 2: 428, 1825) to subordinate (as var. *laevior* DC.) *M. pigra* to the better documented and more certainly identifiable, but posterior *M. asperata* L., a course ratified by Bentham's usage, although illegitimate by the current Code.

Brenan (Fl. Trop. E. Africa, Legum. part 1: 43, 1959) has lectotypified *M. pigra* by Commelijn's plate, without noting its inherent inconsistencies. As Brenan followed Bentham in recognizing only one polymorphic taxon in this group of mimosas, the precise identity of Commelijn's plant did not concern him. On the other hand when *M. asperata* sensu Bentham or *M. pigra* sensu Brenan is divided, the exact identity of the type becomes critical. Having surveyed in detail a very long sequence of specimens assembled from the whole range of *M. pigra* auct. recent. I can assert that the nearest match in leaf-formula for the plants cultivated at Danzig by Breyne and by Commelijn in Holland in the seventeenth century is, on the evidence of the two figures known to Linnaeus, neither the *M. asperata* of Bentham nor the *M. pigra* of contemporary floras but the relatively uncommon species endemic to the shores of the Paraná river and the Plate estuary in Paraguay and Argentina described in detail by Burkart (Darwiniana 8: 94, fig. 10(V), lam. VIII. 1948) as *M. vellosiella* Herter. This identification is not corroborated by the detached pod shown in Commelijn's plate, for its articles are too numerous and too narrow, in these respects agreeing with the fruit of the common *M. pigra* auct. This long hispid many-seeded pod cannot be conspecific with the plant cultivated by Commelijn and certainly cannot be its progeny, as the plant did not flower in cultivation. Possibly it was brought from Brazil by Georg Marcgrave, who had returned to Holland from Maranhão in 1644. There can be no doubt whatever that a pod of this type and the plant that bears it were first described by Marcgrave (Hist. Pl. 2: 74 + fig. on p. 75. 1648) as his second sort of *Caaeo* or *silva d'agoa*, which would later figure largely in the prologue of *M. longisiliqua* Lamk. (see further below). This same type of pod next appeared as a detached element of *Mimosa spinosa secunda . . . siliquis longis hirsutis* Breyne (Exot. Pl. Cent. 37, t. 17. ± 1678), but there was mismatched with a flowering stem of the quite different *M. invisa* Mart. (= *M. diplotricha* Wright ex Sauvalle)..

Obviously a serious disconnection had occurred between these strange long hispid fruits obtained from overseas (but not reproduced in Europe) and not just one but two species of *Mimosa* supposed to have been grown from them.

Accepting Brenan's lectotypification of *M. pigra*, recently confirmed by Wijnands (The botany of the Commelins 151, 1983), one is still obliged to particularize the element in Commelijn's composite *Aeschynomene spinosa quinta* which is to stand as type of the name. Because the foliage must have been drawn from a living plant and as its essential features are corroborated by Breyne's nearly contemporary figure cited in the Linnaean protologue, I hereby designate this as typical of *M. pigra* and consequently reject the detached pod as extraneous. Further, I associate this foliage with *M. vellosiella* sensu Burkart, and propose to transfer the name *M. pigra* to this species. It must be added that *M. vellosiella* Herter (Revista Sudamer. Bot. 6 (5–7); 151, 1940), a substitute for *M. cinerea* Vell. (Fl. Flum. Ic. 11: t, 35, 1835, non Linn.), is in my judgment a taxonomic synonym of *M. elliptica* Benth., so that the species described by Burkart has up till now possessed no correct binomial of its own. Although *M. pigra* in this reassessment turns out to be a species of relatively local dispersal, there is nothing inherently improbable about its appearance in cultivation in Europe a century before Linnaeus. Vessels of Dutch traders and the Hanseatic League had been putting in at the infant city of Buenos Aires from the early years of the XVII century.

2. *MIMOSA ASPERATA* L.

Mimosa asperata Linn. (Syst. Nat. ed. 10, 1312, 1759) was based on two prior descriptions: a) *Aeschynomene spinosa quarta* Breyne (Exot. Pl. Cent. 44, t. 20, 1678), a '*frutex brasiliensis*' illustrated by a single leaf from a plant cultivated at Danzig, and b) *Mimosa caule fruticoso . . . aculeis geminis siliquis radiatis hirsutis* Miller (Fig. Pl. Gard. Dict. t. 182, fig. 3. 1757),'discovered . . . at La Vera Cruz [Mexico] by the late Dr. Houstoun, who sent the seeds to England.' These clearly represent two distinct taxa, the plant of Brazilian origin with 8–12 pairs of pinnae per leaf and recurved aculei on interpinnal segments of the leaf-stalk, equivalent to the widespread '*M. pigra*' of current usage, and the Mexican one, differing in five to six pairs of pinnae and straight interpinnal aculei, instantly identifiable with *M. berlandieri* A. Gray, a species common on the Gulf lowlands of Veracruz. In 1763 Linnaeus (Sp. Pl. ed. 2, 1507) modified and expanded his concept of *M. asperata* by reference to Patrick Browne's *Mimosa* N° 11, a plant brought to Jamaica from Panama and cultivated on the island for ornament and for the curiosity of its sensitive leaves. Brenan (1959, l.c.) incautiously lectotypified *M. asperata* by a plant in the Linnaean herbarium (LINN *1228.32*, annotated by Solander as *M. asperata* but lacking further data; seen in microfiche!). This vigorous specimen cannot have any direct connection with Breyne's cultivated plant, which is not known to have flowered, and is obviously not conspecific with Houstoun's plant from Veracruz. It is in all probability derived from Patrick Browne's herbarium, which Linnaeus had acquired in 1758, just prior to the second edition of *Species Plantarum*, but much later than the protologue of *M. asperata* sens. str. I feel compelled to reject Brenan's lectotypification on these grounds, and to select one of the two sources from which Linnaeus derived the first description of *M. asperata*. The choice lies between the engraving of Breyne's single leaf, very likely the progeny of the long hispid pod associated by Breyne with a plant of *M. diplotricha* Wright (= *M. invisa* Mart.; non Mart. ex Colla), and by Commelijn with *M. pigra*, and Houstoun's fully documented *M. caule fruticoso etc.*, of which the exact indentity and provenance are confirmed by a surviving typotype: *Aeschynomene spinosa siliquis latis hirsutis et articulatis*, Vera Cruz, *Houstoun*, anno 1731 (BM!). The Houstounian element is so far superior in precision that it is the ineluctable lectotype.

CONSEQUENCES

The effects of revised typification of *M. pigra* and *M. asperata* are far-reaching. The equation of *M. pigra* Linn., sens. str., with South American *M. vellosiella* sensu Burkart (non Herter) and of *M. asperata* with Texano-Mexican, Central American and Cuban *M. berlandieri* A. Gray means that neither is available for the very widespread Neotropical and African species called *M. asperata* by Bentham and *M. pigra* by Brenan and all modern floras. While *M. asperata* simply replaces *M. berlandieri*, the choice of a substitute for *M. pigra* auct. lies between competing candidates: *M. longsiliqua* Lamk. (1783); and four specific names proposed by Willdenow in 1805, *M. polyacantha, M. canescens, M. hispida,* and *M. pellita*, examined in turn below:

i) *Mimosa longisiliqua* Lamk. (Encycl. Méth. 1: 21, 1783) was based on two pre-Linnaean antecedents, *Mimosa spinosa secunda . . . siliquis longis hirsutis* Breyne (Exot. Pl. Cent. I: 37, t. 17, ± 1678) and *Caaeo* Marcgrave (Hist. Pl. liber 2: 73. 1648) and a specimen ('v.s.'). Breyne's plate, as noted by Bentham (1875, p. 436), depicts a flowering stem of *M. diplotricha* Wright ex Sauvalle associated with detached pods of *M. pigra* auct. recent. Marcgrave's *Caaeo* consisted of three sorts: a) *herba viva*, described and figured on p. 73, = *Schrankia leptopoda* DC.; b) *silva d'agoa*, described on p. 74 with figure of pod on p. 75 = *M. pigra* auct.; and c) without vernacular name, described at bottom of p. 74 and figured at middle of the same, plainly = *M. polydactyla* H. & B. ex Willd. The particular *Caaeo* to which Lamarck referred is presumably the *silva d'agoa*, for he described the pod of *M. longisiliqua* as measuring ± 12 × 1.6 cm. The Lamarck herbarium (P!) contains three specimens of potential relevance, two of African and one of Brazilian provenance: Sénégal, *Roussillon*, identified as *M. exasperata*, in flower; 2) Thebes, Egypt, *Savigny*, as *M. pigra*? in young fruit; and 3) Brazil, without further data, as *M. pigra*, in flower. Both the African specimens represent *M. pigra* sensu Bentham and Brenan, the Brazilian one *M. elliptica* Bentham, very probably from Rio de Janeiro. None is identified as *M. longisiliqua*, and I was unable to find (in 1985) any specimen so named in other 'herbiers historiques' at Paris. It is impossible to determine which, if any, of the three specimens in the Lamarck herbarium contributed to the protologue of *M. longisiliqua*. Savigny's Egyptian plant is the only specimen that could have suggested the epithet *longisiliqua*, but its fruits fall short of Lamarck's measurements. Certainly there was a large element of *M. pigra* auct. in Lamarck's proposition, but it is so clouded by ambiguity that I think it wise to follow Bentham in treating it as an unredeemable mixture.

ii) *Mimosa polyacantha* Willd. (Sp. Pl. 4: 1034, 1805) was derived from the description and figure of the Abyssinian *ergett el krone* Bruce (Select. Spec. Nat. Hist. 35 + fig., 1790). Its identity with *M. pigra* auct. has gone unchallenged since Bentham's monograph. No specimen exists.

iii) *Mimosa canescens* Willd. (Sp. Pl. 4: 1038, 1805) is based on a specimen of P.E. Isert extant in B-WILLD (no. *19083*, seen in microfiche!), collected in the Dutch colony in equatorial Africa. Like the preceding its identity was established by Bentham and goes uncontested.

iv) *Mimosa hispida* Willd. (Sp. Pl. 4: 1037, 1805) was described from a Bredemeyer collection from near Caracas, Venezuela, extant in B-WILLD (no. *19081*, seen in microfiche!). This represents a distinct local taxon, endemic to north-central Venezuela, distinguished from commonplace *M. pigra* auct. by relatively narrow pods and small lentiform seeds.

v) *Mimosa pellita* H. & B. ex Willd. (Sp. Pl. 4: 1037, 1905) was described from specimens collected by Humboldt & Bonpland near Cumaná in Venezuela, extant in B-WILLD (no. *19081*, fl. and *19082*, fr, seen in microfiche!). The protologue was amplified by redescription and a superb portrait in colour by Kunth (Mimoses 27, t. 9, 1819–20). The plants represent a minor, randomly widespread variant of *M. pigra* auct. marked by especially dense and loose indumentum of setae on young stems and forming pod.

The most effective choice of name will be one based on an extant specimen of known Neotropical provenance that represents unambiguously the most widespread form of the species to which the epithet *pigra* is currently applied. I pass over *M. longisiliqua* as an ambiguous proposition that could only be restored by a lectotypification open to serious challenge. I would eliminate from Willdenow's four names a) *polyacantha* because no herbarium voucher exists, b) *canescens,* because of its African (secondary) source, c) *hispida*, for taxonomic reasons given above. The survivor is *M. pellita* H. & B. ex Willd., the name that I am presently using for *M. pigra* auct.

SUMMARY

In the context of the taxonomy developed for *Mimosa* ser. *Asperatae* Benth., the following nomenclatural equations emerge following retypification of *M. pigra* L. and *M. asperata* L.:

Mimosa pigra Linn., ex parte, lectotypified by the foliage alone, and not the pod, depicted in Commelijn, Horti Med. Amstel. t. 30 = *M. vellosiella* sensu Burkart, Darwiniana 8: fig. 10(V), lam. VIII; non Herter. The species is endemic to the Paraná basin and shores of Mar de Plata in Argentina and Paraguay.

Mimosa asperata Linn., ex parte, lectotypified by description and figure given by Miller, Fig. Pl. Gard. Dict. t. 182, fig. 3 (recapitulated by Houstoun, Reliqu. Houston. 11, t. 20, 1794) = *M. berlandieri* A. Gray, 1859. The species, by many considered only varietally distinct from *M. pigra* auct., has recently been restored to specific rank by Bässler (Feddes Repert. Sp. Nov. 96: 599, 1985). Its range extends from southern Texas through the Pacific and Gulf lowlands of Mexico to Cuba and Nicaragua.

Mimosa asperata sensu Bentham, 1875, p. 437, exclus. lectotyp., and *M. pigra* sensu Brenan, 1959, p. 43, fig. 13, with modified lectotype, = **M. pellita** H. & B. ex Willd.

KEY TO *MIMOSA PIGRA*, *M. ASPERATA*, AND CLOSE KINDRED

1. Stipules firm, densely appressed-setose dorsally and imperceptibly veined externally or, if visibly veined above mid-blade, the nerves 5 or less; extremely common and widespread from tropical Mexico and Cuba S to N Peru and NE Argentina; palaeotropics . . . . *M. pellita*
1. Stipules chartaceous, striately 6–15-nerved dorsally, commonly glabrous, but if setulose the indumentum not concealing the venation.
2. Setae of stems laterally affixed above the base, with a short blunt descending arm; calyx 0.3–0.6 mm, the rim undulate or obscurely pallid-lobulate. Leaf-formula ix–xvi/(25)30–48; aculei of stems and leaf-stalks all recurved; body of pod 7–10 × 1.3–1.6 cm, 12–18-seeded, the articles 4.5–8 mm long, indehiscent, the interseminal septa 0.6–1 mm wide, the valves hispid with vertical setae 1.5–4.5 mm; Orinoco basin in Venezuela and lower Amazon valley in Pará, Brazil . . . . [*M. orinocoënsis* Barneby, in sched.]
2. Setae of stems asymmetrically basifixed, not spurred; calyx 0.6–2 mm, at least 0.7 mm if leaf-formula as above. Aculei and details of pod variable.
3. Leaf-formula x–xviii/30–46; pod when well fertilized 14–24-seeded, the articles individually dehiscent and the seed falling naked; Central America S interruptedly to Bolivia, adapted to permanent high humidity . . . . [*M. tarda* Barneby, in sched.]
3. Leaf-formula iv–viii(ix)/20–40; pod when well fertilized 6–14(15)-seeded, the articles free-falling indehiscent, sealed at each end by the unruptured septa; either of Texas to Central America and Cuba, or of SE Brazil, Paraguay and Argentina, adapted to an annual dry season or to warm-temperate climate or to both.
4. Aculei of stems and leaf-stalks alike perpendicular, straight, broad-based, and early blanched. Corolla strigulose distally; articles of pod 4–7.5 mm long; S Texas to Nicaragua, tropical W Mexico, and Cuba. . . . . . *M. asperata*
4. Aculei of stems recurved or at least distally declined, and brownish except at pallid base, those of leaf-stalks varying from straight to recurved but also brownish.
5. Corolla glabrous or almost so; longer filaments exserted 5–8 mm; articles of pod 3–4.5 mm long; coastal S Brazil southward from S Bahia . . . . *M. elliptica*
5. Corolla strigulose distally; longer filaments exserted ± 2 mm; articles of pod 5–10 mm long; riparian within the Paraguai basin in Paraguay and Argentina . . . . *M. pigra*

NOTE: the term *leaf-formula* denotes *number of pinna-pairs in longer leaves/number of leaflet-pairs in longer pinnae*. Names in square brackets in key are *nomina nuda praecursoria*.

14. THE ADOPTION OF *STOIBRAX* FOR *TRAGIOPSIS* AND *BRACHYAPIUM* (UMBELLIFERAE), AND ITS N–S AFRICAN DISJUNCTION

B. L. Burtt

Royal Botanic Garden, Edinburgh, EH3 5LR

ABSTRACT

Stoibrax Rafinesque (Umbelliferae), based on *Pimpinella dichotoma* L., is the correct name for the homotypic *Tragiopsis* Pomel and for *Brachyapium* (Baillon) Maire. The type of the latter is the S African *Ptychotis didyma* Sonder; reasons for classifying it with the Spanish/N African species (following H. Wolff, who placed it in *Tragiopsis*), rather than with the S African species of *Sonderina* H. Wolff (following Adamson) are given. New combinations necessary in *Stoibrax* are made: *S. didymum* (Sonder), *S. hanotei* (Br.–Bl. & Maire), *S. involucratum* (Maire), *S. pomelianum* (Maire).

Key words: *Brachyapium, Sonderina, Stoibrax*, Umbelliferae.

Stoibrax Rafinesque (1840), based on *Pimpinella dichotoma* L., has already been signalled as the correct name for *Tragiopsis* Pomel (1874) and *Brachyapium* (Baillon) Maire (1932) in *Index Nominum Genericorum* (Farr, Leussink & Stafleu 1979), but it has not yet been widely adopted, nor has its relation to the name *Brachyapium*, which has a South African dimension, been explained.

Tragiopsis Pomel was abandoned by Maire (1932) not because it was a homotypic synonym of *Stoibrax* (that name had not yet come to light), but because it was a later homonym, *Tragiopsis* having been used by Karsten in 1859. A few years previously H. Wolff (1927) had revised *Tragiopsis* Pomel and had cited as a synonym '*Brachyapium* Baillon'. Therefore Maire, when he rejected *Tragiopsis*, simply took up the cited synonym *Brachyapium*. Now, in the first place, Baillon never made a genus *Brachyapium*: he used that name only for a section of *Carum* L. I agree with *Index Nominum Genericorum*, in its attribution of the genus to Maire. However, it is wrongly stated there that a type was not designated and that the name is a superfluous substitute for *Stoibrax*: Baillon's footnote reads 'Le type de cette section est pour nous le *Ptychotis didyma* Sonder. (*Fl. Cap*. II, 538)'. The name *Brachyapium* is valid and legitimate: the view taken here (see below) is that it is a synonym of *Stoibrax*.

H. Wolff had included *Brachyapium* as a synonym of *Tragiopsis* because he had referred *Ptychotis didyma* Sonder to that genus. Maire was correct in taking up the synonym *Brachyapium* because he gave no indication that he was changing H. Wolff's concept of *Tragiopsis*. Without knowledge of *Stoibrax*, Maire's nomenclature was perfectly permissible.

Adamson (1939), however, decided that *Tragiopsis didyma* (Sonder) H. Wolff ought to be transferred to *Sonderina* H. Wolff, the genus H. Wolff (1927) had created for the other four species that Sonder (1862) had placed in *Ptychotis*. (The botanical correctness of this move will be discussed later.) Adamson clearly did not realize that *T. didyma* was the type of *Brachyapium*, for he said that this was the correct name for the Mediterranean *Tragiopsis*. At that time (still ignoring the latent *Stoibrax*) *Tragiopsis* excluding *T. didyma* would have required a new name. It seems unlikely that Mediterranean authors have ever taken cognisance of Adamson's removal of *T. didyma*; if they did, they followed his mistake in using *Brachyapium* for the northern plants (Quézel & Santa 1963, Nègre 1962, Pottier-Alapetite 1979, Smythies 1986).

Merrill (1949, p. 183) brought the long-forgotten *Stoibrax* back into play, but rather curiously he took his classification from Bentham & Hooker (1867, p. 891) where Rafinesque's only species and hence the type, *Pimpinella dichotoma* L., is referred to the genus *Carum*. Merrill made no mention of *Brachyapium*, nor of *Tragiopsis*, which is the name by which this plant had been known for some 60 years before Maire switched to *Brachyapium*. *Stoibrax* is manifestly the earliest name for *Tragiopsis* and must be adopted. The necessary new combinations are made below.

We come back, therefore, to the correct placing of *Ptychotis didyma* Sonder. The chief morphological character on which Adamson (1939) relied for its inclusion in *Sonderina* rather than *Tragiopsis* was that the leaves were pinnately compound, not ternately compound. He then went on: '*P. didyma* differs from *Sonderina* in the fruit being as broad or broader than long and cordate at the base. The fruit approaches in character those of the species of *Brachyapium* [i.e. *Tragiopsis* = *Stoibrax*] but it has less prominent ridges, is less constricted at the commissure and is less wide in proportion to its length. Taking all the characters into consideration the more natural arrangement seems to be to place the species in *Sonderina*. This is supported by the geographical distribution . . . . .'

It is true that *Stoibrax* has the primary division of the leaf ternate, whereas that of *Sonderina* is more distinctly pinnate. H. Wolff was well aware of this, as his generic descriptions show; also he used it as the primary character in his key to *Tragiopsis* to segregate the S African *T. didyma* from the Mediterranean species. Similarly in his key to *Trachyspermum* Link (H. Wolff 1927, p. 8) the first dichotomy is between pinnatisect and ternate or trisect leaved species. Whether a leaf is described as ternatisect or pinnatisect depends largely on whether the lowest pair of segments are notably larger than those next above them. The radical leaves of *Sonderina humilis* (Meisner) H. Wolff come much nearer the ternatisect condition than do those of *S. hispida* (Thunb.) H. Wolff. The character is a weak one for use in defining genera.

The fruit differences that Adamson mentions may be true (I have seen too little material to be sure), but in all three features the descriptive 'less' must mean a very little less. Adamson, however, makes no comment on one of H. Wolff's more important differentials, namely that the lateral ribs are on the margin of the mericarp in *Sonderina*, but not in *Tragiopsis* (i.e. *Stoibrax*). This is true in the material available to me and certainly supports H. Wolff's placement of *T. didyma*, as his figure shows (H. Wolff 1927, p. 96 fig. 7).

Reading Adamson's note, one cannot help feeling that he was chiefly worried that a species endemic to the Cape should be placed in a genus otherwise restricted to N Africa and Spain. Yet a similar disjunction had long been known in a related member of Umbelliferae, the genus *Deverra* DC. (= *Pituranthos* Viv., non Mart.; recently revised by Pfisterer & Podlech (1986)). There are in fact many such examples in different families and at different taxonomic levels: *Ifloga* (Compositae-Gnaphaliineae) is a good one (Hilliard 1981). The whole subject was given a very thorough discussion in a meeting of AETFAT at Munich in 1970 (see Conert 1971, Monod 1971, de Winter 1971). Thus it can be seen that such a disjunction between the Cape and N Africa is in no way sufficiently remarkable to sway taxonomic judgment. The fruit characters indicate that H. Wolff's taxonomy must be upheld and accordingly *Sonderina* (or *Tragiopsis*) *didyma* is here transferred to *Stoibrax*.

What of the other species of *Sonderina*? My preliminary feeling is that the genus is probably too close to *Stoibrax* for it to be maintained. Fruit sections suggest specific differences, but do not indicate two distinct groups. The material I have seen is, however, not yet adequate for a sound judgment and it is best for *Sonderina* to be kept separate until a thorough and critical comparison can be made. The very little cytological information available also sounds a warning note. *Sonderina* (now to be *Stoibrax*) *didyma* has been recorded as having a haploid chromosome number n = 9 (Constance & Chuang 1982, p. 205), whereas Professor Constance informs me (in litt.) that *Sonderina humilis* (Meisner) H. Wolff has n = 8. Unfortunately no counts are available for Mediterranean *Stoibrax*.

Stoibrax Rafinesque, Good Book 52 (1840), reimp. Scad. Gen. Omb. Pl. 52 (Amer. Midl. Nat. reimp. 1913); Merrill, Index Rafinesquianus 183 (1949).
Type: *S. dichotomum* (L.) Rafinesque.

Syn.: *Tragiopsis* Pomel, Nouv. Mat. Fl. Atl. 139 (1874) – nom. illegit., non Karsten (1859). Type as above.
Carum sect. *Brachyapium* Baillon, Hist. Pl. 7: 118 (1879). Type: *Ptychotis didyma* Sonder.
Brachyapium (Baillon) Maire in Bull. Soc. Hist. Nat. Afr. Nord 23: 186 (1932).

Stoibrax dichotomum (L.) Rafinesque, Good Book 52 (1840), reimp. Scad. Gen. Omb. Pl. 52 (Amer. Midl. Nat. reimp. 1913); Merrill, Index Rafinesquianus 183 (1949).

Syn.: *Pimpinella dichotoma* L., Mant. 58 (1767).
Tragiopsis dichotoma (L.) Pomel, Nouv. Mat. Fl. Atlant. 139 (1874) quoad syn. excl. descr. (quae est *S. pomelianum*): H. Wolff in Engler, Pflanzenreich, Umbell.-Apioid.-Amm. Heft 90: 95 (1927).
Tragiopsis scabriuscula Pomel, Nouv. Mat. Fl. Atlant. 139 (1874).
Brachyapium dichotomum (L.) Maire in Bull. Soc. Hist. Nat. Afr. Nord 23: 186 (1932).

Stoibrax didymum (Sonder) B.L. Burtt, **comb. nov.**

Syn.: *Ptychotis didyma* Sonder in Harvey & Sonder, Fl. Cap. 2: 538 (1862).
Trachyspermum didymum (Sonder) Drude in Engler & Prantl, Pflanzenfam. 3(8): 189 (1898).
Tragiopsis didyma (Sonder) H. Wolff in Engler, Pflanzenreich, Umbell.-Apioid.-Amm. Heft 90: 96 (1927).
Sonderina didyma (Sonder) Adamson in J. S. Afr. Bot. 5: 55 (1939).

Although the specific combinations have never been made, this plant is the type species of *Carum* sect. *Brachyapium* Baillon and of the genus *Brachyapium* (Baillon) Maire.

Stoibrax hanotei (Br.-Bl. & Maire) B. L. Burtt, **comb. nov.**

Syn.: *Tragiopsis hanotei* Br.-Bl. & Maire in Bull. Soc. Hist. Nat. Afr. Nord 22: 108 (1931).
Brachyapium hanotei (Br.-Bl. & Maire) Maire in Bull. Soc. Hist. Nat. Afr. Nord 23: 186 (1932); Jahandiez & Maire, Cat. Pl. Maroc 2: 536 (1932).

This species is unknown to me.

Stoibrax involucratum (Maire) B.L. Burtt, **comb. nov.**

Syn.: *Trachyspermum involucratum* Maire in Bull. Soc. Hist. Nat. Afr. Nord 13: 186 (1922).
Tragiopsis involucrata (Maire) H. Wolff in Engler, Pflanzenreich, Umbell.-Apioid.-Amm. Heft 90: 95 (1927).
Brachyapium involucratum (Maire) Maire in Bull. Soc. Hist. Nat. Afr. Nord 23: 186 (1932); Jahandiez & Maire, Cat. Pl. Maroc 2: 537 (1932); Nègre, Petite Fl. Reg. Ar. Maroc Occ. 2: 108, pl. 84 fig. 515 (1962).

Stoibrax pomelianum (Maire) B. L. Burtt, **comb. nov.**

Syn.: *Tragiopsis dichotoma* (L.) Pomel, Nouv. Mat. Fl. Atlant. 139 (1874) quoad descr. excl. basion.
Trachyspermum pomelianum Maire in Bull. Soc. Hist. Nat. Afr. Nord 17: 113 (1926).
Tragiopsis pomeliana (Maire) Maire in Bull. Soc. Hist. Nat. Afr. Nord 20: 23 (1929).
Brachyapium pomelianum (Maire) Maire in Bull. Soc. Hist. Nat. Afr. Nord 23: 186 (1932); Jahandiez & Maire, Cat. Pl. Maroc 2: 536 (1932).

I have not transferred the varieties of this species recognized by Maire. That should await more critical study.

ACKNOWLEDGMENTS

I am grateful to Dr Kit Tan for preparing and sketching sections of the fruits of these plants.

LITERATURE CITED

Adamson, R.S. 1939. Some changes in nomenclature II. J. S. Afr. Bot. 5: 53–58.

Baillon, H. 1879. Histoire des plantes, 7. L. Hachette & Co., Paris. 546 p.

Bentham, G., and J.D. Hooker. 1867. Genera plantarum, 1(3). L. Reeve & Co., London.

Conert, H.J. 1971. The genus *Danthonia* in Africa. Mitt. Bot. Staatss. München 10: 299–308.

Constance, L., and T.I. Chuang. 1982. Chromosome numbers of Umbelliferae (Apiaceae) from Africa south of the Sahara. Bot. J. Linn. Soc. 85: 195–208.

de Winter, B. 1971. Floristic relationships between the northern and southern arid areas in Africa. Mitt. Bot. Staatss. München. 10: 424–437.

Farr, E.R., J.A. Leussink, and F.A. Stafleu, (eds.). 1979. Index nominum genericorum (plantarum). Bohn, Scheltema & Holkema, Utrecht. 1896 p.

Hilliard, O.M. 1981. A revision of *Ifloga* in southern Africa, with comments on the northern hemisphere species. Bot. J. Linn. Soc. 82: 293–312.

Maire, R. 1932. Contributions à l'étude de la flore de l'Afrique du Nord: fasc. 19. Bull. Soc. Hist. Nat. Afr. Nord 23: 163–222.

Merrill, E.D. 1949. Index rafinesquianus. Arn. Arb. Harvard Univ., Jamaica Plain, Mass. 296 p.

Monod, T. 1971. Remarques sur les symétries floristiques des zones sèches nord et sud en Afrique. Mitt. Bot. Staatss. München 10: 375–423.

Nègre, R. 1962. Petite Flore des régions arides du Maroc occidental. Vol. 2. Centre Nat. Recherche Scientifique, Paris. 566 p.

Pfisterer, R., and D. Podlech. 1986. Revision der Gattung *Deverra* DC. (Umbelliferae). Mitt. Bot. Staatss. München 22: 571–604.

Pomel, A. 1874. Nouveaux matériaux pour la flore atlantique. Savy . . . ., Paris. 260 p.

Pottier-Alapetite, G. 1979. Flore de la Tunisie: Apétales-Dialypétales. Ministère de l'Enseignement Supérieur et de la Recherche Scientifique et le Ministère de l'Agriculture. Tunis. 651 p.

Quézel, P., and S. Santa. 1962–1963. Nouvelle flore de l'Algérie, 2 vols. Centre Nat. Recherche Scientifique, Paris. 1170 p.

Rafinesque, C.S. 1840. The good book. [Reimp. Scadiography of 100 genera of Ombelliferous plants. American Midland Naturalist Reprints no. 3, 1913].

Smythies, B.E. 1986. Flora of Spain and the Balearic Islands: checklist of vascular plants: 3. Englera 3(3): 487–882.

Sonder, W. 1862. Umbelliferae, in W.H. Harvey and W. Sonder, Flora Capensis, vol. 2. Hodges, Smith & Co., Dublin. 546 p.

Wolff, H. 1927. Umbelliferae-Apioideae-Ammineae-Carinae, Ammineae novemjugatae et genuinae, in Engler, A. (ed.), Das Pflanzenreich, Heft 90. (IV. 228). W. Engelmann, Berlin. 398 p.

15. TAXONOMIC REVISION OF *MONOCHORIA* (PONTEDERIACEAE)

CHRISTOPHER D.K. COOK

Institut für systematische Botanik, Universität Zürich, Zollikerstrasse 107, CH 8008 Zürich, Switzerland

ABSTRACT

A taxonomic revision of the genus *Monochoria* (Pontederiaceae) is presented with full descriptions of each species with diagnoses, synonyms with typifications, distribution maps and illustrations also containing information on ecology, anatomy, morphology, sporogenesis, chromosomes and applied aspects. Eight species are recognised, four from Asia, two from Africa and two from Australia. One Asian species, *M. vaginalis*, is a weed in ricefields and has extended its range to Europe (Ukraine and Italy), N. America (California) and the Hawaii Islands (Oahu); its status in Australia is uncertain.

Key words: Pontederiaceae, *Monochoria*, rice-field weed.

INTRODUCTION

Since the classical attempts at writing monographs by J.A. and J.H. Schultes (1830) and Solms-Laubach in A. de and C. de Candolle (1883) no recent effort has been made to prepare a world-wide taxonomic revision of *Monochoria*. Two recent local accounts have been written, one for Africa by Verdcourt (1961) and the other on two Australian species by Aston (1985). Both these authors were seriously concerned with problems presented by phenotypic plasticity. Like many aquatic plants *Monochoria* shows extreme plasticity of many characters which, at times, obscures the clear delimitation of species.

This revision is based largely on herbarium material but the work has been helped by field observations in India and cultivation experiments in the botanic garden of the University of Zürich. Specimens were examined from the following herbaria: BM, University of Calicut, DUH, E, G, K, Z, ZT. Photographs were sent by LE.

APPLIED ASPECTS

Monochoria vaginalis is a common and today widespread plant in ricefields and irrigation or drainage ditches. The database of the Aquatic Weed Program of the University of Florida, Gainesville, USA, contains at least 90 items on *Monochoria* and it seems that many control methods have been tried. However, I have found no evidence that *M. vaginalis* or, indeed, any other species of *Monochoria* is a serious weed that endangers man's use of land or water.

In southern and eastern Asia and in India *M. vaginalis* and to a lesser degree *M. hastata* are eaten as vegetables and for this purpose are sometimes cultivated. Usually the leaves and stems are cooked but sometimes in S.E. Asia the inflorescences are eaten raw. The juice of the leaves and sometimes also the roots are used for medicinal purposes and may, depending upon the region, be used for curing boils, stopping toothache or easing asthma. Degener and Degener (1960) considered that *M. vaginalis* was deliberately introduced to Hawaii as a beneficial plant.

SPOROGENESIS

Microsporogenesis has been described by Banerji and Haldar (1942) and macrosporogenesis by Ono (1928) – *M. korsakowii* and *M. vaginalis*, Juliano (1931) – *M. vaginalis*, and Banerji and Haldar (1942) – *M. hastata*. During macrosporogenesis a large micropylar chamber develops giving off a pair of tubular haustorial arms which extend basally and laterally to enclose a column of nucellar tissue. These micropylar processes reach the chalazal end of the embryo sac. As the central nucellar column and the basal chamber degenerate the haustorial arms fill the central space. The chalazal chamber remains free-nuclear in *Monochoria* but becomes cellular before it degenerates in other investigated genera of the Pontederiaceae. In *M. hastata* the synergids exhibit a filiform apparatus, the polar nuclei fuse before fertilization, and the three antipodal cells are ephemeral although their remains persist until late in development. The embryos are central and extend to both poles of the seed.

ANATOMY

The anatomy of the vegetative parts of *M. vaginalis, M. hastata* and *M. cyanea* has been described by Schwartz (1926); he found no anatomical features not found in other genera of the Pontederiaceae. The floral anatomy of *M. vaginalis* and *M. hastata* is described by Singh (1962). The perianth segments are free almost to the base. The stamens remain united at the base to form a short staminal tube. With the exception of *M. cyanea* and *M. australasica*, one stamen is larger than the others; it has an appendaged filament and a blue anther. Glück (1919) considered the appendage which is lateral and somewhat oblique (Fig. 3b) to be of stipular origin. Singh (1962) examined 200 flowers of *M. hastata* and found 75 possessed appendages on both sides of the filament; in this and other species I have never seen more than one appendage. The appendages are clearly without vascular supply and their function is obscure.

Singh (1962) reported that the anthers dehisced introrsely; he seems to have been mistaken as they clearly dehisce laterally; he was also mistaken in considering the placentation to be parietal – it is axile but there are curious cavities between the carpels containing glandular cells. These cavities (well illustrated by Singh, 1962) may represent septal nectaries but from my own studies they seem to remain permanently enclosed within the ovary wall. The pollen from the blue anther appears, even under the electron microscope, to be identical to the pollen of the yellow anthers. Banerji and Haldar (1942) report that both kinds of pollen are equally viable. The details of the pollination biology of *Monochoria* remain to be discovered; *M. korsakowii, M. vaginalis* and *M. hastata* from my own observations are self-compatible.

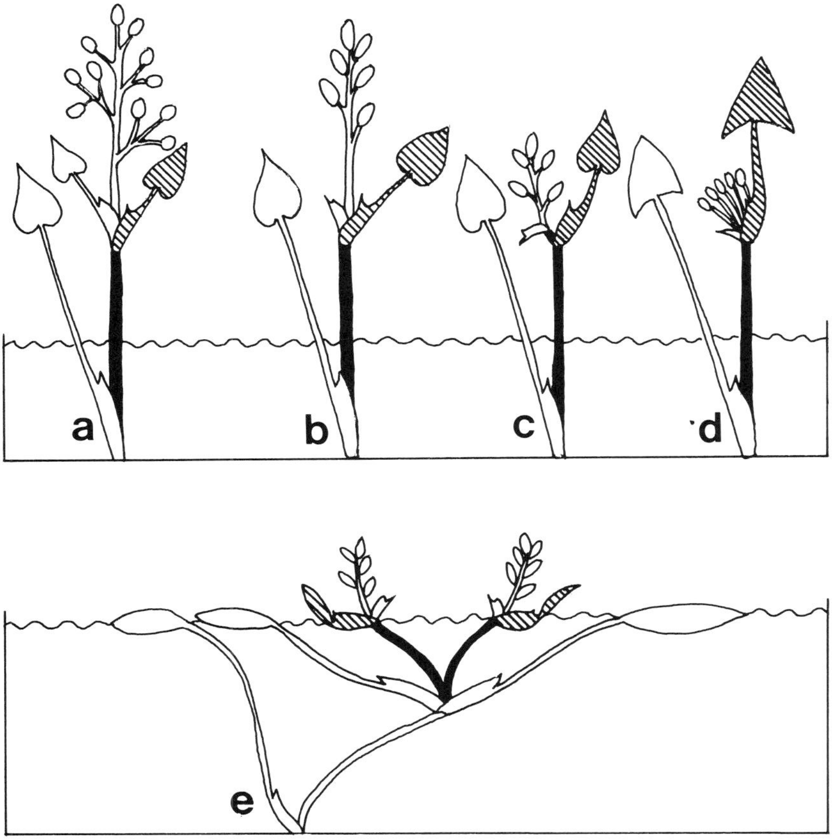

Figure 1. Variation in form of *Monochoria* (inflorescence stalk shaded black, lower spathe shaded with diagonal lines; diagrammatic, not drawn to scale). — a. Both spathes leaf-like, inflorescence dominant (*M. korsakowii*). — b. Upper spathe reduced, inflorescence erect (*M. africana–M. australasica*). — c. Inflorescence reduced (*M. vaginalis*). — d. Lower spathe dominant (*M. hastata*). — e. Floating form as sometimes found in *M. australasica* and *M. cyanea*.

MORPHOLOGY

The growth is sympodial with terminal inflorescences. In *M. korsakowii* the inflorescence is subtended by two leaf-like spathes (bracts or bract with bracteole) illustrated on Fig. 1a and 2b. In the other species of*Monochoria* and the rest of the Pontederiaceae the upper spathe is reduced to a sheath or both upper and lower are reduced to sheaths – hence the term 'spathe' in this family.

The presence of paired leaves below the inflorescence is probably an indication of primitiveness both within the genus *Monochoria* and within the family Pontederiaceae. *Monochoria korsakowii* shows additional primitive charac-

Table 1. *Monochoria* chromosome numbers.

Name (in publication)	Number	Locality	Voucher	Ill.	Reference
M. cyanea					
(*cyanea*)	n = 15	Australia: N. Territory, Tortilla Flats, 1984, *Barrett 1430*	TRT	–	Eckenwalder & Barrett 1986
M. hastata					
(*hastaefolia*)	n = 14	India: suburbs of Calcutta	–	–	Banerji & Haldar 1942
(*hastaefolia*)	2n = 28	no locality given, presumably India: Calcutta	–	+	Majumdar 1953
(*hastaefolia*)	n = 14II	no locality given, presumably India: Calcutta	–	+	Majumdar 1953
(*hastaefolia*)	n = 13II2I	no locality given, presumably India: Calcutta	–	+	Majumdar 1953
(*hastaefolia*)	2n = 28	no locality given, presumably India: Bengal	?	+	Banerji 1974
(*hastaefolia*)	2n = 34	India: Bengal; cult.: Univ. Coll. Calcutta	?	–	Banerji 1974
(*hastaefolia*)	2n = 40	India: Bengal; cult.: Univ. Coll. Calcutta	?	–	Banerji 1974
(*hastaefolia*)	2n = 70	India: Bengal; cult.: Univ. Coll. Calcutta	?	–	Banerji 1974
(*hastaefolia*)	2n = 76	India: Bengal; cult.: Univ. Coll. Calcutta	?	–	Banerji 1974
(*hastaefolia*)	2n = 80	India: Bengal; cult.: Univ. Coll. Calcutta	?	–	Banerji 1974
(*hastaefolia*)	2n = 82	India: Bengal; cult.: Univ. Coll. Calcutta	?	+	Banerji 1974
(*hastaefolia*)	n = 14II	no locality given, presumably India: Bengal	?	+	Banerji 1974
(*hastaefolia*)	n = 40II	India: Bengal; cult.: Univ. Coll. Calcutta	?	–	Banerji 1974
(*hastaefolia*)	n = 41II	India: Bengal; cult.: Univ. Coll. Calcutta	?	+	Banerji 1974
(*hastaefolia*)	n = 42II	India: Bengal; cult.: Univ. Coll. Calcutta	?	–	Banerji 1974
(*hastifolia*)	2n = 28	India: W. Bengal, Hoogly, Dankuni, *14137*	CBLH	–	Sarkar & al. 1980
(*hastaefolia*)	n = 14II	no locality given, presumably India	–	–	Ghosh & Bhattacharya 1981
(*hastaefolia*)	2n = 28	no locality given, presumably India	–	–	Ghosh & Bhattacharya 1981
(*hastata*)	2n = 28	Bangladesh: Joydelpore, 27 March 1975	E	–	Mominul Hoque unpubl.
(*hastata*)	2n = 80	Bangladesh; Sylhet, 3 February 1976	E	–	Mominul Hoque unpubl.
M. korsakowii					
(*korsakowii*)	2n = 52	USSR: Primorskiy Kray, southern part	LE	–	Sokolovskaya 1966
M. vaginalis					
(*vaginalis*)	n = 26II	no locality given, presumably Japan	?	+	Morinaga & Fukushima 1931
(*vaginalis*)	2n = 28	Taiwan: Taipei, *3041*	TAI	–	Hsu 1967
(*vaginalis*)	2n = 28	Taiwan: Taipei, *3184*	TAI	–	Hsu 1967
(*vaginalis*)	n = 40	presumably India	?	?	Sharma, A. & Sarkar, A.K.
(*vaginalis*)	2n = 80	presumably India	?	?	1967–68; fide Moore, D. Reg.

(*plumbaginia*) sic	2n = 52	presumably India	?	?	Veg. 90: 1973
(*vaginalis*)	2n = 52	presumably India	?	?	Krishnappa 1971
(*vaginalis*)	2n = 52	India: Maghalaya, Shillong, September	?	+	Banerji 1974
(*vaginalis*)	2n = 28	different cells of same plant	?	–	Banerji 1974
(*vaginalis*)	2n = 58	different cells of same plant	?	–	Banerji 1974
(*vaginalis*)	2n = 80	different cells of same plant	?	+	Banerji 1974
(*vaginalis*)	2n = 87	different cells of same plant	?	–	Banerji 1974
(*vaginalis*)	2n = 80	India: Meghalaya, Shillong, May	?	+	Banerji 1974
(*vaginalis*)	n = 40II	India: Maghalaya, Shillong, May			
(*vaginalis*)	2n = 72	different cells of same plant	?	+	Banerji 1974
(*vaginalis*)	2n = 74	different cells of same plant	?	–	Banerji 1974
(*vaginalis*)	n = 30	different cells of same plant	?	+	Banerji 1974
(*vaginalis*)	2n = 26	from 2n = 52, Shillong stock, cult.: Univ. Coll. Calcutta	?	+	Banerji 1974
(*plantaginea*)	2n = 52	Nepal, Kathmandu Valley, Bagmati River	?	+	Banerji 1974
(*vaginalis*)	2n = 52	India: Kashmir, Nishat, *Mehra & Pandita 236*	PAN	–	Mehra & Pandita 1978
(*vaginalis*)	n = 26II	India: Kerala, 'different localities'	?	+	Christopher 1983
(*vaginalis*)	2n = 52	India: Kerala, 'all collections'	?	+	Christopher 1983
(*plantaginea*)	n = 12II	India: Kerala	?	+	Christopher 1983
(*plantaginea*)	2n = 24	India: Kerala	?	+	Christopher 1983
(*plantaginea*)	2n = 24	Bangladesh: Joydelpore, 29 November 1975	E	–	Mominul Hoque unpubl.

teristics: it is perennial and it has a dominant and branched inflorescence with cymose lower branches. Evolution within *Monochoria* shows parallel trends in reduction of the inflorescence, reduction of the upper spathe and progressive dominance of the lower spathe – illustrated in sequence in Fig. 1a–1d. In *M. hastata* (Fig. 1d, 4b) the trend has reached the stage where the flowers appear to spring out of an erect petiole. In *M. cyanea* and *M. australasica* the stem is sometimes floating and the sheath of the lower spathe becomes inflated and functions as a float (Fig. 1e) rather like the swollen petioles of *Eichhornia crassipes*.

CHROMOSOMES

The chromosome numbers of*Monochoria* published to date are shown on Table 1 (many are incorrectly cited in the 'standard' abstracting literature). Without checking voucher specimens it is clear that there is considerable variation; there is no simple euploid series but a complex of polyploidy with associated aneuploidy. For example, *M. vaginalis* has $2n = 24$ (× 1), 26(× 1), 28(× 3), 47(× 1), 52(× 7), 58(× 1), 60(× 1), 72(× 1), 74(× 1), 80(× 3) and 87(× 1). *Monochoria hastata* has $2n = 28$(× 6), 34(× 1), 40(× 1), 70(× 1), 76(× 1), 80(× 2), 82(× 1) and 84(× 1). The commonest numbers are $2n = 28$ for *M. hastata* and $2n = 52$ for *M. vaginalis*. In low and high numbers the meiotic configurations are remarkably stable; univalents are very rare and polyvalents are unknown. From published illustrations it appears that the chromosomes are variable in size and shape both within single karyotypes and between different plants.

There is no apparent correlation between chromosome number and morphological or ecological factors. The cytological evidence is of little use in explaining or supporting hypotheses on phyletic relationships within the genus *Monochoria*.

TAXONOMY

Monochoria K.B. Presl, Reliquiae Haenkeanae, 1(2): 127. 1827. Type: *M. hastaefolia* K.B. Presl, nom. illeg., based on *Pontederia hastata* L. (= *Monochoria hastata* (L.) Solms-Laubach).

= *Gomphima* Rafinesque, Fl. Telluriana, 2: 10. 1837. Type: *G. vaginalis* (N.L. Burman) Rafinesque (= *Monochoria vaginalis* (N.L. Burman) Kunth).

= *Limnostachys* F. von Mueller, Fragmenta Phytographiae Australiae, 1(1): 24. 1858. Type: *L. cyanea* F. von Mueller (= *Monochoria cyanea* (F. von Mueller) F. von Mueller).

Subgeneric Classification

Schwartz (1927) placed the genus *Monochoria* close to *Heteranthera* and *Hydrothrix* and used them as a basis for the tribe Heterantthereae. This decision has found universal acceptance and fits the phylogenetic arrangement of Eckenwalder & Barrett (1986). Schwartz split the genus *Monochoria* into two sections (*Eumonochoria* and *Deutomonochoria*) which he based on an incorrect interpretation of the inflorescence structure. Backer (1951) pointed out the mistake made by Schwartz and proposed two new sections. The first he named *Eumonochoria* (which must be called sect. *Monochoria*) for the species with dimorphic stamens, and the second, *Limnostachys* for the species with equal stamens (*M. australasica* and *M. cyanea*). I agree with this decision but find it rather unimportant to split a genus of eight species into two sections.

Nomenclature

Linné in Species Plantarum, 288 (1753) considered the genus *Pontederia* to have three species. The third species, *P. hastata*, is clearly what we call today *Monochoria hastata* in spite of the fact that one of his synonyms 'Carimgola. Rheed. Mal. 11. p. 91. t. 44' is *M. vaginalis*; the original diagnosis, all other citations and the specimens in Linné's herbarium are *M. hastata*. The second, *P. cordata*, is an American species and is considered by Lowden (1973) and others to be the type of the genus *Pontederia*. The first Linnaean species, *P. ovata*, is more problematic; Linné based it on 'Narukila' in Rheede, Hortus Malabaricus, 11: 67. t. 34 (1692); the description and plate are not good but they are generally accepted to represent a member of the Marantaceae, *Phrynium capitatum* Willd. (which presumably should be called *Phrynium ovatum*). The specimen in Linné's herbarium (LINN, Savage Cat., No. 407/1) is *Eichhornia azurea* (Swartz) Kunth; it does not fit Linné's diagnosis – the leaves are not ovate and the flowers are not in heads, also it is native in South America and not in South India. *Monochoria ovata* Kunth is clearly not based on *Pontederia ovata* L. and thus cannot be rejected for nomenclatural purposes.

The following has been described in *Monochoria* but belongs in *Hydrocharis*: *Monochoria dubia* (Blume) Miquel, Fl. Nederl. Indië, Fl. Indae Bataviae, 3(4): 549. 1859 ≡ *Hydrocharis dubia* (Blume) Backer, Handb. Fl. Java, 1: 64. 1925 ≡ *Pontederia dubia* Blume, Enum. Pl. Javae, 1: 33. 1827.

Generic description

Glabrous, erect, ascending or occasionally creeping, emergent or floating annual or perennial aquatics with sympodial growth.

Roots adventitious, fibrous, regularly branched, often blue.

Stems truncate and sometimes corm-like or elongate and creeping, sometimes rhizomatous, simple or sparsely branched.

Leaves radical, spiro-distichous, extremely variable as to shape and size; juvenile leaves simple, scale-like or linear, mostly submerged, usually ephemeral; adult leaves differentiated into petiole and blade; petiole stiff, erect and emergent, or flaccid, inflated and floating; stipules axillary, large, ochrea-like (enveloping the stem and developing leaves), at apex acute or obtuse and often terminating in a filament; blades simple, linear or lanceolate to ovate, cordate, sagittate or hastate, at base cuneate to cordate; margin entire; apex acute or acuminate; venation acrodromous.

Inflorescences terminal, few- or many-flowered, paniculate, raceme- or umbel-like, borne above or below the leaves on naked, elongated, unbranched inflorescence stalks which terminate with 2 opposed spathes; both spathes either similar and leaf-like each with sheath, petiole and blade (only in *M. korsakowii*) or distinctly dimorphic; the lower spathe (bract) leaf-like with sheath, petiole and blade; the upper spathe (bracteole) either scale-like and enclosed within the lower one at anthesis or exposed, when exposed either scale-like and membranous or sheath-like, green and sometimes inflated, also sometimes bearing a filamentous or minute and reduced leaf-like apical appendage; the lower spathe sometimes so dominant (*M. hastata, M. vaginalis*) that the inflorescence appears to burst out of a normal-looking petiole (Fig. 4b).

Flowers bisexual, monomorphic and showy. Perianth spreading at anthesis and usually withering within a day, very nearly actinomorphic, of 6 segments; the segments almost free, very similar in shape and size, blue to white or pinkish, with distinct, green mid-ribs or veins, usually somewhat accrescent, persistent and enveloping the ripe fruit; the lower 3 segments usually somewhat narrower than the upper 3 segments and with somewhat thicker mid-veins.

Stamens 6, all fertile, inserted at the base of the perianth segments; filaments elongate, simple or one with 1 or rarely 2, lateral, obliquely erect teeth; anthers basifixed, 4-sporangiate, dehiscing through 2, apical pores which may later extend downwards, either all alike or 1 larger and blue and the remaining 5 smaller and yellow; pollen monomorphic, 2-celled, ellipsoidal when dry, disulcate with verrucate (tecto-columellate) exine sculpturing.

Gynoecium superior, of 3, united carpels; ovules anatropous, bitegmic, crassinucellar, numerous, borne on axile placentas; endosperm helobial; embryo sac polygonum type; embryony asterad type; style simple, elongate, persisting in fruit; stigma papillose, dry, entire or lobed.

Fruit a loculicidal, ellipsoidal to subglobose or subconical capsule, remaining enclosed within withered perianth segments, normally released as a unit and opening irregularly later.

Seeds numerous, brown to almost black, cylindrical to ovoid or subglobose, with 8–14 longitudinal nerves, ribs or wings, transversely striate; endosperm starchy, embryo central and linear.

Diagnostic Features

Perianth actinomorphic or very nearly so; perianth segments very similar, free almost to base; stamens 6, borne at base of perianth segments, either all alike or 1 large with an appendaged filament and a blue anther, and 5 with simple filaments and yellow anthers; inflorescence subtended by 1 or rarely 2, leaf-like spathes borne on a naked stalk that often superfically resembles the petiole of a leaf; fruit a many-seeded capsule enclosed in perianth segments, dispersed as a unit which opens later.

Distribution

Eight species, all native in the warmer parts of the Old World; 2 in tropical Africa, 2 in Australia and 4 in Asia. One Asian species, *M. vaginalis*, is a weed in ricefields and has been recorded from Europe (Ukraine and Italy), N. America (California), Hawaii Islands (Oahu) and Australia.

MULTIPLE ENTRY KEY

Brackets enclosing a species name denote that the character state is not constant for the species.

1. Perennial with at least 1 cm thick rhizome: *korsakowii, hastata, elata*
2. Petioles (foliage leaves) more than 60 cm long: *hastata, elata*
3. Inflorescence:
   a. paniculate, lower branches 3-flowered: *korsakowii*
   b. racemose, lower branches 2-flowered: (*elata*), (*vaginalis*)
   c. subumbellate: *hastata*
   d. spike-like: *africana, brevipetiolata,* (*cyanea*), *australasica*
   e. with more than 20 flowers: (*korsakowii*), *hastata, elata, africana,* (*brevipetiolata*)
4. Lower spathe:
   a. sheath more than 6 cm long: (*elata*), *australasica*
   b. petiole less than 1 cm long: (*brevipetiolata*), *australasica*
   c. blades sagittate to hastate: *hastata, elata*
   d. blades wider than long (cordate): *korsakowii*
5. Upper spathe:
   a. leaf-like (with petiole and blade): *korsakowii*
   b. sheath less than 2 cm long: *brevipetiolata*
6. Flowerless part of rachis:
   a. less than 1 cm long: *vaginalis, hastata*
   b. longer than the flowering part: *korsakowii*
7. Pedicels more than 1.5 cm long: *vaginalis, hastata, elata*
8. Perianth more than 15 mm long: *korsakowii, elata, cyanea, australasica*
9. All 6 stamens alike: *cyanea, australasica*
10. Yellow anthers less than 2 mm long: (*vaginalis*), *cyanea*
11. Blue anthers more than 5.3 mm long: *hastata, elata*
12. Stigma simple, not wider than style: *vaginalis, hastata, elata*
13. Seeds cylindrical: *korsakowii*
14. Seeds subglobose with 12 or more ribs: *brevipetiolata*

DICHOTOMOUS KEY

1A. Stamens alike; filaments simple (without appendages); anthers yellow, one not much larger than the others (Australia).

2A. Blade of lower spathe with a petiole usually more than 12 mm long; style not exceeding 5.8 mm in length; the flowerless part of the rachis very rarely more than 4 cm long . . . . 7. *M. cyanea*

2B. Blade of lower spathe sessile or with a petiole of up to 8 mm long; style more than 5.5 mm long; the flowerless part of the rachis usually more than 4 cm long . . . . 8. *M. australasica*

1B. Stamens not alike; one filament with an obliquely erect appendage, the remaining five simple, one anther blue, much larger than the remaining 5 yellow ones (widespread).

3A. Inflorescence paniculate (lower branches bearing at least 3 flowers), the flowerless part of the rachis usually longer than the flower-bearing part; leaves usually cordate and wider than long; the upper spathe sometimes bearing a petiole with blade; seeds cylindrical (SE Asia) 1. *M. korsakowii*

3B. Inflorescence spike-like, racemose or sub-umbellate (lower branches simple or with not more than 2 flowers), the flowerless part of the rachis shorter than the flower-bearing part (often remaining within the spathes); leaves linear to ovate, cordate, sagittate or hastate, when cordate then longer than wide; the upper spathe never bearing a petiole with well-developed blade; seeds ovoid to globose.

4A. Mature blades sagittate or hastate; rhizome robust, hard, up to 2 cm diameter; blue anther at least 5 mm long but usually more; inflorescence compact, usually bearing more than 20 flowers (S and SE Asia)

5A. Inflorescence sub-umbellate, the flowers remaining below the leaves; style rarely more than 6 mm long . . . . 3. *M. hastata*

5B. Inflorescence racemose, the flowers overtopping the leaves; style more than 6.5 mm long . . . . 4. *M. elata*

4B. Mature blades linear to ovate or cordate, never sagittate or hastate; rhizome absent or if present then short or slender, soft, less than 0.5 cm diameter; blue anther very rarely more than 5 mm long; inflorescence lax or elongate and spike-like, only spike-like inflorescences bearing more than 20 flowers.

6A. Inflorescences strongly deflexed in fruit, laxly racemose, never elongate and spike-like; flowerless part of rachis usually much less than 2 cm long; pedicels up to 3 cm long, the lower ones sometimes forked (widespread, but not in Africa) . . . . 2. *M. vaginalis*

6B. Inflorescence remaining erect in fruit, elongate and spike-like; flowerless part of rachis usually more than 2 cm long; pedicels less than 1 cm long (Africa).

7A. Petiole of lower spathe at least 5 cm long, usually erect; leaves usually cordate at base, usually at least 6 cm long and 4 cm wide; seeds ellipsoidal to barrel-shaped, ± 0.75 mm long with 10 longitudinal ribs. . . . . 5. *M. africana*

7B. Petiole of lower spathe not more than 4 cm long (usually much less), usually reflexed; leaves cuneate or obtuse at base, rarely more than 6 cm long and 3 cm wide; seeds subglobose, 1.3–1.8 mm long, with 12–14 longitudinal ribs . . . . 6. *M. brevipetiolata*

SPECIES ACCOUNTS

1. Monochoria korsakowii E. Regel & Maack, Tentamen Florae Ussuriensis, in Mém. Acad. Impér. Sci. St.-Pétersbourg, Sér. 7, 4(4): 155, tab. XII. 1861 ≡ *M. vaginalis* var. *γ korsakowii* (E. Regel & Maack) Solms-Laubach in A. de Candolle, Monog. Phaner., 4: 525. 1883. Type: USSR, Ussuri, Ozero Khanka, 'Auf sumpfigen Stellen an den Ufern des Kenga-Sees. Blühet im August', 1857, *P.K. Maack s.n.* (holotype: LE, the specimen with 'teste Rgl' on the label; it is the basis of the plate accompanying the original description; isotypes: K, LE).

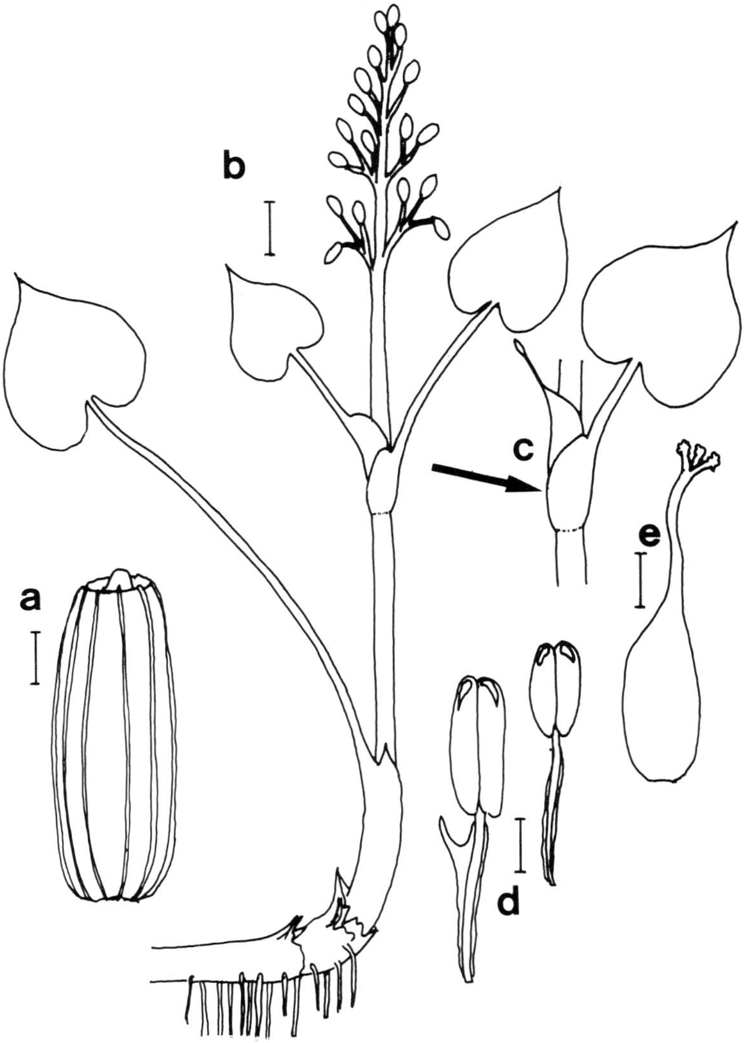

Figure 2. Monochoria korsakowii. — a. Seed (scale = 0.25 mm). — b. Habit (scale = 4 mm). — c. Spathe pair with upper one reduced. — d. Stamens (scale = 2 mm). — e. Young ovary (scale = 2 mm). — a, drawn from *Maire 325* (E); d & e, drawn from *Maire 1001* (E); rest diagrammatic.

= *M. plantaginea* var. *β cordifolia* Franchet & Savatier, Enumeratio Plantarum in Japonia sponte crescentium, 2: 95. 1879. Type: Japan, 'circa Yokoska', September, *Savatier 1217 bis* (holotype: ?P, not found in K).

Perennial or perhaps sometimes annual.

Stem rhizomatous, up to 1 cm diameter or elongate and creeping, with numerous adventitious roots.

Radical leaves either linear, submerged and ephemeral or differentiated into petiole with blade; petioles up to 30 cm or more long, often bright red below; stipules up to *c*. 12 cm long, usually without terminal appendages; blades emergent or rarely floating, rarely lanceolate, usually elliptic to cordate, 4–8 cm or perhaps more long, 1–7 wide.

Inflorescences elongate, below simple and flowerless, above paniculate with (2-) 10–20 (-50 or more) flowers, the lower branches usually bearing at least 3 flowers (Fig. 2b); inflorescence stalks (5-) 15–25 (-40 or more) cm long, usually robust, erect and emergent, rarely flaccid and floating; sheath of the lower spathe (2-) 3–4 (-6) cm long, somewhat inflated; petiole of the lower spathe (1-) 2–7 (-20 cm) long, usually erect; blade of the lower spathe very variable in size, 3–15 cm long, 2–17 cm wide, almost invariably cordate and often wider than long, very rarely ovate; upper spathe (3-) 4–6 (-10) cm long, longer than the sheath of the opposed lower spathe, clearly visible and green (not remaining enclosed by lower spathe); petiole of the upper spathe up to 3 cm long; blade of the upper spathe up to 6.5 cm long and 4 cm wide, cordate to ovate, or reduced and linear or even filamentous (Fig. 2c); flowerless part of the rachis (at late anthesis or in fruit) simple, (3.5-) 6–7 (-15) cm long; flowering part of the inflorescence (2-) 5–6 (-22) cm, mostly shorter than the flowerless part; pedicels up to *c*. 1.5 cm long.

Perianth segments (12-) 15–20 mm long, (6-) 8–12 mm or more wide, not accrescent but persisting and remaining flat in fruit.

Stamens of 2 kinds (Fig. 2d); 5 with simple, 3–5 mm long filaments and yellow, 2.5–3.5 (-4) mm long anthers; 1 with an appendaged, 3–5 mm long filament and a blue, (3.4-) 4–4.5 (-5.2) mm long anther, at least 1/4 longer than the yellow ones.

Fruits ovoid to almost conical, up to *c*. 18 mm long and 9 mm wide at base; style persistent, 5–6 mm long; stigma capitate, distinctly wider than the style (Fig. 2e).

Seeds almost cylindrical, 1.1–1.4 mm long, 0.25–0.5 mm diameter, brown, with 8–10, rather fine, longitudinal veins (Fig. 2a).

Diagnostic Features

Both spathes leaf-like (with petioles and blades); the upper spathe sheath longer than the lower one; inflorescence paniculate, the lower branches bearing 3 flowers; blades cordate, often wider than long.

Distribution

Monochoria korsakowii is a species of eastern Asia. It is found in S.W. China and the northeastern part of China from Shanghai extending northwards through Manchuria and Korea to Sikhote Alin (USSR). In Japan it is widespread on Honshu Island (Map 1). The records from the Ukraine (Laktionov, 1973; I have not personally seen herbarium material) and N. Italy (Vuille, 1985) are referable to *M. vaginalis*.

Notes

Monochoria korsakowii is a very decorative species beautifully illustrated in 'Gartenflora' by Regel (1862). It is rather surprising that it is not in cultivation.

Taxonomically it is very distinct and is the most elaborate and probably the most primitive species in the genus. The possession of two leaf-like spathes and a

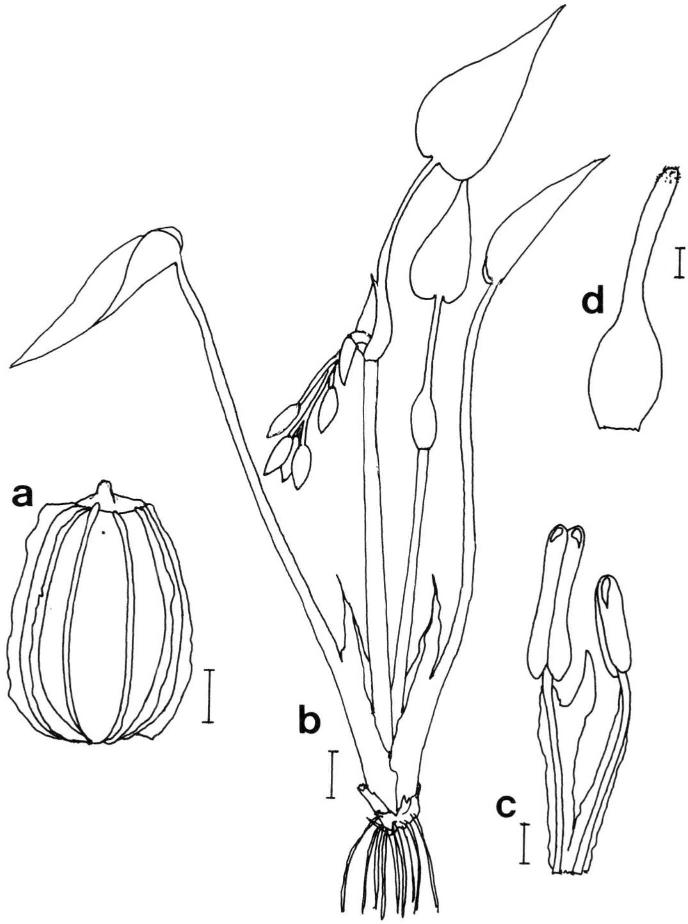

Figure 3. Monochoria vaginalis. a. Seed (scale = 0.2 mm). — Habit (scale = 1 cm). — c. Stamens (scale = 1 mm). — d. Young ovary (scale = 1 mm). — Drawn from *Cook & Frey 4690* (Z).

dominant, branched inflorescence (Fig. 1a) coupled with a perennial rhizome are probably indications of primitiveness. In Japan and occasionally elsewhere the upper spathe sometimes loses the clearly differentiated petiole and blade but it, nevertheless, bears a filamentous appendage distinctly thickened at the apex, illustrated on Fig. 2c. Material collected in Japan cultivated at Zürich has both spathes leaf-like. In Japan it is now a protected species.

2. Monochoria vaginalis (N.L. Burman) Kunth, Enum. Plant., 4: 134. 1843 ≡ *Pontederia vaginalis* N.L. Burman, Fl. Indica, 80. 1768 ≡ *Gomphima vaginalis* (N.L. Burman) Rafinesque, Fl. Telluriana, 2: 10. dated 1836, publ. 1837. Type:

there are two Burman specimens of this species in G; I choose the sheet with the separate label 'Ind. orien. Dr Roxburgh 2' as the lectotype.

= *Pontederia pauciflora* Blume, Enum. Pl. Javae, 1: 32. 1827 ≡ *Monochoria pauciflora* (Blume) Kunth, Enum. Plant., 4: 135. 1843 ≡ *M. vaginalis* var. *pauciflora* (Blume) Merrill, Enum. Philippine Fl. Plants, 1(2): 201. 1923 [comb. illeg., *M. vaginalis* var. *plantaginea* (Roxburgh) Solms-Laubach is cited as a synonym]. Type: Indonesia, Java, 'Crescit in oryzetis humidis Javae occidentalis' (lectotype to be chosen from material in L).

= *Pontederia loureiriana* J.A. & J.H. Schultes, Systema Vegetabilium, 7(2):1145. 1830 ≡ *Monochoria loureiriana* (J.A. & J.H. Schultes) Kunth, Enum. Plant. 4:135, 1843. *M. ?loureirii* sphalm. Type: Based on *Pontederia cordata* Loureiro, non L., no suitable specimen found.

= *Pontederia plantaginea* Roxburgh, [Hort. Bengal., 23. 1814, nom. nud.], Fl. Indica, ed. 2, 2: 123. 1832 ≡ *Monochoria plantaginea* (Roxburgh) Kunth, Enum. Plant., 4: 135. 1843 ≡ *M. vaginalis* var. *β plantaginea* (Roxburgh) Solms-Laubach in A. de Candolle, Monog. Phaner., 4: 524. 1883. Type: no suitable specimen has been found in BM, G or K; Roxburgh plate no. 595 in K is *M. vaginalis* sensu stricto, Roxburgh perhaps based this species on *Plantaginalis stellatae foliis* (see below).

= *Pontederia linearis* Hasskarl, Flora 25(2), Beiblatt 1: 4. 1842 ≡ *Monochoria linearis* (Hasskarl) Miquel, Fl. Nederl. Indië (Fl. Indiae Batavae), 3(4): 549. 1859. Type: Indonesia, Java (to be selected from material in L).

= *Monochoria ovata* (Hooker & Arnott) Kunth, Enum. Plant., 4: 665. 1843 ≡ [*Pontederia ovata* Hooker & Arnott, Botany Beechey's Voyage, 5: 218. 1837; nom. illeg., post-dates *P. ovata* L.]. Type: China, Canton, *Millet* (holotype: E).

= *Monochoria junghuhniana* Hasskarl, Flora 35(8): 115. 1852. Type: Indonesia, Sumatra, 'in niedrigen Strandwiesen bei Samas an der Mündüng des Opak in Sumatra,' *Junghuhn* (holotype: probably destroyed in B).

= *Boottia mairei* Léveillé, Cat. Pl. Yunnan, 131. 1916, Type: China, Yunnan, Dongchuan (Tangdan), 'rizières de la plaine de Tong-Tchouan', alt. 2500 m, August 1912, *E.E. Maire, s.n.* (holotype: E).

– *Plantaginalis stellatae foliis* Plukenet, Almagestrum Botanicum, ed. T. Davis, 2: 299, Tab. 215, Fig. 4. 1769 – a reprint of Phytographica, 3: Tab. 215, Fig. 4. 1692; nom. illeg., pre-Linnaean publication, probably the basis of *Pontederia plantaginea* Roxburgh.

– *Olus palustre* Rumphius, Herb. Amboinense, 6: 178, t. 75. 1750; nom. illeg., pre-Linnaean publication.

– *Pontederia cernua* L., never published, nom. in schaed., LINN, Savage Cat. No. 407/6.

– *Pontederia cordata* Loureiro, non L., Fl. Cochinchinensis, ed. 2, 1:245. 1793; nom. illeg., name used by Linné.

– *Pontederia racemosa* F. Hamilton in Wallich, Cat. No. 5095C, 1831–1832; nom. nud.

– *Pontederia alba* F. Hamilton in Wallich, Cat. No. 5095D, 1831–1832; nom. nud.

– *Pontederia lanceolata* Wallich ex Kunth, Enumeratio Plantarum, 4: 135. 1843; post-dates *P. lanceolata* Nuttall, Genera N. Amer. Plants, 1:216. 1818.

Annual or perhaps sometimes perennating.

Stems truncate or shortly elongate and creeping, rarely rhizomatous.

Radical leaves either linear, submerged and ephemeral or differentiated into petiole with blade, very variable in shape and size; petiole 3–50 cm or more long; stipules up to ± 14 cm but usually less, rarely red, usually brownish or bluish; blades emergent or rarely floating, (2–) 3–7 (–21) cm or perhaps more long, 0.2–4.0 (–10) cm wide, linear or lanceolate to ovate, at base usually cuneate but sometimes truncate or even cordate, at apex acute or acuminate.

Inflorescence very variable in form, from shortly racemose to sub-umbellate or irregular, occasionally branched below but branches bearing not more than 2 flowers, mostly borne below the leaf blades, with (1–) 2–7 (–20) flowers, becoming strongly deflexed in fruit (Fig. 3b); inflorescence stalk erect and emergent, very variable, 2–50 cm long; sheath of the lower spathe (1–) 2–3 (–3.7) cm long; petiole of the lower spathe 0.5–7 cm long, erect or sometimes deflexed; blade of the lower spathe very variable, 1.3–10 cm long, 0.3–5 (–15) cm wide, linear or lanceolate to ovate or cordate (small and terrestrial plants usually have reflexed elongate leaves while robust emergent aquatic plants have erect ovate or cordate leaves); upper spathe 1–3.5 cm long, at apex acute, acuminate or bearing a filamentous appendage; internode between the spathes up to 1.5 cm or rarely more long, strongly deflexed in fruit; flowerless part of rachis usually very short but can reach 2 cm long; pedicels up to 3 cm long but usually much less.

Perianth segments 8–15 mm long, up to ±8 mm wide, green central veins particularly distinct and persisting, somewhat accrescent, persistent and twisted around the ripe fruit.

Stamens of 2 kinds (Fig. 3c); 5 with simple, 4–7 mm long filaments and yellow (1–) 1.5–2.5 (–3) mm long anthers; 1 with an appendaged, 4–6 mm long filament and blue, (1.8–) 1.5–4 mm long anther, 1/4 to 1/3 longer than the yellow ones.

Fruits ellipsoidal to ovoid, 0.8–1.2 mm long, 0.4–0.7 mm diameter; style (2–) 4–5 mm long; stigma papillose, hardly wider than the style (Fig. 3d).

Seeds ovoid to barrel-shaped, (0.5–) 0.7–0.8 (–1.0) mm long, (0.3–) 4.0–6.0 (–6.7) mm diameter, brown, with 8–10, narrow, longitudinal wings (Fig. 3a).

Diagnostic Features

Extremely variable from 3 cm to over 50 cm tall; inflorescence usually racemose, borne below the leaves, strongly deflexed in fruit, few flowered, (1–) 2–7 (–20) flowers; stigma scarcely wider than the style; seeds not exceeding 1 mm long.

Distribution

Monochoria vaginalis is common and widespread. It extends from N.E. Pakistan and N.W. India (Kashmir) southwards through India to Sri Lanka (Ceylon), extending eastwards through China to Japan and the Philippines and southwards through Indo-China, Malaysia and Indonesia (Map 2) to New Guinea and northern Australia (Map 4). It extends into the Pacific to the Fiji Islands. It has become naturalized in Hawaii (Oahu Island) where it is used by the Chinese as a potherb and medicine according to Degener & Degener (1960), in the USA in California where it is confined to ricefields in Butte County according to Mason (1957), in S.W. Russia in ricefields along the northern part of the Black Sea (Chernoye More) according to Laktionov (1972) and in ricefields of N. Italy in Lombardy (Map 5) according to Vuille (1985).

Ecology

Monochoria vaginalis is often gregarious and typically found in inundated places or at the edges of pools, ditches and canals, or in swamps. It is particularly common and almost characteristic of rice fields. Although often cited as a weed of irrigation ditches and paddy fields it is very rarely considered to be harmful. It is usually annual but in permanent water may persist for more than one year. The capsules, enveloped in persistent perianth segments, are liberated as units either on land awaiting the next inundation or on water where they sink and liberate the seeds later. Seeds germinate under water. The seedling develops strap-shaped submerged leaves at first, later floating and emergent leaves develop. The water quality seems to be relatively unimportant, it is commonest in eutrophic water and may be found in distinctly brackish water but has also been seen in almost oligotrophic conditions. It is very rarely found in stable conditions and can be considered to be a stress-tolerant ruderal.

There is no particular flowering period; it flowers as long as the habitat remains wet and the competition with other species is not too high. Anthesis takes place in the morning and by about midday the flowers wither. Usually all the flowers in one inflorescence open on the same day, however, on large and robust plants new flowers may develop for about three consecutive days in a single inflorescence. After fertilization the rachis and pedicels bend downwards and the fruits often develop under water.

Notes

Monochoria vaginalis is extremely plastic in size and shape. In deep water it may develop inflorescence stalks and petioles up to 50 cm or more long. If stranded on land when young the whole plant may be no more than 2–3 cm tall. Large, robust plants tend to have inflorescences bearing 7 or more flowers while small, terrestrially developed plants may have only 1 or 2 flowers. The leaf-shape is also variable. Under stress conditions (excessive competition, shade or cold) the leaf-blades tend to be linear or lanceolate while very robust plants usually have cordate blades, normally they are ovate with somewhat cordate bases. I have found no evidence of clear genecotypic differentiation. Plants at the extremes of the geographic range are smaller and have narrower leaf-blades than most in the centre but in the centre plants under stress resemble those at the ecologically tolerable limits. This plasticity has led to a certain number of 'paper' species with self-explanatory epithets such as *linearis, ovata, pauciflora, plantaginea* and *racemosa.* In natural populations I have seen the perianth colour to vary from white to blue or bluish-pink; this kind of variation, at least, most likely has a genetic basis.

Some plants from the Pacific Region (Fiji, New Guinea, Philippines) are particularly robust with erect petioles and inflorescence stalks and cordate leaves which are often wider than long. They also seem to be somewhat rhizomatous and may be perennial. These plants are slightly transitional to *M. elata* but, except on the island of Hainan, they are distinctly allopatric. On the Fiji Islands these robust erect plants are the only representatives of *Monochoria* present. Without further studies I am reluctant to give them formal taxonomic recognition.

3. Monochoria hastata (L.) Solms-Laubach in A. de Candolle, Monog. Phaner., 4: 523. 1883 ≡ *Pontedena hastata* L., Sp. Pl. 288. 1753. Type: 'India', perhaps Sri Lanka (lectotype designated here: Linn. 407/5).

= *Pontederia dilatata* F. Hamilton in Michael Symes, Account of Embassy to Kingdom of Ava: 475, cum tab. (BM copy, missing in G copy). 1800. ≡ *Monochoria dilatata* (F. Hamilton) Kunth, Enum. Plant., 4: 134. 1843. Type: Burma, 'Rangoon V.', *F. Hamilton* (holotype: BM).

– *Monochoria hastaefolia* C. Presl, Rel. Haenk., 1(2): 128. 1827; nom. illeg., based on *Pontederia hastata* L.

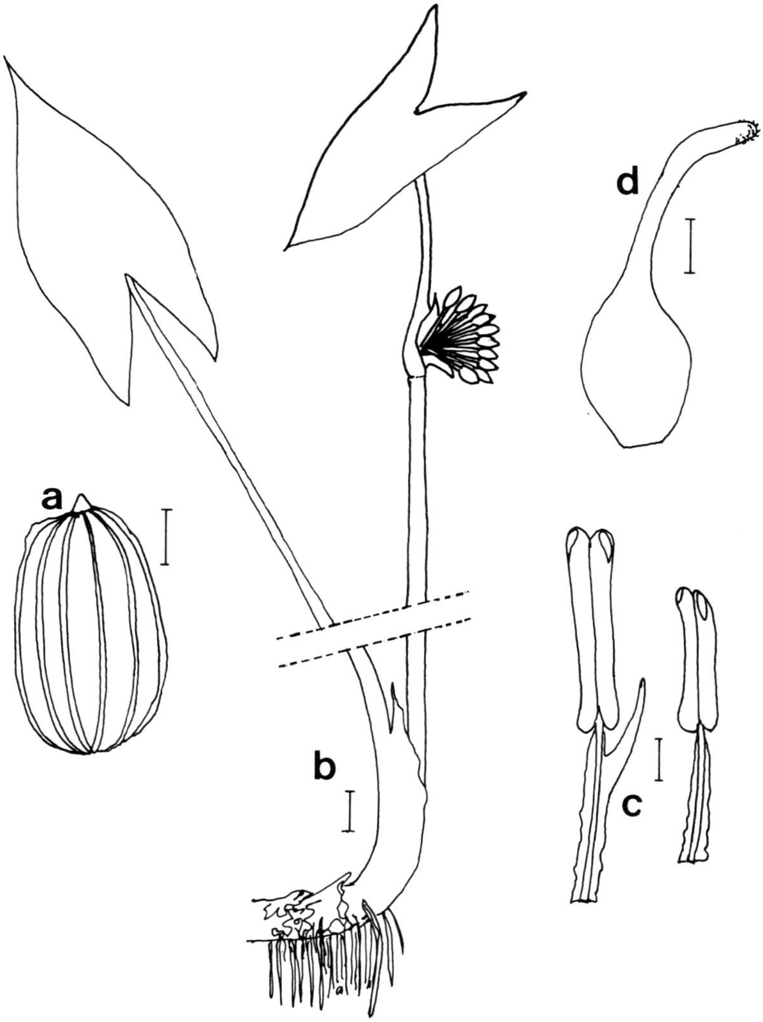

Figure 4. Monochoria hastata. — a. Seed (scale = 0.25 mm). — b. Habit (scale = 3 cm). — c. Stamens (scale = 1 mm). — d. Young ovary (scale = 2 mm). — Drawn from *Cook & Frey 4606* (Z).

– *Pontederia sagittata* Roxburgh, Fl. Indica, ed. 2, 2: 124: 1832. ≡ *Monochoria sagittata* (Roxburgh) Kunth, Enum. Plant., 4: 134. 1843; nom illeg., post-dates *P. sagittata* C. Presl, Rel. Haenk., 1(2). 116. 1827, a synonym of *Pontederia cordata* L. Type: East India, *Roxburgh*: BM, K.

– *Pontederia sagittifolia* Heyne in Wallich, Cat. No. 5093 A, 1831–1832; nom. nud.

Perennial or perhaps sometimes annual.

Stems rhizomatous, strong and robust, up to 30 cm or more long and up to 2 cm diameter, with numerous adventitious roots.

Radical leaves either linear, submerged, ephemeral and absent at anthesis or differentiated into petiole with blade; petioles erect, often robust, usually bright red below, up to 90 cm or more long; stipules up to 30 cm long, often bright red, at apex acute or with a filamentous appendage; blades emergent, up to 20 cm long (measured from point of attachment to the apex), up to 15 cm or more wide, rarely ovate or cordate, mostly sagittate to hastate, the lower lobes rounded to acute or acuminate, the upper lobe acute or acuminate.

Inflorescence sub-umbellate or very shortly racemose, borne below the leaf blades (appearing to burst out of a normal-looking petiole), with (10–) 20–30 (–60 or more) flowers (Fig. 4b); inflorescence stalks robust, red below, erect, up to 65 cm or more long; sheath of the lower spathe sometimes indistinct but usually inflated, (3–) 3.5–5 (–8) cm long; petiole of the lower spathe (3–) 5–10 (–12) cm long, usually erect; blade of the lower spathe emergent, (7–)8–12(–18) cm long, (4–)5–10 (–14) cm wide, sagittate or hastate, the lower lobes acute to acuminate; upper spathe without petiole and blade, green, membranous, (2.5–)3–4(–5) cm long, partly enclosed within and shorter than the lower spathe sheath, margin membranous, apex acute to acuminate; flowerless part of rachis very short, not exceeding 1 cm in length; pedicels arising very close together, up to 3 cm or more long, spreading at anthesis, sometimes becoming deflexed in fruit.

Perianth segments 10–15 (–18) mm long, 7–8 mm wide, somewhat accrescent, persistent and twisted around the ripe fruit.

Stamens of 2 kinds (Fig. 4c); 5 with simple, 5–6 mm long filaments and (3–) 3.5–4 (–4.5) mm long, yellow anthers; 1 with a 5–6 mm long, appendaged filament and (5.3–) 5.5–6.5 mm long, blue anther, 1/4–1/3 longer than the yellow ones.

Fruits ovoid, ± 12 mm long, ± 8.9 mm diameter; style 5–6 mm long; stigma papillose, hardly wider than the style (Fig. 4d).

Seeds ovoid to barrel-shaped, 0.8–1.0 mm long, ± 0.5 mm diameter, brown, with ± 10, narrow, longitudinal ribs (Fig. 4a).

Diagnostic Features

Robust, rhizomatous perennial; inflorescence sub-umbellate, borne below the leaves, usually with more than 20 flowers; leaves widely sagittate or hastate; pedicels up to 3 cm or more long; blue anther (5.3-) 5.5–6.5 mm long; style 5–6 mm long.

Distribution

In India *M. hastata* is common in the Ganges Plain from Punjab to Bangladesh; there are some more or less isolated records from the Deccan Peninsula and the southeast coast; it seems to be rare in the south and absent from the southwest but is present in Sri Lanka. In Southeastern Asia it extends from Assam and northern Burma southwards through the Malay Peninsula extending eastwards through southern China and the Philippines to New Guinea (Maps 3 and 4). It is reported from Australia but I have not seen herbarium material.

Ecology

Monochoria hastata is a large, robust perennial with a thick rhizome. It is characteristically found at the edge of pools, tanks and canals. According to

Banerji & Halder (1942) it attains a height of up to 2–3 m during the rains, adjusting its height with the rise of the water level. The inflorescence and leaf-blades are always borne above the water. In Bengal, according to Banerji & Halder (1942), flowering starts in May and extends to the onset of winter (not later than November). The number of flowers in each inflorescence increases with the advent of the rains and decreases as winter begins. During the winter the above-ground parts die down.

Anthesis takes place in the morning, generally before 08.00, but opening of the flowers may be delayed by cloudy and humid weather. The flowers are self-compatible and largely self-pollinated. The seeds take 10–15 days to develop. Germination takes place under water. Local spread is, however, mostly due to rhizomatous growth and fragmentation; normally it shows a distinctly 'clumped' dispersion.

4. Monochoria elata N.H. Ridley, J. Straits Branch Royal Asiatic Society, No. 79: 99. 1918. ≡ *M. hastata* var. *elata* (N.H. Ridley) Backer, Fl. Malesiana, Ser. 1, 4(3): 258. 1951. Type: Malaya, Kedah, 'Jenum, common in rice fields, flowers blue', 19 November 1915, *Mo. Haniff 1208* (holotype: K; isotypes: BM, K).

Probably a rhizomatous perennial.

Radical leaves erect; petioles 40–60 cm or considerably more long; blade almost linear with an auriculate base to elongate-sagittate or elongate-hastate, ± 10–30 cm long, 0.8–2.0 cm wide, basal lobes elongate to 3 cm long and up to 1 cm wide.

Inflorescences (Fig. 5a) elongate, racemose, with 20 or many more flowers, erect and usually overtopping the leaves; inflorescence stalks exceeding 10 cm, but reaching up to 2 m or perhaps more, erect and stiff with septate pith, partly submerged; sheath of lower spathe 3.5–10 cm or more long; petiole of lower spathe 4–18 cm long; blade of lower spathe elongate-sagittate, 5–10.5 cm long and 0.25–1.3 cm wide, basal lobes elongate, ± 1.2 cm long or blade narrowly triangular with slightly auriculate base, up to 12 cm long and 1 cm wide near base; upper spathe ± 4.5 cm long, membranous, bearing a short filamentous appendage at apex; flowerless part of the rachis very short; flower bearing part of the rachis 8–12.5 cm long; pedicels 0.5–2.0 cm long (mostly long), the lower ones sometimes forked.

Perianth segments up to ± 2 cm long, ± 1.1 cm wide, coiling around the fruit.

Stamens of 2 kinds (Fig. 5b); 5 with simple filaments and yellow, 3.5–6.0 mm long anthers; 1 with appendaged filament and blue, 6–10 mm long anther, at least 1/4 longer than the yellow ones. Fruit ellipsoid, ± 10 mm long, ± 5 mm diameter; style 8–10 mm long (Fig. 5c); stigma 3-lobed, scarcely wider than the style.

Seeds ovoid to barrel-shaped, 0.7–0.8 mm long, 0.3–0.4 mm diameter, with 10, longitudinal, narrow wings (Fig. 5a).

Diagnostic Features

Inflorescence elongate, racemose with 20 or more flowers, erect and overtopping the leaves; pedicels mostly 1–2 cm long, the lower ones sometimes branched; leaves elongate-sagittate or elongate-hastate; blue anthers 6–10 mm long; style more than 8 mm long.

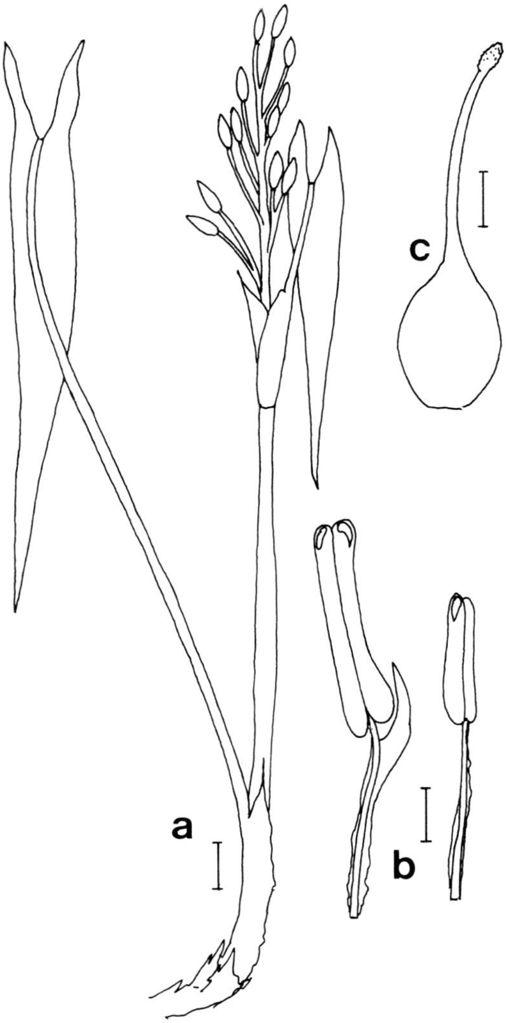

Figure 5. Monochoria elata. — a. Habit (scale = 3 cm). — b. Stamens (scale = 1 mm). — c. Young ovary (scale = 2 mm). — Drawn from *McClure 7650* (E).

Distribution

Monochoria elata extends from central Burma southwards to northern Malaya (Kedah), reaching eastwards through Thailand and Hainan Dao to Guangzhou (Canton) in S. China (Map 1).

Notes

This species is poorly known; we have seen relatively little herbarium material and Ridley's original description is poor and probably partly inaccurate. Ridley mentions its great height, 6–8 feet (±2–2.5 m) but allows it petioles of only 24 inches (*c*. 62 cm) long. Without any comment Backer in Flora Malesiana when he reduces it to a variety of *M. hastata* (above) also reduces its height to 'ca. $1\frac{1}{2}$ m'. Material from Bangkok in Thailand (*Kerr 4497*, BM) has an incomplete inflorescence stalk 185 cm long. Unfortunately leaves with petioles and rhizomes are rarely (if ever) collected.

The floral characteristics and leaf shape certainly indicate affinities to *M. hastata* but the inflorescence has long pedicels and an elongated rachis, a feature sometimes found in *M. vaginalis*. However the inflorescence remains erect and overtops the leaves which are features characteristic of *M. korsakowii*. In spite of the fact that *M. elata* combines features of three species two of which are sympatric, we find no case for considering it to be of hybrid origin.

From the limited herbarium material examined, it looks as if the leaf blades hang from more or less rigid petioles, a feature found in *M. australasica* (the resemblance was noted by Ridley) and *M. brevipetiolata*.

5. Monochoria africana (Solms-Laubach) N.E. Brown, Fl. Trop. Africa, 8: 5. 1901. ≡ *M. vaginalis* var. *δ africana* Solms-Laubach in A. de Candolle, Monog. Phaner. 4: 525, 1883. Type: Sudan, Jur Ghattas, 'Ad Seribah Ghattas territorii Djur,' 22 August 1869, *Schweinfurth 2296* (holotype: destroyed in B; isotype: K).

Annual or perhaps sometimes perennial. Stem truncate, somewhat corm-like or shortly elongate and somewhat rhizomatous.

Radical leaves either linear, up to 40 cm or more long, ephemeral or differentiated into petiole with blade; petioles erect, up to 50 cm or more long; stipules up to about 20 cm long, at apex acute to obtuse with a filamentous appendage; blades elliptic to ovate or cordate, up to ±10 cm long, ±8 cm wide.

Inflorescences elongate, raceme- or spike-like, usually not overtopping the subtending leaves, with (20–) 25–30 (–40) more or less equally spaced flowers (Fig. 6b); inflorescence stalks up to 50 cm or more long, erect and emergent; sheath of the lower spathe (3–) 3.6–4.0 (–4.2) cm long, membranous or somewhat inflated; petiole of the lower spathe (5–) 10–12 (–16) cm long, often reflexed; blade of the lower spathe (5–) 8–9 (–12)cm long, (3–) 5–7 (–10) cm wide, elliptic to ovate or more frequently cordate; upper spathe ±2.5–4.0 cm long, membranous, at least partly hidden within the lower spathe sheath, acute or obtuse at apex with a filamentous appendage; flowerless part of rachis (1.5–) 4–5 (–6) cm long; flower bearing part of rachis (6–) 9–15 (–23) cm long; pedicels up to ± 1 cm long.

Perianth segments 10–14.5 mm long, 3.5–4.5 mm or more wide, somewhat accrescent, persistent and remaining flat along the ripe fruit.

Stamens of 2 kinds; 5 with simple, up to 5 mm long filaments and yellow, (1.8–) 2–4 mm long anthers; 1 with an appendaged (Fig. 6c), up to 5.5 mm long filament and a blue, (3–) 4–5.2 mm long anther, at least 1/4 longer than the yellow ones.

Fruits ellipsoidal to ovoid, ±9 mm long, ±6 mm diameter; style 5.5–6.5 mm at maturity; stigma capitate or lobed, wider than the style (Fig. 6d) .

Seeds ellipsoidal to barrel-shaped, ±0.75 mm, ±0.5–0.6 mm diameter, with 10, narrow, longitudinal ribs or wings (Fig. 6a).

Diagnostic Features

Inflorescences erect, racemose, with 20 or more flowers, usually not overtopping the leaves; leaves usually cordate, usually at least 6 cm long and 4 cm wide; petiole of lower spathe at least 5 cm long, usually erect; seeds ellipsoidal to barrel-shaped, ± 0.75 mm long, 0.5–0.6 mm diameter with 10 longitudinal, narrow ribs or wings.

Distribution

Known from relatively few and widely scattered collections in eastern Africa (Map 5). It extends from southern Sudan southwards to Transvaal.

6. Monochoria brevipetiolata B. Verdcourt, Kirkia 1: 81, t. 9, 10. 1961. Type: Guinea Bissau, 'Portuguese Guinea, Gabú, depressoes alagadas de savana entre Pitche e Canquelifá,' 18 September 1950, *Santo 2777* (holotype: K).

Annual or perhaps sometimes perennial.

Stem truncate and somewhat corm-like or shortly elongate and somewhat rhizomatous.

Radical leaves either linear, up to 35 cm or more long, ephemeral or differentiated into petiole with blade; petioles up to 30 cm or perhaps more long; stipules up to ± 13 cm long, at apex acute with or without a filamentous appendage; blades linear or lanceolate to elliptic or ovate, 2–9 cm long, 0.8–4.0 cm wide, at base usually cuneate but occasionally cordate.

Inflorescences elongate, raceme- or spike-like, usually overtopping the subtending leaves (Fig. 7c), with (5–) 8–15 (–25) more or less equally spaced flowers; inflorescence stalks (8–) 12–25 (–30) cm long, erect and emergent; sheath of the lower spathe (2–) 2.5–3.5 (–4.5) cm long; petiole of the lower spathe (0.7–) 1–2 (–4) cm long, usually reflexed; blade of the lower spathe (2–) 4–6 (–10) cm long, (0.3–) 1–2 (–3) cm wide, lanceolate to elliptic or ovate, at base usually cuneate, rarely cordate; upper spathe ± 2 cm long, membranous, usually completely enclosed by the lower spathe, the apex acute or obtuse, the apical filament sometimes bearing a minute blade; flowerless part of the rachis (2–) 3–5 (–9) cm long; flowering part of the rachis (4–) 6–10 (–15) cm long; pedicels up to ± 0.6 cm long.

Perianth segments 9–14 mm long, up to ± 4 mm wide, somewhat accrescent, persistent and remaining flat along the ripe fruit.

Stamens of 2 kinds (Fig. 7d); 5 with simple, 2.5–3.0 mm long filaments and yellow, (1.5–) 2.5–3.5 (–4.0) mm long anthers; 1 with an appendaged, 2.5–3.0 mm long filament and a blue, 3.5–4.8 mm long anther, at least 1/4 longer than the yellow ones.

Fruits ellipsoidal to ovoid, ± 12 mm long, ± 5 mm diameter; style 3.8–5 (–6) mm long at maturity; stigma capitate, slightly wider than the style (Fig. 7b).

Seeds subglobose to barrel-shaped, 1.3–1.8 mm long, 0.6–1.2 mm wide, with 12–14, narrow, longitudinal ribs or wings (Fig. 7a).

Diagnostic Features

Inflorescences racemose, with usually less than 15 flowers, usually overtopping the leaves; leaves usually cuneate or obtuse but not cordate at base, rarely more than 6 cm long and 3 cm wide; petiole of lower spathe not more than 4 cm long

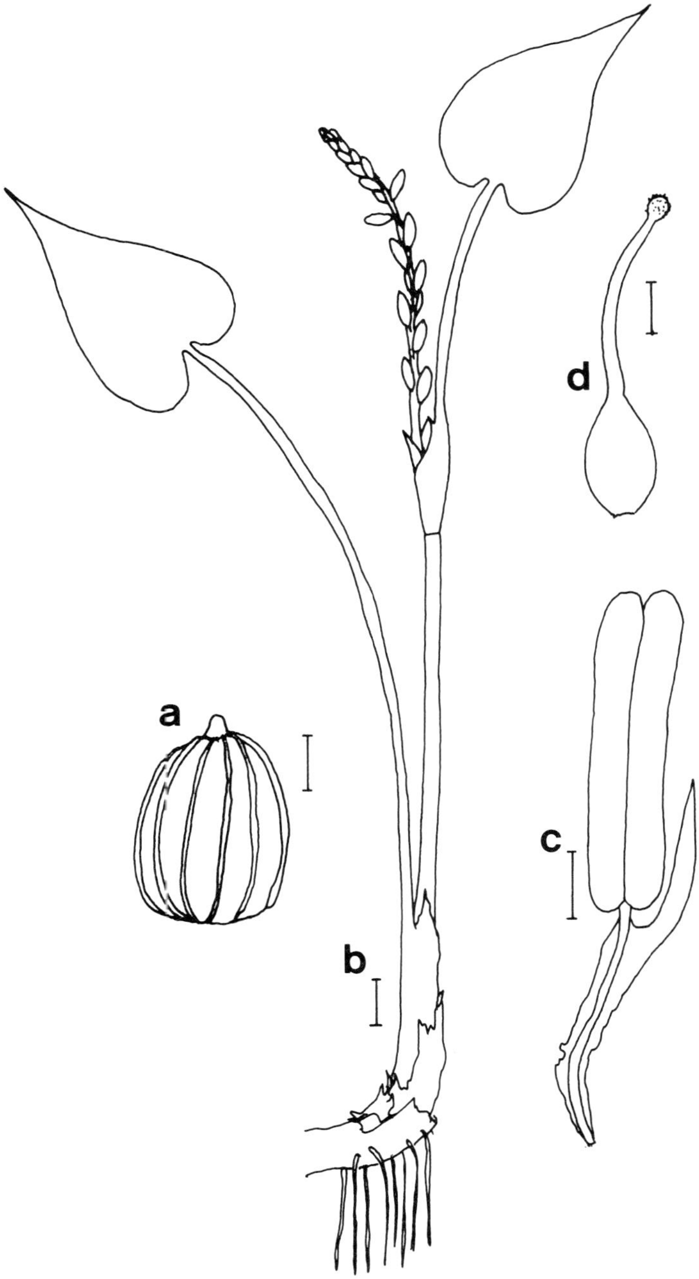

Figure 6. Monochoria africana. a. Seed (scale = 0.25 mm). — b. Habit (scale = 2 cm). — c. Stamen (scale = 1 mm). — d. Young ovary (scale = 2 mm). — Drawn from *Blackmore, Brummitt & Banda 1247* (K).

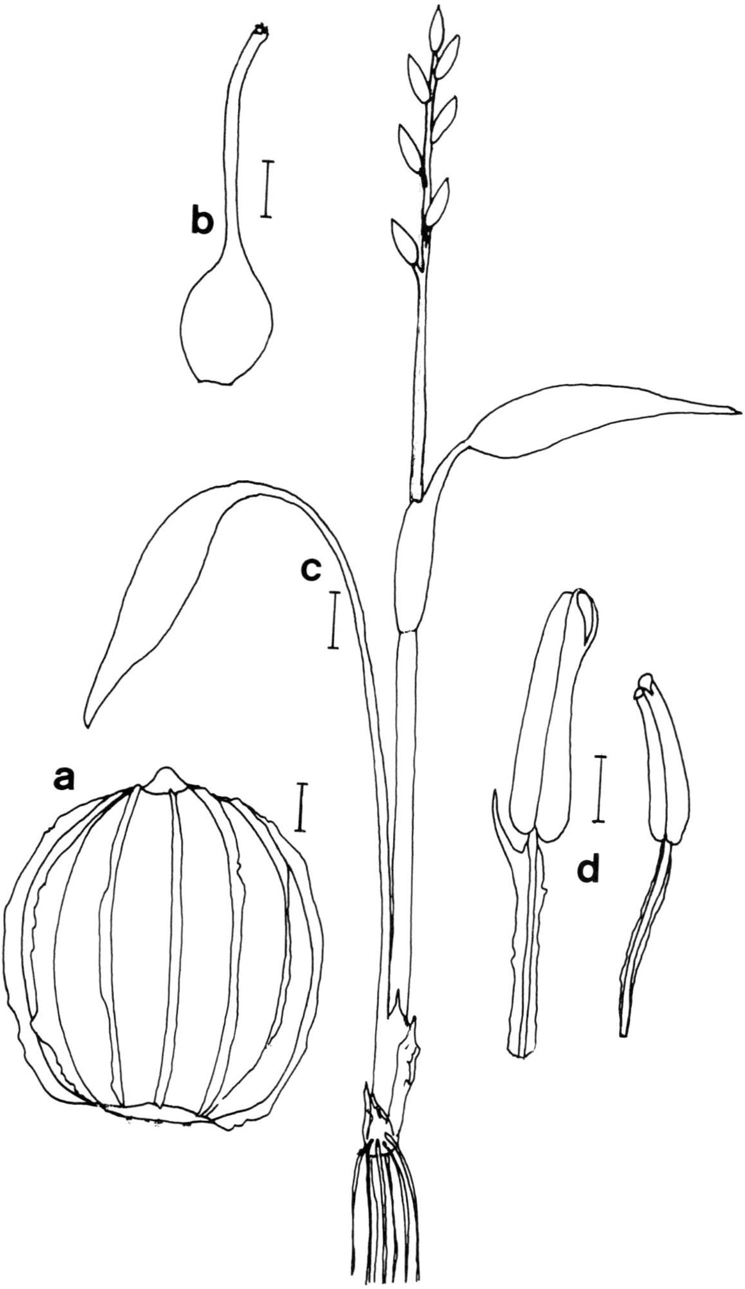

Figure 7. Monochoria brevipetiolata. — a. Seed (scale = 0.2 mm). — b. Young ovary (scale = 2 mm). — c. Habit (scale = 2 cm). — d. Stamens (scale = 1 mm). — Drawn from *Lisowski 51247* (K).

(usually much less), usually reflexed; seeds subglobose, 1.3–1.8 mm long and 0.6–1.2 mm wide with 12–14 longitudinal narrow ribs or wings.

Distribution

Confined to a relatively compact region in tropical West Africa (Map 5). It extends from about 8°N to 14°N, from Senegal to Gambia westwards to the Ivory Coast.

7. Monochoria cyanea (F. von Mueller) F. von Mueller, Fragmenta Phytographiae Australiae, 8 (60): 44. April 1873. ≡ *Limnostachys cyanea* F. von Mueller, Fragmenta Phytographiae Australiae, 1(1): 24. March 1858. Type: Australia, Northern Territory 'In terra Arnhem's Land. Leichhardt. Ad flumen Victoriae' (lectotype: chosen by Aston, Muelleria 6 (1–2): 56. May 1985 – '*Limnostachys cyanea* – Victoria River ferd Mueller' in Mueller's hand on plain blue label – K; syntypes: K, MEL).

Annual or perhaps sometimes perennial.

Stem usually truncate, occasionally somewhat rhizomatous or sometimes one of the lower internodes elongate, up to 20 cm or more long with the upper nodes congested (Fig. 1e & 8b).

Radical leaves either linear, up to 30 cm or more long and 0.4 cm or less wide, ephemeral and submerged, or differentiated into petiole with blade; petioles up to 30 cm or perhaps more long, sometimes inflated and floating otherwise erect and emergent; stipules up to 18 cm long but usually much less, with or without terminal filaments; blades lanceolate to ovate, (1.5–) 4–6 (–14) cm long, (1–) 2–4 (–6) cm wide, floating or emergent, at base usually cuneate but sometimes truncate or rarely cordate, at apex acute.

Inflorescences elongate, raceme- or spike-like, with 1–16 more or less equally spaced flowers, often overtopping the leaves; inflorescence stalks up to 50 cm or more long, either stiff, erect and emergent or flaccid, inflated and floating (Fig. 1e); sheath of the lower spathe (0.8–) 2–3 (–5) cm long, mostly erect and emergent but sometimes floating; petiole of the lower spathe (1.2–) 5–6 (–15) cm long, (0.5–) 2–3 (–7) cm wide, lanceolate to ovate at base cuneate to truncate or sometimes cordate; internode between lower and upper spathes up to 0.5 cm long; upper spathe (1.5–) 2–4 (–5) cm long, membranous, at apex acute or bearing a filamentous appendage; flowerless part of rachis (0.5–) 1.5–2.5 (–4.5) cm long; flower bearing part of rachis (1–) 3–5 (–8.5) cm long; pedicels up to ±1 cm long.

Perianth segments (10–) 12–16 (–20) mm long; 4–6 mm or more wide, somewhat accrescent, persistent and remaining flat along the ripe fruit.

Stamens all alike; filaments simple (Fig. 8c), without appendages, (1.8–) 3–4 (–6) mm long; anthers yellow, (1.2–) 1.6–2.0 (–3.4) mm long, not one obviously larger than the others.

Fruits ellipsoidal, 10–14 mm long, 4–8.5 mm diameter; style (3–) 4 (–5.8) mm long; stigma capitate or 3-lobed (Fig. 8d), distinctly wider than the style.

Seeds ovoid to barrel-shaped, 0.8–1.1 mm long, 0.6–0.7 mm diameter, dark brown to almost black, with 8–10, longitudinal, narrow ribs or wings (Fig. 8a).

Figure 8. Monochoria cyanea. — a. Seed (scale = 0.2 mm). — b. Habit (scale = 1 cm). — c. Stamen (scale = 1 mm). — d. Young ovary (scale = 1 mm). — a, c & d, drawn from *George 15169* (K); b, drawn from *Aston 1983* (K).

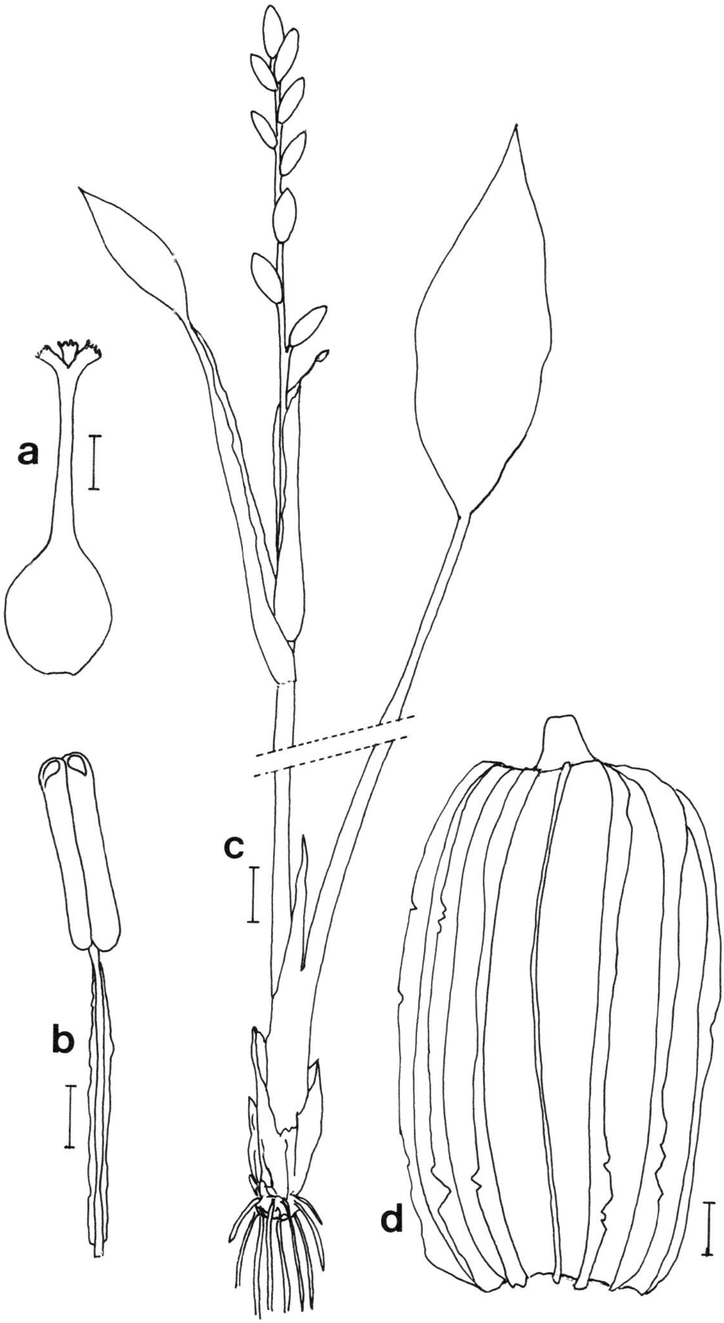

Figure 9. Monochoria australasica. — a. Young ovary (scale = 2 mm). — b. Stamen (scale = 1 mm). — c. Habit (scale = 2 cm). — d. Seed (scale = 0.2 mm). — a, b & c, drawn from *Lazarides & Adams 258* (K); d, drawn from *Allen 81* (K).

Diagnostic Features

Stamens all alike; filaments simple (without appendages); anthers yellow; stigma lobed or capitate and wider than the style; inflorescences racemose, remaining erect in fruit.

Like *M. australasica* but blade of spathe with a petiole usually exceeding 12 mm; style not exceeding 5.8 mm long; the non-flowering part of the rachis rarely more than 40 mm long; leaves usually cordate at base.

Distribution

Widely distributed in northern and western Australia (Map 6). Extending from the mouth of the Fitzroy River in Western Australia eastwards to Queensland and southwards to northeastern New South Wales; it is absent from the northernmost parts of Queensland and Northern Territory.

Ecology

According to Sainty & Jacobs (1981) it is found in stationary or slowly-flowing water to about 40 cm deep in swamps, lagoons or dams. The plants root in mud but can grow out from the margins floating over the water. When in permanent water this species usually behaves as a perennial but in ephemeral waterholes or at the south of its range (New South Wales) it behaves as an annual. It is not considered to be of any economic importance.

8. Monochoria australasica H.N. Ridley, J. Straits Branch Royal Asiatic Society, 79: 100. September 1918. Type: Australia, Northern Territory, 'near Darwin, 8. C.E.F. Allen, Nov. 1914'. Ridley's '8' is '81' and 'Nov.' should be 'February' – he misread 'II' as 'eleven', it is roman 'two' (holotype: near Darwin, February 1914, *C.E.F. Allen 81* – K; isotype: NSW).

Annual or perhaps sometimes perennial.

Stem usually truncate, occasionally somewhat rhizomatous or sometimes one of the lower internodes elongate, up to 20 cm or more long with the upper internodes congested (Fig. 1e).

Radical leaves either linear, up to 50 cm long and 0.6 cm wide, ephemeral and submerged, or differentiated into petiole with blade; petioles up to 30 cm or more long, sometimes floating or submerged and inflated or erect and emergent; stipules up to 26 cm long but usually less, with or without terminal filaments; blades emergent or floating, linear to ovate, (2–) 6–8 (–12) cm long, (0.2–) 1.8–2.2 (–5) cm wide, usually cuneate at base but sometimes somewhat cordate, apex acute.

Inflorescences elongate, raceme- or spike-like, with 2–14, more or less equally-spaced flowers (Fig. 9c); inflorescence stalks up to 30 cm or more long, either stiff, erect and emergent or flaccid, inflated and floating; sheath of the lower spathe (2–) 6–8 (–10) cm long, sometimes inflated and floating otherwise emergent and erect; petiole of the lower spathe (0–) 0.4–0.6 (–0.8) cm long; blade of the lower spathe linear or lanceolate to ovate, cuneate or rarely cordate at base, (2.2–) 6–8 (–17) cm long, (0.2–) 1.6–2.0 (–6) cm wide; internode between lower and upper spathes up to 1.5 cm long; upper spathe (2–)5–6(–8) cm long, membranous, apex acute or bearing a filamentous appendage, sometimes widened and slightly blade-like at the tip; flowerless part of the rachis (3–) 5–10 (–17) cm long; flower-bearing part of the rachis (2–) 7–9 (–15) cm long; pedicels 0–0.3 (–0.5) mm long.

Perianth segments (12–) 18–22 (–26) mm long, 7–9 mm or more wide, somewhat accrescent, persistent and remaining flat along the ripe fruit.

Stamens all alike; filaments simple (Fig. 9b), without appendages, (1.2–) 2–3.5 (–7) mm long; anthers yellow, (2–) 3.5–4.5 (–5.6) mm long, not one obviously larger than the others.

Fruits ellipsoidal, 13–20 mm long, 5–8 mm diameter; style (5.5–) 7–8) (–10 mm long; stigma capitate or 3-lobed (Fig. 9a), distinctly wider than the style.

Seeds ovoid to barrel-shaped (0.8-) 1.0–1.2 (-2.0) mm long, 0.65–0.75 (-1.1) mm diameter, dark brown to almost black, with 8 or 10, longitudinal, narrow ribs or wings (Fig. 9d).

Diagnostic Features

Stamens all alike; filaments simple; anthers yellow; stigma lobed or capitate and wider than the style; inflorescence racemose, remaining erect in fruit.

Like *M. cyanea* but blade of spathe sessile or with a petiole up to 8 mm long; the non-flowering part of the rachis usually more than 40 mm long; leaves usually cuneate at base.

Distribution

It is confined to the northernmost parts of Australia (Map 7), in northwestern Arnhem Land and in the Cape Peninsula of Queensland. It is mostly allopatric to *M. cyanea* but is partly sympatric with *M. vaginalis*.

Notes

Little is known about *M. australasica*. Its ecology is probably like that of *M. cyanea*.

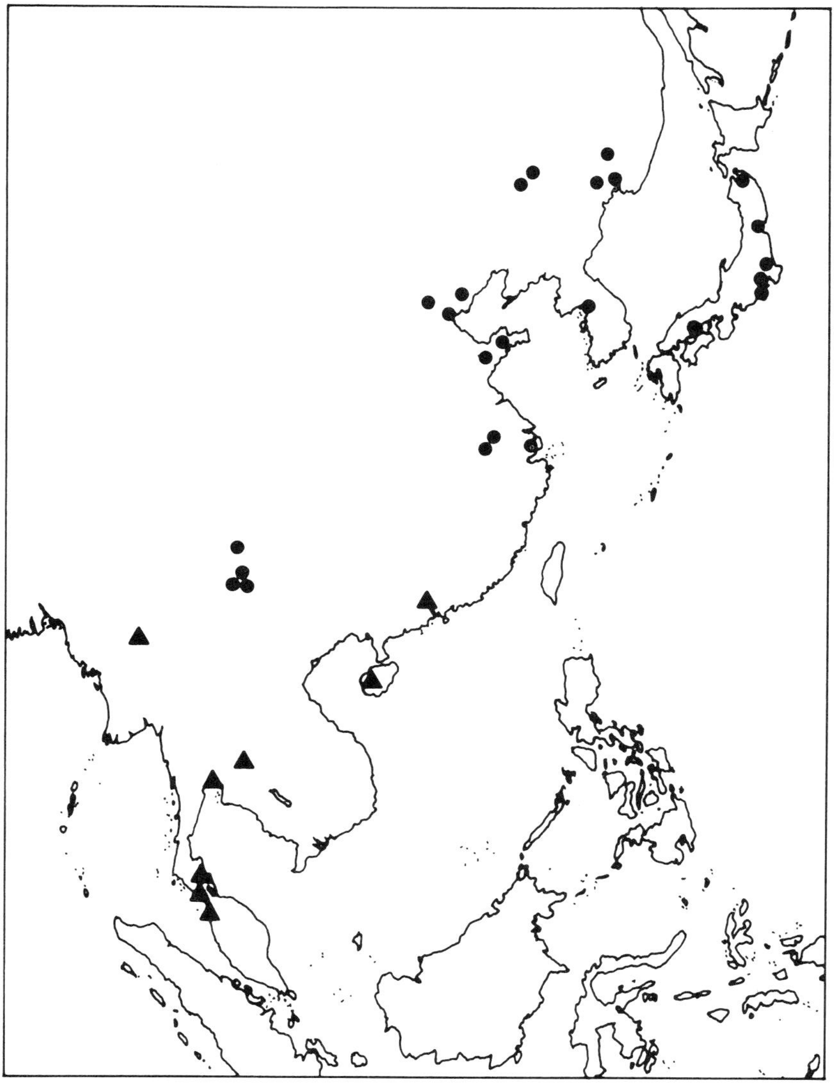

Map 1. Distribution of *Monochoria korsakowii* (circles) and *M. elata* (triangles).

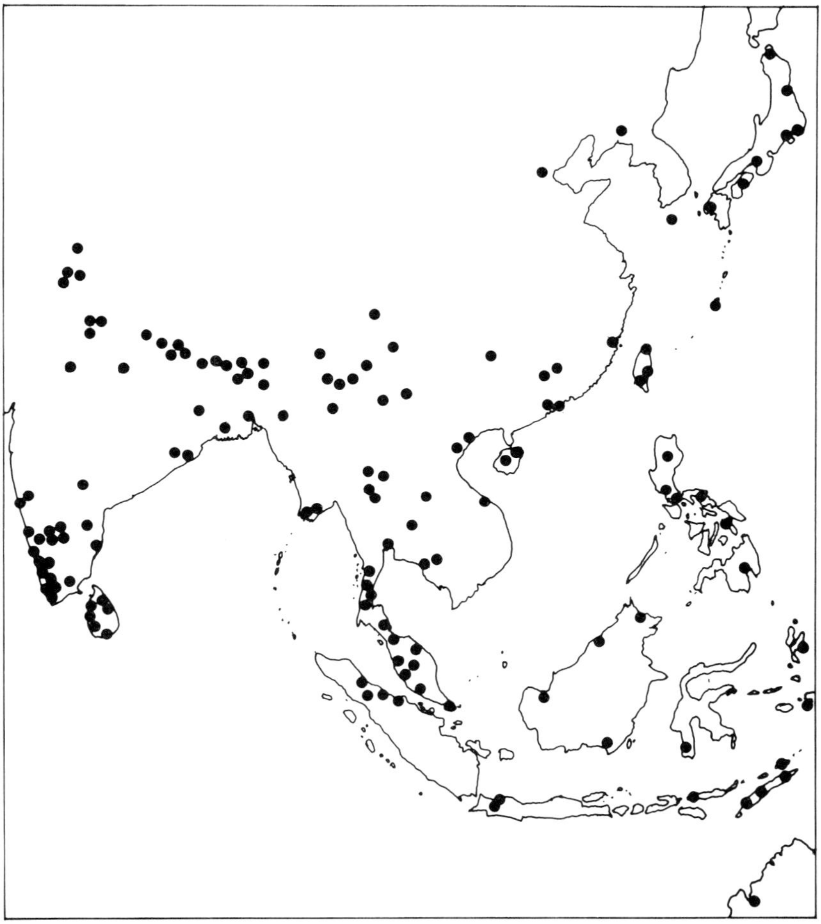

Map 2. Distribution of *Monochoria vaginalis* in SE Asia, India and western Indonesia (see also maps 4 & 5).

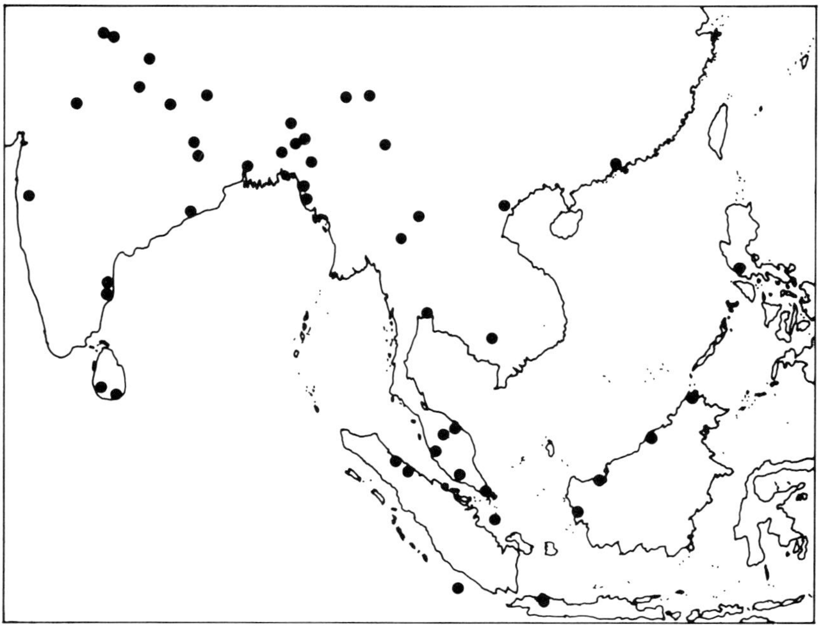

Map 3. Distribution of *Monochoria hastata* (see also map 4).

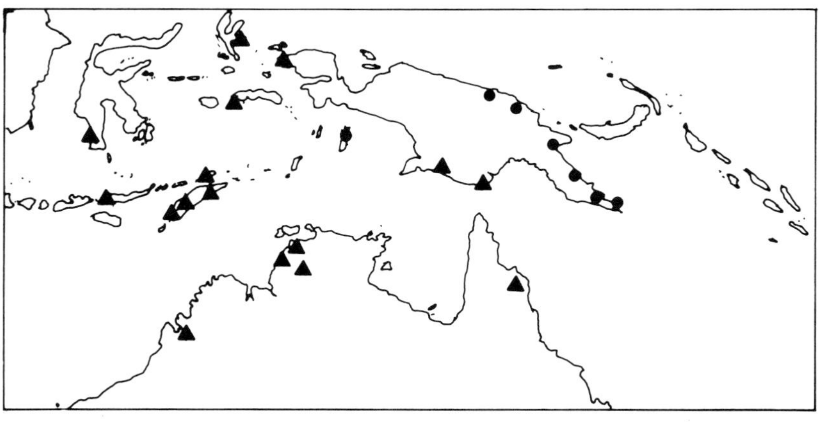

Map 4. Distribution of *Monochoria hastata* (circles) and *M. vaginalis* (triangles); the latter occurs in Fiji – beyond this map.

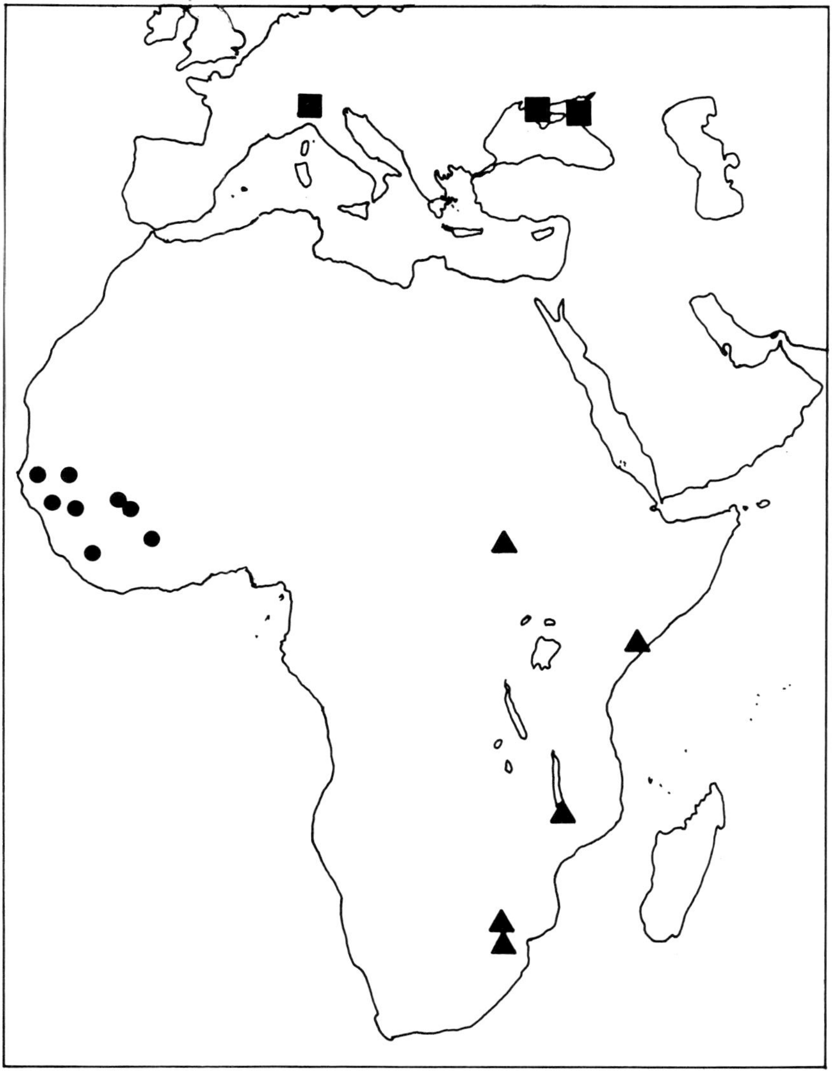

Map 5. Distribution of *Monochoria africana* (triangles), *M. brevipetiolata* (circles) and introduced localities of *M. vaginalis* (squares – I have not seen herbarium material from the Ukraine; the localities in California and Hawaii are beyond this map.)

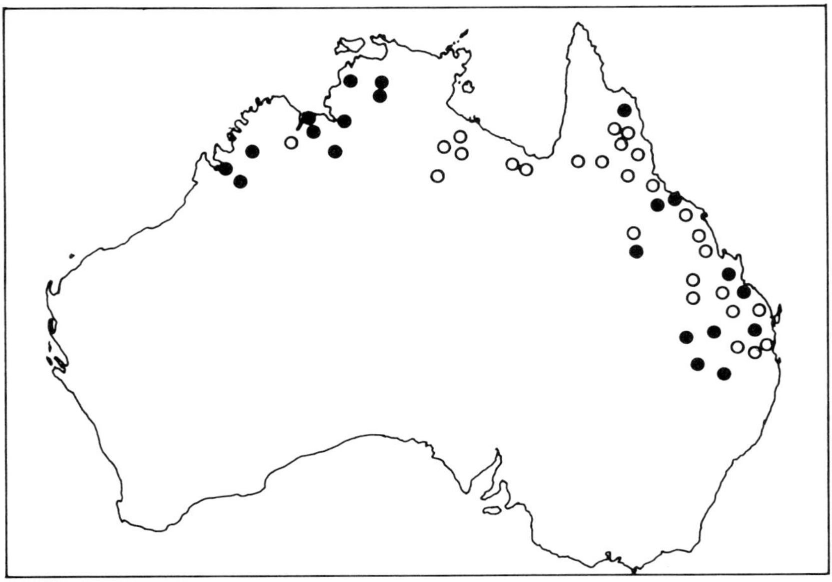

Map 6. Distribution of *Monochoria cyanea* (black circles – based on herbarium material seen by me, hollow circles – based on Aston 1985).

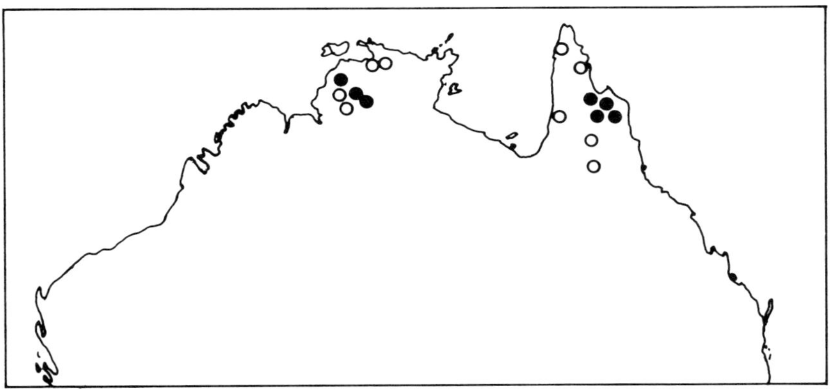

Map 7. Distribution of *Monochoria australasica* (black circles – based on herbarium material seen by me, hollow circles – based on Aston 1985).

ACKNOWLEDGEMENTS

I would like to thank the Fonds National Suisse de la Recherche Scientifique (nr. 3.390–0.83) and the Council of the Georges and Antoine Claraz-Schenkung for financial support. I would like to thank Marc Nicholls for data collecting and processing and Katy Papadopoulos for typing and administrative work. I would also like to thank the following who helped with plant material or specialized knowledge: H.J. Beentje (Nairobi), D.V. Geltmann (Leningrad), Yasuro Kadona (Kobe), Hiroko Kobatake (Okayama), Indu Mehta (Kota, Rajasthan), A.M. Namboodiri (Trivandrum), Victor Ramey (Gainesville) and Norman Robson (London). I also thank the curators of the herbaria listed on p. 149 for allowing me to use their material.

LITERATURE CITED

Aston, H.I. 1985. *Monochoria cyanea* and *M. australasica* (Pontederiaceae) in Australia. Muelleria 6(1–2): 51–57.

Backer, C.A. 1981. *Monochoria. In* C.G.G.T. van Steenis [ed.], Flora Malesiana, Ser. 1, 4(3): 255–259.

Banerji, I., and S. Haldar. 1942. A contribution to the morphology and cytology of *Monochoria hastaefolia* Presl. Proc. Indian Acad. Sci. Ser. B, 16: 91–106.

Banerji, M. 1974. Cytological studies on some members of the family Pontederiaceae. Cytologia 39: 483–491.

Candolle, A. de, and C. de Candolle. 1883. Monographiae Phanerogamarum (Paris) 4: 522–525.

Christopher, J. 1983. Cytology of *Monochoria vaginalis* complex Presl. Cytologia 48: 627–631.

Degener, O., and I. Degener. 1960. Family 63. Pontederiaceae, Genus *Monochoria*. Flora Hawaiiensis, Book 6.

Eckenwalder, J.E., and S.C.H. Barrett. 1986. Phylogenetic systematics of Pontederiaceae. Syst. Bot. 11(3): 373–391.

Ghosh, D.K., and G.N. Bhattacharya. 1981. Study of cytotypes in *Eichhornia* and *Monochoria*. Proc. 68th Indian Sci. Cong., Sect. IV, Botany, Part 3: 74.

Glück, H. 1919. Blatt- und blütenmorphologische Studien. Verlag Gustav Fischer, Stuttgart. 696 p.

Hsu, C.-C. 1967. Preliminary chromosome studies on the vascular plants of Taiwan (1). Taiwania 13: 117–130.

Juliano, B.J. 1931. Morphological studies in the flower of *Monochoria vaginalis*. Phill. Agri. 17: 176 – not seen, *fide* Banerji and Haldar 1942.

Krishnappa, D.G. 1971. Cytological studies in some aquatic angiosperms. Proc. Indian Acad. Sci. Sect. B 73(4): 197–285.

Laktionov, B.I. 1972. Novȳr dlya flory Ukrainȳ vid *Monochoria korsakowii* Regel & Maack. Bot. Žurn. (Moscow & Leningrad) 57(10): 1332–1333.

Lowden, R.M. 1973. Revision of the genus *Pontederia* L. Rhodora 75(803): 426–487.

Majumdar, A. 1953. An investigation on the cytology of *Monochoria hastaefolia* Presl. Caryologia 5(3): 306–312, pl. XXIII.

Mason, H.L. 1957. A Flora of the marshes of California. Univ. Calif. Press, Berkeley and Los Angeles. 877 p.

Mehra, P.N., and T.K. Pandita. 1978. *In* Á. Löve [ed.], IOPB chromosome number reports, LXI. Taxon 27: 375–392.

Morinaga, T., and E. Fukushima. 1931. Chromosome numbers of cultivated plants III. Bot. Mag. (Tokyo). 45(531): 140–145.

Ono, T. 1928. Embryologische Studien an einigen Pontederiaceen. Sci. Rep. Tôhoku Imp. Univ., Ser. 4, Biol. 3: 405–415.

Regel, W. 1862. *Monochoria korsakowii* Rgl. et Maack. Pontederiaceae. Gartenflora 11: 312–313, tab. 374.
Ridley, H.N. 1918. New and rare Malayan plants. J. Straits Branch Royal Asiatic Soc. 79: 99–100.
Sainty, G.R., and S.W.L. Jacobs. 1981. Water plants of New South Wales. Water Resources, New South Wales. 550 p.
Sarkar, A.K., M. Chakraverty, S.K. Das, C.R. Pal, and D. Hazara, 1980. *In* Á. Löve [ed.], IOPB chromosome number reports, LXVII. Taxon 29(2–3): 358.
Schultes, J.A., and J.H. Schultes. 1830. Systema vegetabilium 7(2): 1137–1148. J.G. Cotta, Stuttgart.
Schwartz, O. 1926. Anatomische, morphologische und systematische Untersuchungen über die Pontederiaceen. Beih. Bot. Centralbl. 42(3): 263–320.
Schwartz, O. 1927. Zur Systematik und Geographie der Pontederiaceen. Bot. Jahrb. Syst. 61(2–3); Beiblatt 139: 28–50.
Sharma, A., and A.K. Sarkar. 1967–1968. Annual report of cytogenetics laboratory, department of botany, Univ. Calcutta Res. Bull. 2: 38–48; not seen, *fide* D. Moore in Regnum Veg. 90, 1973.
Simpson, M.G. 1984. Systematics and pollen ultrastructure of the Pontederiaceae. Amer. J. Bot. 71(5/2): 187 (abstract: I also received a long unpublished manuscript).
Singh, V. 1962. Vascular anatomy of the flower of some species of the Pontederiaceae. Proc. Indian Acad. Sci. 56: 339–353.
Sokolovskaya, A.P. 1966. Geograficheskoe rasprostranenie poliploidnych vidov rasteniy (Issledovanie flora Primorskogo Kraya). Vestnik Leningradsk. Univ. Ser. Biol. 1966, 3:93–106.
Verdcourt, B. 1961. The genus *Monochoria* Presl (Pontederiaceae) in Africa. Kirkia 1:80–83, Tab. VII–XII.
Vuille, F.-L. 1985. *Monochoria korsakowii* Regel & Haack (Pontederiaceae): a new plant to the ricefields of N. Italy. Bot. Helvetica 95(2): 323–324.

16. STUDIES IN THE SUBTRIBE *HABENARIINAE* (ORCHIDACEAE) III. TWO NEW SPECIES OF *PERISTYLUS* FROM NEW GUINEA*

Jany Renz

Kirschblütenweg 12, CH-4059 Basel, Switzerland

ABSTRACT

Two new taxa of *Peristylus* Blume are described and illustrated: *P. cymbochilus* Renz and *P. pachyneuroides* Renz, both from Papua New Guinea. Affinities to related taxa are discussed.

Key words: Orchidaceae, *Peristylus*, new speies, *Habenaria*, Papua New Guinea.

INTRODUCTION

It has proved necessary as part of my studies in the genus *Habenaria* Willd. to also consider related genera such as *Peristylus* Blume. The latter has a wide distribution in the Far East, ranging from the NW Himalaya to E Asia, New Guinea, Northern Australia and the Pacific. These plants have been separated on generic level from *Habenaria* mainly because of the development of the stigmata, which are not located on free arms (stigmatic processes), as in *Habenaria*, but firmly adnate throughout their length as convex cushions to the basal edges of the labellum. Other specific characters in their vegetative habit, which are useful for recognizing *Peristylus* as an independent entity, have also recently been worked out by Senghas (1973) and Seidenfaden (1977). Based on these demarcations the two new species described here must be considered as being typical members of *Peristylus*.

TAXONOMY

Peristylus cymbochilus Renz, **sp. nov.** Fig. 1.

Peristylo pachyneuro (Schltr.) P.F. Hunt affinis, a quo caule dissitifoliato, foliis nervis haud incrassatis, labello cymbiformi, calcare multo breviore apice non inflato differt.

Planta gracilis, ad 40–50 cm alta. Caulis ± flexuosus, interdum etiam repens, remote 4-6-foliatus, supra in vaginulis 3–4 adpressis, acuminatis, in bracteas abeuntibus donatus. Folia herbacea, ad 6 × 2 cm, ovato-lanceolata, acuta, sessilia, basi caulem amplectentia, marginibus minutissime denticulato-ciliolata. Inflorescentia usque ad 15 cm longa, ± laxe pluriflora. Bracteae lanceolatae, acuminatae, glabrae, ovarium aequantes vel superantes. Flores minuti, glabri, virides. Sepalum dorsale late ellipticum, obtusum, apice subcucullatum, leviter concavum, 2.3–2.5 × 1.8 mm. Sepala lateralia oblique oblonga, 2.5–3.5 × 1.5 mm. Petala oblique ovata, obtusa, quam sepala paululo breviora. Labellum cymbiforme, integrum, calcaratum, totum cum calcare 3.5 mm longum, explanato subquadratum, trinerve, basi utrinque minute denticulato-gibbosum, 1.8–2.0 mm longum, 1.8 mm latum. Calcar cylindraceum, rectum, obtusiusculum, 1 mm longum, basi 0.8 mm latum. Gynostemium fere 1 mm longum; anthera apice obtusiuscula, canalibus brevibus; stami-

* Part I: Candollea 34:357 (1979); Part II: Pl. Syst. Evol. 155: 319–332 (1987).

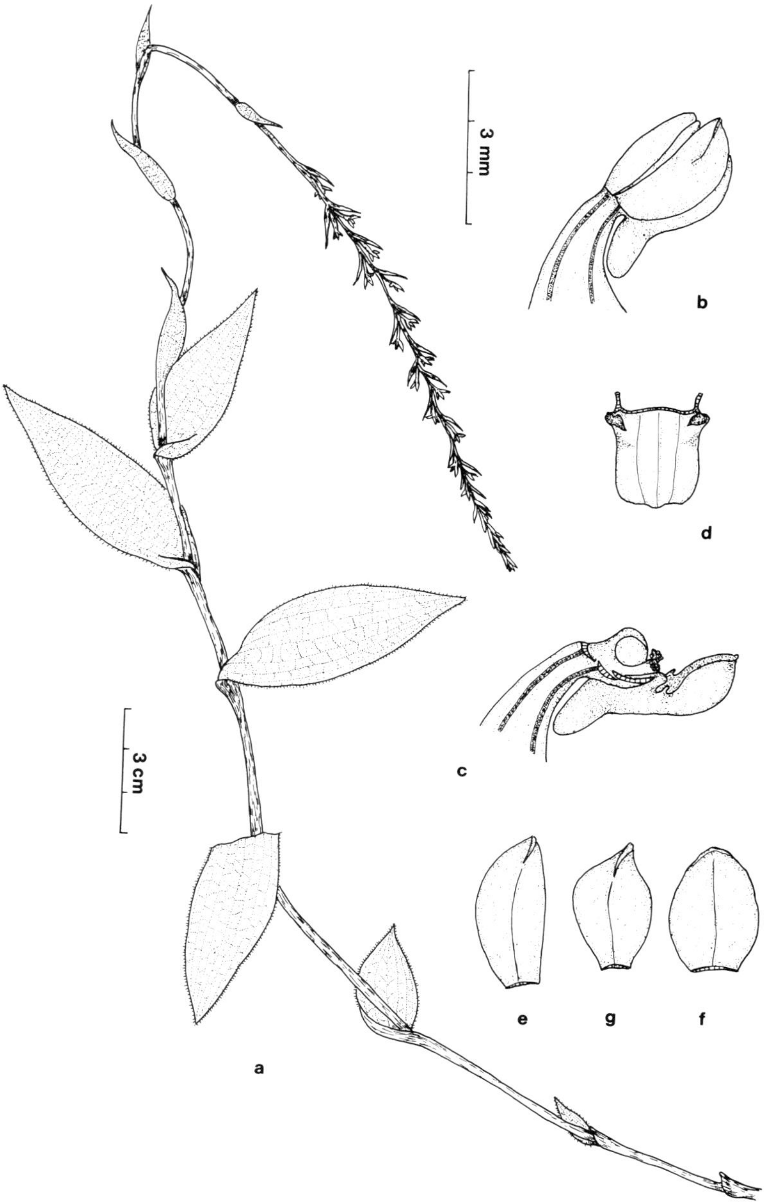

Figure 1. Peristylus cymbochilus Renz (*Renz 12241*). — a. Plant. — b. Opening bud. — c. Column and lip (lateral view). — d. Lip, spread out. — e. Lateral sepal. — f. Dorsal sepal. — g. Petal. — a, with 3 cm scale; b–g, with 3 mm scale.

nodiis oblongis; stigmatibus cum staminodiis et margine basilari labelli adnatis. Ovarium cylindraceum, apice breviter rostratum, 8–11 mm longum.

Plant slender, with a sometimes flexuous, creeping stem, bearing 4–6 evenly spaced leaves and in the upper part a few adpressed, acuminate sheaths. Leaves sessile, amplexicaul, ovate-lanceolate, acute, minutely denticulate-ciliate. Inflorescence narrowly cylindrical, lax, many-flowered, up to 15 cm long. Bracts lanceolate, equalling ovary. Flowers minute, greenish to brownish-green. Dorsal sepal broadly elliptic, slightly concave, up to 2.5 mm long, lateral sepals obliquely oblong, slightly longer and narrower than the dorsal sepal. Petals obliquely ovate. Lip shallowly boat-shaped, entire, when flattened subquadrate, trinerved, with small marginal thickenings near the base on each side, c. 2 mm long and wide. Spur short and straight, 1 mm long. Column c. 1 mm high. Ovary shortly rostrate.

Type: PAPUA NEW GUINEA: Western Highlands Province, S slopes of Mt Hagen above Tomba, 2600–2800 m, rain-forest, 13 Apr. 1978, *J. Renz 12241* (holotype Hb. Renz. Basel).

P. cymbochilus has some peculiar features, but is in habit very similar to *P. pachyneurus* (Schltr.) P.F. Hunt, a widespread plant in the montane forest of New Guinea. It differs from the latter in the arrangement of the rather smooth leaves without clearly prominent veins on the upper side. *P. pachyneurus* also shows some variability in the arrangement of the leaves, which may be clustered near the base of the stem, even prostrate, or condensed in the lower part of the stem. Furthermore the boat-shaped lip is unusual for *Peristylus* and the short, straight, non-inflated spur distinguishes *P. cymbochilus* from *P. pachyneurus*, which has a spur at least three times as long with a sometimes globular inflated tip. The transverse crest on the lip in front of the mouth of the spur, characteristic for *P. pachyneurus*, is absent in *P. cymbochilus*.

J.J. Smith (1914, 1915) described from Western New Guinea (Irian Jaya) a close relative of *P. pachyneurus* as *P. ciliolatus*. In my opinion this plant represents merely an aberrant development of *P. pachyneurus*.

Peristylus pachyneuroides Renz, **sp. nov.** Fig. 2.

Habitu similis *Peristylo pachyneuro* (Schltr.) P.F. Hunt, sed statura strictiore, foliis 2–3, late ovatis, prope basin caulis spisse approximatis, floribus majoribus, labello trilobato bene distinguendum.

Planta 15–30 cm alta. Caulis strictus vel substrictus, prope basin foliatus, infra vaginis 2–3 confertis, arcte amplectentibus, supra vaginis 3–5 lanceolatis, adpressis donatus. Folia approximata, erecto-patentia, ovata, acuta, plana vel conduplicata, basi amplexicaulia, 2–4 cm longa, ad 2.5 cm lata. Inflorescentia anguste cylindrica, densiuscula, usque ad 20 cm longa. Bracteae lanceolatae, ovarium superantes. Flores parvi, glabri, luteo-virides. Sepala lateralia oblique ovata, obtusa, deflexa, apice minute cucullata, 3–4 × 2–2.5 mm; sepalum dorsale late ovatum vel suborbiculare, concavum, 4 × 3.5 mm. Petala sepalo dorsali aequilonga, ovato-triangula, 3 mm lata. Labellum trilobatum, lobis lateralibus ± deflexis, ligulatis, obtusis, ± 3 mm longis, 1 mm latis, lobo intermedio triangulo, acuto, porrecto, lateralibus breviore. Calcar leviter curvatum, cylindricum, dimidio apicali inflatum, apice obtusum, ad 5(–6) mm longum. Gynostemium obtusum, c. 1 mm longum; canalibus antherarum brevissimis; rostelli lobo intermedio humile, triangulo; staminodiis bene evolutis. Ovarium glabrum, subcurvatum, breviter rostratum, 8–12 mm longum.

Small, rather rigid plant with 2–3 short leaves at the base of the stem. Leaves close together, amplexicaul, ovate, up to 4 × 2.5 cm. Inflorescence narrowly cylindrical, rather dense. Bracts lanceolate, exceeding ovary. Flowers small,

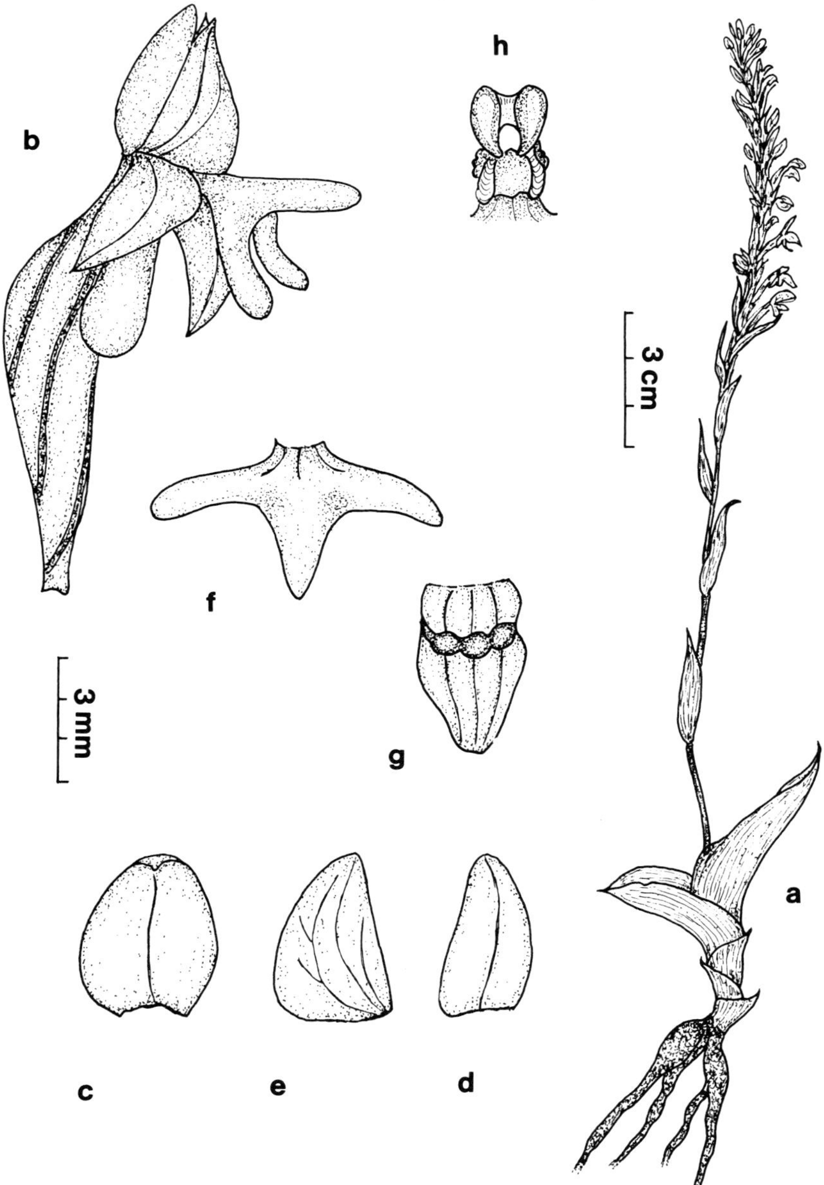

Figure 2. Peristylus pachyneuroides Renz (*Cruttwell 1078*). — a. Plant. — b. Flower. — c. Dorsal sepal. — d. Lateral sepal. — e. Petal. — f. Lip. — h. Column. *Peristylus pachyneurus* (Schltr.) P.F. Hunt. — g. Lip. — a with 3 cm scale; b–h, with 3 mm scale.

yellowish-green. Sepals ovate, up to 4 mm long. Petals equal in length, ovate-triangular, 3 mm wide. Lip 3-lobed, with lateral lobes ligulate and deflexed, 3 × 1 mm, and a somewhat shorter middle lobe. Spur cylindrical, slightly swollen in apical half, apex obtuse, c. 5 mm long. Column with very short anther-canals and well developed staminodes. Ovary glabrous, shortly beaked.

Type: PAPUA NEW GUINEA: Eastern Highlands, Mt. Aniata, near summit, in sparse grassland, 3000 m, 9.7.1959, leg. Rev. Gibson Gisi, in Hb. *N.E.G. Cruttwell no. 1078* (holotype K).

PAPUA NEW GUINEA: Western Highlands, Wabag subdistr., along Wapu River, in treefern grassland, 2900 m, 14.7.1960, *R.D. Hoogland & R. Schodde 7072* (L).

The appearance of this high mountain *Peristylus* somewhat resembles that of *P. pachyneurus*. The upright stem with 2–3 reduced basally arranged leaves are vegetatively distinctive characters. The flowers, especially the labellum, differ very clearly from those of *P. pachyneurus*: their development corresponds with some other New Guinean species of *Peristylus* described by Schlechter (1911, as *Habenaria* sect. *Peristylus*; transferred by P.F. Hunt (1971) to *Peristylus*), occurring either at high altitudes above 2000 m, such as *P. bismarckiensis* (Schltr.) P.F. Hunt, or at lower altitudes between 1100–1300 m, e.g. *P. listeroides* (Schltr.) P.F. Hunt, *P. trianae* (Schltr.) P.F. Hunt, *P. macros* (Schltr.) P.F. Hunt and *P. umbonatus* (Schltr.) P.F. Hunt. All have a somewhat similar 3-lobed lip. As Schlechter's type material is lost, further field studies will be necessary, before a definite conclusion concerning the specific status of the plants from lower altitudes can be made.

ACKNOWLEDGMENT

I am grateful to Dr Kit Tan for her assistance in drawing Figs. 1 & 2.

LITERATURE CITED

Hunt, P.F. 1971. Notes on Asiatic Orchids VI. Kew Bull. 26(1): 171–185.

Schlechter, R. 1911. Die Orchideen von Deutsch Neu Guinea. Feddes Repert. Beiheft 1: 7–10. *In* D.F. Blaxwell 1982: 60–76. The Australian Orchid Foundation. [English translation].

Seidenfaden, G. 1977. Orchid genera in Thailand V. Orchidoideae. Dansk Bot. Arkiv 31(3): 27–64.

Senghas, K. 1973. *Peristylus* Bl., pp. 251–252. *In* R. Schlechter, Die Orchideen [ed. 3], vol. 1(4). Paul Parey, Berlin and Hamburg.

Smith, J.J. 1914. Vorläufige Beschreibungen neuer papuanischer Orchideen. XII. Bull. Jard. Bot. Buitenz. ser. 2(13): 53–75.

——. 1915. Die Orchideen von Niederländisch-Neu-Guinea. Nova Guinea Bot. 12(3): 178, t. 56, 91. E.J. Brill, Leiden.

17. FLORISTIC LINKS AND ENDEMISM IN THE ARMENIAN HIGHLANDS

ELEONORA TZ. GABRIELIAN AND GEORGE M. FAJVUSH

Botanical Institute of the Academy of Sciences of the Armenian S.S.R., Yerevan 63, 375063, U.S.S.R.

ABSTRACT

The Armenian Highlands are a natural biogeographical region of Asia Minor, lying between the Anatolian and Iranian highlands. In spite of the geomorphological entity of this region, its flora is very diverse. Attempts have been made to analyse the endemism and floristic links between separate parts of the Armenian Highlands, as well as between the entire Highlands and the floras of adjacent and more remote areas.

Research showed that the entire flora of the Highlands is not a single common complex; different parts of the region have been subject to dissimilar florogenetic influences. Thus, these different parts should be treated as distinct units of floristic divisions. At the same time, insofar as the Armenian Highlands represent a geomorphological unity, there are whole complexes of taxa, characteristic of the entire Highlands, which show in some features the commonality and similarity of florogenetic processes throughout the territory and the relationships between the floras of Transcaucasia, Anatolia and Iran.

Special attention is given to the analysis of the genus *Centaurea*. It is shown that in Soviet Armenia a greater number of species and sections of this genus is concentrated compared with other regions of South-West Asia; indeed, the concentration is higher than anywhere else throughout the whole area of the genus.

Key words: Armenian Highlands (Soviet Armenia, Turkey, Iran), floristic links, phytogeography, endemism, *Centaurea*.

INTRODUCTION

The Armenian Highlands extend over parts of the U.S.S.R., Turkey and Iran and lie between the highlands of Iran and Asia Minor and between the basin of Transcaucasia and the Mesopotamian lowlands. Their area is about 400,000 km^2. The mountain systems of the Armenian Highlands consist of volcanic as well as folded mountain ridges. The climate is, on the whole, continental. Average annual precipitation is 300–800 mm in the mountains and 150–300 mm on the plains, with maximum rainfall in spring. In its vertical zonation, the vegetation of the Armenian Highlands is characteristic of the Armeno-Iranian province. The lower belts are occupied by semi-desert and phryganoid communities; in the higher belts there are steppes, meadow-steppes and meadows; "thorn-cushion belts" (traganthic vegetation) are widespread at the highest altitudes. Forests are few and, in the main, in more humid habitats (Efremov 1956; Zohary 1973).

Although geomorphologically the Armenian Highlands are an entity, their flora is very diverse, various parts of the Highlands coming under different phytochoria.

The Armenian Highlands are situated at the borders of the Boreal and Ancient Mediterranean subkingdoms of the Holarctic kingdom (Takhtajan 1978, 1986). The greater part of the Armenian Highlands lies within the Irano-Turanian region, almost all of it being in the Armeno-Iranian province. The latter includes the whole of the Armenian subprovince, together with a significant part of the Atropatenian and a small part of the Kurdo-Zagrosian subprovinces. The northernmost part of the Armenian Highlands is included in the Caucasian province of the Circumboreal region. At the same time, as may be seen from the following discussion, there exist floristic complexes characteristic of the Armenian Highlands as a whole.

THE ARMENIAN HIGHLANDS: FLORISTIC COMPOSITION AND ENDEMISM

First of all, let us examine separately the parts of the Armenian Highlands in Turkey, the U.S.S.R. and Iran. The level of floristic wealth of these parts is approximately the same; each contains some 3200–3500 species, of which some 60–70% are common to all three parts, or to two of them (Takhtajan 1954–1987; Davis 1965–1985; Rechinger 1963–1980).

Spectra of large families for all three parts are rather similar. The largest families are Asteraceae, Fabaceae, Poaceae, Brassicaceae, Caryophyllaceae, Lamiaceae, Apiaceae, Scrophulariaceae, Liliaceae and Rosaceae. The importance of any one of the families varies depending on the conditions in a particular region.

In spectra of large genera, differences are also not very great. The first places are occupied by *Astragalus*, *Centaurea*, *Allium*, *Carex*, *Silene*, *Verbascum* and *Veronica*.

Endemism of the Armenian Highlands

Generic endemism. Although generic endemism is very characteristic of the entire Irano-Turanian region, it is particularly strong in its eastern part (Kamelin 1965; Hedge and Wendelbo 1970, 1978). In the Armenian Highlands, the endemic or subendemic genera are: *Peltariopsis*, *Physoptychis*, *Pseudoanastatica*, *Pseudovesicaria* and *Takhtajaniella* (Brassicaceae); *Diplotaenia*, *Smyrniopsis*, *Stenotaenia* and *Szovitsia* (Apiaceae); *Huynhia* (Boraginaceae) and *Callicephalus* (Asteraceae).

Brassicaceae: All three species of *Peltariopsis* grow only in the territory of the Armenian Highlands; *P. grossheimii* is endemic to southern Transcaucasia, *P. planisiliqua* is found in Soviet Armenia, Bitlis and Hakkari (Turkey) and western Iran, while *P. drabicarpa* is a narrow endemic in western Iran.

The monotypic genus *Takhtajaniella* (closely allied to *Alyssum*) is a narrow endemic occurring in mountainous Karabagh.

Pseudoanastatica, near to *Clypeola* and also monotypic, is distributed in southern Transcaucasia and Iran.

The very distinct monotypic physocarpous genus *Pseudovesicaria*, previously considered endemic to the Great Caucasus, was recently found in the Lesser Caucasus in Armenia in the alpine belts of Mt. Aragats at an altitude of 3900 m. A careful search may probably reveal this interesting species in the alpine belt of the Armenian Highlands in regions adjacent to southern Transcaucasia.

The two species of the genus *Physoptychis*, *P. gnaphalodes* and *P. haussknechtii*, are mainly found in all three parts of the Armenian Highlands but extend westwards to Sivas and to the southeast to northern Iraq and northwestern Iran. These beautiful alpine species also have inflated fruits.

Apiaceae: Four genera are endemic. *Smyrniopsis*, according to Tamamschian (1962, 1968), is a monotypic genus occurring in eastern Anatolia, northernmost Iraq, western Iran and southern Transcaucasia; *Diplotaenia* is a monotypic genus found in southeastern Anatolia, northern and western Iran; *Szovitsia* is also monotypic, and occurs in northeastern Anatolia, northwestern Iran and Transcaucasia; *Stenotaenia*, with six species, is distributed in Anatolia, Iran and southern Transcaucasia.

Other families: *Huynhia* Greuter (Syn: *Aipyanthus* Steven, nom. illegit.) is a monotypic genus of the family Boraginaceae, found in Anatolia, Transcaucasia and northern Iran; *Callicephalus* (Asteraceae) occurs in northeastern Anatolia (Çoruh), Transcaucasia, Daghestan and northwestern Iran, and is also monotypic.

Endemism at the species level. Approximately 25–30% of the total flora of the Armenian Highlands is endemic. The family Fabaceae is the richest in endemic species (c. 40%), then follow Asteraceae, Brassicaceae, Caryophyllaceae, Lamiaceae, Liliaceae and Scrophulariaceae. It is noteworthy that the family Poaceae, although it occupies third place as regards total number of species, contains only 25–30% species (c. 10%) which are endemic to the Armenian Highlands. Most grasses of the Highlands are palaeoarctic and ancient mediterranean species have a wide distribution.

Astragalus is the largest genus in the Armenian Highlands, and also has the highest percentage endemism (c. 50–60%). Intensive species formation in this genus is in general peculiar to the Irano-Turanian region and to the Armenian Highlands in particular. Here there are not only many endemic species, but also endemic sections, e.g. Sects. *Acanthophace*, *Adiaspastus*, *Grammocalyx*, *Hololeuce*, *Holophyllus*, *Hymenostegis*, *Macrosemium*, *Ornithopodium*, *Pterosphaerus*, etc. Several *Astragalus* species, hitherto considered endemic to Turkey or Iran, have recently been found in southern Transcaucasia, e.g. *A. eriopodus* (Mirzoeva 1979), *A. campylosema* (Husjan 1982), *A. coarctatus* (Baloyan 1989), *A. bicolor*, *A. cylindraceus* (Arevschatian, in press). Incidentally, establishing Turkish, Iranian or Caucasian endemics must be done with great care (Grossheim 1948; Gabrielian 1961, 1975, 1978; Tamamschian 1968). Besides *Astragalus* the genera *Centaurea*, *Allium*, *Dianthus* and *Verbascum* also have many endemics.

CENTAUREA IN THE ARMENIAN HIGHLANDS

Insofar as the Soviet part of the Armenian Highlands is very closely connected with that in Anatolia, let us continue the analysis performed by Wagenitz (1986) for southwest Asia. Although Wagenitz rightly noted that the method of analysis he used for the distribution of species was rather crude, we have also made use of it in order that our data can be compared with his.

For convenience Wagenitz made use of the grid system used by Davis in *Flora of Turkey* (1965), dividing the entire area into squares each occupying 2° of latitude and longitude.

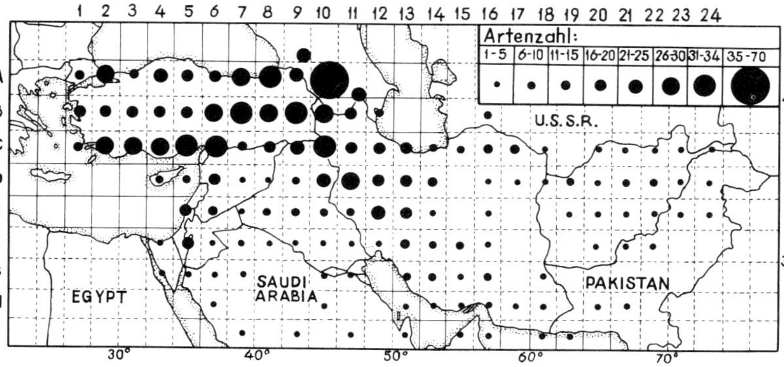

Figure 1. Centaurea in S.W. Asia: species numbers (Artenzahl) per grid square. After Wagenitz (1986), with additions by Gabrielian & Fajvush.

According to Wagenitz (1975, 1986), a great variety of species and sections of the genus *Centaurea* is concentrated in eastern Anatolia and especially where the three countries — Iran, Iraq and Turkey — meet, where 35 species from 17 sections grow. In the parts of eastern Anatolia bordering on Transcaucasia 18–34 species from 13–19 sections are found. Wagenitz predicted that at least as high a concentration of species and sections should also exist in Transcaucasia and nowhere else in the entire area of the genus *Centaurea*. However, the data obtained by us greatly exceed all our expectations. On the tiny territory of Soviet Armenia, occupying less than one grid square, there occur some 70 species of *Centaurea sensu lato* from 25 sections (Fig. 1). On the adjacent territory of Nakhichevan A.S.S.R., 30 species from 20 sections are known. There are even fewer species in southwestern Transcaucasia (Georgia) and eastern Transcaucasia (Azerbaijan). In both areas (Fig. 2) there was the same number — 24 species from 15 sections (naturally the actual species are different).

Similar patterns are shown when the entire flora of these areas is analysed. Estimates of the total number of vascular plant species for the three Transcaucasian republics show that a greater concentration of species (more than 3200) exists in Soviet Armenia (29,965 km^2) in comparison with fairly similar but smaller concentrations in Georgia (4200 species on 76,000 km^2) and Azerbaijan (4000 species on 85,700 km^2). For Soviet Armenia, the ratio of species to total area is unusually high.

Narrow Endemics in Centaurea

Concerning the distribution of narrow endemics in *Centaurea*, according to Wagenitz (1986, Fig. 4 and pp. 13–15), the number of such species is highest in the Cilician Taurus (6 species) and in the adjacent region (5) and also in eastern Anatolia and the Zagros mountains (5 species each). In southern Transcaucasia, however, there are at least 12 (perhaps even more) narrow endemics. Of these, 6 are isolated species without close allies. The rest, although having close relatives, nevertheless have clear distinctive characters. For all endemic species, there are many recent collections.

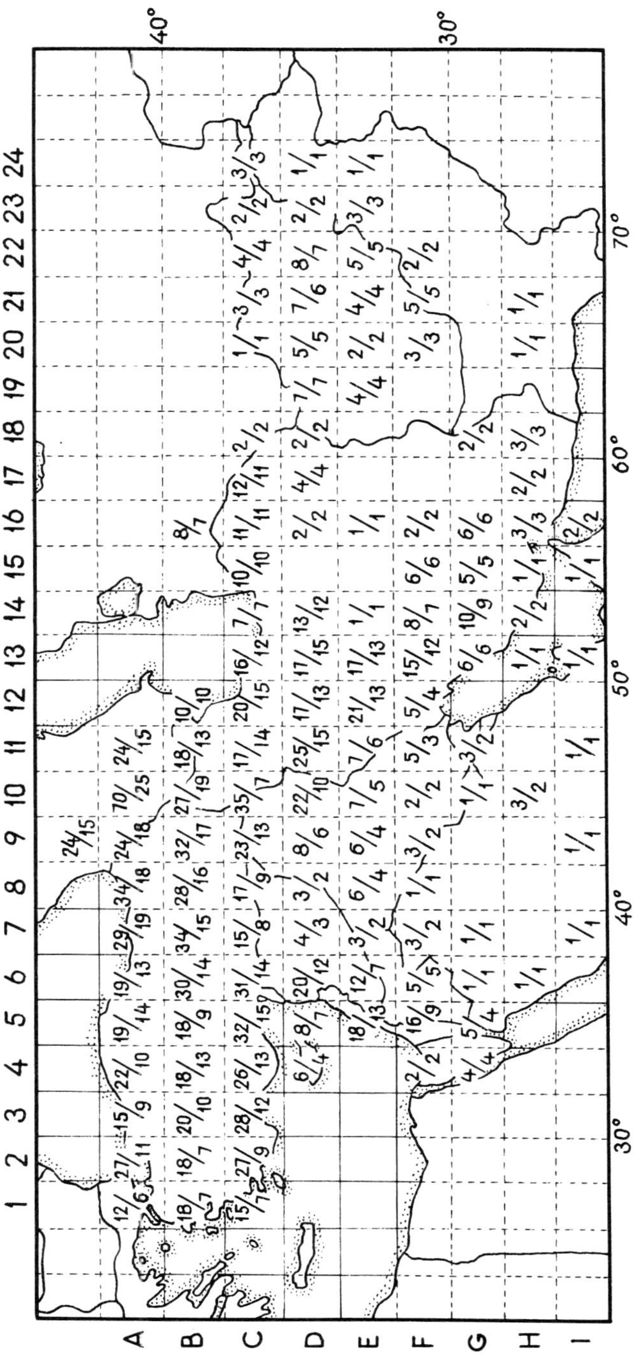

Figure 2. Centaurea in S.W. Asia: number of species/sections per square. After Wagenitz (1986), with additions by Gabrielian & Fajvush.

Species Diversity according to Section

In Southern Transcaucasia the sections of *Centaurea* with the highest number of species are *Acrocentron* and *Cyanus* (7 species each) and *Acrolophus* and *Psephellus*, each having 6 species. Then come *Xanthopsis* and *Phaeopappus*, each having 4 species. Section *Centaurea* is represented by 3 or perhaps 4 species.

Wagenitz (op. cit., 1986, pp. 18–19) writes: "Great diversity in the number of species and sections points to eastern Anatolia and adjacent Transcaucasia, Iran and Iraq as the foremost centre of evolution of this genus. The coherent distribution of most groups and the high degree of endemism seem to indicate speciation in relatively recent times and *in situ*. The mountains of S. Turkey have been a very active (secondary) centre of species formation".

It must be added that in Transcaucasia, besides an unusual diversity of different infrageneric taxa and species in *Centaurea*, in many different groups there has also been much infraspecific evolution. As a result, besides the great number of endemic species, there are numerous endemic taxa in lower ranks.

Caryology, Endemism and Seismicity

The abovementioned is to some extent explained by the detailed caryological researches of Ts. Tonian and co-workers (Tonian 1968a, b, 1971, 1972, 1980; Avetisian and Tonian 1975; Tonian and Mekhakian 1975). They studied most of the Armenian representatives of the subtribes Centaureinae and revealed great variability in both basic chromosome number (x = 8, 9, 10, 11, 12, 13, 14, 15) and also in the somatic chromosome number (2n = 16, 18, 20, 24, 26, 28, 30, 32, 40, 44, 54, 60). In addition, they disclosed the correlation between ploidy and the nature of the sexine structure of the pollen grains; ploidy and size of pollen grains; ploidy and the combination of various morphological features, sizes and quantities of stomata, and many other such relationships. Six polyploid series were established. Some sections are characterised by one basic number (Sect. *Chartolepis*, Sect. *Acrolophus* and Sect. *Phaeopappus*, all with x = 9), while other infrageneric taxa have two or several (Sect. *Cyanus*: x = 8, 10, 11, 12; Sect. *Microlophus*: x = 8, 9). In a number of species they discovered B-chromosomes, e.g. *C. phaeopappoides* (3B), *C. carduiformis* (1B) and *C. behen* (3B).

We would like here to draw the attention of botanists to a paper by the Soviet zoologist-geneticists N.N. Vorontsov and E.A. Lyapunova (1984), who revealed the existence of a correlation between zones of high seismic activity with regions of high chromosome variability. It is common in certain localities that unstable karyotypes are found in mountainous regions where the effects of isolation and of biotope mosaic patterns are more pronounced than on the plains, bringing about a change in chromosomes. According to the authors, the following hypotheses may be postulated.

1. The simplest hypothesis is that in the seismic regions more often than in the non-seismic ones, the degree of isolation is changed due to landslides, changes of river-beds and mud-flows, thus creating particularly favourable conditions for the fixation of new chromosome variants and increasing the rate of speciation.
2. Chromosome variation can be caused by mutagenic factors accompanying earthquakes, such as gamma radiation, increased concentration of radioactive

isotopes of radon (Rn) in water and occurrence of salts of heavy metals; these result in an increased rate of chromosome aberrations.

To briefly summarise the geomorphogenesis of the Armenian Highlands, the peaks of Zangezur and other mountain ridges were formed in the Upper Pliocene, c. 2 million years ago (Vardaniantz 1943). Our mountains, as well as the entire Armenian Highlands, are relatively young and have high seismicity. It would be interesting to study the species of plants growing in regions with various degrees of seismic activity as the factors which bring about chromosome variation in animals perhaps also affect vegetation.

It should be added that the highly seismic Meghri region is very mountainous, while plains of the intra-montane basins such as Ararat and Shirak have a similar degree of high seismicity and are also characterised by intensive diversification, speciation and high endemism.

Let us now return to *Centaurea*. We have noted above the wide variability in chromosome number in the Armenian species. The fans of microspecies in a whole series of groups, as well as in the taxa of higher rank in the subtribe Centaureinae, are undoubtedly the result of not only the factor of isolation of small populations forming a complex mosaic of biotopes in mountain regions, but also of increasing chromosomal mutagenesis, perhaps due to the high seismicity of the region.

A phenomenon very characteristic of the genus *Centaurea* is the occurrence of a high percentage of local endemic subgenera, sections and species, taxonomically isolated from other groups. We consider these to be the result of saltation (macro-evolutionary mutagenesis). According to Takhtajan (1983), "The basic and more usual type of speciation corresponds to the dominant microevolutionary model. But at least many if not the majority of isolated species, especially edaphic endemics have saltatory origin. The same is true about many monotypic neoendemic genera." The neglected, endemic monotypic genus *Chrysopappus*, described by Takhtajan in 1938 from south-east Anatolia, has no close relatives and may serve as an example.

Similar polyploid species such as *Centaurea erivanensis* (2n = 32), *C. behen* (2n = 36 + 3B), *C. huetii* (2n = 40 + 2B), *C. pulcherrima* (2n = 36), *C. glastifolia* (2n = 36), *Serratula biebersteiniana* (2n = 60), *Cheirolepis kotschyi* subsp. *persica* (2n = 44) and others, also have clearly saltatory origin and have successfully persisted. Wagenitz (1986), in contrast noted that a large number of endemic *Centaurea* species are known only from their type locality and some have not been collected for more than a hundred years, and quite possibly have become extinct.

OTHER GENERA OF PHYTOGEOGRAPHICAL INTEREST

No less interesting from a phytogeographical point of view is the genus *Cousinia*, the area of distribution of which almost completely lies within the Irano-Turanian region. The maximum variability of species in this genus is in central Iran, where some 400 species are concentrated (Rechinger 1972, 1986; Agakhanianz 1981). It may undoubtedly be confirmed that the Iranian Highlands is the original primary centre of speciation of this genus. However, as the study of the distribution of all species of *Cousinia* has shown, there are secondary centres of speciation within peripheral areas (Cherneva 1974). Two such centres are located on the Armenian

Highlands (Huber-Morath 1972: Tamanian & Fajvush 1987; Takhtajan & Tamanian, 1988; Tamanian, in press). It should be noted than in Soviet Armenia the highest number of species is in the south — the Meghri and Zangezur floristic regions. The most active speciation has been in series *Ilicifolia* and *Decipiens* of section *Cirsioideae* (*Cousinia erivanensis, C. iljinii, C. lomakinii*) and in section *Cynaroideae* (3 new species with affinities to *C. macrocephala* from south Armenia will be described by Tamanian).

Very curious is the genus *Onosma* (Boraginaceae), about 60 species of which occur in the Armenian Highlands, about 40 of which are distributed in the southern part of the Highlands, 25% of which are endemic to the whole area of the Highlands. On the other hand, speciation of the genus *Pyrus* (Rosaceae) has been most active in the northern part of the Armenian Highlands, especially the Lesser Caucasus, where many narrow endemics occur.

The greater number of species of the genera mentioned above are neoendemics; in that case all these genera are distinguished by major centres of speciation for the entire Irano-Turanian region as well as for the Armeno-Iranian province.

DISTRIBUTION TYPES

All the species of the flora of the Armenian Highlands may be divided into a few large categories according to distribution type. We are naturally more interested in narrow-endemic and endemic species and also relict species with very strong disjunctive areas.

Narrow Endemics

Narrow endemics form the first, rather large category. This group of species is extremely interesting from taxonomic, floristic and phytogeographical points of view. An abundance of such species in any genus clearly indicates active speciation in a comparatively 'young' group. To list only half of the narrow endemics of the Armenian Highlands would occupy considerable space; for that reason only a few of the most interesting and characteristic ones will be mentioned. The following are endemic to Soviet Armenia: *Acantholimon gabrieljanae, Alchemilla heteroschista Alyssum hajastanum, Galium sevanense, Isatis arnoldiana, I. sevangensis, Scrophularia olgae, Stipa × gegarkunii* (endemics of the Lake Sevan basin); *Carthamus tamamschjanae, Gypsophila takhtadzhanii, Milium transcaucasicum, Minuartia daralaghezica, Pyrus browiczii, P. × daralagezii, Ribes armenum, Stenotaenia daralaghezica* (endemics of Dareleghis); *Acantholimon fedorovii, Astragalus megricus, A. ordubadensis, Colutea komarovii, Iris grossheimii, Linaria megrica, Scrophularia takhtajanii, Verbascum erivanicum* (endemics of the Meghri and Ordubad regions); *Centaurea hajastana, C. takhtajanii, C. tamanjanae* (endemics of Shirak) and many others.

Narrow endemics of the Turkish and Iranian parts of the Armenian Highlands include, for example, *Acantholimon spirizianum, Isatis bitlisica, Marrubium vulcanicum* (Bitlis), *Arenaria angustipetala, A. davisii, Cousinia vanensis* (environs of Lake Van), *Astragalus hakkiaricus* (Hakkari), *Galium tortumense, Verbascum transcaucasicum* (northeastern Turkey, the latter also in Transcaucasia); *Cousinia harazensis*, *C. stroterolepis*, *Echinops elbursensis*, *E. iranschahrii*, *Scorzonera luristanica* (northwestern and western Iran) and many others.

Narrow endemic species, reflecting intensity of speciation in a given territory and demonstrating the uniqueness of its flora, serve as necessary evidence for the division of floristic regions but do not show sufficiently clearly the links between adjacent floras. From this point of view, rather greater interest is presented by comparatively narrow species, components of two or more bordering floras.

Narrow Species Linking Bordering Floras

If the Armenian Highlands are represented as a triangle, the apices of which lie in Turkey, Iran and the Soviet part, it is possible to single out groups of species, characteristic of two parts taken separately. It must be remembered that these three parts are divided between the Armenian and Atropatenian subprovinces of the Armeno-Iranian and Caucasian provinces. Many species or sections from different genera with Mediterranean or Irano-Turanian distributions have their easternmost, westernmost or northernmost representatives in Armenia.

"Armenian" species. Thus, groups of "armenian" species may be singled out, which are distributed in the Turkish and Soviet parts of the Armenian Highlands. The species of these groups are very numerous and include interesting taxa such as *Acantholimon araxanum*, *A. caryophyllaceum*, *A. glumaceum*, *Alcea karsiana*, *A. sophiae*, *Allium karsianum*, *Allochrusa versicolor*, *Asperula affinis*, *Astragalus szovitsii*, *A. torrentum*, *Campanula massalskyi*, *Carduus navaschinii*, *Catabrosella araratica*, *C. fibrosa*, *Cousinia brachyptera*, *Diarthron magakjanii*, *Festuca karsiana*, *Hedysarum elegans*, *Heracleum trachyloma*, *H. transcaucasicum*, *Hypericum formosissimum*, *Oenanthe sophiae*, *Peucedanum zedelmeyerianum*, *Prangos arcis-romanae*, *Ranunculus obesus*, *Sameraria glastifolia*, *Scilla rosenii*, *Scrophularia macrobotrys*, *S. nachitschevanica*, *Verbascum hajastanicum* and others. Among the endemics of the Armenian Highlands in practically all large families a significant part consists of "armenian" species; many of them have an important role in vegetation formation.

"Atropatenian" species. No less in number are the "atropatenian" species distributed in the Iranian and Soviet parts of the Armenian Highlands. Among them there are also numerous species, interesting from their taxonomic and florogenetic as well as phytocoenological points of view, e.g. *Acantholimon karelinii*, *Allium derderianum*, *A. materculae*, *Astragalus grammocalyx*, *A. urmiensis*, *Campanula bayerniana*, *Centaurea leuzeoides*, *C. phaeopappoides*, *Corydalis persica*, *Cousinia chlorocephala*, *C. macrocephala*, *Euphorbia woronowii*, *Heliotropium szovitsii*, *Iris elegantissima*, *I. lycotis*, *Lactuca takhtadzhianii*, *Leontice armena*, *Limonium fischeri*, *Pimpinella confusa*, *Pseudoanastatica dichotoma*, *Salsola cana*, *S. macera*, *Scilla mischtschenkoana*, *Scrophularia zvartiana*, *Teucrium taylorii*, *Verbascum megricum*, etc.

Species of the Turkish and Iranian parts. A group of species which are less numerous but also very interesting is distributed only in the Turkish and Iranian parts of the Armenian Highlands. Most of them are concentrated close to the border between these countries, in the regions between Lakes Van and Urmia.

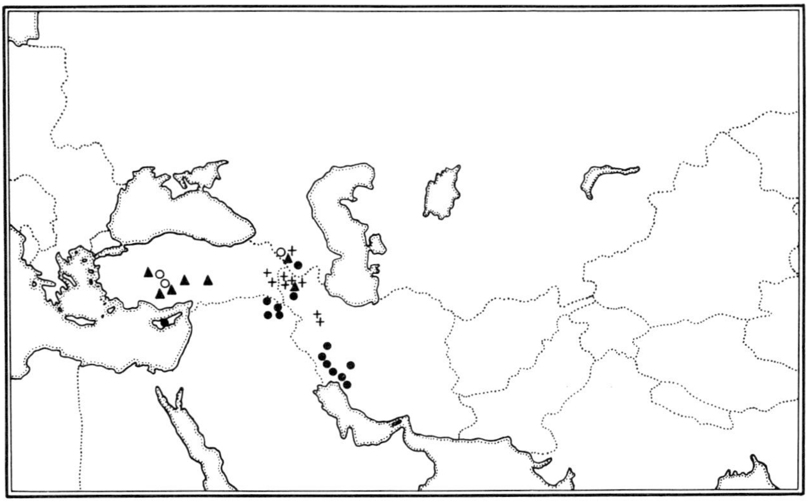

Figure 3. Distributions of species in S.W. Asia: ● *Chaenorhinum gerense.* + *Tomanthea carthamoides.* ▲ *Thesium compressum.* ○ *Linum seljukorum.*

Particularly noteworthy are *Acantholimon curviflorum*, *Alcea excubita*, *Alchemilla hessii*, *Astragalus capito*, *A. sphaeranthus*, *A. siliquosus*, *Centaurea vanensis*, *Cephalaria hirsuta*, *Cousinia urumiensis*, *Diplotaenia cachrydifolia*, *Dracocephalum aucheri*, *Elymus gentryi*, *Erigeron daenensis*, *Heracleum lasiopetalum*, *H. persicum*, *Isatis kotschyana*, *Oxytropis kotschyana*, *Ranunculus crymophilus*, *Salvia macrochlamys* and *Silene pungens*, but many others could be cited.

"Armeno-atropatenian" species. However, a more interesting group of species of those distributed over all three parts of the Armenian Highlands, are typical "armeno-atropatenian" species. They are rather numerous and reflect the more interesting aspects of florogenesis on the Armenian Highlands. These species include *Acantholimon bracteatum*, *Acanthophyllum mucronatum*, *Alcea flavovirens*, *Allochrusa bungei*, *Arenaria blepharophylla*, *A. szowitsii*, *Astragalus paradoxus*, *A. saganlugensis*, *A. shelkovnikovii*, *Campanula karakuschensis*, *Eryngium wanaturi*, *Hedysarum formosum*, *Hypericum armenum*, *Linaria pyramidata*, *Onobrychis altissima*, *O. atropatana*, *O. subacaulis*, *Onopordum armenum*, *Salvia hydrangea*, *S. limbata*, *Scrophularia amplexicaulis*, *S. atropatana*, *S. ilwensis*, *Sorbus luristanica*, *Szovitsia callicarpa*, *Tomanthea carthamoides*, *Veronica microcarpa* and others.

Relict and Disjunctive Species

Extremely great interest is attached to those species with strong disjunctive and unusual areas (Fig. 3). Most of them are relicts, preserved from previous eras and revealing ancient floristic links. Let us discuss some of these species. *Chaenorhinum gerense* is a very interesting species found in southern Iran, north Iraq, Crete and in Soviet Armenia (Dareleghis); *Microcnemum coralloides* has a very disjunct distribution, growing in Spain, central Anatolia and Soviet Armenia (environs of Yerevan); *Linum seljukorum* is found on saline marshes in central Anatolia and in the environs of Yerevan; *Citrullus colocynthis*, a Saharo-Sindian species, was recently discovered in Armenia (Meghri); *Sorbus hajastana*, a species until recently considered endemic to Soviet Armenia, has recently been discovered on the northern slope of the main Caucasus ridge; *Thesium compressum* and *Falcaria falcarioides* grow in the saline marshes of central Anatolia, in Soviet Armenia and northwestern Iran; *Lycium anatolicum* from Inner Anatolia, and the southern part of Soviet Armenia; *Didymophysa aucheri*, which occurs in the eastern Caucasus, the Armenian Highlands, Iraq and Middle Asia; and *Nectaroscordum tripedale*, which has a very disjunct distribution, being known in small populations on the Great Caucasus and Soviet Armenia, northwest Iran, northern Iraq and eastern and southwestern Anatolia.

When speaking of the ancient floristic links of the Armenian Highlands, connections between the latter and Daghestan cannot be neglected. As Tumadjanov (1966) showed, beginning in the Miocene, dispersion of xerophytic Antasian species on the Great Caucasus continued to take place, being particularly intense under the dry continental climatic conditions of the Middle Pliocene. The route for the Antasian xerophytic floras to the Lesser and Great Caucasus might have been open from the end of the Miocene (Sarmat-Meotis) when the Caucasus became a large peninsula of Asia Minor. At that time there was no high mountain relief in the Caucasus. This intensive penetration of xerophytes continued in the Middle Pliocene as well.

The evidence for this is provided by such halo-xerophytic relicts as *Salsola dendroides*, *S. ericoides*, *Nitraria schoberi*, *Reaumuria hypericoides*, etc. of the ancient saltwort desert of Daghestan found by Tumadjanov.

During neotectonic elevations in the Upper Pliocene the territory of mountainous Daghestan was isolated by high mountain ranges. The migration of species ceased owing to these natural barriers and the formation of local endemics with narrow ranges, such as *Convolvulus ruprechtii*, *Salvia daghestanica*, *Hordeum daghestanicum*, *Salsola daghestanica* (very close to *S. canescens*) and others, began. Besides these, many other plants have been preserved until the present time and grow in separate isolated populations in the ancient desert of Daghestan and also in the arid regions of the Armenian Highlands, as for example *Centaurea caspia*, *Colpodium versicolor*, *Saccharum ravennae*, *Melampyrum chlorostachyum*, *M. mulkijanianii*, *Scleropoa rigida*, *Stipa arabica*, *Verbascum nudicaule* etc. An interesting example confirming these ancient links is the discovery on the Sevan ridge, the high mountainous relief of which did not exist until the Upper Pliocene, of the Antasian species *Convolvulus calvertii* (Gabrielian 1980). Later mighty mountain rises and other geomorphological transformations brought about the isolation of the Antasian xerophytic flora, the disappearance of many and the preservation of some elements as relicts in different refugia. One of these, evidently, is the gorge where *C. calvertii* was discovered, a relic of the ancient xerophytic flora which existed before the epoch of the neotectonic rise. Thus, the discovery of the Antasian species *C. calvertii* on the Sevan ridge (in comparison with its closely related vicariad, *C. ruprechtii*) shows the ancient floristic links between the desert of mountainous Daghestan and southern Transcaucasia and Antasia.

It should certainly be noted that not all species interesting from a phytogeographical and taxonomic point of view have been included in the above mentioned framework. For example, a very interesting group of species, distributed in Iraq and Iran, penetrated eastern Anatolia, and in the north reached southern Transcaucasia (mainly to the Meghri and Zangezur floristic regions of the Armenian S.S.R. and the Nakhichevan A.S.S.R.): *Anchonium elichrysifolium*, *Convolvulus commutatus*, *Physoptychis gnaphalodes* and *Vavilovia aucheri*. Another example is the species *Dracocephalum botryoides*, which grows in the eastern Caucasus, Daghestan, on two of the highest peaks of Soviet Armenia (Aragats and Kaputdjukh) and in Iran and elsewhere.

THE ARMENIAN HIGHLANDS: INFLUENCES OF ADJACENT FLORAS

Till now we have spoken of similarities in different floristic regions of the Armenian Highlands. Let us now turn to the differences.

As is known, any flora is not unchanging. The process of florogenesis never stops. In time every flora is enriched by certain elements and loses others. Naturally, the greatest contribution to the enrichment of any specific flora is brought about by those neighbouring or adjacent. Thus the flora of the Armenian Highlands is under the strong influence of its surrounding territory. It goes without saying that different parts of such a vast territory undergo dissimilar influences from those in surrounding regions. Let us briefly pause to reveal these influences in various parts of the Armenian Highlands.

Influence of the Euxine Flora

In the northwestern part of the Armenian Highlands the influence of the Euxine flora is rather strongly evident. It is here that the *Fagus orientalis* — *Corylus colurna* forests, so characteristic of the Euxine province (Davis 1985), are best represented and birch forests with *Betula medwediewii* and *B. litwinowii* distributed. South of the Pontus mountains, there are insurmountable barriers for many Euxine species (e.g. many *Hieracium* species); nevertheless, some Euxine or Hyrcano-Euxine elements have penetrated into the Armenian Highlands, e.g., *Campanula lactiflora*, *C. pontica*, *Chamaesciadium acaule*, *Dianthus carmelitarum*, *Geranium ptilostemon*, *Hypericum orientale*, *Pimpinella rhodantha*, *Rhamnus imeretinus*, *Scaligera tripartita* and *Teucrium hircanicum*, together with many relatively mesophytic boreal and often Mediterranean species. Another means for the penetration of boreal species into the territory of the Armenian Highlands is provided by the Caucasus. Evidently, it was in just that way that Euro-Siberian species such as *Bellis perennis*, *Chrysosplenium alternifolium*, *Lychnis flos-cuculi* and others migrated to Soviet Armenia (Husjan 1977; Gabrielian and Husjan 1980).

Influence of Mediterranean and Central Anatolian Floras

The main route for migration of Mediterranean and central Anatolian species to the Armenian Highlands is from the west, e.g., *Alyssum huetii*, *Atraphaxis grandiflora*, *Haplophyllum cappadocicum*. In this case the overwhelming majority of the

Anatolian and Mediterranean species do not go further east than the 'Anatolian Diagonal' (Davis 1971, 1985; Ekim and Güner 1986). However, a few Mediterranean species are widely distributed all over the Armenian Highlands. It must be noted here that in the Zangezur floristic region of Armenia, the largest grove of *Platanus orientalis* in the Caucasus exists, with an undergrowth of *Periploca graeca*, and numerous trees of *Juglans regia*.

Influence of Antasian and Mesopotamian Flora

The most xerophytic Antasian (and also Mesopotamian) species penetrate into the Armenian Highlands from the south and southwest. In distribution many of them could pass only over the Kurd Mountain in the Hakkari region and Van (e.g. *Asplenium haussknechtii, Nigella unguicularis, Ranunculus diversifolius, R. myosuroides*), but a rather large group of species of the "Asia Minor–Caucasian" distribution type grow from Syria and Iraq to the Great Caucasus (often extending to mountainous Daghestan). These include, for example, *Campanula aucheri, C. collina, Cephalaria gigantea, Corydalis alpestris, Dianthus cretaceus* and *Poa iberica*.

Influence of the Caucasian Flora

In the north of the Armenian Highlands (especially in the Soviet Armenian flora) the Caucasus has a very great influence. Naturally, the majority of the Caucasian species are concentrated in northern Armenia (e.g., *Agasylis latifolia, Astragalus calycinus, Astrantia trifida, Cerastium holosteum, Galanthus alpinus, Gypsophila tenuifolia, Verbascum formosum, Viola caucasica*), although many reach south to a significant degree, extending their distribution to Syria and Iraq and in the southeast to the Zagros Mountains in Iran. These groups of species reflect ancient floristic relations of the Ciscaucasian steppes with southern Transcaucasia. Thus, only recently, *Papaver bracteatum* was discovered in the southern part of Armenia; this is an extremely beautiful large-flowered species of poppy known only from Ciscaucasia, East Anatolia and northwest Iran.

Influence of the Iranian Flora

In the east and southeast of the Armenian Highlands the influence of the Iranian flora, in particular the Khorassan and Kurdo-Zagrossian subprovinces, is rather strongly revealed. In spite of its tremendous floristic diversity, there are rather many Atropateno-Khorassanian (*Euphorbia marschalliana, Acanthophyllum mucronatum*), Atropateno-Kurdo-Zagrossian (*Allium viride, Reseda microcarpa, Rosularia elymaitica, R. persica*) and even Armeno-Atropateno-Kurdo-Zagrossian (*Echinops pungens, Onosma sericeum, Paracaryum strictum*) species (Sagatelian 1981).

CONCLUSIONS

The flora of the Armenian Highlands does not represent a united floristic complex; it is completely right and natural to treat the separate parts of the various regions of the floristic divisions as distinct entities. In that case, it is quite possible, as shown by Takhtajan (1978), that for the Armenian and Atropatenian 'subprovinces' this rank is somewhat too low and should be considered instead at the level of 'province'. On the other hand, as long as the Armenian Highlands are a geomorphological unit, a whole complex of species exists, characteristic of the Armenian Highlands as a whole.

Secondly, all parts of the Armenian Highlands are under the mutual influence, on the one hand, as evidence by species connected, let us say, with Soviet and Turkish Armenia, or species growing in Soviet Armenia and northwestern Iran. On the other hand, each section of the Armenian Highlands undergoes the strongest influence from the flora bordering it. It is very difficult to forecast, at present, the future of florogenesis but two variations may be suggested connected with general changes in the climate. If there were a fall of temperature and a simultaneous rise in humidity, i.e., a general decrease of the Continental climate, then evidently the influence of the Caucasian and Euxine floras (in a boreal and wide sense) would increase; in that case, a wiping out of specially strong differences between the Armenian and Atropatenian subprovinces might occur. If general aridization were to continue and the continentality of the climate as a whole increased, the influence of the central Anatolian and Iranian flora would become stronger, and therefore an even greater division into parts of these sections of the Armenian Highlands might be expected.

ACKNOWLEDGMENT

We wish to express our gratitude to Dr Kit Tan for kindly editing this paper.

LITERATURE CITED

Agakhanianz, O.E. 1981. Arid mountains of U.S.S.R. Mysl', Moscow, 271 p.

Avetisian, E.M., and Tz. R. Tonian. 1975. Morphologia pyl'zy i čisla chromosom nekotorych vidov podtriby *Centaureinae* Less., pp. 45–49. *In* E.M. Avetisian [ed.], Palynologia. Akad. Nauk Armjan. S.S.R., Yerevan.

Baloyan, S.A. 1989. Analys alpijskoj flory Priaraxinskyh hrebtov. *In* A.M. Barseghjan [ed.], Flora, rastitel'nost' i rastitel'nie resursy Armjan. Vol. 12. Akad. Nauk Armjan. S.S.R., Yerevan.

Cherneva, O.W. 1974. Kratky analis geografičeskogo rasprostranenia vidov roda *Cousinia* Cass. Bot. Žurn. (Moscow & Leningrad) 59: 183–191.

Davis, P.H. [ed.]. 1965–1985. Flora of Turkey and the East Aegean Islands, vols. 1–9. Edinburgh University Press, Edinburgh.

——. 1971. Distribution patterns in Anatolia with particular reference to endemism, pp. 15–27. *In* P.H. Davis, P.C. Harper and I.C. Hedge [eds.], Plant Life of South-West Asia. Botanical Society of Edinburgh, Edinburgh.

——. 1985. Why is the flora of Turkey interesting and important? Kew Mag. 2(4): 357–367.

Efremov, J.K. 1956. Armenian Highlands, pp. 88–188. *In* D.L. Armand et al., Zarubezhnaja Asia. Uchredgiz, Moscow. 608 p.

Ekim, T., and A. Güner. 1986. The Anatolian Diagonal: fact or fiction? Proc. Roy. Soc. Edinburgh 89B: 69–77.

Gabrielian, E.C. 1961. The genus *Sorbus* in Turkey. Notes Roy. Bot. Gard. Edinb. 22: 483–496.

——. 1975. On floristic connections of the Caucasus and the Balkans with particular reference to the genus *Sorbus*, pp. 216–222. *In* D. Jordanov et al. [eds.], Problems of Balkan flora and vegetation. Bulgarian Academy of Sciences, Sofia.

——. 1978. The genus *Sorbus* in Western Asia and the Himalayas. Akad. Nauk Armjan S.S.R., Yerevan. 263 p.

——. 1980. Rod *Convolvulus* L. v Armenii [The genus *Convolvulus* L. in Armenia.] Biol. Žurn. Arm. 33(1): 73–77.
——, and K.E. Husjan. 1980. Novye i redkie rody i vidy iz Severnoj Armenii. Biol. Žurn. Arm. 33(5): 535–537.
Grossheim, A.A. 1948. Rastitel'nyi pokrov Kavkaza. Moskovskoe obshchestvo estestvoispitateley prirody, Moscow. 267 p.
Hedge, I.C. 1986. Labiatae of South-West Asia: diversity, distribution and endemism. Proc. Roy. Soc. Edinburgh 89B: 22–35.
——, and P. Wendelbo. 1970. Some remarks on endemism in Afghanistan. Israel J. Bot. 19: 401–417.
——, and ——. 1978. Patterns of distribution and endemism in Iran. Notes Roy. Bot. Gard. Edinb. 36: 441–464.
Huber-Morath, A. 1972. Die anatolischen Arten der Gattung *Cousinia* Cass. Ber. Schweiz. Bot. Ges. B 82: 223–268.
Husjan, K.E. 1977. Novyj rod *Chrysosplenium* L. v Armenii. Učenye zapiski Er. Gos. Univ. 3 (135).
——. 1982. Novye florističeskie nachodki is Severnoj Armenii. Učenye zapiski Er. Gos. Univ. 2 (156).
Kamelin, R.V. 1965. O rodovom endemisme flory Srednej Asii. Bot. Žurn. Moscow & Leningrad) 50: 1702–1710.
Mirzoeva, N.V. 1979. *Astragalus* (*A. eriopodus*) — a new species of U.S.S.R. flora. Biol. Žurn. Arm. 32(6): 586–587.
Rechinger, K.H. [ed.]. 1963–1980. Flora Iranica. Lfg. 1–143. Akademische Druck- u. Verlagsanstalt, Graz.
——. 1972. Compositae–Cynareae I: *Cousinia*. *In* K.H. Rechinger [ed.], Flora Iranica Lfg. 90. Akademische Druck- u. Verlagsanstalt, Graz. 329 p. + 184 plates.
——. 1986. *Cousinia*: morphology, taxonomy, distribution and phytogeographical implications. Proc. Roy. Soc. Edinb. 89B: 45–58.
Sagatelian, A.A. 1981. Geographical analysis of the flora of Mehri district in Armenian S.S.R. Bot. Žurn. (Moscow & Leningrad) 66: 650–661.
Takhtajan, A.L. 1938. *Chrysopappus* Takht. - novyj rod semejstva Compositae. Mater. Gerb. Bot. Inst. Komarova Akad. Nauk. S.S.S.R. 7(12): 274–276.
—— [ed.]. 1954–1987. Flora Armenii, vols. 1–8. Akad. Nauk Armjan. S.S.R., Yerevan.
——. 1978. Florističeskie oblasti Zemli. Nauka, Leningrad. 247 p.
——. 1983. Macroevolutionary processes in the history of Plant World. Bot. Žurn. (Moscow & Leningrad) 68: 1593–1603.
——. 1986. [Translated by T.J. Crovello, ed. A. Cronquist]. Floristic regions of the world. University of California Press, Berkeley. xxii + 522 p.
——, and K.G. Tamanian. 1988. *Cousinia gabrieljanae* (Asteraceae)—novyj vid iz Armenii. Bot. Žurn. (Moscow & Leningrad) 73: 1609–1612.
Tamamschian, S.G. 1962. Pierre Edmon Boissier, 1810–1885 (For the 150th anniversary of his birth and 75th anniversary of his death). Bot. Žurn. (Moscow & Leningrad) 47: 144–154. [In Russian.]
——. 1968. On some relationships in the floras of Transcaucasia, Anatolia, and Iran. Notes Roy. Bot. Gard. Edinb. 28: 201–208.
Tamanian, K.G. (In press). Three new for the Soviet Union species of the genus *Cousinia* (Asteraceae). Bot. Žurn. (Moscow & Leningrad).
——, and G.M. Fajvush. 1987. Phyto-geographical analysis of the Armenian species of the genus *Cousinia*. Biol. Žurn. Arm. 40(6): 464–469.
Tonian, Tz. R. 1968a. Čisla chromosom nekotorykh vidov roda *Centaurea* L. Biol. Žurn. Arm. 21(5): 86–96.
——. 1968b. Čisla chromosom nekotorykh rodov triby Centaureinae Hoffm. Biol. Žurn. Arm. 21(8): 69–78.
——. 1972. Correlation between some corriological and morphological features

of representatives of Centaurinae Less. Subtribes. Biol. Žurn. Arm. 25(11): 47–50.

——. 1980. Relation between chromosome number and some morphological signs of Centaurinae Less. representatives. Biol. Žurn. Arm. 33(5): 552–554.

——, and A.K. Mekhakian. 1975. Ploidnost' i rasmery pyl'zevych zeren u vidov podtriby Centaurinae Less., pp. 50–56. *In* E.M. Avetisian [ed.], Palynologia. Akad. Nauk Armjan. S.S.R., Yerevan.

Tumadjanov, I.I. 1966. Ancient desert in the mountainous part of Daghestan. Bot. Žurn. (Moscow & Leningrad) 51: 784–791.

Vardaniantz, L.A. 1943. Postpliozenovaja istoria Kavkazsko-Černomorsko-Kaspijskoj oblasti. Akad. Nauk Armjan. S.S.R., Yerevan. 148 p.

Vorontsov, N.N., and E.A. Lyapunova. 1984. Explosive speciation in seismic active regions. Chromosome Today 8: 279–294.

Wagenitz, G. 1975. Floristic connections between the Balkan peninsula and the near East as exemplified by the genus *Centaurea*, pp. 223–238. *In* D. Jordanov et al. [eds.], Problems of Balkan flora and vegetation. Bulgarian Academy of Sciences, Sofia.

——. 1986. Centaurea in South-West Asia: patterns of distribution and diversity. Proc. Roy. Soc. Edinb. 89B: 11–21.

Zohary, M. 1973. Geobotanical foundations of the Middle East. Gustav Fischer Verlag, Stuttgart, and Swets & Zeitlinger, Amsterdam. 2 vols., 739 p.

18. FORM, FUNCTION AND ADAPTATION IN THE GRASS PALEA, WITH SPECIAL REFERENCE TO *BROMUS*

P.M. Smith

Department of Botany, University of Edinburgh, The King's Buildings, West Mains Road, Edinburgh

ABSTRACT

The range of form of the grass palea, generally regarded as a conservative taxonomic feature, is discussed, along with its morphological homologies and supposed functions. An investigation of palea form and function is reported in *Bromus* L. Considerable adherence of palea and caryopsis occurs in many species of this genus. The 'glueing' seems to be an active part of the normal developmental process. The glued areas are associated mainly with the palea keels, and it is argued that this is important in dispersal, but also in protecting the caryopsis mechanically and by means of enclosed air spaces. Where atypical palea form exists it is shown that alternative protection is provided via transference of function.

Key words: morphology, form and function, palea, Gramineae, *Bromus*, dispersal.

INTRODUCTION

The palea of the grass floret is relatively seldom used as a source of taxonomic characters. Certainly it has been much less generally studied than the lemma. It may be regarded as a fairly conservative feature. The study of palea form and function here reported is the first fruit of a wider study of adaptive radiation in grass floret and spikelet structure. Paleas, it was hoped, might reveal, under close study, rather more of interest than appears at first sight. Perhaps paleas have been unjustly neglected by taxonomists and evolutionists.

GENERAL FORM OF PALEAS

The palea is typically a fragile scale only a few cells thick, its adaxial surface lying against one side of the caryopsis (see Fig. 1A). The lemma lies against the other side of the caryopsis. It is thus one of the two components of the grass *anthoecium*. Paleas usually are two-veined, 'binerved' structures. The nerves are near the margins which are infolded on to the adaxial side, forming 'keels' containing the veins. Long hairs (cilia) are usually associated with the veins. Fig. 1B illustrates a typical palea. The apex of the palea may be truncate or acute, often it is irregularly notched. Between and beyond the veins the tissue is colourless: the palea is commonly translucent. On either side of the veins there is typically a narrow green strip, 2–10 cells wide, of chloroplast-containing cells. Above these strips the epidermis contains rows of stomata.

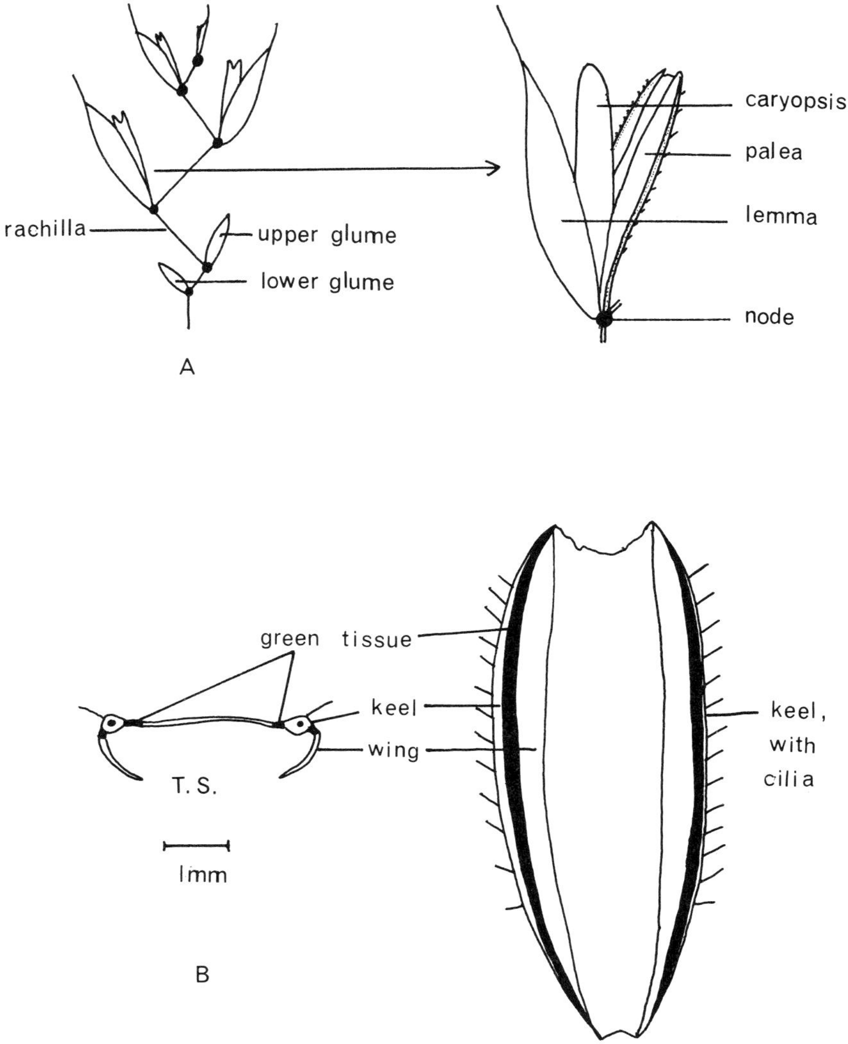

Figure 1. — A. Diagram of a grass spikelet, showing six main nodes of the rachilla, and an enlargement to show the parts of a single floret. — B. A view of the palea of *Bromus hordeaceus* in T.S., and from the adaxial (grain-opposed) side.

Variation in palea form involves mainly size, texture, the number of veins, the form and number of cilia on the veins, and the indumentum of the two surfaces. Palea shape is rather unvarying, but there are some strikingly unusual forms. Mature paleas are generally as long as or a little longer than the mature caryopsis, and somewhat shorter than the lemma. In a relatively few cases the mature grain

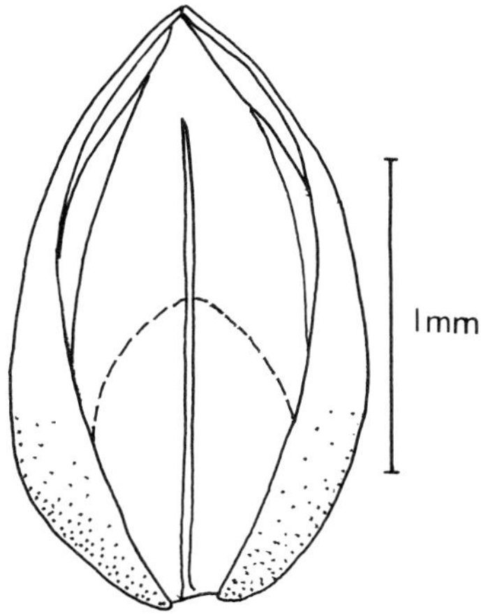

Figure 2. Diagram of a fertile floret of *Panicum plicatum* from the palea side. The young grain is outlined within. Note the enclosure of the palea margins by those of the lemma, and the development of some central vascularisation in the palea.

exceeds the length of the palea. Rarely the palea is very small (e.g. *Andropogoneae*); and in some species it is completely absent (e.g. *Alopecurus*). Paleas are at first much shorter than the lemma, and grow in tandem with the ovary. Though usually the delicate scales previously described, paleas are sometimes very robust — thick and hard or leathery. Such is the case when they mimic the mainly protective function of the lemma, when protective glumes are absent or very small, and the palea side of the caryopsis is not protected by the lemma of a succeeding floret. An example of this transference of function is *Panicum* (Fig. 2).

Palea veins can be numerous, in some bamboos for instance, and it is not uncommon in more advanced genera for there to be one or two minor veins, parallel and in between the main marginal ones. These veins are normally very short and generally lack accompanying green tissue. Three equal size veins are found in the palea of *Leersia*, though here it is discharging a glume function. The lateral (or marginal) veins usually stop short of the apex of the palea, or may enter the irregular apical cusps, on either side of a terminal sinus. Rarely they are emergent as short bristles.

Cilia on the marginal veins may be few and scattered (as in *Bromus*, Fig. 3) or numerous and dense (*Melica*). They may be long or short. They are probably never completely absent, and even modest magnification reveals that the most obvious cilia on the veins are merely the longest cohort of a series of cilia on either side of the vein, in the region of the green cells. There is variation within as well as between genera and it can have some taxonomic significance (Fig. 4).

Cilia (essentially macro-hairs) may be rather widely distributed on either or both surfaces of the palea, or the latter may be completely glabrous.

The most striking exceptions to the general palea morphology described above, which are hence the variations to consider first in any general theory of palea form

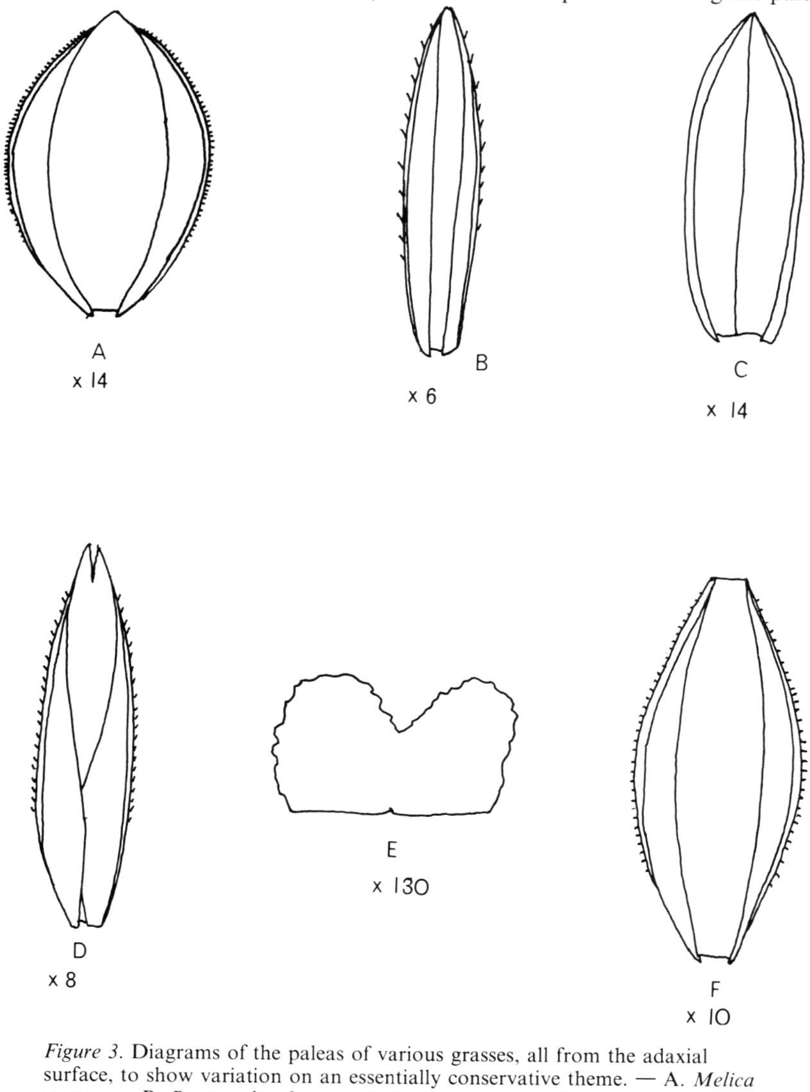

Figure 3. Diagrams of the paleas of various grasses, all from the adaxial surface, to show variation on an essentially conservative theme. — A. *Melica nutans*. — B. *Bromus diandrus*. — C. *Leersia oryzoides* (3-veins). — D. *Avena fatua*. — E. *Digitaria ischaemum* (vestigialised). — F. *Sieglingia decumbens*.

and function, are found in species where the palea is either not a single scale, but is two, or where the single organ is deeply bifid. In *Streptochaeta* the 'palea' consists of two structures (Fig. 5B). Here and in *Ampelodesmos*, Stebbins (1974) has reported that two primordia are involved, each with its own procambial strand. In *Streptochaeta*, these initials produce separate structures; in

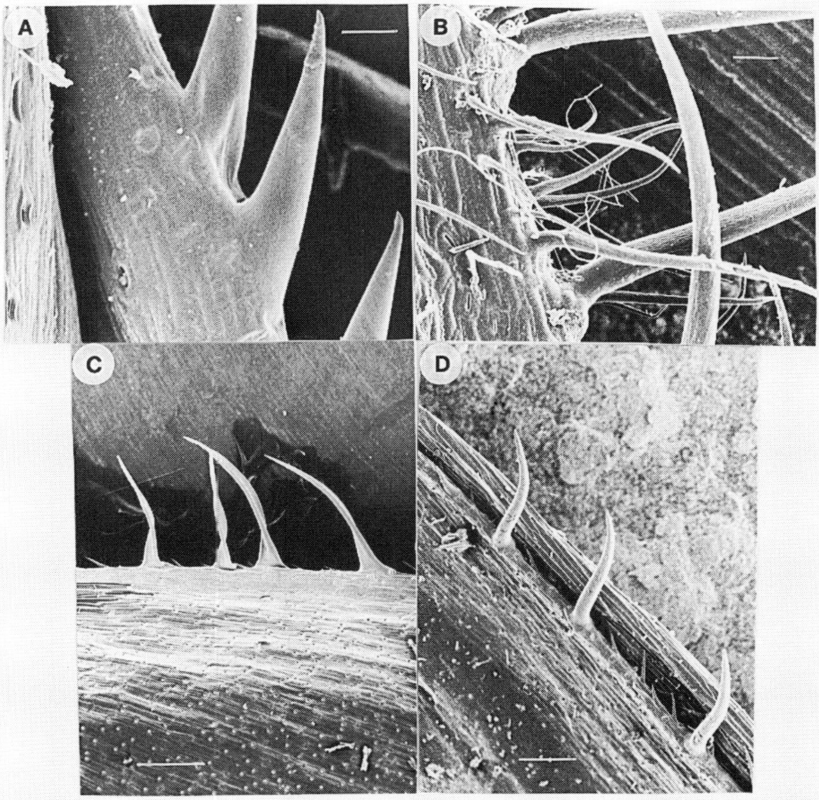

Figure 4. Diversity of palea cilia in *Bromus*. — A. *B. lanceolatus*. — B. *B. secalinus*. — C. *B. commutatus*. — D. *B. squarrosus*. — Note the stomata in the green strip along the palea keel, shown most clearly in A and B. Bar scales, respectively, 20, 50, 200, 100 μm.

Ampelodesmos they fuse at a later stage producing a single bifid structure (Fig. 5A). Both *Streptochaeta* and *Ampelodesmos* have other characters deemed to be relatively primitive in grasses, both showing, for example, a relatively high degree of floral trimery. In *Bromus*, a comparatively advanced genus in terms of *Ampelodesmos* and *Streptochaeta*, and only very distantly related to either, there is a single species, *B. interruptus*, which has a deeply bifid palea (Fig. 5C).

ORIGIN AND MORPHOLOGICAL NATURE OF THE PALEA

Arber (1934), Philipson (1935) and Stebbins (1974) have discussed palea origins critically. It is virtually certain that the palea is a leaf homologue of some kind.

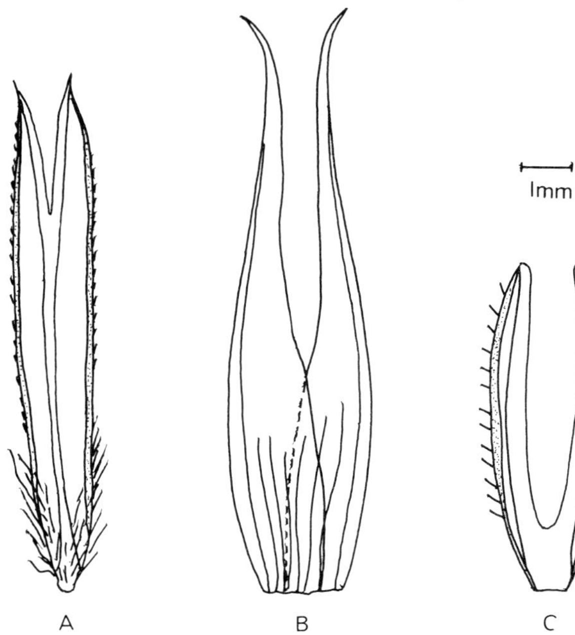

Figure 5. Dual or bifid palea structures of certain grasses. — A. *Ampelodesmos mauritanicus*. — B. *Streptochaeta spicata*. — C. *Bromus interruptus*. — Adaxial appearance shown, with 1 mm scale.

Most have seen it as a version of the prophyll (Arber 1934; Gould and Shaw 1983; Clayton and Renvoize 1986). Prophylls are the structures produced on grass stems, primary or lateral, before the leaves (Fig. 6). Though not always present, they are very common. Prophylls usually have two, more or less equal vascular bundles, near their lateral margins. Effectively they are juvenile leaves of distinctive form. Prophylls sometimes have prominent infolded marginal wings, forming keels. Keels are often hairy. The marginal wings are sometimes fused on the far side of the stem (Fig. 6). Some of them, at least, look very much like a vegetative version of the palea.

Proponents of the prophyllar nature of the palea point to this overall resemblance, and also to the binary nature of the very first emergent aerial structure from the embryo stem, the coleoptile. This structure, itself of disputed morphological nature (Eames 1961) is a tube-like structure with two opposite vascular bundles.

Stebbins (1974) proposes that the palea arises from two fused lodicules. Lodicules are generally accepted to be the vestigialised perianth segments of the grass flower, which structure is of course sited between the palea and the lemma. Usually two lodicules are produced, often fused at their base. Stebbins suggests a second lodicule fusion — producing the palea. Lodicules have a hydraulic

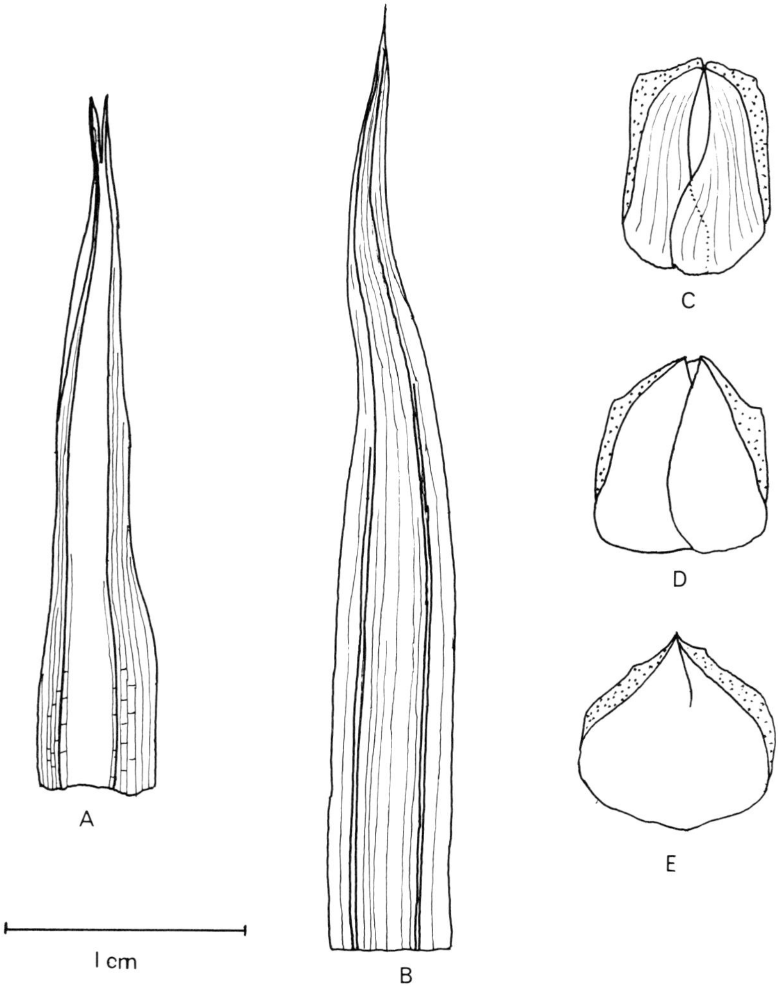

Figure 6. Prophylls, from the adaxial side. — A. Branch prophyll of *Diplachne fusca*. — B. Branch prophyll of *Cymbopogon nardus*. — C. Bud prophyll of *Chimonobambusa quadrangularis*. — D. Bud prophyll of *Semiarundinaria fastuosa*. — E. Bud prophyll of *Bambusa multiplex*. — C, D and E show a thin (stippled) membrane arising from the prophyll keels in a position homologous with that of palea keel cilia. [C, D and E from McClure, F.A. (1966) The Bamboos, a fresh perspective (*Harvard University Press*), after Takenouchi].

function, forcing apart the bases of palea and lemma so exposing or releasing the reproductive organs during anthesis. Primitive grasses have four to six lodicules which are considered remnants of a two-whorled trimerous perianth.

Though the lodicules can be larger and more scale-like in primitive grasses (some bamboos for instance), there are no obvious examples of structures clearly intermediate in form between lodicules and paleas. There are numerous well-documented cases of structures intermediate between lodicules and stamens (the next undeniably floral whorl). Arber (1934) illustrates and extensively discusses these 'stamen-lodicules'. On this observation, stamens and lodicules might be crudely estimated to have a similar 'evolutionary distance' from leaves (their ancestor-organs), while paleas may have a different one, involving a fusion between "leaf-descendants" of a different order, and probably of different function.

Though the 'fused-lodicules' theory may not be correct, the duality of the origin of paleas, or indeed of prophylls, cannot seriously be doubted, though Hubbard, in his general theory of spikelet evolution (1948), suggested an origin from a single bracteole. Except for some primitive species, paleas, prophylls and coleoptiles share the feature of a pair of distant (opposite where they enter the vascular system of the stem) main vascular bundles, with relatively undifferentiated tissue in between and they are commonly bifid at least at the apex. The veins terminate separately, not converging at all, as they often do in lemmas and foliage leaves, and seem as unconnected evolutionarily as they are in fact. Paleas arise from separate procambial initials which usually fuse later, but do not always (*Streptochaeta* etc.). They are usually bifid at the apex. Stebbins (1974) is surely right to point out that syncotyly, the fusion of two cotyledons into one, probably in their petiolar areas, is the most likely origin of monocotyledony itself. A subsequent further fusion of leaf initials producing prophylls and paleas would seem likely, especially in parts of the stem where internodes were short, if there were appropriate, perhaps similar selection pressure, i.e. some parallelism of function.

The palea is a characteristic organ, produced to function, if it functions at all, at a *late* stage in the life cycle of the plant. Conversely, the prophyll is an early structure presumably discharging an early function. In the palea, we therefore seem to see the late appearance of a juvenile structure, if the homologies are correctly interpreted here. Though the coleoptile is an earlier feature of the growing grass plant than the prophyll, it has a more complex function, not just being meristem-protective and basally absorptive, but also incorporating light-piping features (Mandoli and Briggs 1982).

Arber (1934) sought to explain the characteristic form of the palea by what she called 'the pressure theory'. She suggested that, in the reproductive apex, space limitations and pressure on the developing palea forced it into its intriguing form. On this basis, the two vascular bundles and keels became prominent as and where they do because that is where space for them exists in the bud. Though the pressure is released later, the palea remains fixed in its early form. Though not explicitly stating this, Arber thus suggests that palea form results from generations of selection for smallness and simplicity, a classic case of vestigialisation. She does not have much to say about palea function.

Though plants are, to some extent, constrained by the organ-systems they inherit, over the generations one might reasonably suppose that, depending on apex-architecture, grasses could have evolved vestigial paleas with two vascular bundles or six, with two or three keels or twenty, or perhaps very often with no palea at all. Thomasson (1985) shows that paleas had disappeared in *Nassella*

(*Stipeae*) by the Miocene. Yet grass paleas are still common, and have, over the family as a whole, a remarkable consistency of form. The 'pressure theory' does not attract: most paleas do not seem to me to be mere vestiges. Stebbins (1974) rejects Arber's views on developmental grounds.

The function of the palea has been accepted for a long time to be essentially protective. The developing and mature grain is enclosed between the two scales of the anthoecium. Usually the palea is an important part of this protective coat. The protection offered could be from desiccation, from insect predation or from contact with decay-causing micro-organisms, all in the early stages of development. In maturity, protection from grazing damage is largely by the lemma, which often toughens as it dries out, and may almost completely envelop the grain *and* palea. Larval insects penetrate mature *Bromus* grains usually through the palea or around its margin.

Commonly the lemma and palea are part of the dispersal unit. Numerous dispersal-related features of lemmas — inflations, awns, hairs, pointed calli, etc. — are well-known. It would be curious if the form of the palea was not in some way also connected to the vital dispersal and ecesis stages of the plant's life cycle. Thomasson (1985) shows how the lemma margins interlock in the abaxial groove formed between the keels of the palea in *Piptochaetium*, *Leersia* and the Miocene fossil genus *Archaeoleersia*. Less strong mechanical interlocking occurs, for example, in *Panicum*. This adaptation may usefully preserve the integrity of the anthoecium against shocks and abrasion.

Protection of the grain is also necessary if the route of dispersal includes a sojourn in a herbivore gut. The palea, if not too grossly damaged, forms part of an inert layer on the outside of the grain, of cellulose or sometimes partly of silica. Lastly, there is the phase of survival and germination while buried in soil, during which paleas may discharge some useful, perhaps protective function.

PALEA FORM AND FUNCTION IN BROMUS

When mature *Bromus* caryopses are freed from their anthoecia it is obvious that there is a considerable bond between the pericarp wall and both the lemma and the palea. It is difficult to detach the caryopsis free from pieces of lemma and/or palea. This firm adhesion has been mentioned by Dore and McNeill (1980). Though it is a notable feature in *Bromus*, it is found to some extent in other grasses, e.g. *Festuca*, where the lemma/palea architecture is similar. The *Bromus* anthoecium seems to be stuck to the grain surface quite firmly. Fig. 7 shows a typical *Bromus* floret from the palea side.

In previous chemotaxonomic investigations of *Bromus* seed proteins (e.g. Smith 1972) the initial procedure in preparing the protein extracts has been to 'de-husk' the grains. No mechanical method or apparatus, manual or electrical, has been found to achieve an acceptable threshing operation. Instead, the florets have been soaked in 50% sulphuric acid for 30 minutes, after which the husks (lemmas and paleas), are freed and can be separated cleanly by agitation in liquid. Evidently the bond is a substantial one.

These observations led to the investigation of palea form and function in *Bromus* here reported. Results with other genera will be reported separately.

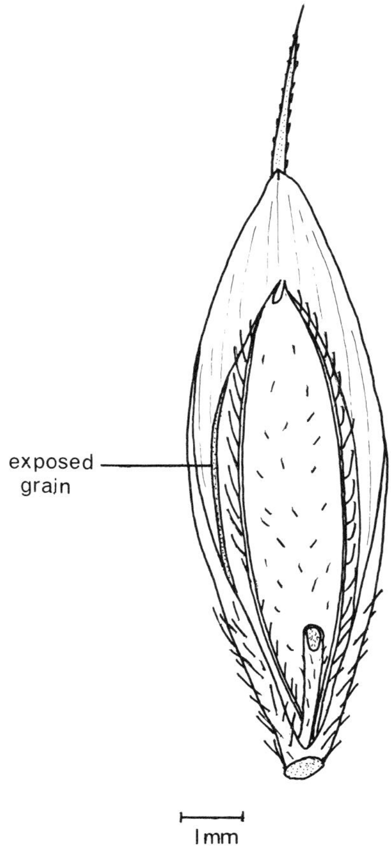

Figure 7. Diagram of floret of *Bromus latiglumis*, from the palea side.

MATERIALS AND METHODS

Young and mature florets of *Bromus* species were manually dissected and the previously attached, now separated surfaces (adaxial lemma surface, adaxial palea surface and the opposed caryopsis surfaces) were examined by eye, and by light and scanning electron microscopy. Transverse sections of young and mature grains were studied. Mature florets were similarly examined after 50% sulphuric acid treatment for 30 mins and, after soaking in water, in EDTA solution (2%) and a pectinase preparation for two hours. The pectinase treatment was with *Aspergillus* pectinase (100 units per ml) (Sigma Chemical Co.) in a medium buffered at pH4.

RESULTS

When a lemma is manually separated from the mature grain there is tissue damage to the lemma and to the grain (Fig. 8). Damage is slight or absent when the grain is young. The damage is spread over the entire width of the grain, and most of the opposing lemma surface. Sheets of lemma epidermis are found stuck to the caryopsis, and pieces of the outer pericarp are found attached to the lemma. The grain surface is generally pockmarked. The adhesion, as shown by frequency of damage on separation, is greatest between the lower half of the opposed surfaces of lemma and grain.

When the opposed grain/palea surfaces are inspected after manual separation, tissue damage of a similar kind is seen on both, but it is largely confined to the vein areas of the palea and the opposing surface of the grain. Two parallel lines of adhesion seem to exist essentially along the keel areas of the palea (Fig. 9). In between the keels, and especially clearly at the base, the palea is generally free from the grain.

Similar observations have now been made for twenty species of *Bromus*, in several sections of the genus.

Manual separation of lemmas and paleas from grains after soaking in water remained difficult and incomplete. Tissue damage of the type described above was still evident. Much less separation difficulty was found after soakings in acid, EDTA or pectinase. Tissue damage was slight or absent after these treatments.

Sections of developing and mature grains (see Fig. 10) revealed that the outermost pericarp cell layers are at first mostly green, probably actively photosynthesising, but that when the grain is mature the surface layer is hard, rough, brown and amorphous, often with irregular superficial patches of still recognisable, but dead, cells.

Light-microscopic and naked-eye inspection of undamaged surface areas of the grain reveal an apparent surface encrustation of whitish, often glistening material.

DISCUSSION

Mechanism of Adhesion

It seems inescapable that there is an efficient glueing mechanism at work in the adhesion between the *Bromus* anthoecium and grain, which supplements or replaces any other mechanical grip which the anthoecium develops.

The glueing process seems to be related to the presence, in the lemma and palea, of living green tissue. The lemma is green and alive along its entire internal surface, while the palea is green and evidently alive only along the keels. This parallels the distribution of glueing. Glueing may conceivably occur only at contact points, in which case the late survival of keel cells is unexplained.

The glueing takes place only when the grain has reached its maximum size: at about this time the green tissue of the lemma, palea and grain rapidly senesces.

The most obvious plant glue responsible would appear to be pectin, or a pectin product. Pectin glues are increasingly believed to be implicated in plant tissue developmental changes (Jarvis 1984; Fry 1986). It may be that the quite rapid, active 'senescence' changes in the superficial green tissues of lemma and palea involve the production of a pectin-gel cementing property. The outer cells of the pericarp wall senesce differently: most of them are quite unrecognisable as cells at maturity. The glistening white material visible on the pericarp may be the pectin glue itself.

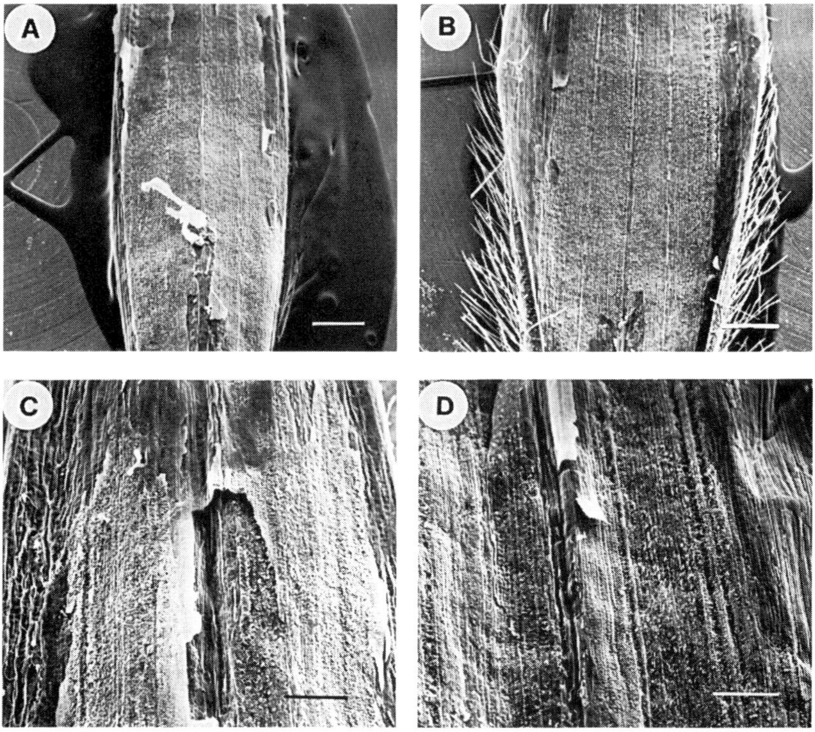

Figures 8 and 9. Adhesion of ripe grain to lemma and palea in *Bromus latiglumis*.

Figure 8. Lemma surface/grain surface adhesion (SEM photographs). — A. Lemma-opposed surface of grain, after separation from lemma (adaxial epidermis of lemma is stuck to grain). — B. The grain-opposed surface of the lemma removed as described. Most of the (adaxial) epidermis has been stripped off. — C. Close-up of part of the specimen in A, showing lemma adaxial epidermis stuck on to the grain, and a central area where the grain surface is exposed and part of it lifted off. — D. Close-up of part of B. The cell wall outlines of broken cells of the epidermis and sub-epidermal mesophyll can be seen. Bar scales, respectively, 500, 500, 250, 250 μm.

Pectin gels and glues can be destroyed by pectolytic enzymes, like those used in part of this investigation. Pectinase treatment considerably reduced the adhesion. So too did treatment with EDTA. Calcium ions, which are necessary for the formation and preservation of pectin lattices, are chelated, hence effectively removed, by EDTA. The acid treatment would effectively hydrolyse pectin. These findings strongly suggest a pectin glue mechanism.

There would appear to be a closely controlled and correlated "senescence physiology" of lemma, palea and pericarp, resulting in an effective glueing procedure when the grain has reached its final size relative to those of the anthoecial organs. The senescing pericarp produces a pectin product: the nearby green,

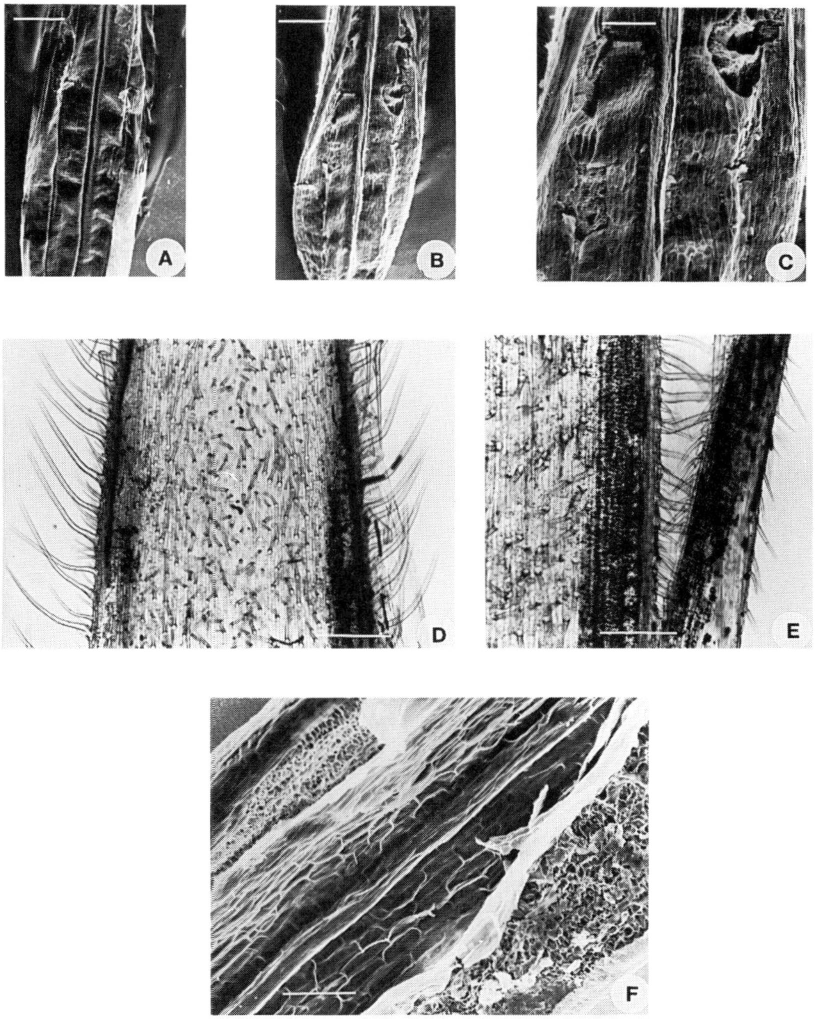

Figure 9. Palea surface/grain surface adhesion. — A. Adaxial palea surface after removal from grain. Note the grain surface material adhering to the keel areas (SEM picture). — B. The grain surface after the palea (in A) was removed. Surface tissue has been removed along the palea keel-opposed areas (SEM picture). — C. Close-up of B. Sub-pericarp and testa layers of the grain are revealed where the superficial layers (layers a + b in Fig. 10) have been pulled off (SEM picture). — D, E. Light-microscope photographs of *Bromus latiglumis* paleas. The palea was pulled off the grain and mounted in water. Damaged cells along the keels have filled with air: other cells remain intact. E. shows a palea which ruptured along the damaged keel area during removal. — F. SEM photograph of the palea-opposed grain surface from which the palea has been removed. Grain surface damage is along the keel-opposed areas. The central surface is intact. Bar scales, respectively, 800, 800, 2000, 500, 800, 100 μm.

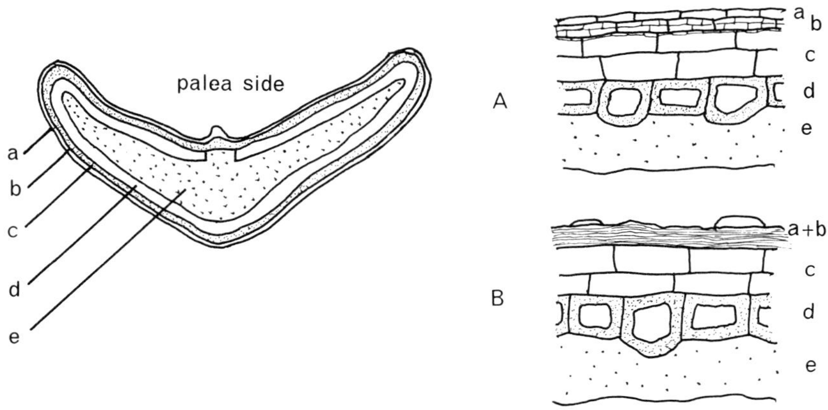

Figure 10. Diagram of transverse sections of grains of *Bromus cappadocicus*. — A. Tissue layers in the unripe, green stage. — B. Tissue layers of the ripe, brown grain. — a. Pericarp epidermis. — b. Sub-epidermal, chloroplast containing cells. — c, d. Nucellar cells, the inner characteristically thick-walled. — e. Endosperm. N.B. The layer of tiny cells beneath b (in A), which is often difficult to discern, is the testa. The cells in layer c frequently become disorganised and thick-walled. Layers a and b senesce into the adhesive layer described in the text.

living cells of the lemma and palea either convert it into a glue, or merely act as pressure points to an adhesive surface.

Adhesion in Relation to Palea Form and Function in Bromus

A glue joint between the keels and grain comprises a mechanically efficient means of ensuring a durable seal attaching the outer margins of the palea to the structure it is deemed to be protecting. The wings of the palea are glued near the vein like the flaps of an envelope: their free edges wrap around the edge of the grain. In *Bromus*, this glued area is overlapped in the lower half of the floret by the lemma, but not usually above (see Fig. 7). Physical interlocking of lemma and palea (of the type outlined above in *Piptochaetium*) is thus weak or lacking — the anthoecium is agape. The glue may therefore be an important means of attaching the caryopsis to its dispersal apparatus, usefully supplementing the basal attachments of the palea, caryopsis and lemma to the rachilla node.

It remains to be explained why the palea glueing mechanism does not occur over the whole opposing surface of the caryopsis, i.e. why other, central vascular bundles with associated living green tissues do not exist, reinforcing the adhesion at the marginal area. As reported above, the central tissue of the mature palea is largely free from the (grooved) opposing grain surface, and in the area of the embryo it forms an almost dome-shaped membrane. Immersion in de-gassed water for a prolonged period shows that an air-filled gap persists along the central palea/grain area, and in the embryo area, beneath the dome. This would seem to ensure a local oxygen supply perhaps for continuity of germination in periodically water-logged soils, and perhaps a barrier to acid or digestive enzyme damage during passage through a herbivore gut.

The apical end of the central air-channel is, in *Bromus*, closed off by an unusual, bilabiate, hairy appendage arising from the top of the caryopsis: the surface-tension properties of the hairs seem to resist the ingress of water.

The longer cilia on the keels of the palea close the gap between the protective anthoecial layer of the lemma and that of the palea. They are best developed in the central and upper area of the keel, at the position where the caryopsis edge is most often exposed. Though these cilia may help to contain an air reserve it is more likely that they frustrate entry by small insects. If this is so there may turn out to be a relationship between cilia size/number, the size/shape of caryopses and the degree of lemma/palea overlap. Preliminary observations suggest that this is so in *Bromus*.

An argument that the characteristic morphology of paleas comprises a useful protection, and/or part of the dispersal and ecesis adaptations must consider the cases of non-glued, absent, dual and bifid paleas.

The obvious and probably quite realistic interpretation of palea absence (or vestigialisation) is that in these cases the anthoecial function is wholly discharged by wrap-around lemmas (e.g. *Nassella*, *Alopecurus*) or by transference of anthoecial function to glumes. Non-glued paleas may be accounted for in the same way, or by the existence of other means whereby the lemma and palea are firmly clamped round the grain.

It will be a matter for further enquiry to determine how the divided or dual floral axis prophylls of *Streptochaeta* and *Ampelodesmos* relate to the view of palea function advanced here. In *Ampelodesmos* the degree of palea division may not be functionally significant at all. In *Streptochaeta*, both Soderstrom (1981) and Clayton and Renvoize (1986) doubt that the dual structure is homologous with the palea: certainly it is functionally important in the remarkable dispersal system which has evolved there. In *Bromus interruptus*, however, the deeply bifid palea form (Fig. 5) lends itself to a fairly ready interpretation. The glistening pectin-glue is produced by the caryopsis. The two lobes of the palea consist of the keels and wings, which are firmly glued to the caryopsis as in other *Bromus* species. The air gap over the basal, embryo area is still intact. But no physical protection to the largest part of the caryopsis surface is offered, because there is no opposing palea tissue. Very few of the florets of this heterodiasporous species (now probably extinct in the wild) are dispersed singly. Most of them are dispersed as a 4- or 5-partite, synaptospermous unit of tightly congested florets, comprising most of the spikelet. The inner caryopsis surface of one floret is thus covered and protected by the abaxial hairy lemma surface of the floret above it. The protection function of the palea has been transferred, and the palea somewhat vestigialised.

CONCLUSIONS

The relative conservatism in palea form probably implies consistent functions in caryopsis protection and ecesis, which may sometimes include a glue-based adherence property. The extent and extremity of this property remain to be established, but it is not peculiar to *Bromus*. The chemical nature of the glue is almost certainly pectinoid: this is currently under further investigation.

Where the protection/ecesis functions are transferred to one or more other organs, or there is heterodiaspory, or caryopsis form is diverse, palea variations can be expected, and many will generate taxonomically useful characters. It is not possible to argue that paleas are mere vestigial organs: they continue to discharge functions. There is a probable homology of coleoptiles, prophylls and paleas, which discharge similar functions at early, middle and late stages of the grass life-cycle, viz. meristem or primordium protection. Of the three, prophylls are least specialised in form and function.

Whenever the dispersal biology and adaptive radiation of grass taxa are being considered, palea form and function should not be overlooked.

ACKNOWLEDGMENTS

I thank G.L. Stebbins for stimulating me to think about paleas, Fatima Sales and Matiur Rahman for supplying me with provoking references and plant material, Grace Forrest and Bill Foster for enterprising technical assistance. This investigation was supported by S.E.R.C. grant no. GR/B/83957, for which I am grateful.

LITERATURE CITED

Arber, A. 1934. The Gramineae: a study of cereal, bamboo and grass. Cambridge University Press, Cambridge. 480p.

Clayton, W.D., and S.A. Renvoize, 1986. Genera Graminum, grasses of the World. Kew Bull. Addit. Ser. XIII. Her Majesty's Stationery Office, London. 389p.

Dore, W.B., and J. McNeill. 1980. Grasses of Ontario. Agriculture Canada, Ottawa, Monogr. 26. Hull, Quebec. 566p.

Eames, A.J. 1961. Morphology of the angiosperms. McGraw-Hill, New York, Toronto and London. xiii + 518p.

Fry, S.C. 1986. Cross-linking of matrix polymers in the growing cell walls of angiosperms. Ann. Rev. Pl. Physiol. 37: 165–186.

Gould, F.W., and R.B. Shaw. 1983. Grass systematics. Texas A & M University Press, College Station, Texas. xiv + 397 p.

Hubbard, C.E. 1948. Gramineae, pp. 284–348. *In* J. Hutchinson, British Flowering Plants. P.R. Gawthorn, London.

Jarvis, M.C. 1984. Structure and properties of pectin gels in plant cell walls. Plant, Cell and Environment 7: 153–164.

Mandoli, D.F., and W.R. Briggs. 1982. Optical properties of etiolated plant tissues. Proc. Natl. Acad. Sci. USA 79: 2902–2906.

Philipson, W.R. 1935. The development of the spikelet in *Agrostis canina*. New Phytologist 34: 421–436.

Soderstrom, T.R. 1981. Some evolutionary trends in the Bambusoideae (Poaceae). Ann. Missouri Bot. Gard. 68: 15–47.

Smith, P.M. 1972. Serology and species relationships in annual bromes (*Bromus* sect. *Bromus*). Ann. Bot. (London) 36: 1–30.

Stebbins, G.L. 1974. Flowering plants: evolution above the species level. Belknap, Harvard, and Edward Arnold, London. xviii + 399p.

Thomasson, J.R. 1985. Miocene fossil grasses. Possible adaptation in reproductive bracts (lemma and palea). Ann. Missouri Bot. Gard. 72: 843–851.

19. NESTS OF HARVESTING ANTS: A PREFERRED PRIMARY HABITAT OF WILD *BETA VULGARIS* IN ISRAEL

AVINOAM DANIN

Department of Botany, The Hebrew University, Jerusalem, Israel

ABSTRACT

In Israel, the common Mediterranean synanthropic weed *Beta vulgaris* L. is adapted to a kind of symbiotic relationship with harvesting ants. The ants collect the synaptospermic diaspores with 1–7 seeds and fail to obtain most of the seeds enclosed by the hardened sepals. The diaspores are then planted in the refuse zone together with rich organic material and soil particles removed from the nest. The seeds in the diaspores are protected from fire and also from granivores in the nest and its close vicinity. The wild beet grows successfully in the nutrient-rich refuse zone and competes successfully with the desert annuals. It is a successful weed and synanthropic ruderal plant.

Key words: Harvesting ants, ant nests, *Beta vulgaris*, primary habitat, Israel.

INTRODUCTION

Beta vulgaris L. (Chenopodiaceae), the progenitor of the cultivated beet (Bailey 1925; D. Zohary 1983), grows in synanthropic habitats around the Mediterranean (Zohary, 1966). Its taxonomic position is not fully agreed upon. Zohary (1966) regarded our taxon as *B. vulgaris* L. subsp. *maritima* (L.) Arc., whereas Greuter, Burdet and Long (1984, p. 297) consider it as *B. maritima* L. and do not record any *B. vulgaris* (*sensu stricto*) from Israel or its adjacent countries. According to D. Zohary (1983) and our own observations, *B. vulgaris* develops in slightly saline heavy soils under sub-Mediterranean conditions. The successful growth of the plant on clayey soils near nests of harvesting ants in the Jordan Valley is spectacular. Harvesting ants influence soil properties and vegetation near their nests in many areas (Cole 1932, Golley and Gentry 1964, Wight and Nichols 1966, Koyumdjisky et al. 1967, King 1977, Buckley 1982, Beattie 1985, Rissing 1986). Most of these authors found increased soil fertility near the nests, better aerated soil, and increased production of phytomass by the natural vegetation.

The aim of the present study is to show the impact of harvesting ants on soil fertility; to correlate species distribution, species diversity and phytomass production with ant-induced soil conditions, and to present the life strategies of *Beta vulgaris* that are important to its selection of primary and secondary habitats.

MATERIALS AND METHODS

The study was conducted in the Jordan Valley near Yafit, 25 km north of Jericho, Israel (35°27′40″E/32°02′30″N); altitude 250 m below sea level. The climate is Mediterranean with an average annual rainfall of 200 mm (Dorfman, 1981), falling mainly between October and April. The geological substratum is the Lisan Formation (Begin et al., 1974).

Eleven pairs of ant-induced sites and between-nest control sites were chosen. The controls ("a" sites) were about 0.5–1 m away from the sharp boundaries of the nest-induced vegetation ("b" sites). Sites 1–6 were on Alluvial Arid Brown Soil and sites 7–11 were on Calcareous Serozem (Dan et al., 1975). In addition, sites were sampled in a small wadi on the alluvial soil. Sampling of plants took place on February 20, 1987 when most species had ripe fruits. All the aerial parts of the plants were harvested in plots of 50 × 50 cm. One plot was harvested in each of the following habitats: refuse zone of 11 ant nests, control sites between nests, and 3 wadi sites.

The harvested plant material was separated into its constituent species. The number of individuals of each species was counted, then all specimens of each species were oven-dried for 48 hours at a temperature of 90°C and separately weighed.

Plant names and their chorotypes are according to Feinbrun-Dothan (1978, 1986) and Zohary (1966, 1972).

Sörensen's index of similarity (s) was used to compare pairs of floristic lists:

$$s = 2c \times 100/a + b$$

a = number of species in the first list
b = number of species in the second list
c = number of species common to both lists

The dry weight and number of individuals of each species were used to calculate species diversity (e^H, Hill 1973). The species that occurred in more than 50% of their habitat plots (i.e. the ant nests, control sites, and the wadi sites) were grouped. The dry weight of each of these species was calculated as a percentage of the total dry weight of all the species in the sample.

Phytogeographical analyses of floristic lists were made by calculating the frequency of each chorotype.

The coefficient of linear regression was calculated for vegetation variables (i.e., phytomass production, number of species, species diversity calculated by phytomass, and by number of individuals) and environmental variables (i.e., nutrient content of the soil and electrical conductivity).

Waste material disposed of by the ants in the refuse zone was collected on June 8, 1987 around a few nests on the alluvial soil. These included sites b3 and b6 that were sampled in the spring. The contents of some 40 cm^3 of waste material from each refuse zone was analyzed. The number of seeds and holes in the diaspores were counted in all diaspores of *B. vulgaris* and *Pteranthus dichotomus* Forsskål found in the 40 cm^3.

Soil was sampled on the day of harvesting, at depths of 0–5 cm and 5–10 cm in most sites (excluding 2 of the wadi sites). The nitrate content of the soil was analysed by the use of Szechrome NAS Reagent (Szekely et al. 1972). Phosphorus and potassium contents were analysed by the flame photometer method (Richards 1954). The electrical conductivity (EC) of extracts of saturated soil was used to estimate total soil salinity.

Environment and Vegetation

The site is in the Mediterranean territory of Isreal where bi-regional species

Figure 1. A nest of harvesting ants surrounded by tall annual herbs with *Beta vulgaris* dominant — Jordan Valley near Pazael (site b3).

(Mediterranean-Irano-Turanian) are the second most frequent chorotype (Danin and Plitmann 1987). A phytogeographical analysis of an area of 225 km^2, including the study sites in its centre, comprised 477 species of which 34.8% were Mediterranean, 15.9% bi-regional Mediterranean-Irano-Turanian, and 11.5% Saharo-Arabian. An analysis of the 74 species that occurred in all the 25 plots studied showed: Mediterranean: 23%, Mediterranean-Irano-Turanian: 23%, Saharo-Arabian: 24%. This relatively high proportion of Saharo-Arabian species is probably due to the occurrence of halophytes and is discussed further later.

The two areas studied were dominated by annual plants; a few scattered semi-shrubs of the halophyte *Salsola damascena* Botsch. (*Salsola vermiculata* L. sensu M. Zohary, 1966) occur here and there. This is an indicator of the low salinity of the soils (Danin, 1978). The dominant annuals in patches with higher salinity are *Mesembryanthemum nodiflorum* L., *Pteranthus dichotomus*, *Aizoon hispanicum* L. and *Limonium thouinii* (Viv.) O. Kuntze. The dominants in leached sites out of wadis are *Stipa capensis* Thunb., *Plantago ovata* Forsskål, and *Reboudia pinnata* (Viv.) O.E. Schulz. Most of these annual plants are 5–20 cm high.

There are circles or oval-shaped concentrations up to 4 m across (Fig. 1) of 100–150 cm high plants of *Beta vulgaris*, *Chrysanthemum coronarium* L., and *Malva parviflora* L. Most of these concentrations are nests of the harvesting ant *Messor semirufus* (E. Andre) Santschi subsp. *ebeninus* Forel (Avidov 1968). Some of the tall plants are also concentrated along small depressions.

RESULTS

Soil Properties

The nutrient content and conductivity of the topsoil in the sites is presented in Table 1. The mean nitrate content of the ant-affected soils was 48.1 times higher

Table 1. Means and standard errors of nutrient content (ppm) and electrical conductivity (EC, mmho/cm) of soils in the study area.

site	depth (cm)	NO_3^-	± SE	P	± SE	K	± SE	EC	± SE
a1–a6	0–5	6.8	1.8	6.5	1.4	10.9	2.5	1.2	0.5
a1–a6	5–10	6.4	1.7	4.7	1.4	10.8	2.2	7.6	2.9
b1–b6	0–5	326.9	13.4	36.4	3.6	35.9	3.5	2.5	0.9
b1–b6	5–10	140.5	14.2	12.7	2.3	10.1	2.2	1.6	1.1
a7–a11	0–5	14.7	3.6	12.1	1.6	74.5	7.3	6.7	2.7
a7–a11	5–10	36.2	7.8	4.2	0.7	99.3	7.4	15.9	4.0
b7–b11	0–5	256.0	13.5	49.8	4.7	135.2	8.8	2.1	1.1
b7–b11	5–10	308.5	21.1	20.5	3.2	70.4	4.9	1.5	1.0
13	0–5	255.0		32.8		67.6		2.9	
13	5–10	124.0		16.0		26.2		1.2	

than that of the control (Table 1). In 21 of the 22 sample pairs that were compared, the phosphorus was higher in the ant-induced sites; the mean was 5.6 times higher near the nests than in the controls. In 15 pairs there was higher content of potassium and in 7, it was lower in the ant-induced sites than in the controls; the mean was 3.3 times higher than the control at the topsoil near the ant nests. In addition to plant parts and soil grains in the refuse zone of the ant nests there were dead ants, other dead insects, and the shells of land snails. In most cases the nutrient concentration in the topsoil (0–5 cm) was higher than that at a depth of 5–10 cm; this is also clear in the mean values (Table 1).

The electrical conductivity (EC in Table 1) was relatively low in the top layer of the control sites as compared with that of the soil at a depth of 5–10 cm. The ant-induced sites (b) had much lower conductivity values than in the control sites; the mean values (Table 1) were 4.75 times lower in sites 1–6 and 10.6 times lower in sites 7–11. Topsoil in the ant-induced sites had slightly higher values of EC than that in the 5–10 cm layer.

Flora and Phytomass

Of the 77 plant species found in the harvested samples 18 species occurred only in the control areas, 13 occurred only on the ant nests, and 4 only in the wadi sites. The most common species that occurred in more than 50% of the plots are shown in Table 2. In this table the percentage of presence (P) in their habitat group of samples is noted. The average contribution of these species to the total dry weight of each plot (W) is presented as well. The mean percentage of dry weight of the most frequent species (Table 2) indicates the importance of *Aizoon hispanicum* and *Pteranthus dichotomus* in the control areas and their rare occurrence or absence from the nests and wadi. The dominant plants of the nest sites are *Beta vulgaris* and *Malva parviflora*.

The mean Sörensen's index of similarity (± SE) for individual ant nests and their neighbouring control site was 30.2% ± 2.7% (n = 11). The mean index of similarity for the control plots of sites 1–6 was 56.6% ± 3.1, and for the ant plots in these sites it was 42.1% ± 3.2.

A phytogeographical analysis was conducted for the list of species of each site.

Table 2. Percentage (P) of species which occurred in more than 50% of the plots in each habitat, and their average weight (W), expressed as a percentage, in each plot where they were present.

No. of plots	Control sites (11)		Ant nests (11)		Wadi sites (3)	
	P	W	P	W	P	W
Aizoon hispanicum L.	91	30.4	36	0.4	66	0.6
Anagallis arvensis L.	73	0.5	36	0.7	100	6.1
Beta vulgaris L.	64	6.6	100	44.3	100	57.6
Echium judaeum Lacaita	9	3.6	73	25.7	33	4.1
Emex spinosus (L.) Campd.			27	0.8	100	5.0
Erodium malacoides (L.) L'Hér.	9	1.1	36	0.6	100	1.6
Eruca sativa Miller	27	5.6	55	9.8		
Limonium thouinii (Viv.) O. Kuntze	73	8.1	9	1.4		
Malva parviflora L.	18	8.8	73	11.7	100	10.6
Medicago laciniata (L.) Miller	64	1.2	27	1.4	33	0.2
Plantago coronopus L.	55	2.0				
Plantago ovata Forsskål	91	14.6				
Pteranthus dichotomus Forsskål	73	22.9				
Reboudia pinnata (Viv.) O. Schulze	55	4.4	36	5.1		
Spergula fallax (Lowe) E.H.L. Krause	55	4.0	45	0.9	66	1.2
Stipa capensis Thunb.	91	11.9	45	7.3	33	0.2

All the Saharo-Arabian, Irano-Turanian and bi-regional Saharo-Arabian-Irano-Turanian species were grouped as "desert species". All the Mediterranean, bi-regional Mediterranean-Irano-Turanian, and tri-regional Mediterranean-Irano-Turanian-Euro-Siberian species were regarded as "mesophytic species". The rest constitute the remaining percentage. The results are:

1. The control sites had 50% desert species and 31% mesophytic species.
2. The ant nests had 26.8% desert species and 60.3% mesophytic ones.
3. The wadi had 17.3% desert species and 72.3% mesophytic ones.

Values of phytomass production, number of species, and species diversity of each plot are presented in Table 3. The highest mean values of phytomass production per plot was in the ant nests (226.6 g), 4.66 times higher than the control; the wadi had phytomass 4.32 times higher than the control. Species number and species diversity were higher in the control, lower in the wadi and the lowest in the ant nests.

The results of linear regressions between the vegetation variables and environmental ones are presented in Table 4. The only significant relationship is that of phytomass production with topsoil fertility. The three variables related to the number of species per plot are either not in linear relationship with any of the five variables tested, or their significance is too low.

Relationship of Beta vulgaris with Ant Nests

The means and standard error dry weight of an individual plant of *B. vulgaris* in the 7 control sites where it occurred was 0.20 ± 0.14 g; in the 11 ant nest sites it

Table 3. Means and standard errors of phytomass, species number, and species diversity in various habitats.

	Control (n = 11)		Ant nests (n = 11)		Wadi (n = 3)	
	mean	SE	mean	SE	mean	SE
Phytomass $g/0.25\,m^2$	48.6	5.1	226.6	8.8	210.4	5.8
Number of species/$0.25\,m^2$	15.5	2.2	13.5	2.0	14.3	2.3
Species diversity (e^H)	5.35	1.7	3.8	1.3	4.1	1.1

was 8.36 $\pm$ 3.06 g (41.8 times higher than in the control); in the 3 wadi sites it was 4.43 $\pm$ 1.44 g (23.1 times higher than in the control).

Synaptospermic diaspores of *B. vulgaris* contain 1–7 seeds. Of the 141 seeds in the previous year's diaspores from site b3, 4.2% were eaten; of 236 seeds in the older diaspores 17.8% were missing. In site b6, 0.6% of the 318 seeds in the previous year's diaspores were eaten and 35.6% of the 104 older ones.

Synaptospermic diaspores of *Pteranthus dichotomus* contain 1–5 seeds. Here most diaspores collected in the refuse zone were those of the previous year. Of the 351 seeds in diaspores from site b3, 88% were removed by ants; in site b6, 86% of the 702 seeds were taken.

DISCUSSION

Beta vulgaris is regarded as a synanthropic plant in the east Mediterranean area by several authors (Post, 1932–1933; Eig, Zohary and Feinbrun, 1948; M. Zohary, 1966, D. Zohary, 1983). The principal hypothesis that I wish to examine in the present paper is that this plant is pre-adapted to synanthropic and other nutrient-rich environments by being adapted to live with harvesting ants.

My findings show that soil conditions near the nests are improved as compared to the surrounding soils. This is in agreement with other areas as reviewed by Beattie (1985). The chemical improvement is indicated by the increased amounts of macronutrients. The physical improvement is apparent by the enhanced leaching of the soil near the ant nests. One of the principal sources for nutrients is the organic material disposed of by the harvesting ants in the refuse zone (Avidov 1968, Offer 1980). The amounts of nitrates and phosphorus in the soil of the wadi, where phytomass is high, are closer to those of the nest than to the amounts in the control areas. However, it seems that the higher values of macronutrients near the nests may be related to the function of the refuse zone as a graveyard. The enhanced leaching of the wadi, where large amounts of water are conducted is well known (Danin 1983). The comparable salinity at depths of 5–10 cm near the nests is probably a result of good aeration of the ground in the tunnels and pedoturbation caused by the ants (Baxter and Hole 1967, Koyumdjisky et al. 1967, Offer 1980). The improvement of soil conditions matches the development of vegetation. Phytomass production had significantly increased

Table 4. Results of linear regressions between vegetation variables and environmental ones. Soil variables refer to the layer of 0–5 cm. Variables: Dep. = dependent variable; DN = species diversity calculated by number of individuals; DW = species diversity calculated by dry weight; EC = electrical conductivity; Indep. = independent variable; N = concentration of nitrate; n.s. = not significant; P = concentration of phosphorus; K = concentration of potassium; r = coefficient of linear regression; s = number of species in the plot; Sig. = level of significance; W = dry weight of the plot. Number of plots = 22.

Dep.	Indep.	r	Sig.
W	N	0.709	0.0001
W	P	0.454	0.05
W	K	0.444	0.05
W	EC	−0.133	n.s.
s	N	−0.063	n.s.
s	P	−0.060	n.s.
s	K	−0.139	n.s.
s	EC	−0.369	0.05 < Sig. < 0.1
s	W	−0.209	n.s.
DW	N	−0.129	n.s.
DW	P	−0.181	n.s.
DW	K	−0.121	n.s.
DW	EC	−0.226	n.s.
DW	W	−0.392	0.05 < Sig. < 0.1
DN	N	−0.286	n.s.
DN	P	−0.277	n.s.
DN	K	−0.228	n.s.
DN	EC	−0.388	0.05 < Sig. < 0.1
DN	W	−0.0933	n.s.

in the study area with increased soil fertility. The phytogeographical spectrum of the nests' vegetation has more mesophytes than in the control area.

Although in close proximity, the flora of the nests has a low index of similarity with that of the control sites. This value is higher for the flora of the control sites as compared with each other, or for the ant nests with each other. This floristic distance is a result of the impact of edaphic conditions on the nests' flora because seeds of plants from the control area can easily arrive at the ant nests. This is exemplified by the seeds of the halophyte *Pteranthus dichotomus* that are found in the refuse zone. No adult plants of the latter were in the harvested plots although it was of major importance in the vegetation of the control plots. The flora of the nests successfully competes with the halophytes in the leached and fertile soil.

Contrary to the good correlation between species diversity and soil salinity, found in the saline and leached soils 32 km south of this area (Danin 1976), the relations here are much more complicated. Several control sites had low species diversity resulting from their high salinity (as reported by Danin 1976); ant nests near these sites have relatively high species diversity. Other control sites which were less saline had higher species diversity as expected. However, the nearby nests had relatively low species diversity resulting from overtopping of most species by successful species such as *Beta vulgaris*. The increase in phytomass and the reduction in species diversity on the nest parallel change induced by fertilization of poor sandy soils in Devon, England with the macronutrients discussed above (Willis and Yemm, 1961; Willis 1963).

Seeds of *B. vulgaris* are dispersed by ants in synzoochory or dyszoochory (van der Pijl 1972). The term myrmecochory as applied by Sernander (1906) cannot be used in this case since there are no elaiosomes or other ant-attractants on the diaspores. The seeds are protected from the ants by the indurated sepals. Of the many beet diaspores collected by the ants, very few are consumed. Most of the seeds are disposed of in the nutrient-rich sites of the refuse zone. The wild beet reaches its optimal vegetative and reproductive development in the refuse zone of ant nests. The case of *B. vulgaris* certainly follows the "nutrient hypothesis" (Beattie 1985) of plant-ant relationships. However, it is achieved here by synzoochory and not by the better documented myrmecochory. Although *B. vulgaris* also thrives in the wadis, it seems that the ant nests function as a safer habitat for the beet. Fires that can easily affect the beet plants amongst uninterrupted vegetation in the wadi have less access to the isolated 'islands' of vegetation (Fig. 1) near the ant nests. Beet diaspores that are stored in the nest before being planted in the refuse zone are thus protected for some time from possible fires and granivores. This adaptation is in agreement with the "fire-avoidance hypothesis" (Berg 1975, Beattie 1985).

The pre-adaptation of *B. vulgaris* to the ant nest habitat enabled it to become an important weed in agricultural areas and in ruderal synanthropic habitats. These habitats became important only in the last millennia (Pignatti 1978, 1983, Zohary 1973).

ACKNOWLEDGMENTS

I thank Prof. U. Plitmann for discussing ant-plant relationships during my research, Miss Navah Eisland and Miss Morit Danin for technical assistance, and the staff of the soil-water laboratory, Ministry of Agriculture, Jerusalem, for the soil analyses.

LITERATURE CITED

Avidov. Z. 1968. Harvesting ants in Israel. Sifriath Hassadeh, Tel Aviv (Hebrew). 87 p.

Bailey, L.H. 1925. The standard cyclopedia of horticulture. Macmillan, New York.

Baxter, F.P., and F.D. Hole. 1967. Ant (*Formica cinerea*) pedoturbation in a prairie soil. Proc. Soil Sci. Soc. of Amer. 31: 425–428.

Beattie, A.J. 1985. The evolutionary ecology of ant-plant mutualisms. Cambridge University Press, Cambridge.

Begin, B.Z., A. Ehrlich, and Y. Nathan. 1974. Lake Lisan, the Pleistocene precursor of the Dead Sea. Geological Survey of Israel, Report GSI/29/85, Jerusalem. 18 p.

Berg, R. 1975. Myrmecochorous plants in Australia and their dispersal by ants. Austral. J. Bot. 23: 475–508.

Buckley, R.C. 1982. Ant-plant interactions: a world review, pp. 111–141. *In*: R.C. Buckley [ed.], Ant-Plant interactions in Australia. Dr. W. Junk, The Hague.

Cole, A.C. 1932. The ant, *Pogonomyrmex occidentalis* Cr., associated with plant communities. Ohio J. Sci. 32: 10–20.

Dan, J., D.H. Yaalon, H. Koyumdjisky, and Z. Raz, 1975. The soils map of

Israel (1:500,000). Ministry of Agriculture, Bet Dagan, Israel.
Danin, A. 1976. Plant species diversity under desert conditions. I. Annual species diversity in the Dead Sea Valley. Oecologia 22: 251–259.
——. 1978. Species diversity of semishrub xerohalophyte communities in the Judean Desert of Israel. Israel J. Bot. 27: 66–76.
——. 1983. Desert vegetation of Israel and Sinai. Cana, Jerusalem. 148 p.
——, and U. Plitmann. 1987. Revision of the plant geographical territories of Israel and Sinai. Plant Syst. Evol. 150: 43–53.
Dorfman, Z. 1981. Rainfall map. Ministry of Agriculture, Tel Aviv, Israel.
Eig, A., M. Zohary, and N. Feinbrun. 1948. Analytical Flora of Palestine. Hamagdir, Jerusalem (Hebrew). 514 p.
Feinbrun-Dothan, N. 1978. Flora Palaestina. Part 3. Israel Academy of Sciences and Humanities, Jerusalem. 481 p.
——. 1986. Flora Palaestina. Part 4. Israel Academy of Sciences and Humanities, Jerusalem. 463 p.
Golley, F.B., and J.B. Gentry, 1964. Bioenergetics of the southern harvester ant *Pogonomyrmex badius*. Ecology 45: 217–225.
Greuter, W., H.M. Burdet, and G. Long. [eds.] 1984. Med-Checklist 1. Editions des Conservatoire et Jardin botaniques de la Ville Genève, i–xvi, 330 p., xvii–c.
Hill, M.O. 1973. Diversity and evenness: a unifying notation and its consequences. Ecology 54: 427–432.
King, T.J. 1977. The plant ecology of ant-hills in calcareous grasslands. I. Patterns of species in relation to ant-hills in southern England. J. Ecology 65: 235–256.
Koyumdjisky, H., H. Tsaban, and J. Offer. 1967. The influence of ant-nests on soil fertility in natural pasture. Ktavim 17: 195–208 (Hebrew).
Offer, J. 1980. The ecology of ant populations of the genus *Messor* and their influence on the soil and flora in pasture. Ph.D. thesis, The Hebrew University, Jerusalem. (Hebrew).
Pignatti, S. 1978. Evolutionary trends in Mediterranean flora and vegetation. Vegetatio 37: 175–185.
——. 1983. Human impact in vegetation of the Mediterranean basin, pp. 151–161. *In* W. Holzner, M.J.A. Werger and I. Ikusima [eds.], Man's impact on vegetation. Dr. W. Junk, The Hague.
Pijl, L. van der. 1972. Principles of dispersal in higher plants. 2nd ed. Springer-Verlag, New York. 162 p.
Post, G.E. 1932–1933. Flora of Syria, Palestine and Sinai. ed. 2 by J.E. Dinsmore. Beirut. 2 vols., 658 & 928 p.
Richards, L.A. [ed.] 1954. Saline and alkali soils. United States Salinity Laboratory Staff, Agriculture Handbook No. 60, Washington D.C. 160 p.
Rissing, S.W. 1986. Indirect effects of granivory by harvester ants: plant species composition and reproductive increase near ant nests. Oecologia 68: 231–234.
Sernander, R. 1906. Entwurf einer Monographie der europaeischen Myrmekochoren. Kongl. Svenska Vetenskapsakad. Handl. 41: 1–410.
Szekely, E., E. Bandel, and M. Flitman. 1972. Volumetric determination of primary arylamines and nitrates using an internal indicator system. Talanta 19: 1429–1435.
Wight, J.R., and J.T. Nichols. 1966. Effects of harvester ants on production of saltbush community. J. Range Manage. 19: 68–71.
Willis, A.J. 1963. Braunton Burrows: the effects on the vegetation of the addition of mineral nutrients to the dune. J. Ecol. 51: 353–374.
——, and E.W. Yemm. 1961. Braunton Burrows: mineral nutrient status of the dune soils. J. Ecol. 49: 377–390.
Zohary, D. 1983. Wild genetic resources of crops in Israel. Israel J. Bot. 32: 97–127.

Zohary, M. 1966. Flora Palaestina. Part 1. Israel Academy of Sciences and Humanities, Jerusalem. 364 p.

——. 1972. Flora Palaestina. Part 2. Israel Academy of Sciences and Humanities, Jerusalem. 489 p.

——. 1973. Geobotanical foundations of the Middle East. Gustav Fischer Verlag, Stuttgart. 2 vols., 739 p.

20. THE EMBRYOLOGY AND SYSTEMATIC POSITION OF *RHABDODENDRON* (RHABDODENDRACEAE)

HIROSHI TOBE

Biological Laboratory, Yoshida College, Kyoto University, Kyoto 606, Japan

AND

PETER H. RAVEN

Missouri Botanical Garden, P.O. Box 299, St. Louis, Missouri 63166, U.S.A.

ABSTRACT

The embryology of *Rhabdodendron*, one of the genera whose systematic position is most controversial at the present time, is studied and compared with its alleged relatives Phytolaccaceae (Caryophyllales), Rutaceae (Rutales), and Rosales. The main embryological features of *Rhabdodendron* are mostly plesiomorphic, and include: mature pollen 2-celled; ovule campylotropous, bitegmic and crassinucellate; ovular archesporium multi-celled; embryo sac of a unique variety of the Polygonum-type; endothelium not formed; endosperm of the Nuclear-type; exotesta not palisadal; and mature seed containing a small amount of endosperm. Comparisons on the basis of embryological data available indicate that *Rhabdodendron* hardly agrees with Phytolaccaceae, but that it agrees with Rutaceae and Rosales sensu Dahlgren, respectively, in many characters, although it is clearly distinct from each of them in a few, mainly common, characters. The embryological features of *Rhabdodendron*, taken together with other evidence, suggests that *Rhabdodendron* can best be placed near or in Rosiflorae as a distinct family, but that it should not be considered a member of the order Rosales sensu Dahlgren.

Key words: embryology, Rhabdodendraceae, *Rhabdodendron*, Rosales, Rutaceae.

INTRODUCTION

Rhabdodendron, comprising 3 species confined to South America (Prance 1972), is currently one of the most controversial taxa of dicots with respect to its systematic position and affinities. Historically, it has been included in Chrysobalanaceae or Rosaceae sensu lato (including Chrysobalanaceae) as '*Lecostemon*' (e.g., Bentham 1853; Hooker 1867), in Rutaceae (Gilg and Pilger 1905; Huber 1909; Engler 1931; Scholz 1964; Takhtajan 1969; Thorne 1983; see also Puff and Weber 1976), in Phytolaccaceae (Record and Hess 1943), or in its own family Rhabdodendraceae (Airy Shaw 1966; Prance 1968, 1972). In recent years, *Rhabdodendron* has mainly been treated as a separate family, sometimes as a member of Rosales sensu lato or sensu stricto (Cronquist 1981, 1983; Dahlgren 1983), Rutales (Takhtajan 1980), or Caryophyllales (= Centrospermae) (Prance 1968, 1972).

The position of *Rhabdodendron* has been considered by a number of authors, some of whom exclusively investigated a certain anatomical structure(s) or chemicals. Its wood anatomy was examined by Record (1933), Milanez (1943), Metcalfe and Chalk (1950), and Prance (1968); except for Milanez, who suggested a close resemblance with Rutaceae, the other authors considered it to be near Phytolaccaceae. The leaf anatomy of *Rhabdodendron* was examined by Prance (1968) and Puff and Weber (1976), who also studied nodal anatomy; in these characters, *Rhabodendron* appears to be more similar to Rutaceae than to Phytolaccaceae. Sieve element plastids were investigated by Behnke (1976) and Behnke and Dahlgren (1976), and pollen ultrastructure by Behnke (1977); these investigators concluded that *Rhabdodendron* has no structures characteristic of Caryophyllales (including Phytolaccaceae). Puff and Weber (1976) examined some reproductive structures, including floral and ovule morphology and chromosomes, and concluded that *Rhabdodendron* was related to Rutaceae. More recently ellagic-acid derivatives were analyzed by Wolter Filho et al. (1985); this work suggested that chemically *Rhabdodendron* fits into the Rosiflorae or even better into Myrtiflorae.

Despite those intensive studies, we are still far from a consensus concerning the affinities of *Rhabdodendron*. The purpose of this paper is to present the embryological features of *Rhabdodendron*, which have been little studied so far (see Cronquist 1981, 1983), and to use these findings to evaluate its systematic position.

MATERIALS AND METHODS

One of the 3 species of the genus, *Rhabdodendron macrophyllum* (Spruce ex Bentham) Huber, was examined in this study. The material was collected at Manaus, Brazil and fixed in FAA (voucher specimen: *B. Nelson 1290*, NY). For observations, 83 buds and flowers and 24 fruits in various stages of development were sectioned longitudinally, and 8 buds and flowers were sectioned transversely. Preparations of microtome sections were made following a technique described elsewhere (Tobe and Raven 1983a).

OBSERVATIONS

Anther and Microspores

The anther is tetrasporangiate (Fig. 4). The wall structure prior to maturity comprises four cell layers; i.e., an epidermis, an endothecium, one middle layer, and a tapetum (Fig. 1). The middle layer shares its developmental origin with the tapetum (Fig. 1), and therefore wall formation conforms to the Monocotyledonous-type (see Davis 1966, p. 10). During maturation, the middle layer collapses, while the endothecium develops fibrous thickenings (Fig. 3). The cells of the epidermis, which are tanniferous throughout all stages of development, are persistent (Figs. 1–4). The tapetum is glandular; its cells become two-nucleate before degeneration (Fig. 2). For these reasons, the mature anther wall is composed of the tanniferous epidermis and the fibrous endothecium (Figs. 3 and 4).

Meiosis in the microspore mother cells is accompanied by simultaneous cytokinesis. The arrangement of microspores of the resultant tetrad, on the basis of the examination of 50 selected tetrads, is usually (90%) tetrahedral and very occasion-

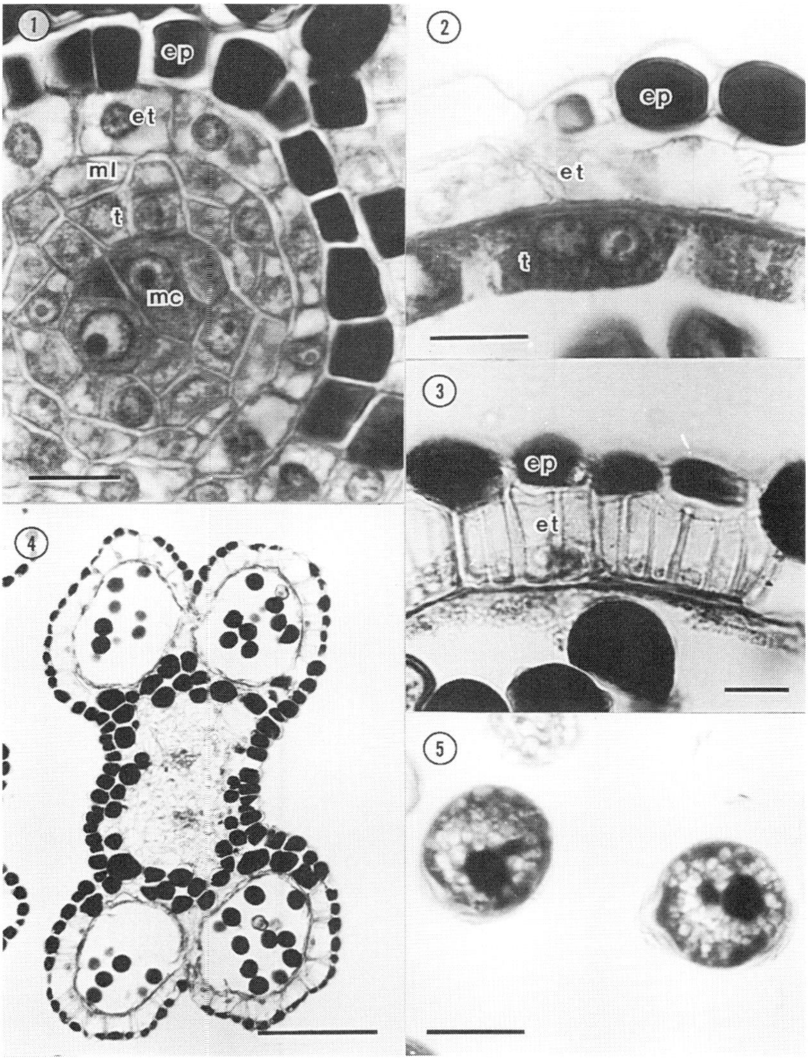

Figures 1–5. Development of anthers and microspores. — 1. Transverse section (TS) of young microsporangium showing the wall structure. — 2. TS of older microsporangium. — 3. TS of mature microsporangium showing the fibrous endothecium. — 4. TS of mature anther. — 5. Section of mature pollen grains. Note that grains are two-celled at the time of shedding. Abbreviations: ep, epidermis; et, endothecium; mc, microspore mother cell; ml, middle layer; t, tapetum. Scales, respectively, 10, 10, 10, 100, 10 μm.

ally (10%) decussate. Pollen grains are two-celled when shed (Fig. 5), contrary to the results of Puff and Weber (1976, p. 211), who described 'most pollen grains' as 'trinucleate.' We confirmed the two-celled nature of mature pollen grains using both microtome-sectioned and acetocarmine-stained pollen grains.

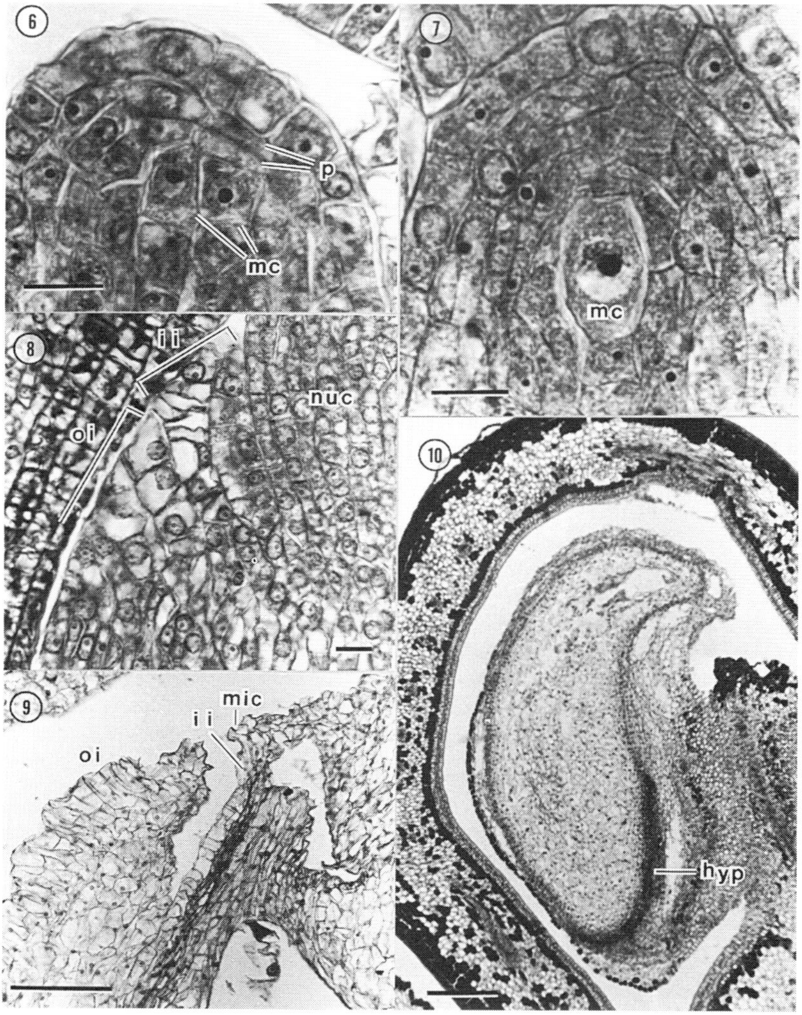

Figures 6–10. Structure of ovule and nucellus. — 6. Longitudinal section (LS) of young ovule with plural youngest megaspore mother cells. — 7. LS of older ovule with a megaspore mother cell. — 8. LS of young ovule showing an early indication of bitegmy. — 9. LS of mature ovule. Note that a micropyle is formed by the inner integument alone. — 10. LS of mature ovule showing its campylotropous orientation. Abbreviations: hyp, hypostase; ii, inner integument; mc, megaspore mother cell; mic, micropyle; nuc, nucellus; oi, outer integument; p, parietal cell. Scales, respectively, 10, 10, 10, 100, 200 μm.

Megagametophyte and Nucellus

To investigate whether the archesporium is one-celled or multi-celled, we prepared sections of some young ovules. In these, however, we could not easily distinguish archesporial cells because there are no clear morphological and histological differences between them and the surrounding cells. However, since we observed plural (usually 2 or 3) megaspore mother cells in the older ovules (Fig. 6), we concluded that the archesporium of *Rhabdodendron* is multi-celled. Certainly, more than two parietal cell-layers are present above those youngest megaspore mother cells (Fig. 6), and therefore the ovule is crassinucellate. Only one megaspore mother cell functions (Fig. 7), undergoing meiosis to give rise to a tetrad of megaspores (Figs. 11 and 12). The megaspore tetrads are linear, with the chalazal megaspore functional (Figs. 13 and 14).

Subsequently, by three successive nuclear divisions, the functional megaspore develops into a two- (Fig. 15), four- (Fig. 16), and eight-nucleate embryo sac (Fig. 17). Of the four chalazal nuclei in the eight-nucleate embryo sac, two nuclei soon degenerate without developing into antipodal cells. At that time, one of the remaining four micropylar nuclei moves toward the chalazal end and is positioned close to one of the two non-degenerated chalazal nuclei, forming 'polar nuclei,' while another non-degenerated chalazal (= 'antipodal') nucleus remains as it was. Consequently, an organized embryo sac is composed of six nuclei or cells; that is, an egg cell and two synergids at the micropylar end, and two 'polar nuclei' and one 'antipodal' nucleus at the chalazal end (Fig. 18). The three nuclei at the chalazal end are always surrounded by relatively dense cytoplasm, which also contains many starch grains (Fig. 18). We confirmed these features of the organized embryo sac in all seven of the mature ovules that we examined except one; in that one, the embryo sac lacked all antipodal cells. *Rhabdodendron* thus has a monosporic Polygonum-type embryo sac, which, however, has a unique six-nucleate or six-celled structure.

In the megaspore mother cell stage, the nucellus is small, with the megaspore mother cell surrounded by nucellar tissue four or five cells thick. During the development of the embryo sac, however, the nucellus enlarges by cell divisions so that the functional megaspore would be relatively deeply buried in the nucellus (Fig. 13). The nucellar tissue is persistent and remains as one to two cell layers in the mature seed (Fig. 22). An endothelium, which is usually combined with an early disintegrating nucellar tissue, is not formed. The nucellar apical cells do not divide periclinally to form a nucellar cap. At the mature embryo sac stage the nucellus is curved, even though Puff and Weber (1976, p. 207) described it as 'non-arched,' and a conspicuous hypostase is formed (Fig. 10).

Integuments and Ovular Orientation

Puff and Weber (1976, p. 207) described *Rhabdodendron* as having only one integument, and their report has been cited subsequently (e.g., Cronquist 1981, p. 586). Despite this, the ovule of *Rhabdodendron* is clearly bitegmic. Two integuments are initiated as a single common primordium in the young ovule, a pattern of development that probably misled Puff and Weber. For this reason, the bitegmy of the ovule cannot be recognized until the ovule develops into an early megaspore mother cell stage, when the common primordium shows bifurcation at its tip (Fig. 8). Even at the mature ovular stage, the bitegmic condition is restricted to the micropylar part and never extends wholly toward the chalazal end (Fig. 13). The inner integument is initially two cell-layers thick (Fig. 8), whereas the outer one is

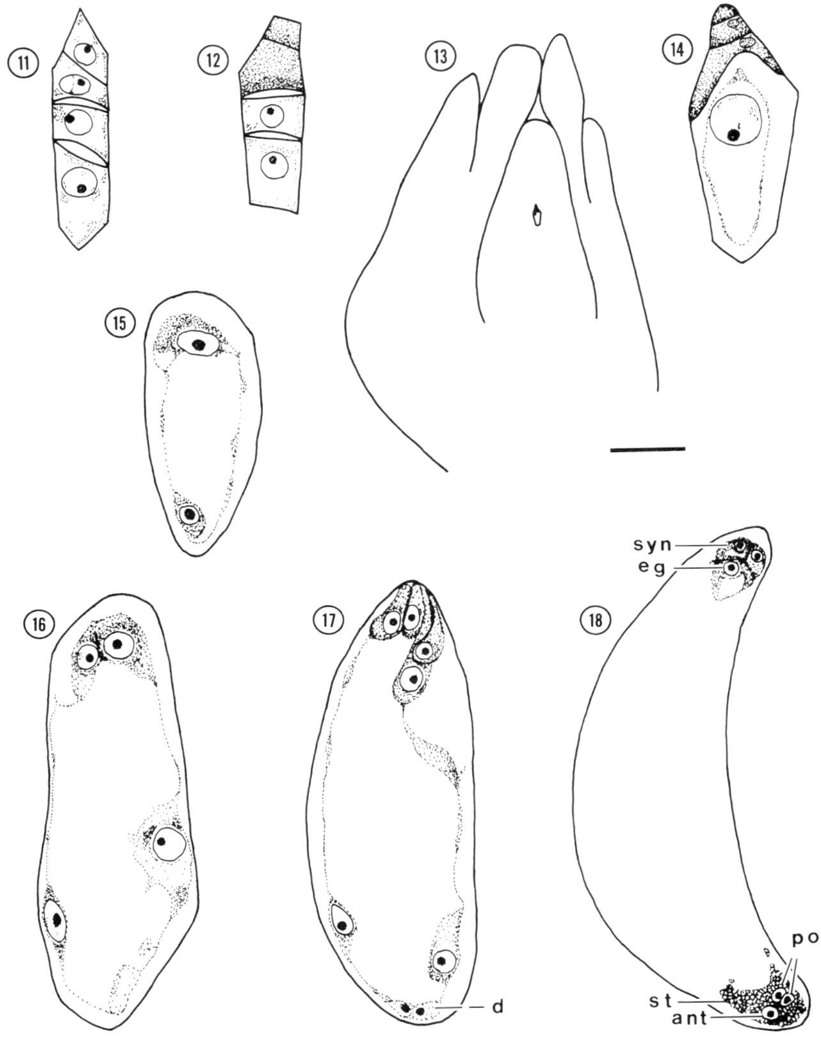

Figures 11–18. Development of embryo sac. — 11 and 12. Tetrad of megaspores; two upper megaspores are degenerating in Fig. 12. — 13 and 14. Functional megaspore; Fig. 13 also shows that the bitegmic condition is restricted to the micropylar part. — 15. Two-nucleate embryo sac. — 16. Four-nucleate embryo sac. — 17. Eight-nucleate embryo sac. Note that two of the chalazal nuclei are degenerating. — 18. Organized embryo sac, which lacks antipodal cells, instead with two polar nuclei and one antipodal nucleus at the chalazal end. Abbreviations: ant, antipodal nucleus; d, degenerating nuclei; eg, egg cell; po, polar nucleus; st, starch grains; syn, synergid. Scales, respectively, 10, 10, 10, 100, 20, 20, 20, 50 μm.

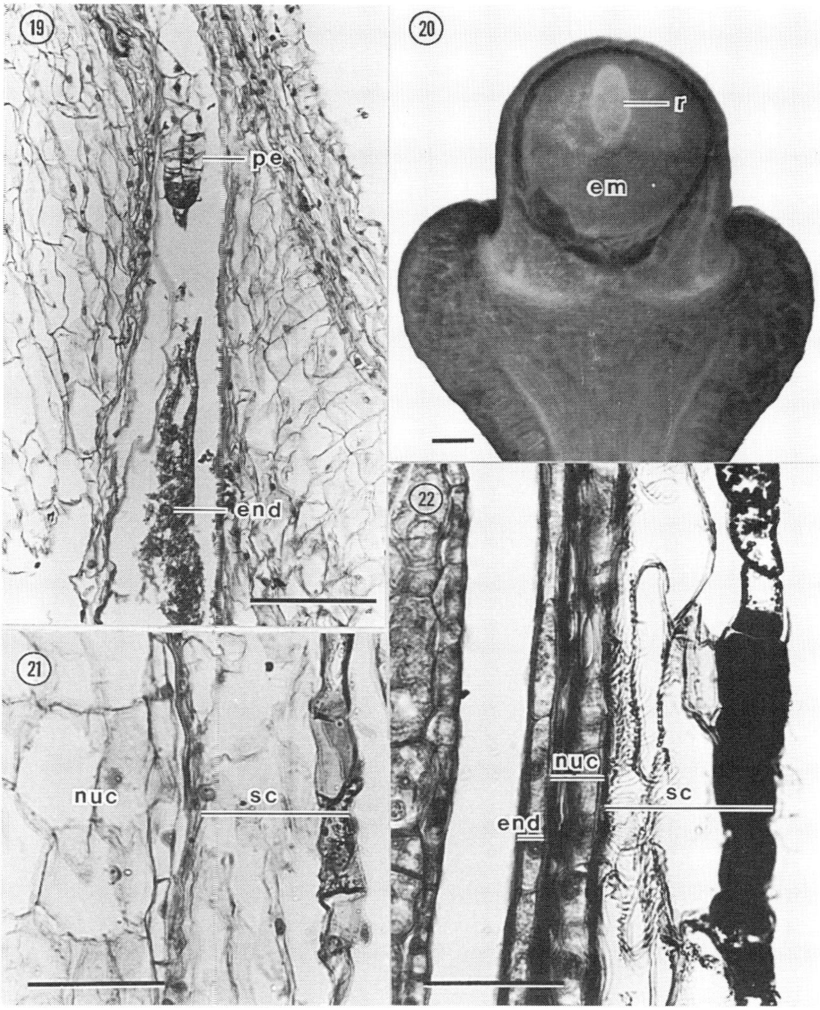

Figures 19–22. Structure of seed and seed coat. — 19. Longitudinal section (LS) of young seed containing a proembryo and free endosperm nuclei. — 20. LS of mature fruit. — 21. LS of young seed coat. — 22. LS of mature seed coat. Note that a small amount of endosperm and of nucellar tissue are persistent. Abbreviations: em, embryo; end, endosperm; nuc, nucellus; pe, proembryo; r, radicle; sc, seed coat. Scales, respectively, 100, 1000, 50, 50 μm.

much thicker. The inner integument is multiplicative particularly at its upper (i.e., micropylar) part and secondarily increases in thickness, whereas the outer integument does not multiply so extensively. Consequently, when the ovule is mature, the upper part of the inner integument is five cell-layers thick, and the outer integument four to five cell-layers thick. The micropyle is formed by the inner integument alone (Fig. 9).

The ovule of *Rhabdodendron* is always campylotropous (more strictly, 'hemi-campylotropous' [Bocquet and Bersier 1960]; Fig. 10), as reported by Prance (1972, p. 11). Puff and Weber (1976, p. 207), described the ovule as hemianatropous, and their report was cited by Cronquist (1981, p. 586). Since the ovular orientation changes from hemitropy to campylotropy as the ovule matures, it seems that the observation by Puff and Weber was probably made on the basis of immature ovules.

Obturator

An obturator is not formed in *Rhabdodendron*.

Fertilization, Endosperm and Embryo

Fertilization is porogamous. Endosperm formation is of the Nuclear-type (Fig. 19). At early free endosperm nuclear stages, the nuclei are located only on the chalazal side and are not present around the proembryo (Fig. 19). The mature seed contains a small amount of endosperm, which is one cell-layer thick (Fig. 22).

We did not examine embryogenesis in detail, but did observe the developing proembryo, which has a fairly long, multiseriate suspensor (Fig. 19). The proembryo eventually develops into a nearly globular embryo (Fig. 20). In a longitudinally sectioned mature embryo, we clearly observed the position of the radicle, which develops upwards (Fig. 20). We could not distinguish the individual cotyledons in the mature embryo, which interestingly is chlorophyllous.

Mature Seed and Seed Coat

The mature seed of *Rhabdodendron* is globular and lacks appendages such as arils or wings.

The seed coat is formed over most of the seed by the unitegmic part of the ovule, rather than by the bitegmic part, which is restricted to the area near the micropyle. Throughout its development, the constituent cells of the seed coat generally do not collapse and are still evident at maturity (compare young seed coat [Fig. 21] with mature seed coat [Fig. 22]). At maturity, the epidermal cells of the seed coat are more or less tangentially enlarged and tanniferous, and cells of the three to four underlying layers develop into short tracheoidal cells with helical wall thickenings.

DISCUSSION

As a result of the present study, nearly all important embryological features of *Rhabdodendron* are available for use in comparing this enigmatic genus with other groups. The embryological features of *Rhabdodendron* may be summarized as follows (see also Table 1).

Anther tetrasporangiate; anther wall development of the Monocotyledonous-type; anther epidermis tanniferous and persistent; endothecium fibrous; anther tapetal cells 2-nucleate; mature pollen two-celled at the time of shedding.

Ovule campylotropous (not ana- or hemitropous), bitegmic and crassinucellate; a small amount of nucellar tissue persistent in the mature seed; outer integument four to five cell-layers thick, and inner integument two to five cell-layers thick; micropyle formed by the inner integument alone; endothelium not formed; conspicuous hypostase present; nucellar cap not formed; obturator absent. Ovule archesporium multi-celled, with one of the archesporial cells undergoing meiosis to form a linear tetrad of megaspores; chalazal megaspore developing into a variation of the Polygonum-type embryo sac, which nearly always lacks two of the three antipodals and instead contains two polar nuclei and one antipodal at the chalazal end; starch grains present in the embryo sac.

Fertilization porogamous; endosperm formation of the Nuclear-type; mature seed containing a small amount of endosperm; embryo in mature seed straight, round-shaped and chlorophyllous. Seed non-arillate; seed coat formed by one-layered tanniferous exotesta composed of somewhat tangentially enlarged cells and the three or four underlying cell-layers composed of short tracheoidal cells with helical wall thickenings.

Table 1 presents comparisons of *Rhabdodendron* with its alleged relatives on the basis of selected embryological characters. Some selected literature references, from which embryological data for Phytolaccaceae, Rutaceae, and Rosales were obtained, are also provided in Table 1.

Comparison with Phytolaccaceae (Caryophyllales)

Embryologically *Rhabdodendron* resembles Phytolaccaceae in many features, and especially in its campylotropous ovules. Campylotropous ovules are common to all Caryophyllales, but are also known to occur in many unrelated groups. Importantly, *Rhabdodendron* has neither 3-celled pollen grains at the time of shedding nor a palisadal exotestal seed-coat structure composed of radially elongated epidermal cells, both of which are key characteristics not only of Phytolaccaceae, but also of all other Caryophyllales (see Brewbaker 1967, for pollen cell number condition distribution; Corner 1976, for seed coat structure of Caryophyllales).

In addition, there are several other embryological differences between *Rhabdodendron* and Phytolaccaceae. For example, the anther wall development is of the Monocotyledonous-type in *Rhabdodendron*, but of the Basic-type in Phytolaccaceae; a hypostase is present and conspicuous in *Rhabdodendron*, but is not found in Phytolaccaceae; *Rhabdodendron* lacks both a nucellar cap and an obturator, which are ubiquitous in Phytolaccaceae; *Rhabdodendron* has a unique variety of the Polygonum-type embryo sac, whereas Phytolaccaceae has a typical Polygonum-type embryo sac; and the embryo in the mature seed is chlorophyllous in *Rhabdodendron*, but non-chlorophyllous in Phytolaccaceae.

In its embryological characteristics, therefore, *Rhabdodendron* differs substantially and fundamentally not only from Phytolaccaceae but from Caryophyllales as a whole. Behnke (1976, 1981) showed that *Rhabdodendron* has P-type sieve elements plastids (i.e., PIcs-type) that are not identical with those found in Caryophyllales including Phytolaccaceae (i.e., PIIIcf-type). According to Puff and Weber (1976), moreover, *Rhabdodendron* does not appear to have a centrifugal androecium development, which is characteristic of Caryophyllales. Thus, the evidence from embryology and other sources suggests strongly the conclusion that

Table 1. A comparison of some selected embryological characters in *Rhabdodendron* and its alleged relatives.

Characters	*Rhabdodendron*	Phytolaccaceae	Rutaceae	Rosales sensu Dahlgren
Anther and microspores				
No. sporangia	4	4	4	4
Anther wall development	Monocotyledonous	Basic	?	?
Anther epidermis	Persistent	?	?	?
Endothecium	Fibrous	Fibrous	Fibrous	Fibrous
Tapetum	Glandular	Glandular	Glandular	Glandular
Tapetal cell	2-nucleate	2- or multi-nucleate	2- or multi-nucleate	2- or multi-nucleate
Cytokinesis in meiosis	Simultaneous	Simultaneous	Simultaneous	Simultaneous
Mature pollen	2-celled	3-celled	2- or 3-celled	2-celled
Megagametophyte and nucellus				
Archesporium	Multi-celled	Multi-celled	Multi-celled	Multi-celled
Pattern of embryo sac formation	A variety of Polygonum type	Polygonum type	Polygonum type	Polygonum type
Antipodal cells	Ephemeral	Ephemeral or persistent	Ephemeral	Ephemeral or persistent
No. nuclei in mature embryo sac	6	5 or 8 (or more)	5	5 or 8 (or more) (with irregularities)
Starch grains in embryo sac	Present	Absent	Present	Present
Nature of nucellus	Crassinucellate	Crassinucellate	Crassinucellate	Crassinucellate
Nucellar cap	Not formed	Formed	Formed	Formed
Hypostase	Present	Absent	Present	Present
Endothelium	Not formed	Not formed	Not formed	Not formed
Nucellar tissue in mature seed	Remain	Remain	Remain	Remain
Integuments and ovular orientation				
No. integuments	2	2	2 (rarely 1)	1 or 2
Micropyle formation	ii	ii	Both, ii only or oi only	Both, or ii only
Thickness of integuments	ii, 2–5 cells thick; oi, 4–5 cells thick	ii, 2–4 cells thick; oi, 5–8 cells thick	ii, 2–5 cells thick; oi, 3–6 cells thick	ii, 2–6 cells thick; oi, 3–14 cells thick
Ovule orientation	Campylotropous	Campylotropous	Ana-, hemi-, or amphitropous	Ana-, hemi- or campylotropous

Obturator				
Presence or absence	Absent	Present	Present	Present
Endosperm and embryo				
Endosperm formation	Nuclear	Nuclear	Nuclear	Nuclear
Endosperm in mature seed	Present	Present	Present	Present
Embryo	Straight, and chlorophyllous	Curved and non-chlorophyllous	Straight or curved, and chlorophyllous	Straight or curved, and non-chlorophyllous
Seed and seed coat				
Appendage on seed	Absent	Sometimes arillate	Sometimes arillate	Absent
Characteristic feature of mature seed coat	Exotestal cells somewhat tangentially enlarged and tanniferous; underlying cells short-tracheoidal	Exotesta palisadal	Varied	Varied
Selected references		Brewbaker (1967); Corner (1976); Davis (1966); Kajale (1944, 1954); Lewis (1905); Mauritzon (1934); Woodcock (1925)	Bacchi (1943); Banerjee and Pal (1958); Boesewinkel (1977, 1978); Boesewinkel and Bouman (1978); Brewbaker (1967); Corner (1976); Davis (1966); Desai (1962) ; Johri and Ahuja (1956, 1957); Mauritzon (1935); Narayana (1963); Yakovlev and Zhukova (1980)	Brewbaker (1967); Corner (1976); Davis (1966); Harrold (1935); Hjelmqvist (1956, 1957, 1959a, b, 1962); Juliano (1931); Mauritzon (1939); Nordborg (1967); Pratt and Einset (1955); Schneider (1953); Webb (1902)

Rhabdodendron is not related to Phytolaccaceae, and supports the exclusion of *Rhabdodendron* from Caryophyllales, as recent authors have mainly concluded (e.g., Puff and Weber 1976; Takhtajan 1980; Cronquist 1981; Thorne 1983; Dahlgren 1983).

Comparison with Rutaceae (Rutales or Sapindales)

Rutaceae are diverse in some of the embryological features. For example, different members of the family have both two- and three-celled pollen grains; furthermore, ana-, hemi- to amphitropous ovules are also found in Rutaceae; and the seed coat structure is highly diversified. Because of this extensive variability, most of the embryological features of *Rhabdodendron* are found in Rutaceae, the primary differences being the lack of both a nucellar cap and an obturator in *Rhabdodendron* and its unique variety of the Polygonum-type embryo sac, whereas Rutaceae have a typical Polygonum-type embryo sac. These differences are sufficiently impressive that they seem clearly to contradict the placement of *Rhabdodendron* in Rutaceae, as suggested by Puff and Weber (1976). In addition, *Rhabdodendron* has the PIcs-type sieve element plastids, which are never found in Rutaceae, where S-type plastids are ubiquitous (Behnke 1976, 1981). Furthermore, in contrast with the tri- or unilacunar nodes in Rutaceae, *Rhabdodendron* has a multilacunar node. In view of all these fundamental differences, there does not appear to be a case for a direct relationship between *Rhabdodendron* and Rutaceae.

Comparison with Rosales

With respect to the position of *Rhabdodendron*, Cronquist (1981, 1983) suggests a broadly defined Rosales, and Dahlgren (1983) a narrowly defined Rosales (although Dahlgren placed *Rhabdodendron* [as a family, Rhabdodendraceae] in that order with some doubt). As mentioned earlier, *Rhabdodendron* was once assigned to Chrysobalanaceae, which was traditionally treated as a subfamily or tribe of Rosaceae; Cronquist suggested that Chrysobalanaceae and Rhabdodendraceae might indeed be related, placing both in his broadly defined order Rosales. However, our recent embryological study (Tobe and Raven 1984) showed that Chrysobalanaceae have characteristic features (e.g., tenuinucellate ovules and the formation of an endothelium) in common with Thealean families (particularly with Theaceae), and we consequently proposed that the family be transferred to Theales (Dahlgren and Thorne [1984, p. 687] also briefly referred to the validity of our suggested transfer). Embryological evidence seems clearly and unequivocally to contradict any suggestion of relationship between *Rhabdodendron* and Chrysobalanaceae, therefore it seems unnecessary to compare these two groups further. The following comparison is made with Rosales sensu stricto (Dahlgren 1983), from which Chrysobalanaceae have been excluded.

As shown in Table 1, Rosales as a whole have a certain diversity in embryological characters, yet it is evident that the order can be characterized by a certain combination of embryological features. These embryological features of Rosales include most of those of *Rhabdodendron*, with some notable exceptions. *Rhabdodendron* differs from other families in Rosales in lacking both the nucellar cap and the obturator, and in having chlorophyllous embryos whereas Rosales have non-chlorophyllous embryos. Considering further that *Rhabdodendron* differs from Rosales (with S-type plastids) in having PIcs-type sieve element plastids (Behnke 1976, 1981), we conclude that *Rhabdodendron* should not be included in Rosales sensu stricto even if there is some degree of relationship between the groups.

Position of Rhabdodendron

The preceding discussion clearly indicates that *Rhabdodendron* is distinct from all of the families or groups with which it has been considered to be related. This justifies the treatment of *Rhabdodendron* as a separate family, Rhabdodendraceae, as proposed by Prance (1968, 1972) and accepted by some subsequent authors.

The embryological features of *Rhabdodendron* are mostly plesiomorphic and therefore of limited value for determining its affinities. The combination of embryological features that characterizes *Rhabdodendron* is approximately equivalent to that found in Rosiflorae (particularly Rosales sensu Dahlgren) and some Rutiflorae (particularly Sapindales, Rutales, and Polygalales). Character states shared by *Rhabdodendron*, some Rosiflorae, and some Rutiflorae include: anther tetrasporangiate; endothecium fibrous; tapetum glandular; pollen 2-celled when shed; ovule bitegmic and crassinucellate; ovule archesporium basically multi-celled; embryo sac basically of Polygonum-type; starch grains common in embryo sacs; endosperm formation of the Nuclear-type; and mature seed with scanty or copious endosperm and a persistent nucellar tissue. Since most of these shared character states are apparently plesiomorphic, they may not indicate a close relationship between the groups involved; rather, they seem to suggest that the groups are at a similar evolutionary stage with respect to their embryological features.

Among the other kinds of evidence that have been used to elucidate the systematic position of *Rhabdodendron*, the nature of the pollen exine sculpturing suggest a rutaceous affinity for *Rhabdodendron* (Behnke 1977), and Puff and Weber (1976), on the basis of some vegetative and reproductive anatomy, argued for a very close relationship between these groups. In contrast, Cronquist (1981, 1983) concluded that, except for the shared secretory cavities, there is no compelling evidence for concluding that *Rhabdodendron* is allied even to Rutales (or 'Sapindales'), much less to Rutaceae itself. Cronquist (1981, p. 586) pointed out that the simple leaves, perigynous flowers, numerous stamens, single carpel with a basal style and long-decurrent stigma, and the absence of a nectary disk of *Rhabdodendron* are all discordant in Rutaceae. The recent analysis of ellagic-acid derivatives by Wolter Filho et al. (1985) likewise suggests that *Rhabdodendron* agrees with Rosiflorae or even better with Myrtiflorae, and differs fundamentally from Rutales. Myrtiflorae are generally considered to have been derived directly from Rosiflorae, and such an order as Myrtales, a representative group of Myrtiflorae, is well defined by a distinct combination of wood anatomical and embryological characters (van Vliet and Baas 1984; Tobe and Raven 1983b; see also Dahlgren and Thorne 1984) that are not shared by *Rhabdodendron*. Possession of P-type sieve element plastids always has been one of the keys to suggest a distinct position of *Rhabdodendron*. According to Behnke (1981), 'Rosidae' have close to 30%

P-type plastids and 70% S-type. Of special interest is the fact that PIcs-type plastids of the sort found in *Rhabdodendron* are common in Magnoliiflorae (Behnke 1971, 1981). This accords with Behnke's conclusion that P-type plastids are primitive, and were present in the common ancestor of the angiosperms.

In conclusion, considering resemblances and differences in embryological and other characters between *Rhabdodendron* and other groups, *Rhabdodendron* seems best placed in or near Rosiflorae, but not within the order Rosales sensu Dahlgren. *Rhabdodendron* certainly merits treatment as a distinct family. The possession of many plesiomorphic embryological character states, as well as of PIcs-type sieve element plastids, seems to suggest that *Rhabdodendron*, like some other Rosiflorae, retains a number of features inherited directly or indirectly from Magnoliiflorae. Additional embryological and other detailed studies of Rosiflorae and Rutiflorae would probably help to clarify the relationships of *Rhabdodendron* more precisely than we have been able to do on the basis of the available evidence.

ACKNOWLEDGMENTS

The work was supported by grants from the U.S. National Science Foundation to Peter H. Raven. We are grateful to Bruce W. Nelson for collecting the excellent embryological material that was used in this study, and to Tokushiro Takaso for his assistance in drawing the figures.

LITERATURE CITED

Airy Shaw, H.R. 1966. J.C. Willis' A dictionary of the flowering plants and ferns. 7th ed. Cambridge University Press, Cambridge. 1214 p.

Bacchi, O. 1943. Cytological observations in *Citrus*. 2. Megasporogenesis, fertilization, endosperm, and polyembryony. Bot. Gaz. 105: 221–225.

Banerjee, I., and S. Pal. 1958. Studies in the embryology of *Feronia limonia* Swingle. Bull. Bot. Soc. Bengal 12: 18–28.

Behnke, H.-D. 1971. Sieve-tube plastids of Magnoliidae and Ranunculidae in relation to systematics. Taxon 20: 723–730.

——. 1976. Sieve element plastids of *Fouquieria*, *Frankenia*, and *Rhabdodendron*. Taxon 25: 265–268.

——. 1977. Zur Skulptur der Pollen-Exine bei drei Centrospermen (*Gisekia*, *Limeum*, *Hectorella*), bei Gyrostemonaceen und Rhabdodendraceen. Plant Syst. Evol. 128: 227–235.

——. 1981. Sieve-element characters. Nord. J. Bot. 1: 381–400.

——, and R. Dahlgren. 1976. The distribution of characters within an angiosperm system. 2. Sieve-element plastids. Bot. Not. 129: 287–295.

Bentham, G. 1853. Notes on two little-known genera connected with the South American flora. Hooker's J. Bot. Kew Gard. Misc. 5: 289–296.

Bocquet, G., and J.D. Bersier. 1960. La valeur systématique de l'ovule: Développements tératalogiques. Arch. Sci. 13: 475–496.

Boesewinkel, F.D. 1977. Development of ovule and testa in Rutaceae I. *Ruta*, *Zanthoxylum*, and *Skimmia*. Acta Bot. Neerl. 26: 193–211.

——. 1978. Development of ovule and testa in Rutaceae III. Some representatives of the Aurantioideae. Acta Bot. Neerl. 27: 341–354.

——, and F. Bouman. 1978. Development of ovule and testa in Rutaceae II. The unitegmic and pachychalazal seed of *Glycosmis* cf. *arborea* (Roxb.) DC. Acta Bot. Neerl. 27: 69–78.

Brewbaker, J.L. 1967. The distribution and phylogenetic significance of binucleate and trinucleate pollen grains in the angiosperms. Amer. J. Bot. 54: 1069–1083.

Corner, E.J.H. 1976. The seeds of dicotyledons. Cambridge University Press, Cambridge. Vol. 1. 311 p.

Cronquist, A. 1981. An integrated system of classification of flowering plants. Columbia University Press, New York. 1262 p.

——. 1983. Some realignments in the dicotyledons. Nord. J. Bot. 3: 75–83.

Dahlgren, R. 1983. General aspects of angiosperm evolution and macrosystematics. Nord. J. Bot. 3: 119–149.

——, and R.F. Thorne. 1984. The order Myrtales: circumscription, variation and relationships. Ann. Missouri Bot. Gard. 71: 633–699.

Davis, G.L. 1966. Systematic embryology of the angiosperms. John Wiley & Sons, New York. 528 p.

Desai, S. 1962. Cytology and embryology of the Rutaceae. Phytomorphology 12: 178–184.

Engler, A. 1931. Rutaceae. Pp. 187–359. *In* Engler, A. & K. Prantl [eds.], Die Natürlichen Pflanzenfamilien, 2. Aufl., 19a. Engelmann, Leipzig.

Gilg, E., and R. Pilger. 1905. Rutaceae. Pp. 152–154. *In* Pilger, R. (ed.), Beiträge zur Flora der Hylaea nach den Sammlungen von E. Ule. Verh. Bot. Ver. Prov. Brandenburg 47.

Harrold, T.J. 1935. Comparative study of the developing and aborting fruits of *Prunus persica*. Bot. Gaz. 96: 505–520.

Hjelmqvist, H. 1956. The embryology of some South African *Alchemilla* species. Bot. Not. 1956: 21–32.

——. 1957. The apomictic development in *Malus sieboldii*. Bot. Not. 1957: 455–467.

——. 1959a. Studien über Embryologie und Variabilitat bei einigen *Aphanes*-Arten. Bot. Not. 1959: 17–64.

——. 1959b. On the embryology of two *Malus* hybrids. Bot. Not. 1959: 459–464.

——. 1962. The embryo sac development of some *Cotoneaster* species. Bot. Not. 1962: 208–236.

Hooker, J.D. 1867. Rosaceae. *In* Martius, C.F.P. von, Flora Brasiliensis 14(2): 1–76.

Huber, J. 1909. Materiaes para a Flora Americana. VII. Plantae Duckeanae austro-guyanenses. Bol. Mus. Paraense Hist. Nat. 5: 424–431.

Johri, B.M., and M.R. Ahuja. 1956. Development of endosperm and nucellar polyembryony in *Aegle marmelos* Correa. Curr. Sci. 25: 162–164.

——, and ——. 1957. A contribution to the floral morphology and embryology of *Aegle marmelos* Correa. Phytomorphology 7: 10–24.

Juliano, J.B. 1931. Floral morphology of *Lyonothamnus floribundus*. Bot. Gaz. 91: 426–438.

Kajale, L.B. 1944. A contribution to the embryology of the Phytolaccaceae. 1. Studies in the genus *Phytolacca*. Patna Univ. J. 1: 10–21.

——. 1954. A contribution to the embryology of the Phytolaccaceae. 2. Fertilization and the development of embryo, seed and fruit in *Rivina humilis* and *Phytolacca dioica*. J. Indian Bot. Soc. 33: 206–225.

Lewis, J.F. 1905. Notes on the development of *Phytolacca decandra*. Johns Hopkins Univ. Circ. 178: 34–42.

Mauritzon, J. 1934. Ein Beitrag zur Embryologie der Phytolaccaceen und Cactaceen. Bot. Not. 1934: 111–135.

——. 1935. Über die Embryologie der Familie Rutaceae. Svensk Bot. Tidskr. 29: 319–347.

——. 1939. Contributions to the embryology of the orders Rosales and Myrtales. Acta Univ. Lund 35: 1–121.

Metcalfe, C.R., and L. Chalk. 1950. Anatomy of the dicotyledons. 2 vols. Clarendon Press, Oxford. 1500 p.

Milanez, F.R. 1943. Anatomia das principais madeiras brasileiras das Rutaceae. Rodriguésia 7: 5–22.

Narayana, L.L. 1963. A note on the embryology of a few Rutaceae. Curr. Sci. 32: 516–517.

Nordborg, G. 1967. Embryological studies in the *Sanguisorba minor* complex (Rosaceae). Bot. Not. 120: 109–119.

Prance, G.T. 1968. The systematic position of *Rhabdodendron* Gilg & Pilg. Bull. Jard. Bot. Nat. Belg. 38: 127–146.

——. 1972. Rhabdodendraceae. Flora Neotropica, monograph no. 11. Hafner Publ. Co., New York. 22 p.

Pratt, C., and I. Einset. 1955. Development of the embryo sac in some American blackberries. Amer. J. Bot. 42: 637–645.

Puff, C., and A. Weber. 1976. Contributions to the morphology, anatomy and karyology of *Rhabdodendron* and a reconsideration of the systematic position of the Rhabdodendronaceae. Plant Syst. Evol. 125: 195–222.

Record, S.J. 1933. The woods of *Rhabdodendron* and *Duckeodendron*. Trop. Woods 33: 6–10.

——, and R.W. Hess. 1943. Timbers of the New World. Yale University Press, New Haven, CT. 640 p.

Schneider, G.W. 1953. Megagametogenesis and embryology in a diploid and an aneuploid apple. Amer. J. Bot. 40: 196–203.

Scholz, H. 1964. Rutales. Pp. 262–276. *In* H. Melchior (ed.), A. Engler's Syllabus der Pflanzenfamilien. 2. Borntraeger, Berlin.

Takhtajan, A.L. 1969. Flowering plants: origin and dispersal. Oliver & Boyd, Edinburgh. 310 p. (Translated from Russian by C. Jeffrey.)

——. 1980. Outline of the classification of flowering plants (Magnoliophyta). Bot. Rev. (Lancaster) 46: 225–359.

Thorne, R.F. 1983. Proposed new realignments in angiosperms. Nord. J. Bot. 3: 85–117.

Tobe, H., and P.H. Raven. 1983a. The embryology of *Axinandra zeylanica* (Crypteroniaceae) and the relationships of the genus. Bot. Gaz. 144: 426–432.

——, and ——. 1983b. An embryological analysis of the Myrtales: its definition and characteristics. Ann. Missouri Bot. Gard. 70: 71–94.

——, and ——. 1984. An embryological contribution to the systematics of the Chrysobalanaceae. I. Tribe Chrysobalaneae. Bot. Mag. (Tokyo) 97: 397–411.

Vliet, G.J.C.M. van, and P. Baas. 1984. Wood anatomy and classification of the Myrtales. Ann. Missouri Bot. Gard. 71: 783–800.

Webb, J.E. 1902. A morphological study of the flower and embryo of *Spiraea*. Bot. Gaz. 33: 451–460.

Wolter Filho, W., A.I. Da Rocha, M. Yoshida, and O.R. Gottlieb. 1985. Ellagic-acid derivatives from *Rhabdodendron macrophyllum*. Phytochemistry 24: 1991–1994.

Woodcock, E.F. 1925. Observations on the morphology of the seed in *Phytolacca*. Pap. Michigan Acad. Sci. 4: 413–417.

Yakovlev, M.S., and G.Y. Zhukova. 1980. Chlorophyll in embryos of angiosperm seeds, a review. Bot. Not. 133: 323–336.

21. THE LAST DAHLGRENOGRAM. SYSTEM OF CLASSIFICATION OF THE DICOTYLEDONS

GERTRUD DAHLGREN

Department of Systematic Botany, Ö. Vallgatan 20, S-22361 Lund, Sweden

ABSTRACT

A diagram illustrating a modified classification of the dicotyledons, sensu Dahlgren 1980 is presented. The greatest changes are found in the superorders Theanae, Malvanae, Rutanae and Cornanae. A survey of the families and orders of the dicotyledons is added.

Key words: angiosperm classification, dicotyledons, Magnoliidae.

INTRODUCTION

A revised system of classification of the angiosperms was presented by my husband, Rolf Dahlgren, in 1980. Subsequent research, in particular on the orders Myrtales (Dahlgren and Thorne 1984) and Rhizophorales and associated taxa (Dahlgren 1989) convinced him that certain changes should be made in the diagram, with the result that some families and orders now form new constellations. He also found support for this in connection with symposia, where these groups were treated from different aspects (Behnke 1982, 1984; Tobe and Raven 1983; Keating 1984; Juncosa and Tomlinson 1989). A few weeks before his tragic death in a traffic accident in February 1987 he sketched a modified bubble diagram of the dicotyledons, which would reveal phylogenetic relationships better than that of 1980. However, he had not written down any comments and explanations. I have now analysed the changes he made and I present the diagram together with some compilatory notes and an example of a character state illustrated in the diagram. A list of the families and orders of dicotyledons is also appended. Since the ending '-florae' for superorders does not apply to plant groups other than angiosperms, I have changed to '-anae' in the interests of nomenclatural uniformity.

THE SYSTEM OF CLASSIFICATION

The system of classification of the dicotyledons presented here (Figs. 1–2) differs from the 1980 classification in the aspects presented below, the greatest changes being found in the superorders Theanae, Malvanae, Rutanae and Cornanae.

Theanae

Theanae now comprises four orders: Theales, Lecythidales, Dilleniales and Paeoniales, the last two having formerly been treated in Malvanae.

Lecythidales has previously been placed as a family (Lecythidaceae) in Myrtales

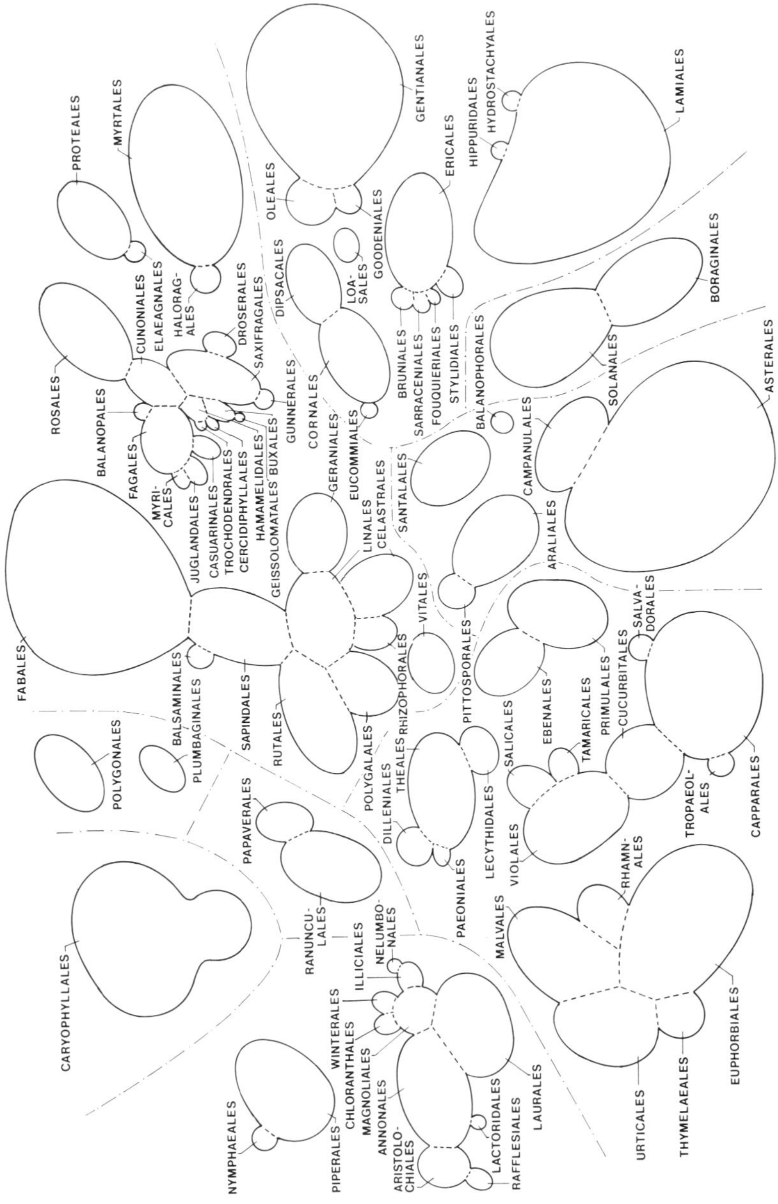

Figure 1. Classification of dicotyledons with the superorders illustrated two-dimensionally as clusters of orders. Only names of orders are given.

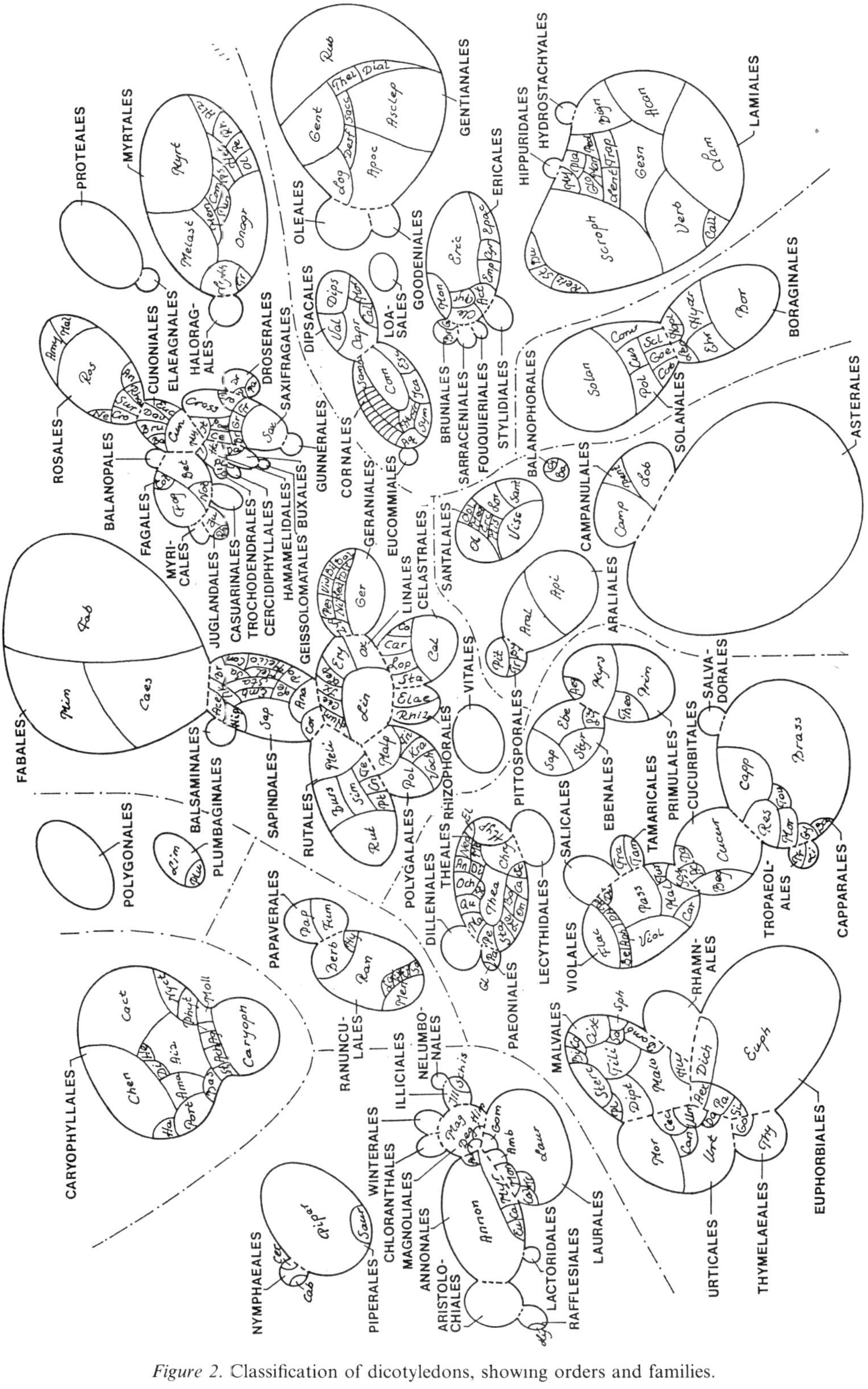

Figure 2. Classification of dicotyledons, showing orders and families.

(e.g. Prance and Mori 1977, 1978) or in Theales (Dahlgren 1975, 1980, Thorne 1976) or has been treated as a separate order (Cronquist 1968). Dahlgren and Thorne stressed the similarities between Lecythidaceae and families in Theales with regard to anatomical characters (lack of internal phloem, vessels without vestured pitting, presence of wedge-shaped phloem rays), morphological characters (such as alternate leaf arrangement and polymerous centrifugally developing androecium) and embryological characters (tenuinucellate ovules, lack of endosperm in the ripe seed). They suggested a position in Theales treating it as a suborder. Lecythidales is here treated as a separate order in Theanae.

In the 1980 classification it was stated that Dilleniales is less clearly allied to Malvales, and Paeoniales was with some reservation placed together with Dilleniales in Malvanae. It is now found that they are better placed in Theanae.

Chrysobalanaceae, previously in Rosales, are included in Theales. The chief reason for not placing them in Rosales (Dahlgren and Thorne 1984) is the occurrence of foliar sclereids and the syncarpous character of the gynoecium when this consists of 2 or 3 carpels. In many morphological characters and in the occurrence of ellagic acid the family resembles both thealean and myrtalean families; in embryological features it resembles families in Theales. Perhaps Chrysobalanaceae can be put forward as a connecting link between Theales and Myrtales.

Droserales has now been moved to Rosales.

Malvanae

Of the 9 original orders of Malvanae (Paeoniales, Dilleniales, Malvales, Urticales, Rhamnales, Elaeagnales, Plumbaginales, Thymelaeales, Euphorbiales), 5 remain, viz. Malvales, Urticales, Euphorbiales, Thymelaeales and Rhamnales. Plumbaginales now forms a superorder, Plumbaginanae, placed next to Polygonanae, with which they share many character states. As an order in Malvanae its position was uncertain: in embryological features it shows affinities with Euphorbiales, in some morphological characters with Rhamnales and Elaeagnales. Elaeagnales is a rather isolated group that has now been transferred to Proteanae, mainly on the basis of anatomical but also morphological similarities. Its place there, however, seems somewhat doubtful.

Rutanae

Fabales and Celastrales are placed in Rutanae, Fabales having previously been treated as a superorder, Fabanae, and Celastrales as an order in Santalanae. Geraniales has been split up into Geraniales and Linales. Rhizophorales has been transferred from Myrtanae and includes Elaeocarpaceae (from Malvales). Dahlgren (1987) studied 'rhizophoraceous' character states in a series of families or orders that were presumed to be related to Rhizophoraceae. He found that Elaeocarpaceae and Rhizophoraceae agree in alkaloid chemistry, many floral and fruit characters and embryology. They should be treated as two distinct but related families together forming an order. Further analyses have shown that Celastrales, a variable group difficult to define, as well as families of Geraniales-Linales are fairly closely related to Rhizophorales and they seem now all to fit in this constellation in Rutanae. Anisophylleaceae, formerly included in the family Rhizophoraceae, are now a separate family which on the basis of gross morphology and chemical characters have been placed in Rosales sensu lato in Rosanae.

Cornanae

Cornanae have been split up into two superorders, Ericanae and Cornanae sensu stricto, Ericanae with the old Cornanae orders Fouquieriales, Ericales and Sarraceniales. The position of Fouquieriales is still uncertain. Stylidiaceae have been removed from Cornales and given the rank of order in Ericanae. Furthermore Bruniaceae from Cunoniales (Rosanae) are treated here as the order Bruniales, emphasis being preferably laid on chemical similarities.

Orders left over from the old Cornanae are Cornales and Eucommiales. A newly transferred order is Dipsacales (from Gentiananae) with the families Dipsacaceae, Valerianaceae and Caprifoliaceae. Amoeboid tapetum and a special type of nectary in the corolla tube are two autapomorphies in Dipsacales, as circumscribed here, since *Sambucus* and *Viburnum*, usually included in Caprifoliaceae, are here treated as families in Cornales – they differ in the type of tapetum and nectary. Cornales has become a large, and somewhat heterogeneous, order comprising 32 families. Of these, Roridulaceae and Diapensiaceae were formerly treated in Ericales and Menyanthaceae in Gentianales. In Menyanthaceae endosperm formation is cellular, whereas in all Gentianales, in this circumscription, it is nuclear.

Some other changes to be observed:

Magnolianae. – The families Winteraceae, Chloranthaceae and Lactoridaceae in Magnoliales are treated as orders.

Violanae. – The families Cucurbitaceae, Achariaceae, Begoniaceae and Datiscaceae of Violales together now form the new order Cucurbitales. Tropaeolales, previously in Rutanae, has been raised to an order here.

Rosanae. – Cercidiphyllales now forms a new order with Cercidiphyllaceae and Eupteleaceae, previously families in Trochodendrales.

Vitanae. – The order Vitales, Santalanae, has been given the rank of superorder, Vitanae, with a single order, Vitales.

Lamianae. – Lamiales and Scrophulariales have been united into a single order, Lamiales.

In addition to these major changes some minor changes are also presented. In many orders small families are recognized, which previously were treated as subfamilies or genera, whereas in other orders families have disappeared being now included in other families.

ENDOSPERM HAUSTORIA
AN EXAMPLE OF A CHARACTER STATE ILLUSTRATED IN THE DIAGRAM

A survey of the embryological literature has been made with the purpose of using patterns of distribution of embryological character states for clarifying relationships in the dicotyledons at and above family level. The result of this investigation will be published in a symposium volume of *Aliso*, but one character state, endosperm haustoria, has been broken up and selected as an example of a character illustrated in the diagram (Fig. 3). The pattern of distribution will be discussed in terms of phylogeny.

Endosperm haustoria occur in scattered groups of the dicotyledons. They can be divided into those that develop at the chalazal end of the endosperm only (termed *chalazal endosperm haustoria*) and those that develop both at the micropylar and the chalazal ends (here termed *terminal endosperm haustoria*). Rarely, haustoria develop at the micropylar end only (*micropylar haustoria*). Micropylar haustoria are recorded in some Ochnaceae and Marcgraviaceae (Theales), and in certain taxa in Epacridaceae (Ericales), Acanthaceae (Lamiales) and Boraginaceae (Boraginales). They will not be further commented on here. The distribution of the two main types and their phylogenetic significance varies conspicuously.

Chalazal endosperm haustoria have a very scattered distribution in the dicotyledons. They are usually combined with nuclear endosperm formation and occur, for example, in isolated genera of Nymphaeaceae (Nymphaeales), Lauraceae (Laurales), in several genera of Mimosaceae and Fabaceae (Fabales), in Linaceae (intermediate endosperm formation; Linales), in many genera of Cucurbitaceae (Cucurbitales) where they are often well developed and aggressive, in a few genera of Myzodendraceae and Santalaceae (Santalales), Icacinaceae (Cornales) and Proteaceae (Proteales), and in several genera of Boraginaceae (Boraginales). This pattern of distribution is disparate and gives little guidance to phylogeny above family level.

On the other hand it is striking that terminal endosperm haustoria have a concentrated distribution in the dicotyledons. With few exceptions they are combined with cellular endosperm formation. They are found in Rutanae: Balsaminaceae (intermediate endosperm formation; Balsaminales); Loasanae: Loasaceae (Loasales); Ericanae: practically all taxa except the Sarraceniales and Fouquieriales and Actinidiaceae and Pyrolaceae in Ericales; Lamianae: nearly all Lamiales; and in Asteranae: Campanulales.

Since Balsaminaceae differs from the other groups in a number of characters the terminal endosperm haustoria in this family have presumably evolved independently. Furthermore in Campanulales endosperm formation differs from that in other groups, suggesting an independent evolution, which is also supported by chemical differences.

The Ericanae, Lamianae and Loasanae resemble one another in embryological characters as well as in other characters, e.g. in the production of iridoids and thus probably form a monophyletic unit. Somewhat doubtful in this constellation are the Loasanae, however, which are seco-iridoid taxa. Other plant groups in which seco-iridoids are produced (Cornanae, Gentiananae) lack endosperm haustoria.

Stylidiaceae, which have aggressive endosperm haustoria, do not fit well in the Cornanae, where they have previously been placed, and are here placed as a separate order (Stylidiales) in Ericanae. Within Ericales there are two families, viz. Actinidiaceae and Pyrolaceae in which endosperm haustoria are lacking, although in all other characteristics they agree well with the other families. Incipient endosperm haustoria occur in Clethraceae and Empetraceae.

In Lamiales terminal endosperm haustoria occur in all families. Growth at the chalazal end of these haustoria may, however, be strongly retarded where a

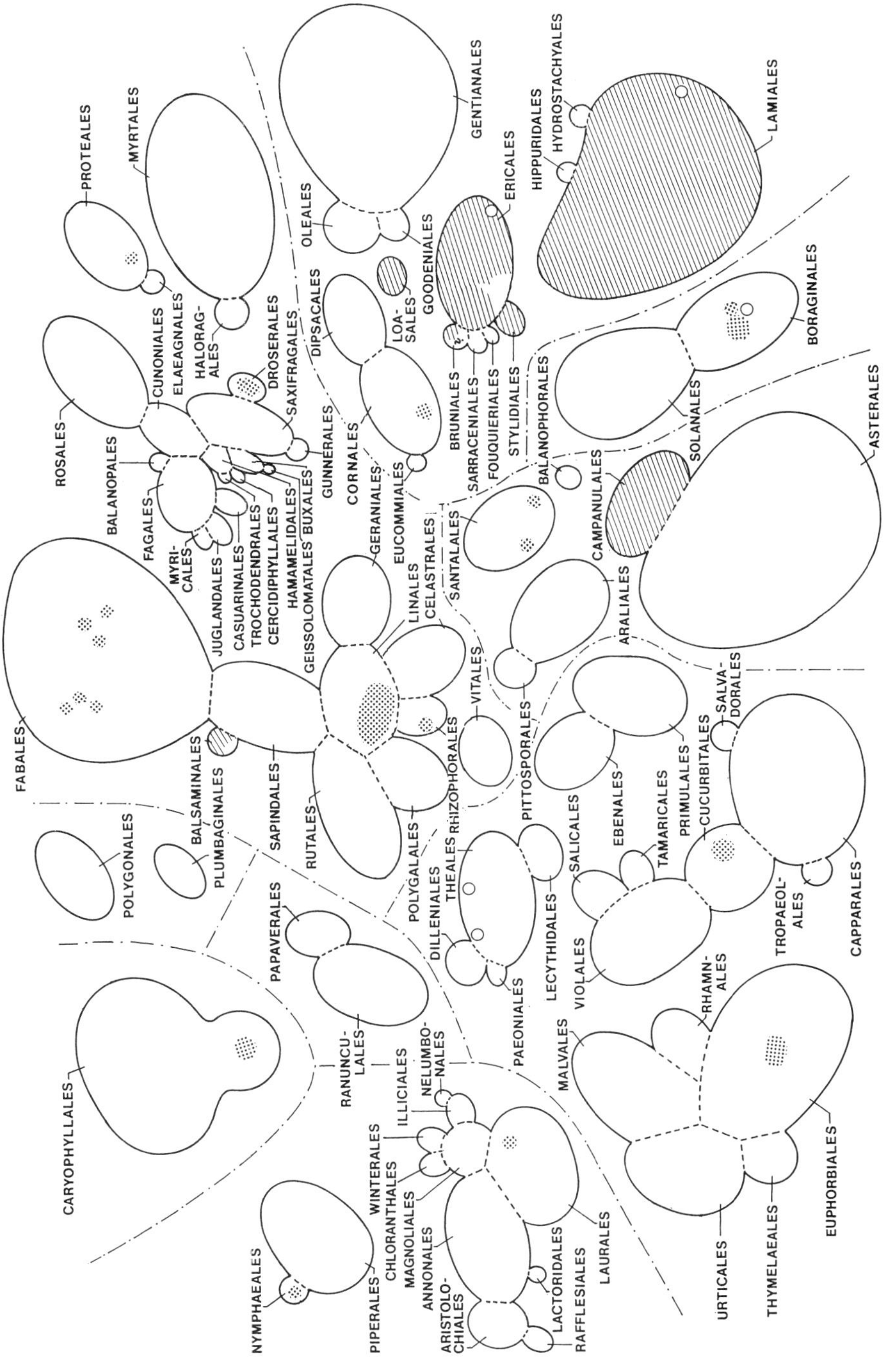

Figure 3. Presence of endosperm haustoria. Terminal: hatched; chalazal: dotted; Micropylar: rings.

well-developed hypostase is present as in Retziaceae and Buddlejaceae and certain genera of Scrophulariaceae. No endosperm haustoria are developed in Hippuridales. In older literature and also in Davis (1966) it has been reported that Hydrostachyales have suspensor haustoria, but according to Rauh and Jäger-Zürn (1966) they have terminal endosperm haustoria similar to the type found in other Lamianae. Thus, this character supports the position of Hydrostachyales in Lamianae, while the lack of terminal endosperm haustoria in Hippuridales could be an indication that this order might be better placed in Cornanae where such haustoria do not occur.

SYSTEM OF CLASSIFICATION OF DICOTYLEDONS

(some small families may not have been formally described)

Magnolianae

Annonales: Annonaceae, Myristicaceae, Eupomatiaceae, Canellaceae, Austrobaileyaceae
Aristolochiales: Aristolochiaceae
Rafflesiales: Rafflesiaceae, Hydnoraceae
Magnoliales: Degeneriaceae, Himantandraceae, Magnoliaceae
Lactoridales: Lactoridaceae
Winterales: Winteraceae
Chloranthales: Chloranthaceae
Illiciales: Illiciaceae, Schisandraceae
Laurales: Amborellaceae, Trimeniaceae, Monimiaceae, Gomortegaceae, Calycanthaceae, Lauraceae
Nelumbonales: Nelumbonaceae

Nymphaeanae

Piperales: Saururaceae, Piperaceae
Nymphaeales: Cabombaceae, Nymphaeaceae, Ceratophyllaceae

Ranunculanae

Ranunculales: Lardizabalaceae, Sargentodoxaceae, Menispermaceae, Kingdoniaceae, Circaeasteraceae, Ranunculaceae, Hydrastidaceae, Berberidaceae
Papaverales: Papaveraceae, Fumariaceae

Caryophyllanae

Caryophyllales: Molluginaceae, Caryophyllaceae, Phytolaccaceae, Achatocarpaceae, Agdestidaceae, Basellaceae, Portulacaceae, Stegnospermataceae, Nyctaginaceae, Aizoaceae, Halophytaceae, Cactaceae, Didiereaceae, Hectorellaceae, Chenopodiaceae, Amaranthaceae

Polygonanae

Polygonales: Polygonaceae

Plumbaginanae

Plumbaginales: Plumbaginaceae, Limoniaceae

Malvanae

Malvales: Sterculiaceae, Plagiopteridaceae, Bixaceae, Cochlospermaceae, Cistaceae, Sphaerosepalaceae, Sarcolaenaceae, Huaceae, Tiliaceae, Dipterocarpaceae, Bombacaceae, Malvaceae
Urticales: Ulmaceae, Moraceae, Cecropiaceae, Barbeyaceae, Cannabaceae, Urticaceae
Euphorbiales: Euphorbiaceae, Simmondsiaceae, Pandaceae, Aextoxicaceae, Dichapetalaceae
Thymelaeales: Gonystylidaceae, Thymelaeaceae
Rhamnales: Rhamnaceae

Violanae

Violales: Flacourtiaceae, Berberidopsidaceae, Aphloiaceae, Physenaceae, Passifloraceae, Dipentodontaceae, Peridiscaceae, Scyphostegiaceae, Violaceae, Turneraceae, Malesherbiaceae, Caricaceae
Cucurbitales: Achariaceae, Cucurbitaceae, Begoniaceae, Datiscaceae
Salicales: Salicaceae
Tamaricales: Tamaricaceae, Frankeniaceae
Capparales: Capparaceae, Brassicaceae, Tovariaceae, Resedaceae, Gyrostemonaceae, Batidaceae, Moringaceae
Tropaeolales: Tropaeolaceae, Limnanthaceae
Salvadorales: Salvadoraceae

Theanae

Dilleniales: Dilleniaceae
Paeoniales: Glaucidiaceae, Paeoniaceae
Theales: Stachyuraceae, Pentaphylacaceae, Marcgraviaceae, Quiinaceae, Ancistrocladaceae, Dioncophyllaceae, Nepenthaceae, Medusagynaceae, Caryocaraceae, Strasburgeriaceae, Ochnaceae, Chrysobalanaceae, Oncothecaceae, Scytopetalaceae, Theaceae, Bonnetiaceae, Hypericaceae, Elatinaceae
Lecythidales: Lecythidaceae

Primulanae

Primulales: Myrsinaceae, Aegicerataceae, Theophrastaceae, Primulaceae, Coridaceae
Ebenales: Sapotaceae, Styracaceae, Lissocarpaceae, Ebenaceae

Rosanae

Trochodendrales: Trochodendraceae, Tetracentraceae
Cercidiphyllales: Cercidiphyllaceae, Eupteleaceae
Hamamelidales: Hamamelidaceae, Platanaceae, Myrothamnaceae

Balanopales: Balanopaceae
Fagales: Nothofagaceae, Fagaceae, Corylaceae, Betulaceae
Juglandales: Rhoipteleaceae, Juglandaceae
Myricales: Myricaceae
Casuarinales: Casuarinaceae
Buxales: Buxaceae, Daphniphyllaceae, Didymelaceae
Geissolomatales: Geissolomataceae
Cunoniales: Cunoniaceae, Baueraceae, Brunelliaceae, Davidsoniaceae, Eucryphiaceae
Saxifragales: Saxifragaceae, Francoaceae, Greyiaceae, Brexiaceae, Grossulariaceae, Iteaceae, Cephalotaceae, Crassulaceae, Podostemaceae
Droserales: Droseraceae, Lepuropetalaceae, Parnassiaceae
Rosales: Rosaceae, Neuradaceae, Malaceae, Amygdalaceae, Anisophylleaceae, Crossosomataceae, Surianaceae, Rhabdodendraceae
Gunnerales: Gunneraceae

Proteanae

Proteales: Proteaceae
Elaeagnales: Elaeagnaceae

Myrtanae

Myrtales: Psiloxylaceae, Heteropyxidaceae, Myrtaceae, Onagraceae, Trapaceae, Lythraceae, Combretaceae, Melastomataceae, Memecylaceae, Crypteroniaceae, Oliniaceae, Penaeaceae, Rhynchocalycaceae, Alzateaceae
Haloragidales: Haloragidaceae

Rutanae

Sapindales: Coriariaceae, Anacardiaceae, Leitneriaceae, Podoaceae, Sapindaceae, Hippocastanaceae, Aceraceae, Akaniaceae, Bretschneideraceae, Emblingiaceae, Staphyleaceae, Melianthaceae, Sabiaceae, Meliosmaceae, Connaraceae
Fabales: Mimosaceae, Caesalpiniaceae, Fabaceae
Rutales: Rutaceae, Ptaeroxylaceae, Cneoraceae, Simaroubaceae, Tepulanthaceae, Burseraceae, Meliaceae
Polygalales: Malpighiaceae, Trigoniaceae, Vochysiaceae, Polygalaceae, Krameriaceae
Geraniales: Zygophyllaceae, Peganaceae, Nitrariaceae, Geraniaceae, Vivianiaceae, Ledocarpaceae, Biebersteiniaceae, Dirachmaceae, Balanitaceae
Linales: Linaceae, Humiriaceae, Ctenolophonaceae, Ixonanthaceae, Erythroxylaceae, Lepidobotryaceae, Oxalidaceae
Celastrales: Stackhousiaceae, Lophopyxidaceae, Cardiopteridaceae, Corynocarpaceae, Celastraceae
Rhizophorales: Rhizophoraceae, Elaeocarpaceae
Balsaminales: Balsaminaceae

Vitanae

Vitales: Vitaceae

Santalanae

Santalales: Olacaceae, Opiliaceae, Loranthaceae, Medusandraceae, Misodendraceae, Eremolepidaceae, Santalaceae, Viscaceae

Balanophoranae

Balanophorales: Cynomoriaceae, Balanophoraceae

Aralianae

Pittosporales: Pittosporaceae, Tremandraceae, Byblidaceae
Araliales: Araliaceae, Apiaceae

Asteranae

Campanulales: Pentaphragmataceae, Campanulaceae, Lobeliaceae
Asterales: Asteraceae

Solananae

Solanales: Solanaceae, Sclerophylacaceae, Goetzeaceae, Convolvulaceae, Cuscutaceae, Cobaeaceae, Polemoniaceae
Boraginales: Hydrophyllaceae, Ehretiaceae, Boraginaceae, Lennoaceae, Hoplestigmataceae

Ericanae

Bruniales: Bruniaceae, Grubbiaceae
Fouquieriales: Fouquieriaceae
Ericales: Actinidiaceae, Clethraceae, Cyrillaceae, Ericaceae, Empetraceae, Monotropaceae, Pyrolaceae, Epacridaceae
Stylidiales: Stylidiaceae
Sarraceniales: Sarraceniaceae

Cornanae

Cornales: Garryaceae, Alangiaceae, Nyssaceae, Cornaceae, Roridulaceae, Davidiaceae, Escalloniaceae, Helwingiaceae, Toricelliaceae, Aucubaceae, Aralidiaceae, Diapensiaceae, Phellinaceae, Aquifoliaceae, Paracryphiaceae, Sphenostemonaceae, Symplocaceae, Icacinaceae, Montiniaceae, Columelliaceae, Alseuosmiaceae, Hydrangeaceae, Sambucaceae, Viburnaceae, Menyanthaceae, Adoxaceae, Dulongiaceae, Tribelaceae, Eremosynaceae, Pterostemonaceae, Tetracarpaeaceae
Eucommiales: Eucommiaceae
Dipsacales: Caprifoliaceae, Valerianaceae, Dipsacaceae, Morinaceae, Calyceraceae

Loasanae

Loasales: Loasaceae

Gentiananae

Goodeniales: Goodeniaceae
Oleales: Oleaceae
Gentianales: Desfontainiaceae, Loganiaceae, Dialypetalanthaceae, Rubiaceae, Theligonaceae, Gentianaceae, Saccifoliaceae, Apocynaceae, Asclepiadaceae

Lamianae

Lamiales (= Scrophulariales): Retziaceae, Stilbaceae, Buddlejaceae, Scrophulariaceae (incl. Orobanchaceae and Selaginaceae), Myoporaceae, Globulariaceae, Plantaginaceae, Lentibulariaceae, Pedaliaceae, Trapellaceae, Martyniaceae, Gesneriaceae, Bignoniaceae, Acanthaceae, Verbenaceae, Lamiaceae, Callitrichaceae
Hydrostachyales: Hydrostachyaceae
Hippuridales: Hippuridaceae

LITERATURE CITED

Behnke, H.-D. 1982. Sieve-element plastids of Cyrillaceae, Erythroxylaceae and Rhizophoraceae: Description and significance of subtype PV plastids. Plant Syst. Evol. 141: 31–39.

——. 1984. Ultrastructure of sieve-element plastids of Myrtales and allied groups. Ann. Missouri Bot. Gard. 71: 824–831.

Cronquist, A. 1968. The evolution and classification of flowering plants. Houghton Mifflin Co., Boston, and Thomas Nelson & Sons, London. xi + 396 p.

Dahlgren, R. 1975. A system of classification of the angiosperms to be used to demonstrate the distribution of characters. Bot. Not. 128: 119–147.

——. 1980. A revised system of classification of the angiosperms. Bot. J. Linn. Soc. 80: 91–124.

——. 1989. Rhizophoraceae and Anisophylleaceae: summary statement, structure and relationships. Ann. Missouri Bot. Gard. 75: 1259–1277.

——, and R.F. Thorne. 1984. The order Myrtales: Circumscription, variation, and relationships. Ann. Missouri Bot. Gard. 71: 633–699.

Davis, G.L. 1966. Systematic embryology of the angiosperms. John Wiley & Sons, New York. 528 p.

Juncosa, A.M., and P.B. Tomlinson. 1989. A historical and taxonomic synopsis of Rhizophoraceae and Anisophylleaceae. Ann. Missouri Bot. Gard. 75: 1278–1295.

Keating, R.C. 1984. Leaf histology and its contribution to relationships in the Myrtales. Ann. Missouri Bot. Gard. 71: 801–823.

Prance, G.T., and S.A. Mori. 1977. What is *Lecythis*? Taxon 26: 209—222.

——, and ——. 1978. Observations on the fruits and seeds of neotropical Lecythidaceae. Brittonia 30: 21–33.

Rauh, W., and T. Jäger-Zürn. 1966. Zur Kenntnis der Hydrostachyaceae. 1. Teil. Blütenmorphologische und embryologische Untersuchungen an Hydrostachyaceen unter besonderer Berücksichtigung ihrer systematischen Stellung. Springer Verlag, Heidelberg. 115 p.

Thorne, R.F. 1976. A phylogenetic classification of the Angiospermae. Evol. Biol. 9: 35–106.

Tobe, H., and P.H. Raven. 1983. An embryological analysis of Myrtales: its definition and characteristics. Ann. Missouri Bot. Gard. 70: 71–94.

22. PLANTSMEN AND POTTERY

JENNIFER LAMOND

Royal Botanic Garden, Edinburgh EH3 5LR, Scotland

ABSTRACT

An affinity between plantsmen and ceramics is noted and the contributions of Carl Linnaeus, Theodor Holmskiold, Lewis Weston Dillwyn, John Wedgwood, Joseph Hooker, George Maw, Christopher Dresser, Arthur Herbert Church, Thomas Sprague, Ronald Copeland, Peter Davis, Oleg Polunin and Patrick O'Hara are discussed in relation to this affinity.

Key words: pots, botanists, links, plantsmen, ceramics.

INTRODUCTION

Enquiries from the public come to staff at the Royal Botanic Garden, Edinburgh, in many and varied forms. One which involved the identification of a plant depicted on a Wedgwood plate and tracing its source has opened up new fields far removed from the original query. It has led to an introduction to the wealth of botanical illustrations published during the last century, to the artists, to the plants themselves, to their intrepid collectors and to the anecdotes and myths surrounding them. Botanists and growers, familiar names, are now seen in a new light as an affinity has emerged between them and ceramics often in a more than superficial way.

CARL LINNAEUS (1707–1778)

The trail can start, as does modern botanical nomenclature, with Linnaeus, Wilfred Blunt's 'Compleat Naturalist'. The art of making true porcelain similar to that known from China since medieval times was not discovered in Europe until early in the eighteenth century when it spread from the Meissen factory in Germany to all the major factories of the continent. In 1741 Linnaeus was sent to survey the Baltic islands of Öland and Gotland with among his instructions a plea to seek clay suitable for porcelain manufacture. His search was unsuccessful and perhaps he felt that the standards of the east could never be attained in Europe for he fell under their spell and in 1752 his pupil Pehr Osbeck returned from China bearing a set of Chinese porcelain which Linnaeus had commissioned (Blunt, 1971). These delicate pieces, decorated with the eponymous *Linnaea borealis*, can be seen in the Linnaeus museum at Uppsala (Baer & Lack, 1979, p. 33). A trivial connection this perhaps but it serves as an introduction to some much closer affinities.

THEODOR HOLMSKIOLD (1732–1794)

One of Linnaeus' pupils was Theodor Holmskiold. Retzius (1791, pp. 16, 31) named the attractive verbenaceous genus *Holmskioldia* after him and elsewhere referred to his 'Beata Ruris', a two volume treatise on the fungi of Denmark, as the most brilliant work which had appeared up to that time (Lind, 1913, p. 12). He is still held in high regard among mycologists. Many of his names are still extant and he is credited as being the first to observe the entomogenous nature of *Cordyceps militaris* (Ainsworth, 1976, p. 25). He became professor of natural history and medicine at the College of Søro and later director of the Botanic Garden in Copenhagen.

However he did not confine his talents to the botanical and medicinal field but was a jack-of-all-trades. He was not only postmaster general in Copenhagen, Lord Chamberlain of the royal household and a privy councillor but also manager of the Royal Porcelain factory.

In 1762 the first part of Flora Danica was published under the auspices of Georg Christian Oeder, a German by birth, who with the patronage of the Danish king, Frederick V, had begun a systematic and exhaustive study of the Danish flora. This massive work continued throughout the reign of five kings and was not finished until 1883. During his tenure of the Copenhagen factory Holmskiold was instrumental in having some of the illustrations in Flora Danica reproduced on ceramic ware. By 1790 over one thousand designs were available from which to choose. This was done with care and sensitivity and resulted in one of the finest examples of botanical porcelain known. Could Holmskiold ever have envisaged that nearly 200 years later many of the pieces would still be in production? (Stearn, 1983).

JOHN WEDGWOOD (1766–1844)

Meanwhile in Britain 1804 saw the formation of the Horticultural Society of London, the forerunner of the Royal Horticultural Society (Fletcher, 1969). One of the founders was John Wedgwood and there is no need to stress the ceramic connection here! After the death of his father Josiah, John was with the Staffordshire firm for a short time during which a botanical series of designs was introduced in 1810. These made use of such works as Curtis' Botanical Magazine, Andrew's Botanist's Repository, Salisbury's Paradisus Londinensis, Redouté's Les Liliacées, Roxburgh's Plants of Coromandel and Sowerby's English Botany. The series was produced throughout the nineteenth century as earthenware, and later porcelain, but only the original prints, published before 1810, were used. No catalogue of sources seems to exist but it is possible that all were available in John Wedgwood's private library. On leaving the firm he devoted much of the rest of his life to horticultural and botanical pursuits.

LEWIS WESTON DILLWYN (1778–1855)

Lewis Weston Dillwyn (Fig. 2) was born in 1778, the son of a Quaker businessman who had settled in London although the family originated in the Welsh border country. He is best known, perhaps, for his work on British Confervae (Figs. 3 &

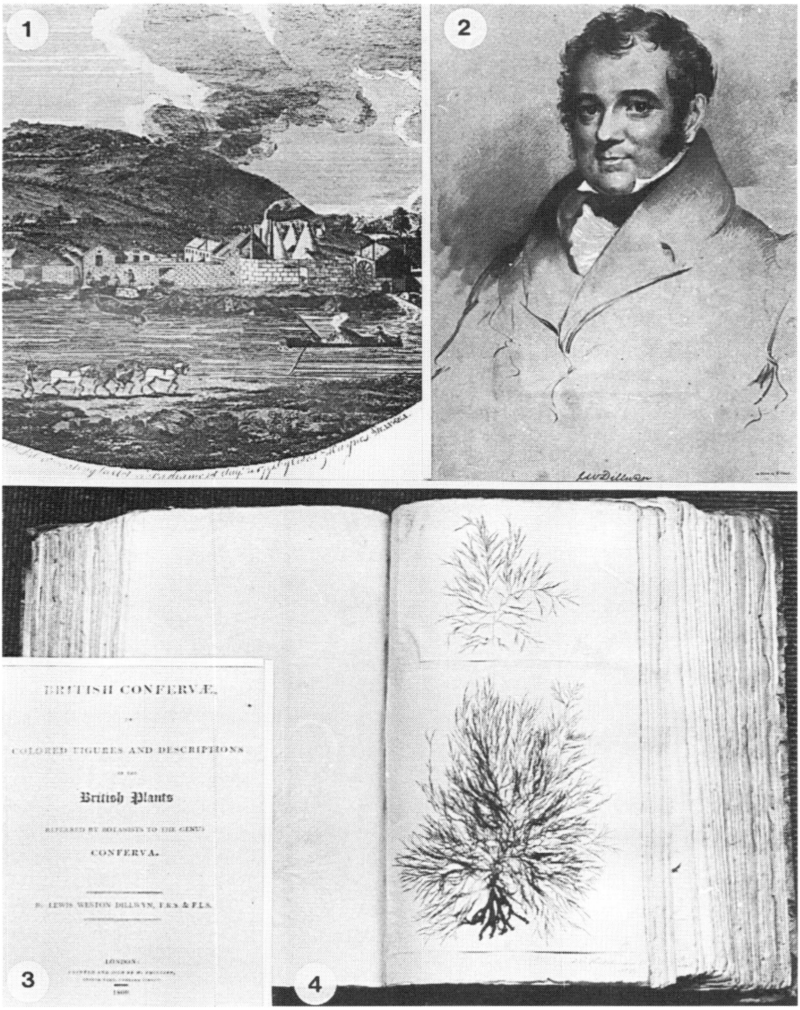

Figures 1–4. Lewis Weston Dillwyn. — 1. Engraving of Cambrian Pottery, Swansea, c. 1791. — 2. Portrait. — 3. Title page of British Confervae. — 4. Hortus Siccus. (1–2 courtesy Swansea Museum, 3–4 courtesy National Museum of Wales, Cardiff).

4) which he published between 1802 and 1814 but many other botanical publications are to his credit (Hyde, 1963). J.E. Smith named the Australian leguminous genus *Dillwynia* 'in honour of my friend whose scientific labours concerning *Conferva* as well as his knowledge in other abstruse parts of botany merit such a memorial' (Smith, 1805, p. 510). However by this time Dillwyn was settled in Swansea where his father had purchased a controlling interest in the

Cambrian pottery on his behalf. Completely inexperienced in what was to be his bread-and-butter for over the next thirty years Dillwyn was initially trained by George Haynes who as manager had, over the previous twelve years, built up the pottery to a high standard enlarging and improving the buildings and employing craftsmen of skill and repute. One was Thomas Rothwell, an engraver who was responsible for many detailed views of Swansea and its environs among which was that of the pottery complex from the river Tawe (Fig. 1). Dillwyn himself appointed William Weston Young who had also prepared many of the illustrations for British Confervae (Hughes, 1983, p. 3).

Dillwyn had married Mary Llewelyn, daughter of John Llewelyn of Penllergaer near Swansea and settled at Sketty Hall also near the city where he became absorbed in growing exotic shrubs from India and Australia. His father-in-law owned an estate just north of Swansea and had already established a garden there which flourished until shortly before the second world war. Later, red clay from this estate was used at the pottery for the production of 'Etruscan Ware' much of which was designed by Dillwyn's daughter-in-law (Armistead & Grant Davidson, p. 10). The construction of an orchid house at Penllergaer is the subject of the first article in the first volume of the Journal of the Horticultural Society of London in 1846 (Llewelyn, 1846). Despite the popularity and challenge of growing exotic orchids in the Victorian age comparatively few are ever depicted on pottery. Why?

For several years Dillwyn kept up a lively correspondence with Sir Joseph Banks (Dawson, 1958). Many subjects were covered including the discovery of skeletons, sizes of cockerels in Carmarthenshire, polydactylism, copper coins, design of rat traps, purchase of merino sheep, not to mention traditional methods of cooking salmon in Lancashire. Banks was rather scathing about the Swansea ware. He found it 'thicker and heavier than the French, in facture too glassy and its glaze too floating'. However he continued to encourage its production hoping for an improvement!

Eventually Dillwyn entered politics serving in various local capacities before becoming Mayor of Swansea. For nine years he represented Glamorgan in parliament (Jackson, 1888, p. 91).

JOSEPH DALTON HOOKER (1817–1911)

Dillwyn was a friend and correspondent of both W.J. Hooker and his son Joseph. Neither Hooker needs any botanical introduction, their achievements are legion. It is of interest however to note that for relaxation Joseph Hooker turned to pottery and in a letter to Charles Darwin he wrote (Huxley, 1918, p. 77), 'By the way – now don't despite me – I am collecting Wedgwoods, simply and solely because they are pretty and I love them'. These 'Wedgwoods' were medallions – Gladstone also collected them and the two corresponded – made in black basalt or jasper ware of blue and white or other colours, c. 2″ × 1–1¾″ and commemorating poets, painters, philosophers, physicians, princes and statesmen from historical times to Hooker's own day. At the Royal Botanic Gardens, Kew, there can still be found not only medallions of Linnaeus and of William Hooker but also, in the director's house, specially designed panels of ferns incorporated in a marble mantelpiece surrounding a central panel portraying the likeness of Sir

William Hooker. The memorial plaques to the two Hookers in St Anne's Church on Kew Green are also of Wedgwood jasper ware (Allan, 1967 facing p. 241).

It was not just the medallions which appealed to Hooker. He refers to the functional aspect of Wedgwood ware and describes at some length the merits of the tea-pots which neither spurt, nor dribble, nor lose their lids, nor scald the knuckles. 'The same care *in adapting to purpose* is displayed in every pot, jug, plate or other article that he made. You may pile plate upon plate of old Wedgwood from floor to ceiling and the whole forms a rigid column; and you never can spin one of his plates in another' (Huxley, 1918, p. 360).

Joseph Hooker was a great traveller. He had been on the *Erebus* in the Antarctic with Ross and had explored Sikkim and Nepal where his published journals prove him to be, like his father, an artist of no mean ability. He travelled in the Rockies, in Syria and Palestine and on more than one occasion in Morocco where he was accompanied by his friend, George Maw.

GEORGE MAW (1832–1912)

It is just over a hundred years since Maw's monograph on the genus *Crocus* was published (Maw, 1886). George Maw (Fig. 8), the subject of a recent paper by Brian Mathew (1986), was by trade a tile maker, running with his brother the Benthall works near Ironbridge in Shropshire where the production of decorative tiles proved a lucrative business feeding the fashion of the mid-Victorian era (Fig. 9). They were used in churches, stations, hospitals, municipal buildings, schools and colleges, even in warships and, more modestly, in private houses. When Joseph Hooker built his house at Sunningdale in 1882 tiles from the Maw factory graced its floors. At its most productive over twenty million tiles were being made at the factory in a year, royalty and aristocracy at home and abroad were among the

Figure 5. Headed notepaper of Maw's Benthall factory. (Taken from Maw letter to John Hutton Balfour; Royal Botanic Garden Edinburgh archive: Balfour letters vol. 9, no. 152).

MAW AND CO.'S PATENT IMPERISHABLE TERRA COTTA PLANT MARKERS AND GARDEN TALLIES, a cheap and permanent substitute for wooden and zinc Tallies, for the labelling of plants in pots and in the open border, and for suspending to Roses, Orchids, Ferns, Fruit Trees, &c. These Labels, unlike Wood and Zinc, combine perfect durability with a lasting legible inscription, and form an elegant adjunct to the Garden and Conservatory for the naming of Horticultural and Botanical Collections. They are also invaluable for the labelling of Wine Cellars, and for employment in any position where other materials would be perishable by damp.

Commended by the Editor of the "Gardeners' Chronicle," Dr. Hooker, the Director of the Royal Gardens, Kew, and the Floral Committee of the Royal Horticultural Society.—(See "Gardeners' Chronicle," July 11th, 1868.)

Specimens sent post free on Application.

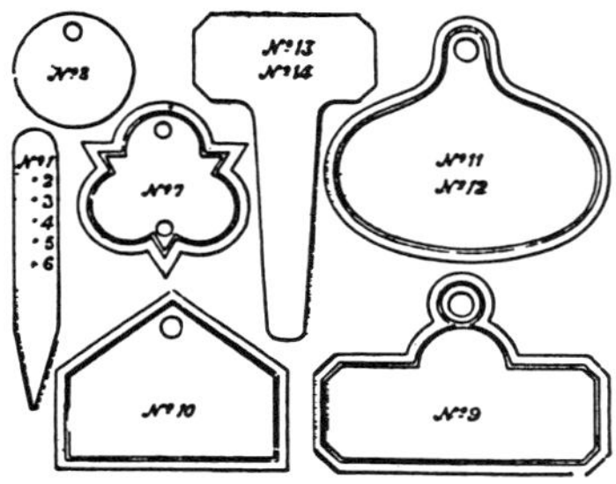

SIZES IN CREAM COLOUR OR BUFF.

No. 1	3½ × ½ inch	(Upright)
No. 2	4½ × ⅝ "	"
No. 3	5½ × ¾ "	"
No. 4	6 × ⅞ "	"
No. 5	7 × ⅞ "	"
No. 6	8 × 1 "	"
No. 7	2¼ × 2 "	(Suspending)
No. 8	1½ in. in diameter	"
No. 9	3½ × 2½ inches	"
No. 10	2¾ × 2¼ "	"
No. 11	3½ × 3 "	"
No. 12	5 × 4½ "	"
No. 13	2 × 4 "	(Upright)
No. 14	3 × 5 "	"

And other larger sizes mounted on Iron shanks for Arboretums and Botanical Gardens.

6 Can be seen in use at the Royal Gardens, Kew, the Gardens of the Royal Horticultural and Royal Botanic Societies, &c.

Figure 6. Catalogue extract of types of ceramic labels produced at Maw's factory, c. 1875. (Courtesy J. Benthall).

clients and lavish catalogues were printed advertising the wares. Agents were appointed in Europe, Canada, USA, South Africa, Australia, New Zealand, India, Japan and China and the firm won many international awards (Fig. 5) (Austwick & Austwick, 1980).

Maw's home was at Benthall Hall at Broseley near Much Wenlock in Shropshire, a fine late sixteenth century house built for the Benthall family and now owned by the National Trust. Maw travelled a great deal in Europe, North Africa, the Near East and North America (Nelmes & Cuthbertson, 1932) combining detailed observations on tile designs and uses, clay deposits and rock formations with the study and collection of plants. Correspondence in the archive at the Royal Botanic Garden, Edinburgh shows that both seeds and herbarium specimens were distributed from these journeys. In his own garden at Benthall Hall he grew the large collection of *Crocus* used in the preparation of his fine monograph. Wilfred Blunt in his Art of Botanical Illustration (Blunt, 1950, p. 238) remarked that 'His monograph, the result of ten years of inquiry, was the most complete work of its kind that had been published on any genus. The plates of this work are marvels of comprehensive detail and put to shame those of many more skilled draughtsmen. From the botanical point of view this work is a landmark; and Ruskin in an unpublished letter described the drawings . . . as "most exquisite . . . and quite beyond criticism"' (Fig. 10). The bulbs were grown in the same meticulous fashion, coded and labelled with permanently fired ceramic labels (Fig. 6) some of which still turn up at Benthall Hall today where Mr James Benthall is

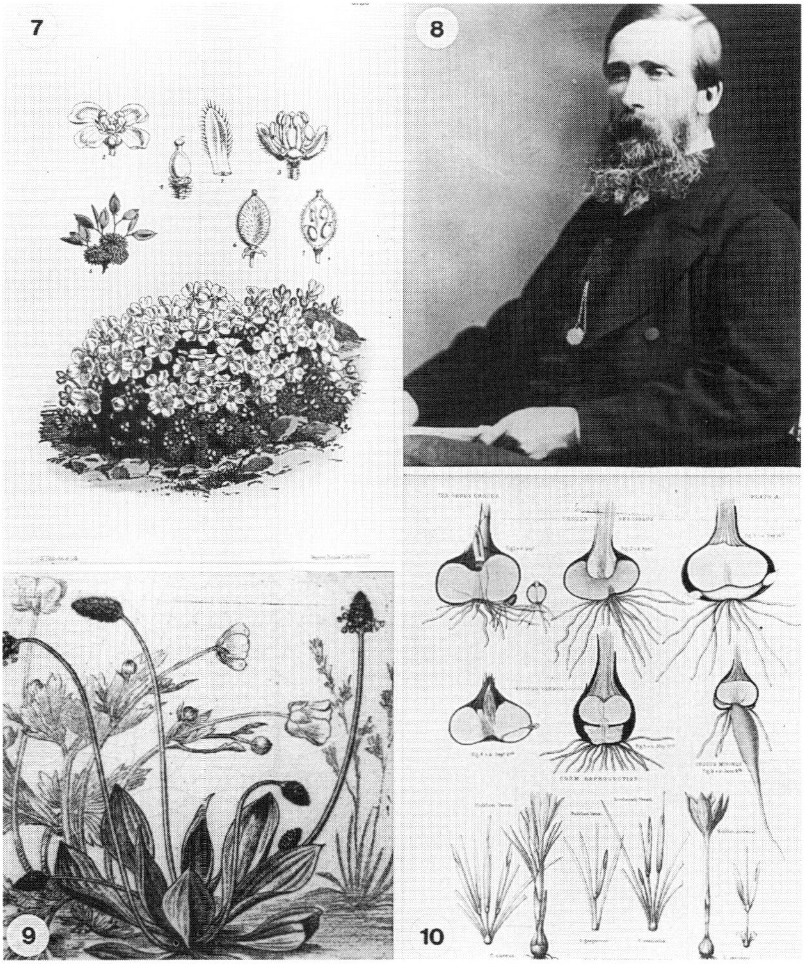

Figures 7–10. George Maw. — 7. *Draba mawii* (Bot. Mag. t. 6186, 1875). — 8. Portrait (courtesy Royal Botanic Gardens, Kew). — 9. Maw tile decorated with *Plantago lanceolata* (courtesy Tony Herbert). — 10. The Genus Crocus, plate A.

hoping to build up a collection of plants connected with George Maw, both his introductions and those such as *Saxifraga maweana*, *Chrysanthemum mawii* or *Draba mawii* that bear his name (Fig. 7). Maw's efforts were rewarded when Joseph Hooker dedicated a volume of the Botanical Magazine to him 'as a tribute to the value of your exertions in introducing hardy herbaceous plants into English gardens' (Hooker, 1874). One of the best known of these introductions and which graces our spring gardens is the invasive *Chionodoxa forbesii*, Glory of the Snow, known to gardeners until recently as *C. luciliae* (Meikle, 1970, 1984). It is likely that some of the botanical subjects on tiles are also Maw's work and one design which may never have reached production incorporated both *Crocus* and *Chionodoxa*, genera almost synonymous with George Maw!

CHRISTOPHER DRESSER (1834–1904)

Another designer of ceramic tiles, this time for the Minton factory at Stoke-on-Trent, was Christopher Dresser (Durant, 1972). However he is perhaps best known in the ceramic world for his designs for the Linthorpe and Ault potteries, distinctive classical shapes, stylised plant forms and glowing rich glazes being his hallmarks. His ideas on the importance of function in design 'Utility must precede beauty' mirror Hooker's 'Adaption to purpose'. Dresser's talents were not restricted to pottery: fabrics, carpets, wallpapers, glass and metal-work all fell heir to his skills. He also developed a technique for nature printing which he patented in 1855. He visited Japan in 1876–77 and studied traditional architecture and methods of manufacture of metal-work, lacquer and ceramics (Dresser, 1882). In the course of visits to about seventy potteries he met potters whose families a generation later, were to befriend the pioneer Cornish studio potter, Bernard Leach.

This was, however, a turnaround of career for earlier he had held professorships of botany at several institutions and is the author of several botanical books and papers. A member of the Botanical Society of Edinburgh, he had been awarded Doctor of Philosophy by the University of Jena 'in consideration of services he has rendered to botanical science' and in 1861 had been elected FLS. Letters in the archive of the Royal Botanic Garden, Edinburgh suggest that he was not, however, on good terms with John Lindley and as a result he was unable to publish through the Linnean Society. His contact with John Hutton Balfour, Regius Keeper of the Garden at Edinburgh ensured that his papers were accepted and published by the Botanical Society of Edinburgh. Conservationists will be interested in another subject in the Balfour letters where Dresser states 'the system of plundering our orchids is barbarous and intolerable: when got to London they only die'. This in 1859!

ARTHUR HERBERT CHURCH (1834–1915)

Although strictly a chemist and professor of chemistry at the Royal Agricultural College, Cirencester, from 1863–1879, Professor Sir Arthur Herbert Church (not to be confused with Arthur Henry Church, Reader in Botany at Oxford) did publish on crop plants and diseases and amassed a valuable collection of botanical water-colour drawings eventually given by his widow to the Royal Botanic Gardens, Kew (Anon., 1916) so that he can thus legitimately be classified as a plantsman.

However his main career was at the Royal Academy where as well as teaching duties he delved deeply into the chemistry of pigments (A.P.L., 1915). Interested in all branches of the arts and an avid collector he is author of two respected books on English porcelain and English earthenware. It would be a pity to miss the chance to air his views on the botanical decoration of pottery 'Admit that naturalistic flowers are a legitimate decoration for shaped vessels, they certainly seem out of place sometimes, as when we find a full-blown rose, dewdrops and all, in the midst of the gravy on our plate' (Church, 1894, p. 12).

THOMAS ANTHONY SPRAGUE (1877–1958)

Thomas Anthony Sprague was born and educated in Edinburgh, training in botany under Isaac Bayley Balfour. He collected plants in Venezuela and Colombia at the end of last century before joining the staff at Kew in 1900 and eventually becoming Deputy Keeper of the herbarium. A recent pen picture of Sprague in the decade before the second world war conjures up the image of an energetic, respected taxonomist as adept at instilling enthusiasm into students as dealing with nomenclatural niceties or editing *Hooker's Icones Plantarum* to a meticulously high standard (Burtt, 1986). Remembered for his well-ordered, acute and logical mind when applied to his work on the Loranthaceae, he was also an authority on early botanical books and on the development of the science of botany over the centuries. A glance at the list of his botanical publications on subjects ranging from African plant studies and itineraries of South American travellers to nomenclature will show the wide variety of his interests and knowledge (Bullock, 1960). This sense of logic and feeling for history found satisfaction again in his extensive and detailed studies on New Hall porcelain which he collected for twenty-five years. In a series of articles in *Apollo* (Sprague, 1949–1959) his discerning eye picked out detail after detail as he compared and contrasted cream jugs and coffee cups, tea-pots and sucriers. Pieces were described as 'globose', 'ellipsoid', 'ventricose', 'compressed obconical' and who but a botanical taxonomist could write of a tea-pot 'foot short, 7–8 mm high; shoulder comparatively flat, spout fluted halfway up ... handle with or without an upturned cusp ...'. All that is missing is the Latin description: perhaps even that still lurks unpublished in a dark corner of the Kew herbarium!

RONALD R.J. COPELAND (1884–1958)

The western coasts of the British Isles are washed by the warm waters of the Gulf Stream lulling them into the belief that they can support an exotic vegetation. So they can in all but the severest of winters and one area well endowed with subtropical gardens is the Duchy of Cornwall, the south-westernmost tip of England. Here amongst a bevy of 'Tre's – tre is a Celtic word signifying house or home and also turns up frequently in Brittany and Wales – lies Trelissick, twenty-five acres laid out in its modern form by the Copeland family from 1937 until 1955 and still their home though now under the aegis of the National Trust. The garden contains a splendid collection of choice trees and shrubs: *Magnolia*, *Eucalyptus*, *Nothofagus*, *Hoheria*, over one hundred species or cultivars of *Hydrangea* but above all *Rhododendron* and *Camellia* of which some cultivars originated in the garden (Thomas, 1979).

Ronald Copeland was chairman of the family company W.T. Copeland and Sons Ltd, manufacturers of Spode china. He frequently brought floral specimens from Trelissick to the factory at Stoke-on-Trent and these were reproduced on the ware. A large number of *Rhododendron* subjects, starting in 1943 with *R. augustinii*, were depicted on 10" plates later extended to dessert plates; subsequently a complete range of tea and coffee sets was produced followed by full dinnerware. Several species and cultivars were portrayed including *R. batemanii*, *R. hodgsonii*, *R. fortunei* and *R. albrechtii*, 'Cowslip', 'Alison Johnstone' and 'Gwylt King'. Camellias also were used and more recently his son, Spencer Copeland, now residing at Trelissick, was personally responsible for the reproduction of *Camellia* 'Donation' (S. Copeland pers. comm.). Plants also came from other gardens and a set of six plates of rhododendrons from Bodnant in North Wales was included in the January 1987 Sotheby's exhibition 'The Glory of the Garden' when they were illustrated as number 500 of the accompanying catalogue.

PETER HADLAND DAVIS (b. 1918)

At about the same time as the garden at Trelissick was being planned a young man was standing, half a world away, amazed and captivated by the blue domes of Islam. Peter Davis (Fig. 12) first went to the Levant before the second world war and visited Turkey in 1938. Seeing the need for a modern Flora, he determined to make this his main research activity and many expeditions followed which resulted in a magnificent herbarium collection of Turkish plants. During his period as lecturer, reader and professor in taxonomic botany at the University of Edinburgh and supported by a succession of loyal and diligent assistants, he has published the complete Flora of Turkey in ten volumes including a supplement in just over twenty years (Figs. 11 & 13!). As a recent reviewer has remarked (Barneby, 1983), 'The botanical community is deeply indebted to Professor Davis, who initiated a noble project, orchestrated its complex realization with single-minded zeal, wrote a substantial portion of the text and contributed enormously to the raw material of specimens on which it is based'. In addition, during this period, dedication to the needs of students has resulted in the publication of two indispensable textbooks, 'Principles of angiosperm taxonomy' with V.H. Heywood in 1963 and 'The identification of flowering plant families' with J. Cullen in 1965.

However, since the fifties when his acquisitive eye first fell on the jolly, colourful earthenware produced from 1882 until 1930 in Kirkcaldy another project has been brought to fruition when in 1986 a monograph on Wemyss ware (pronounced Weems) was published jointly with Robert Rankine (Figs. 14 & 15). In the introduction to a catalogue for the Scottish Arts Council's exhibition of Wemyss in 1971 Peter Davis wrote (Davis, 1971, p. 3–4): 'Wemyss ware is the most distinctive and decorative of the late Victorian–Edwardian wares made in Scottish potteries. The pots are not so much decorated as the "canvas" for vigorous, three-dimensional, underglaze painting. The meretricious, niggling aspects of Victorian decoration were set aside by this back-to-nature style that threw open the windows of bedrooms and drawingrooms to the world of the garden and farmyard outside – to fruits and flowers and chanticleer. This wedding of peasant art and relative sophistication was fresh enough to please but not bold enough to shock the upper middle-class market at which it was originally aimed ... Although Wemyss ware was exuberantly decorative much of it was useful if not very durable. Wemyss could have played a considerable part in a country dowager's day. With its aid she could have taken tea in bed, peed and washed, curled her hair and

Figures 11–15. Peter Davis. — 11. Flora of Turkey (courtesy Edinburgh University Press). — 12. Portrait (courtesy The Scotsman, Edinburgh). — 13. Cartoon by the late Paul Aellen, Basel, 1964. — 14. Wemyss Pottery (courtesy R. Eudall and Scottish Arts Council). — 15. Wemyss monograph (courtesy Scottish Academic Press).

powdered her nose, tucked into porridge and honey for breakfast, arranged the flowers, watered the plants in their *jardinières*, dealt with her correspondence with the aid of inkstands and paper-racks and taken a lavish afternoon tea. Finally, having chosen a suitable candlestick, she tottered up to bed; if it wasn't her day she may have tripped over a pig door-stop and soaked her poor old feet in a Wemyss foot-bath. It may not have been a very useful day but, with breakages common, it was not unprofitable for the Fife pottery'. Many a Turkish botanist has spent a bemused evening amidst this overwhelming collection but has never failed to be carried along with Peter's infectious enthusiasm. With his naughty sense of humour and enviable turn of phrase it has been an exasperating delight working with him, travelling with him but above all learning from him in the last thirty years!

OLEG POLUNIN (1914–1985)

One of Peter Davis's early travelling companions in Turkey in the fifties was Oleg Polunin. Together they covered Western Anatolia, Lycia, Pamphylia, Cilicia and the Amanus to the area around Lake Van and the Cilo Dağ in the south-east. The journeys resulted in over 4000 collections including many new taxa such as *Ranunculus poluninii* P.H. Davis, *Primula davisii* W.W. Smith and *Iris pamphylica* I.C. Hedge. Already a veteran of botanical expeditions in Nepal, the Turkish expeditions were followed by others to Kashmir, Iraq, Lebanon and to the Karakorum. Since then Oleg Polunin became well known to a new generation of travellers for his sparkling leadership of holiday groups in the East Mediterranean and the Himalayas and for his magnificent series of field guides: 'Flowers of the Mediterranean' written with Anthony Huxley in 1965 was followed by 'Flowers of Europe' (1969), 'Flowers of SW Europe' (with Bill Smythies, 1973), 'Flowers of Greece and the Balkans' (1980), 'Flowers of the Himalaya' (with Adam Stainton, 1984) and others published posthumously. These, as well as forming practical guides aiding identification of plants, have stressed time and again the need for their conservation. For thirty-four years he was biology master at Charterhouse School where his influence in infusing into his pupils a love of the natural sciences was great and permanent (Akeroyd, 1986).

Twenty-five years ago he took up pottery as a hobby, his interest probably fired by his godmother Katharine Pleydell-Bouverie, an early pupil of Bernard Leach. Once a week, first at Charterhouse and then at Farnham College of Art, he would repair to the wheel. His wife writes (pers. comm.), 'It was his day of relaxation, he liked the physicalness of potting after sitting at his desk writing. He was a very practical potter, he made things to be used – usually in stoneware in which to do our cooking. All our friends and relations are well supplied with Oleg's casseroles, mugs, bowls, dishes of every sort – plus a few decorative pots for plants!' He would probably have agreed with Patrick O'Hara, who follows, that plants and pots are a logical association. They also shared a deep concern about the threat to wildlife and in their different ways have made practical contributions to conservation. Oleg Polunin was one of the contributors to a 'Folio of Quotations' accompanying an exhibition at the United Nations in Geneva in 1984. The exhibition, 'Secret Gardens of the World', showed original sculptures in porcelain by Patrick O'Hara while the proceeds all went to support plant conservation efforts managed by the International Union for the Conservation of Nature and Natural resources and the United Nations High Commission for Refugees.

PATRICK O'HARA (b. 1936)

'Few are the media and rarer still are the artistic skills that have captured the spirit of plants and butterflies with such an expression of dedication, love and genuine feeling as those evident in the extraordinary porcelain sculptures created by Patrick O'Hara'. So wrote William Davis and Richard Evans Schultes (1984) in the foreword of the catalogue to that same exhibition in Geneva. On the southern shores of Ireland in the little village of Currabinny near Carrigaline Patrick O'Hara creates his unique sculptures with equal concern both for scientific accuracy and aesthetic appeal. This is, however, the last stage of the process which may have taken several years of careful preparation. Having begun by studying and portraying the plants of his local countryside he now travels the world with particular habitats and species in mind making drawings and notes on colours, sizes and communities so detailed that they alone are necessary to create the design and eventually capture in porcelain a little bit of alpine limestone bright with gentians and daphne, or the heat beating from the desert rock home of *Opuntia compressa*. He has been exhibiting flower sculptures since 1972 and his work – he has now depicted well over 500 species – is in museums and private collections worldwide.

Like Christopher Dresser, however, his career has changed direction. Patrick O'Hara has no formal art training but read botany, zoology and geology at Reading before working for ten years as an agricultural adviser. The two disciplines are not divorced but happily fused, a formal training combining with natural talent to produce works of art which cannot fail to accentuate the fragile state of much of our heritage.

CONCLUSION

Botanists with close ceramic ties still abound! Roger Kemp, a discerning collector with a penetrating knowledge of ceramics particularly of early British creamware with a special love for Swansea, has taught genetics in the Botany Department of the University of Edinburgh for more than twenty years. Tony Herbert, now an adviser on historic tiles and architectural ceramics, started off with a degree in botany from Bristol. Walter Lack of the Botanical Garden and Museum at Berlin-Dahlem has been responsible in conjunction with Winfried Baer for two papers on botanical designs on porcelain (Lack & Baer, 1978, 1986) and for the organisation of what must have been the finest display of botanically decorated ceramics ever seen as part of the tercentenary of the Berlin Botanic Garden in 1979. He wrote the foreword for the catalogue produced at the time (Baer & Lack, 1979), itself an indispensable reference book crying out for an English edition!

Is there then an affinity between plantsmen and ceramics or is it just coincidence? Certainly, similar links could be found between plantsmen and music or painting but have individuals made such major contributions to both fields? Is there a reason for the relationship? Patrick O'Hara feels that it is a logical association, both subjects being completely apolitical. Answers at present remain obscure: perhaps a symposium (which the dictionary defines as a drinking party) should be convened to explore them! Meanwhile thinking in terms of plants or pots let Joseph Hooker continue, and have the last word – 'They afford me pleasure – *voilà tout*'.

ACKNOWLEDGMENTS

For help in various ways I thank James Benthall, Bill Brydon, B.L. Burtt, Spencer Copeland, George Gibb, Ross Eudall, Tony Herbert, Michael Isaac, Roger Kemp, Patrick O'Hara, Roy Perry, Lorna Polunin, Graham Stuart Thomas, Roy Watling and Paddy Woods. The Scottish Arts Council gave permission to quote from their Wemyss catalogue for which I am grateful. All unacknowledged figures are taken from material at the Royal Botanic Garden, Edinburgh and are the work of Ken Grant to whom particular thanks are due for his preparation of the photographs.

LITERATURE CITED

Ainsworth, G.C. 1976. Introduction to the history of mycology. Cambridge University Press, Cambridge. xi + 359 p.

Akeroyd, J. 1986. Oleg Vladimirovich Polunin 1914–1985. Watsonia 16: 105–107.

Allan, M. 1967. The Hookers of Kew 1785–1911. Michael Joseph, London. 273 p.

Anon. 1916. Sir Arthur Church's collection of botanical drawings. Kew Bull. 1916: 162–168.

'A.P.L.' 1915. Sir A.H. Church, K.C.V.O., F.R.S. *Nature* 95: 399–400.

Armistead, K.M., and W.J. Grant-Davidson. Undated. Catalogue Kildare S. Meager bequest Swansea pottery. Glynn Vivian Art Gallery, Swansea. 48 p.

Austwick, J. and B. 1980. The decorated tile: an illustrated history of English tile-making and design. Pitman House, London. 160 p.

Baer, W., and H.W. Lack. 1979. Pflanzen auf Porzellan 1750–1850. Botanischer Garten und Botanisches Museum Berlin-Dahlem, Berlin. 125 p.

Barneby, R.C. 1983. Book review of P.H. Davis, Flora of Turkey, volume 7. Notes Roy. Bot. Gard. Edinb. 41: 95–96.

Blunt, W. 1950. The art of botanical illustration. Collins, London. xxxi + 304 p.

——. 1971. The compleat naturalist: a life of Linnaeus. Collins, London. 256 p.

Bullock, A.A. 1960. Dr Thomas Archibald Sprague, 1877–1958. Taxon 9: 93–102.

Burtt, B.L. 1986. Thomas Archibald Sprague, 1877–1958. The Golden Bough 8: 1–3.

Church, A.H. 1894. English porcelain. Chapman and Hall, London. xiii + 99 p.

Davis, P.H. 1971. Introduction. *In* Wemyss ware: catalogue to exhibition. Scottish Arts Council, Edinburgh. 20 p.

Davis, W.A., and Schultes, R.E. 1984. Foreword. *In* Patrick O'Hara, Secret gardens of the world. Patrick O'Hara, Currabinny. 20 p.

Dawson, W.R. [ed.]. 1958. The Banks letters. Trustees of the British Museum, London. xlii + 965 p.

Dresser, C. 1882. Japan: its architecture, art and manufactures. Longmans, Green and Company, London. xi + 467 p.

Durant, S. 1972. Christopher Dresser 1834–1904. *In* R. Dennis and J. Jesse, catalogue to exhibition. Fine Art Society, London. 38 p. (unpaged).

Fletcher, H.R. 1969. The story of the Royal Horticultural Society. Oxford University Press, London. xii + 564 p.

Hooker, J.D. 1874. Dedication to George Maw. Bot. Mag. ser. 3, 30.

Hughes, P. 1983. Welsh China. Amgueddfa Genedlaethol Cymru (National Museum of Wales), Cardiff. 28 p.

Huxley, L. 1918. Life and letters of Sir Joseph Dalton Hooker Volume 2, John Murray, London vii + 569 p.

Hyde, H.A. 1963. Lewis Weston Dillwyn as a botanist. South Wales and Monmouth Rec. Soc. Trans. 5: 6–8.
Jackson, B.D. 1888. Lewis Weston Dillwyn. *In* L. Stephen [ed.], Dictionary of national biography 15: 90–91.
Lack, H.W., and W. Baer. 1978. Ein 'botanisches' Porzellanservice aus Berlin fur Kaiserin Josephine. Willdenowia 8: 235–259.
——, and ——. 1986. Pflanzen aus Kew auf Porzellan aus Berlin. Willdenowia 16: 285–312.
Lind, J. 1913. Danish fungi as represented in the herbarium of E. Rostrup. Gyldendalske Boghandel - Nordisk Forlag, Copenhagen. 650 p.
Llewelyn, J.D. 1846. Some account of an orchideous house constructed at Penllergare [sic], South Wales. J. Hort. Soc. London 1: 5–8.
Mathew, B. 1986. George Maw and his monograph. The Kew Magazine 3: 186–189.
Maw, G. 1886. A monograph of the genus *Crocus*. Dulau & Co., London. viii + 326 + xx p. + index.
Meikle, R.D. 1970. *Chionodoxa luciliae*, a taxonomic note. J. Roy. Hort. Soc. 95: 21–24.
——. 1984. *Chionodoxa forbesii*. *In* P.H. Davis [ed.], Flora of Turkey 8: 225–226.
Nelmes, E., and W. Cuthbertson. 1932. George Maw. *In* Curtis's Botanical Magazine dedications 1827–1927. Bernard Quaritch, London. xxi + 400 p.
Retzius, A.J. 1791. Observationes Botanicae 6. Crusius, Leipzig. 68 p. + 3 tabs.
Smith, J.E. 1805. Remarks on the generic characters of the decandrous papilionaceous plants of New Holland. Ann. Bot. (London) 1: 501–512.
Sprague, T.A. 1949a. Hard paste New Hall porcelain, part 1. Apollo 49: 165–167.
——. 1949b. Op. cit. part 2. l.c. 50: 16–18.
——. 1950a. Op. cit. part 3. l.c. 52: 51–54.
——. 1950b. Op. cit. part 4. l.c. 52: 109–112.
——. 1959. New light on polychrome Caughley. l.c. 69: 14–18.
Stearn, W.T. 1983. The Flora Danica, its history and illustrations. *In* M. & M. Rossler, Plant portraits from the Flora Danica 4–7. The Mendip Press, Bath. 12 p. + 12 plates.
Thomas, G.S. 1979. Gardens of the National Trust. Weidenfeld and Nicolson, London. 292 p.

23. AN UNPUBLISHED FLORA OF CONSTANTINOPLE FROM THE SECOND HALF OF THE XIX CENTURY

KAZIMIERZ BROWICZ

Institute of Dendrology, Polish Academy of Sciences, 62-035 Kórnik near Poznań, Poland

ABSTRACT

A review is made of a manuscript, stored in the Library of the Polish Academy of Sciences in Krakow, which is a study by Wladyslaw Jablonowski (1841–1894) entitled 'Materyaly do Flory Konstantynopola i jego okolic' (Materials on the Flora of Constantinople and its environs), written in the years 1882–1886. It contains a physiographic survey of the region, on both the European and Asiatic coasts of the Bosporus, and lists of species compiled according to their occurrence in various types of sites although unfortunately the herbarium collections on which these lists were based do not appear to have survived. Information is also given on the activities of Jablonowski in Turkey where he first went in 1866 as an army physician.

Key words: Flora of Turkey, of Constantinople (Istanbul), of Bosporus.

In the library of the Polish Academy of Sciences in Krakow, in the Manuscripts Division, under no. 1946 there is stored a manuscript by Wladyslaw Jablonowski entitled 'Materyaly do Flory Konstantynopola i jego okolic' (Materials on the flora of Constantinople and its environs). Who was Jablonowski, how did he come to be in Constantinople, when and whence came his floristic interests and what is the value of this work? These are the questions I shall try, at least partially to answer.

Three times in the XIX century a wave of Polish refugees settled in the Ottoman Empire, particularly in Constantinople. This happened after the sad end of two Polish uprisings in 1831 and 1863 and after the fall of the Hungarian uprising in 1849, in which many Poles also participated. Some of them spent the rest of their days in Constantinople, frequently taking an active part in the cultural life of the town, leaving their mark behind them. They included military men, engineers building roads and railways, architects, physicians, lecturers in academic schools, cartographers and painters. Some entered administration, particularly in the Turkish army: those embracing Islam attained high positions, Michal Czajkowski (Sadyk Pasha) for example commanded a regiment of Ottoman Cossacks, gen. Jozef Bem (Murat Pasha) was an administrator in Aleppo and Konstanty Borzecki (Mustafa Celalettin Pasha) wrote a book published in 1869 in Constantinople entitled 'Les Turcs anciens et modernes', where he traced the historic and ethnic links between the Ottoman Turks and the Turks from central Asia; he was also chief of the Cartographic Division in the Turkish General Staff. A permanent remnant of this period is the Polish village of Adam-Pol (today known as Polonezköy) lying on the Asiatic shore of the Bosporus, at the foot of Alem Daği. Established in 1842 and originally an agricultural-cum-military community, a small colony of the grandchildren of these first settlers still survives today.

Wladyslaw Jablonowski was one such military emigrant. He was born in 1841 in the Russian part of partitioned Poland. First he studied at the Mathematical-Natural Sciences Faculty in Kiev University, and then went to Italy and attended lectures on descriptive anatomy in Genoa. After returning to Poland he took part in the 1863 uprising against the Russians with the rank of captain. When this failed he took up medical studies at Krakow University (in the part of Poland under Austria) practising in the surgical department of a local hospital. He graduated in 1865, but in the beginning of 1866, as a result of a police decision he had to leave the country. He arrived in Constantinople in March 1866, at the age of 25 and remained in exile for the rest of his life. He died in 1894 in Burgas, in Bulgaria.

Though educated not as a botanist but as a physician, Jablonowski writes in the introduction to his 'Flora' that botany had been his hobby ever since his days in Poland. It is possible that these floristic interests were influenced by his friend, Antoni Rehman, a geographer, geomorphologist, and botanist, known for his travels to southern Africa, the Crimea and in the Caucasus. Jablonowski had known him in Krakow and in later years they again met several times in Vienna. It was Rehman who introduced Jablonowski as a member of the Vienna Zoologisch-botanischen Verein.

His first contact with the vegetation of the vicinity of Constantinople, both wild and cultivated, inspired him to conduct botanical studies. There and then he decided to write the 'Flora of the vicinity of Istanbul' (see 'Pamietniki Jablonowskiego', Memoirs of Jablonowski, published as recently as 1967 based on the manuscript no. 1280 in the Library of the Polish Academy of Sciences in Krakow). He was enchanted by the vegetation of the hills near the town, and by finding a magnificent specimen of 'arum' (probably *Arum italicum* Miller). To realize his project, however, he had to wait, for in November 1866, after passing an exam under a special medical commission of the Turkish Army, he had to leave Constantinople as a military surgeon, moving to various parts of the Ottoman Empire. He was stationed in Baghdad, Mosul, Karbala and Sulaljmaniah in Iraq, in Penjwin on the Iraqi–Persian border, in Hamadan and Kermanshah in Persia, in Batumi in the Caucasus, in Trabzon, Diyarbakir, Urfa in Anatolia, Aleppo in Syria and on the Balkan peninsula (Albania, Montenegro, Macedonia and in Greece). He acted also as the physician for a marble and boracite quarry in Sultan Cair on the Sea of Marmara and was engaged in the establishment of the balneological sanatorium in Bursa. Appointed by the Turkish authorities as a delegate to the International Sanitary Commission in the East, stationed in Constantinople, he travelled to areas endangered by plague and cholera. As a result of this activity he obtained high Turkish and Persian distinctions and also a personal appreciation from the Sultan and the Shah. In 1888 he went to Bulgaria as a quarantine doctor.

During his many travels, in spite of the pressures of his profession Jablonowski continued his botanical and archaeological studies, gathering among other things a substantial herbarium.

Returning frequently to Constantinople he concentrated primarily on the study of the flora of this fascinating town and its vicinity: that is stated several times in his 'Memoirs'. He started to edit the text of his botanical work in the years 1882–1886. Careful preparation is indicated by the accumulation of appropriate literature, the list running to 45 titles among which are the works of Boissier, Grisebach, Jaubert & Spach, Sibthorp & Smith, Marschall von Bieberstein, Delile, Jacquin, Endlicher, Willdenow, Reichenbach, Parlatore, Redouté, De Candolle and many others. He does not list the work of Tchihatcheff (Asie Mineure III Botanique), although in the text of his 'Flora' he frequently quotes it, so that it was certainly known to him. This list of literature indicates clearly that Jablonowski was able to identify plants that he saw and collected.

During each stay in Constantinople he undertook several floristic trips around the town on foot or on horseback both on the European and Asiatic sides. Additionally, as a physician, he had frequent access to estates of Turkish noblemen (he was even called to treat ladies in harems!) permitting him to become acquainted with cultivated ornamental plants: often these were exotic, their presence quite surprising. He devoted much attention to these cultivated plants in the latter part of his 'Flora'.

Jablonowski's manuscript consists of 262 sheets, 27 × 21 cm in size, each divided into two parts or pages, written in a fine but very elegant hand.

The introduction is most important for the understanding of the presentation of the floristic data. It has 52 pages (among which one page with a list of plants from Syria is accidentally interpolated). It is composed of (a) a general survey of the area under study (20 pages); (b) characteristics of the climate (8 pages); (c) geological details (13 pages); (d) a list of literature (5 pages) and (e) a list of localities along the Bosporus, divided into European and Asiatic, the list including 169 names (5 pages).

Jablonowski, who had some cartographic training (while in Iraq he introduced some additions and corrections to W. Kiepert's 'Karte des turkischen Reiches in Asien'), describes in great detail the landscape around Constantinople, taking particular note of the courses of valleys, features of the sea coast and the presence of various types of water, both standing and running. All of this is discussed from the point of view of the formation of plant communities. He delimits the area under study in the north along the shores of the Black Sea, more or less from Lake Terkos in the west to Riva in the east, in the west from Lake Terkos through the valley of the Karasu to Büyük Çekmece Bay and along the coast to Silivri, in the south along the shores of the Sea of Marmara from Silivri in the west to Tusla in the east and in the east from Riva, through Alem Daği-Çatal Dağ-Aydos Daği to Pendik and Tusla. Also included in the region is the Princes Islands archipelago (Kizil Adalar).

In his description Jablonowski points out the differences in the vegetation between the European and Asiatic shores of the Bosporus. He enthuses about the plants of Alem Daği and the beautiful valleys of the Kutschuk-Gok-Su and Bujuk-Gok-Su (the so-called Sweet Waters of Asia), about the Beykos wood at the foot of Alem Daği and the belt of laurels in that region the importance of which for the Asiatic shores of the Bosporus he compares to that of the Belgrad Forest on the European side. He describes the Princes Islands, particularly Halki

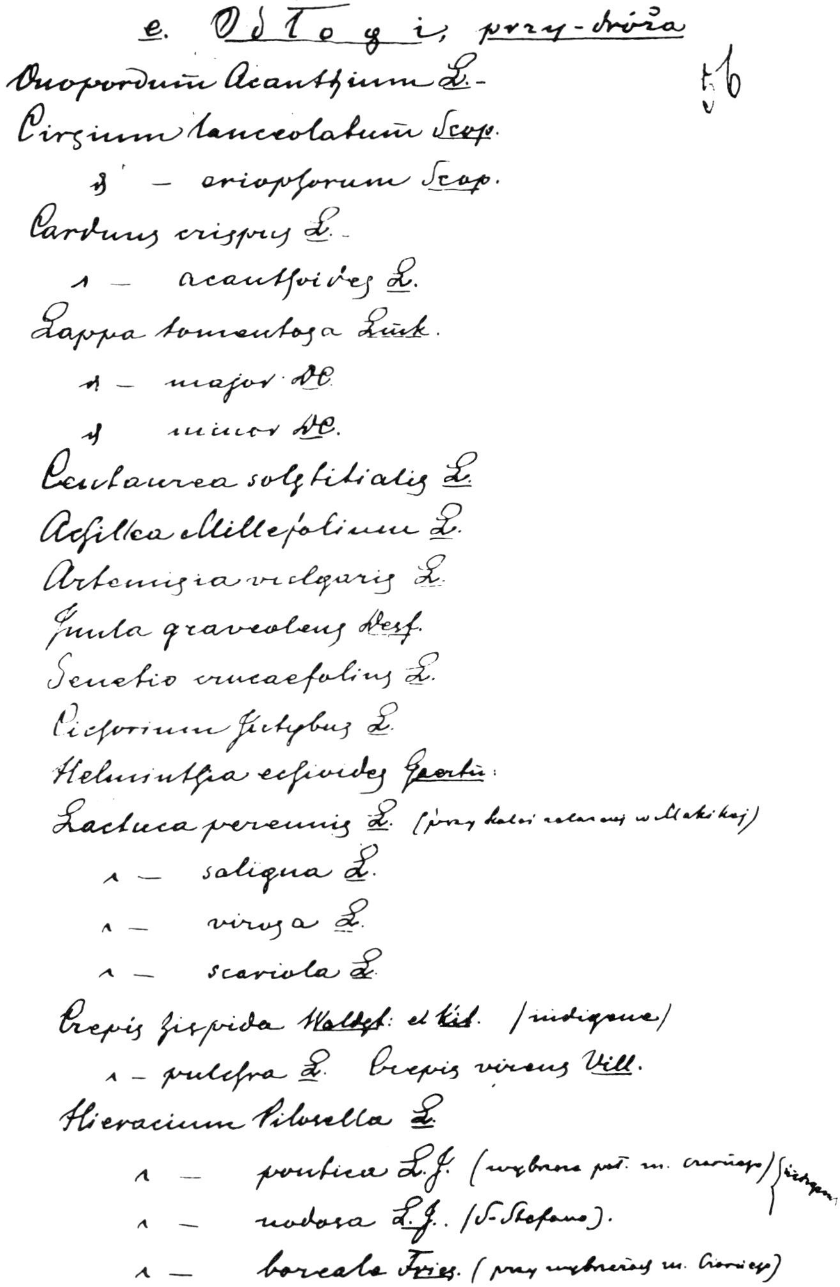

e. Odłogi, przy-dróża

Onopordum Acanthium L. 56

Cirsium lanceolatum Scop.

" — eriophorum Scop.

Carduus crispus L.

" — acanthoides L.

Lappa tomentosa Lmk.

" — major DC.

" minor DC.

Centaurea solstitialis L.

Achillea Millefolium L.

Artemisia vulgaris L.

Inula graveolens Desf.

Senecio erucaefolius L.

Cichorium Intybus L.

Helminthia echioides Gaertn.

Lactuca perennis L. (przy kolei żelaznej w Makrikoj)

" — saligna L.

" — virosa L.

" — scariola L.

Crepis hispida Waldst. et Kit. (indigene)

" — pulchra L. Crepis virens Vill.

Hieracium Pilosella L.

" — pontica L.J. (wybrzeże [illegible] m. [illegible]) [illegible]

" — nodosa L.J. (S-Stefano).

" — borealis Fries. (przy wybrzeżu m. [illegible])

Figure 1. Fallow land and roadsides – beginning of the list, part 4.

(Heybeli) and Prinkipo (Büyükada). He had also visited small uninhabited islands lying at the mouth of Tusla Bay where he found a rich occurrence of species of *Sedum*. He even reached a group of small, rocky islands located at the northern tip of the Bosporus, near the European shore, opposite Rumelifeneri, where he found only mosses and lichens and one species of fern.

Rather interesting are Jablonowski's opinions about the changes taking place in the vicinity of Constantinople as a result of frequent fires, rapid development of the town, establishment of new communication lines with the use of explosives in the process, and also the cutting of forests. Writing the introduction to his 'Flora' Jablonowski reports that after 16 years stay in Turkey, there had been, in the vicinity of Constantinople, considerable changes in the occurrence of certain species, as well as the appearance of new species on disturbed sites. He also remarks on the Princes Islands, where one of the causes of change had been the settlement of foreigners and the establishment of ornamental gardens in place of natural plant communities.

'The materials' are arranged according to ecological requirements rather than on a systematic basis. They consist of lists of plants occurring in regions with similar sites without, however, detailing the locations. Only in exceptional cases, apparently when mentioning rare species or those of particular interest to himself, the exact location is given. The whole work, apart from the Introduction, has been divided into the following 23 parts:

1. Hills and dry meadows – 25 pages including 6 pages of text. The list contains 479 species. In a short note added to the text Jablonowski mentions *Liquidambar orientalis* Miller on the Asiatic shore of Bosporus, between Aydos Daği and Bulgurlu, and also from Prinkipo Island. This may refer to cultivated plants: *L. orientalis* is not mentioned from this part of Anatolia in Davis, 'Flora of Turkey' vol. IV.
2. Meadows in valleys – 16 pages including 8 pages of text and 306 species.
3. Cultivated fields and post harvest stubbles – 15 pages with 6 pages of text and 219 species.
4. Fallow land and roadsides – 18 pages with 3 pages of text and 384 species (Fig. 1).
5. Cemeteries – 7 pages with 2 pages of text and 127 species.
6. Old walls, ruins, hollows – 10 pages with 1 page of text and 227 species. The list includes *Utricularia vulgaris* L. and *U. minor* L., found in moats near old forts. These species are not mentioned for this part of Turkey in Davis, 'Flora of Turkey' vol. VI.
7. Vegetable and fruit gardens – 4 pages with 76 species. Jablonowski mentions *Viscum album* L. as occurring here on fruit trees.
8. Gardens – 7 pages and 158 species.
9. Cypress forests (probably this concerns plantations of *Cupressus sempervirens* L.) – 4 pages and 102 species.
10. Rivers and riversides – 15 pages with 4 pages of text and 278 species. Jablonowski cites here from Kutschuk (Küçük) Cekmece Bay *Myricaria germanica* (L.) Desv. This species is at present not known from the European part of Turkey. It is possible that Jablonowski erroneously described some species of *Tamarix*,

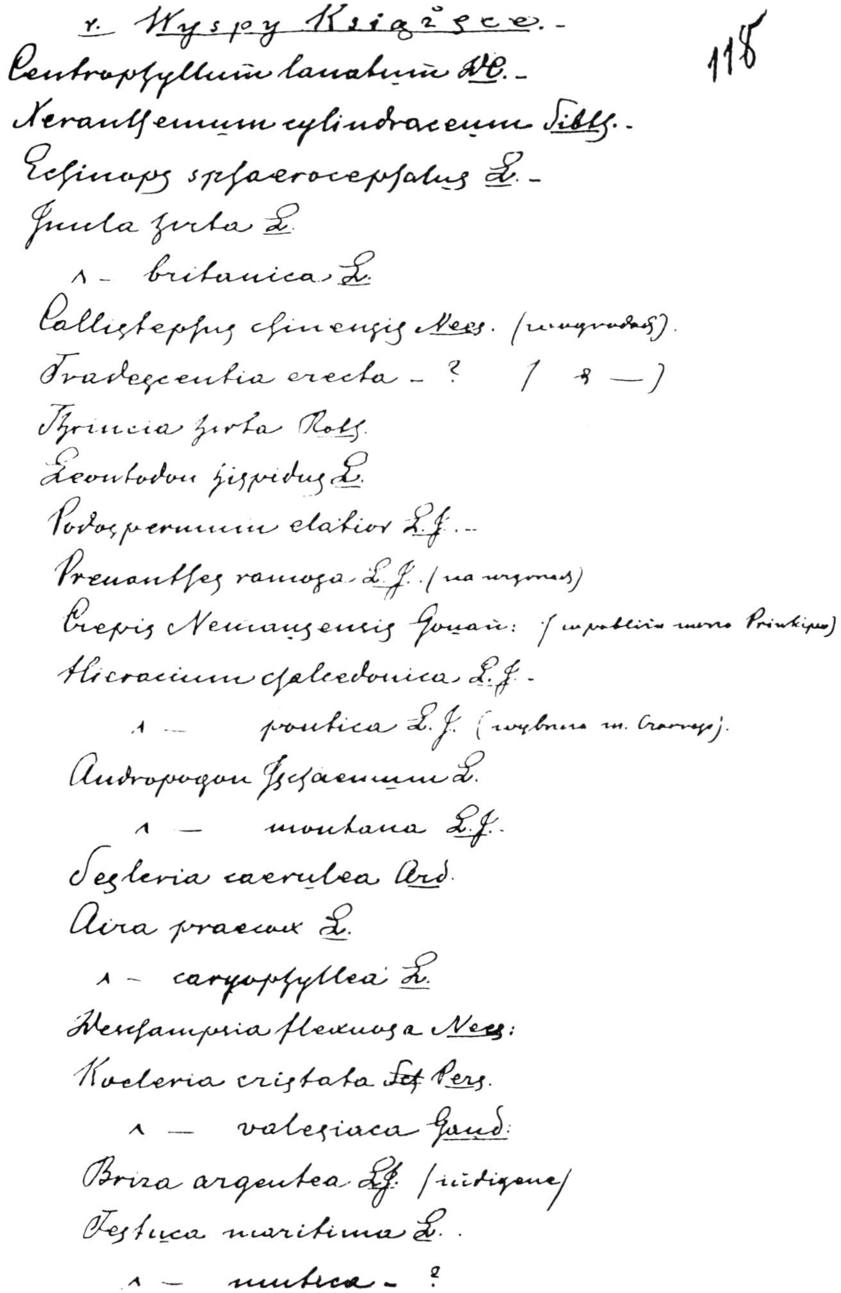
r. Wyspy Książęce.- 118
Centrophyllum lanatum DC.-
Xeranthemum cylindraceum Sibth.-
Echinops sphaerocephalus L.-
Inula hirta L.
" – britanica L.
Callistephus chinensis Nees. (w ogrodach).
Tradescantia erecta – ? (" –)
Thrincia hirta Roth.
Leontodon hispidus L.
Podospermum elatior L.f.--
Prenanthes ramosa L.f. (na wzgórzach)
Crepis Nemausensis Gouan: ([illegible] Prinkipo)
Hieracium chalcedonica L.f.-
" – pontica L.f. ([illegible]).
Andropogon Ischaemum L.
" – montana L.f.
Sesleria caerulea Ard.
Aira praecox L.
" – caryophyllea L.
Deschampsia flexuosa Nees:
Koeleria cristata [illegible] Pers.
" – valesiaca Gaud.
Briza argentea L.f. (indigena)
Festuca maritima L.
" – mutica – ?

Figure 2. Princes Islands – beginning of the list, part 16.

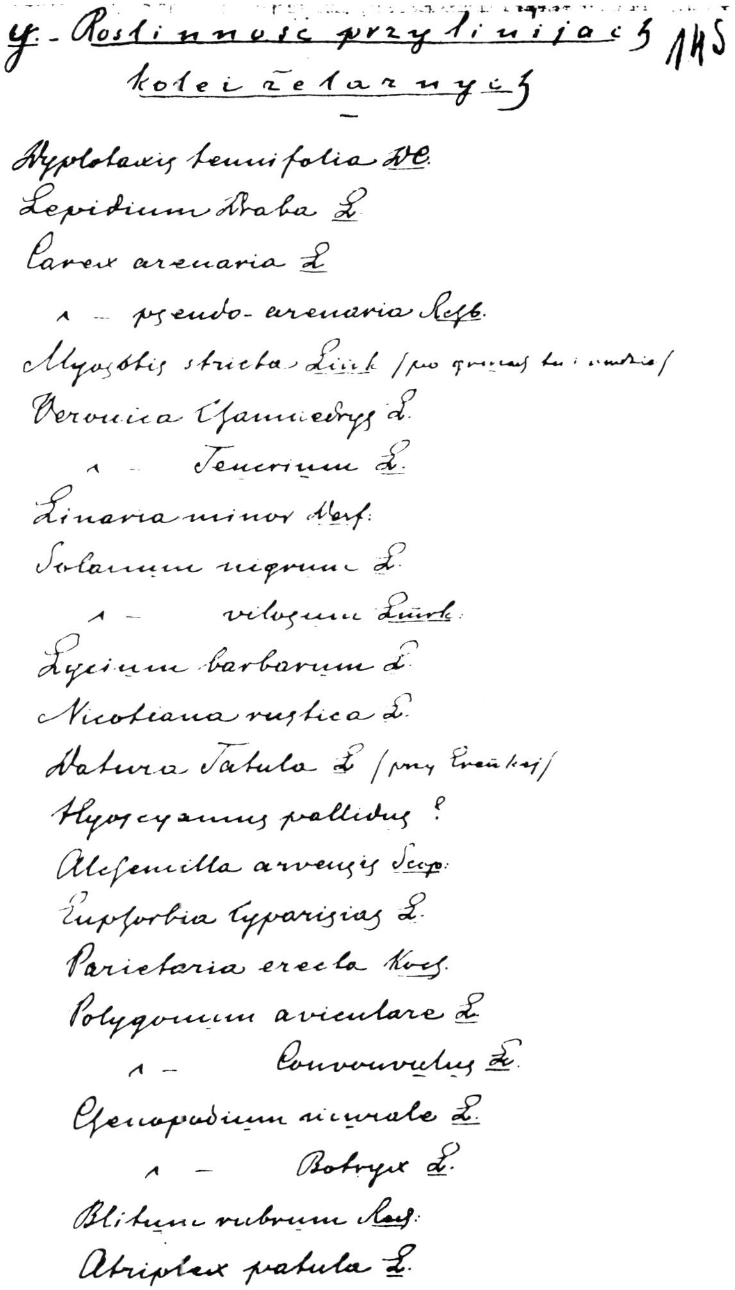

4. - Roślinność przy liniach kolei żelaznych 145

Dyplotaxis tenuifolia DC.
Lepidium Draba L.
Carex arenaria L.
" — pseudo-arenaria Rchb.
Myosotis stricta Link (po [illegible])
Veronica Chamaedrys L.
" — Teucrium L.
Linaria minor Desf.
Solanum nigrum L.
" — villosum Link.
Lycium barbarum L.
Nicotiana rustica L.
Datura Tatula L. (przy [illegible])
Hyoscyamus pallidus ?
Alchemilla arvensis Scop.
Euphorbia Cyparissias L.
Parietaria erecta Koch.
Polygonum aviculare L.
" — Convolvulus L.
Chenopodium murale L.
" — Botrys L.
Blitum rubrum Rchb.
Atriplex patula L.

Figure 3. Plants along railway lines – beginning of the list, part 21.

although he must have known *M. germanica* from Poland, where it grows along rivers in the southern part of the country.

11. Wet meadows, lakes and ponds – 11 pages including 2 pages of text and 223 species.

12. Forests (primarily Belgrad Forest) – 14 pages with 2 pages of text and 295 species.

13. Pastures – 12 pages with 4 pages of text and 187 species.

14. 'Bodiaki' (communities of thistles) and forest shrub thickets – 15 pages with 4 pages of text and 254 species.

15. Heaths – 1 page, no scientific description but *Calluna vulgaris* (L.) Hill is mentioned.

16. Princes Islands – 15 pages of text and 387 species (Fig. 2).

17. Shoreline vegetation (a coastal belt of the Black Sea and the Sea of Marmara) – 22 pages and 515 species. It is worth mentioning here that Jablonowski includes in the list of species of these communities *Pilularia globulifera* L. from the Asiatic side in the vicinity of the so-called Sweet Waters of Asia. *Pilularia* is not mentioned at all in Davis, 'Flora of Turkey' vol. I.

18. The laurel belt – 5 pages and 77 species. Already in the 'Memoirs' Jablonowski had mentioned that during a trip to the vicinity of Beykos, in the first year of his stay in Turkey, he had noted the variability within this species, promising himself that if he were to identify some new variety of *Laurus nobilis* L., he would name it after his friend Antoni Rehman. This he does when describing the laurel belt though unfortunately *Laurus nobilis* L. var. *Rehmanii* is without any diagnosis.

19. Mosses, algae and lichens – no data except the title page.

20. Vineyards – 5 pages and 84 species.

21. Plants along railway lines – 3 pages and 54 species (Fig. 3).

22. Exotic plants (cultivated in gardens) – 93 pages. According to the author there are about 1400 species. This part of the work, however, is not consistently presented. It has the appearance of being composed of two lists, species which grew in Constantinople and its vicinity and excerpts from literature about various species which Jablonowski expected to find.

23. Colonial plants and their utilisation – 31 pages of text. This chapter, however, has no connection with the flora of Constantinople. These are notes of the author about the origin, history of cultivation and utilisation of various exotic plants such as *Vanilla*, *Angraecum*, *Cananga*, *Palmae*, *Musa*, *Ginkgo*, *Chrysophyllum*, *Dracaena*, *Boehmeria*, *Paulownia*, *Passiflora* and others.

A contemporary evaluation of Jablonowski's manuscript is very difficult without the herbarium specimens which would confirm or contradict the identifications of plants included in his lists of species. However, such a herbarium did exist and Jablonowski mentions the fact that he is compiling it several times in his 'Memoirs'. From these 'Memoirs' we also learn that in September 1892 he presented a herbarium of 300 species to Archbishop Menini in Filippol (now Plovdiv in Bulgaria). We do not know, however, whether this was his full herbarium or only a part. On the other hand J. Fijalek, who prepared the 'Memoirs' for publication, includes in a footnote the information, presumably based on sources known only to him and possibly from existing correspondence of Jablonowski, that a part of the archaeological and ethnographic collections and the whole of the herbarium collections were deposited in Vienna (but where exactly?). Possibly they were placed there by Jablonowski himself as he was a member of the Zoologisch-botanischen Verein. Dr A. Polatschek, from the Naturhistorisches Museum (Botanische Abteilung) conducted a search on my behalf in Vienna, both in the Museum and at the University, but unfortunately without success. Before this Rechinger (1938), who studied the flora of Constantinople in such great detail, does not mention the herbarium of Jablonowski, and certainly when preparing materials for his work he must have made a detailed search of all Vienna herbaria. Could the herbarium have been lost altogether? Or maybe it still rests forgotten in some Viennese institution? It is also worth mentioning that H. Czeczott (1939), who worked on the flora of Constantinople and its vicinity and wrote about it on pages 129–135 of her monograph on the flora of northern Anatolia, does not mention Jablonowski and apparently knew nothing of his activities in the area.

However, even without a herbarium Jablonowski's work still has some value since it permits the comparison of data contained in it with later publications. Excluding the fragmentary data contained in Forsskål (1775) or Grisebach (1843–45) it is the first flora of the area. It is only towards the end of the nineteenth century that Aznavour (1897) published his first notes concerning this flora: these were supplemented several times in the years 1899–1913 and were based primarily on his own herbarium collections made between 1885 and 1913 and fortunately well preserved at Geneva. However, he did not meet Jablonowski, because the latter in 1885 was in Iraq and Persia, sent there by the International Sanitary Commission in connection with an epidemic of plague: he returned to Constantinople only in late November of that year. One could perhaps only dream that should these two men have met they could have prepared a flora of Constantinople that would have predated by 50 years that produced by Rechinger.

'Materyaly do flory Konstantynopola i jego okolic' of Jablonowski is one more example of how the flora of that region inspired people of many nationalities to conduct botanical research. Today when we have such abundant documentation as that compiled by Rechinger (1938), Yaltirik (1963), Webb (1966) or Davis (1965–1988) the value of this manuscript is only historical. Jablonowski did not have the luck to see the fruits of his efforts published. His work remained in archive for a century and only now has been discovered and for the first time reviewed. It permits, however, the inclusion of the name of Wladyslaw Jablonowski in the list of eminent botanists that have worked at the point where Asia meets Europe, in the Bosporus region in Turkey. Jablonowski certainly deserves this.

LITERATURE CITED

Aznavour, G.V. 1897. Notes sur la flore des environs de Constantinople. Bull. Soc. Bot. France 44: 164–177.

Czeczott, H. 1939. A contribution to the knowledge of the flora and vegetation of Turkey. Feddes Repert. Beih. 107: 1–282.

Davis, P.H. (ed.). 1965–1988. Flora of Turkey and the East Aegean Islands, Vols 1–10. Edinburgh University Press, Edinburgh.

Forsskål, P. 1775. Flora aegyptiaco-arabica. Hauniae.

Grisebach, A.H.R. 1843–1846. Spicilegium florae rumelicae et bithynicae Vols. 1 & 2. Brunsvigae.

Jablonowski, W. 1967. Pamietniki z lat 1851–1893 (in Polish – elaborated by J. Fijalek). Ossolineum Press. Wroclaw-Warszawa-Krakow. 561 p.
Rechinger, K.H. 1938. Enumeratio Florae Constantinopolitanae. Feddes Repert. Beih. 98: 1–73.
Yaltirik, F. 1963. Belgrad Orman vejetasjonun floristik analizi. Ist. Üniv. Orman Fak. Derg. Ser. A, 13: 69–80.
Webb, D.A. 1966. The flora of European Turkey. Proc. Roy. Irish Acad. 65, Sect. B, no. 1: 1–100.

24. THE *OLEA EUROPAEA* COMPLEX

P.S. GREEN AND G.E. WICKENS

Royal Botanic Gardens, Kew, Richmond, Surrey, TW9 3AB

ABSTRACT

Olea europaea L. (including var. *europaea* and var. *sylvestris*, syn. var. *oleaster*) is considered, as a species, to include not only the cultivated olive but also at infraspecific rank its close wild relatives from Asia and Africa: subsp. *cuspidata*, subsp. *laperrinei* and subsp. *cerasiformis*. Morphological, karyological, anatomical, palynological and biochemical (flavonoid) evidence is considered. A full synonymy is presented.

Key words: Olive, *Olea europaea*, flavonoids, taxonomy, synonymy.

INTRODUCTION

Olea europaea L., the olive, is a plant of great historic importance, as well as of major economic value. As a cultivated plant its point of origin is lost in the mists of time. It could have arisen in the 'fertile crescent', the Nile delta (Newberry, 1937) or the eastern Mediterranean area, where the remains of olive endocarps have been found in some of the earliest archaeological excavations (Chalcolithic Teleilat Ghassul, 3700–3500 B.C.; Zohary and Spiegel-Roy, 1975). Olive stones from the third millenium B.C. have also been reported from Bronze Age Lachish, Israel (Helbaek in Tufnell, 1958), and Tell Soukas, Syria (Renfrew, 1973).

Today *Olea europaea* is particularly associated with countries bordering the Mediterranean Sea and has in fact been referred to by phytogeographers as one of the species which characterises the 'Mediterranean Region'. An exact match for the domesticated olive cannot be found wild in any country today and seed taken from the cultivated plant produces not the parent olive but a form which reverts towards the 'wild type'. There is a divergence in opinion as to whether the 'wild' olive, generally referred to as var. *oleaster*, has originated by naturalization from cultivated plants (Turrill, 1951) or whether the wild forms are indigenous in the area of the Mediterranean (Zohary and Spiegel-Roy, 1975). Whichever is correct we believe the cultivated olive to be an ancient cultigen selected by man in the earliest days of plant domestication. The Pliocene records of *Olea europaea* var. *oleaster* from Bulgaria by Stojanoff and Stefanoff (1929: 90) and Stefanoff and Jordanoff (1935: 74) are based on leaf impressions. In our opinion there is considerable doubt as to whether these impressions have been correctly identified.

When Linnaeus (1753) proposed the name *Olea europaea* he described only one other species, *O. capensis*, differing widely in its foliage, but also, although not at the time specified, in its terminal inflorescence. With the botanical exploration of Asia and Africa, however, plants were discovered which clearly belonged to *Olea* and in fact closely resembled the cultivated olive except in their fruits. Because *O. europaea* was such an important economic and historic plant, botanists felt constrained to recognize these later discoveries as different species. Furthermore, as they became known from different regions, these related wild plants were redescribed. Yet when the variation exhibited by all the species of *Olea* is analyzed it is clear that some plants from three different continents would, if approached with an open mind, be seen to belong to the same entity. In fact, in many cases it is impossible to refer certain members of the genus *Olea* to a traditional species unless the origin of the specimen is known. Furthermore – the 'acid-test' in taxonomy – it proves impossible to construct a practical key for identification of the so-called species for, morphologically, they are more or less indistinguishable and no reliable character has been found. Wild specimens from Pakistan have, on geographical grounds, been called *O. ferruginea*, others from Africa *O. chrysophylla* or *O. africana*, while those from Greece, for example, have been referred to *O. oleaster* or *O. europaea* var. *sylvestris*. Here it is worth noting that Linnaeus in 'Hortus Cliffortianus' (1737: 4) described two species of *Olea*, one from southern Europe ('Olea foliis lanceolatis, ramis teretiusculis') and one from Africa ('Olea foliis lanceolatis, ramis tetragonis'), while in his subsequent 'Species Plantarum' (1753: 8) he treated both as *O. europaea*. The whole *O. europaea* complex extends from the Canary Islands and Madeira westwards across the Mediterranean and south-west Asia to the Sino-Himalayan region, and south through eastern Africa to South Africa (Figs. 1 and 2). We believe that all the variation is assignable to one species, which should be called *Olea europaea*.

Recently Mazzolani and Altamura Betti (1979) have published a review of what they call the 'cycle of *Olea chrysophylla*', including a long list of taxonomic synonyms. Unfortunately, the material they were reviewing and their concepts are clearly discordant, so we cannot agree with their findings. They reduce to synonymy (p. 131) several names which belong to species not closely related to *O. chrysophylla*: *O. woodiana* Knobl., *O. listeriana* Sim, *O. mackenii* Harv., and *O. schliebenii* Knobl., all of which belong to the quite different Section *Ligustroides* Benth. and Hook. f. They also include *O. neriifolia* Li, a member of yet a different group of species which, according to Johnson (1957), should constitute a separate genus, *Tetrapilus* Lour. Furthermore, in their discussion they refer to *O. excelsa* Ait. as belonging, through affinity, to the same 'forms-cycle' as *O. chrysophylla*, yet this species is today generally and correctly classified in the genus *Picconia* DC. (Johnson, 1975; do Amaral Franco in Tutin *et al.*, 1972; Kunkel, 1975). Similar confusion reigns in their later paper on *O. chrysophylla* (Mazzolani and Altamura Betti, 1980). Nowhere do they show whether they have studied the types of the taxa involved.

In a third paper Mazzolani and Altamura Betti (1981) continue to confuse species belonging to two different sections of *Olea*, stating in their abstract that *O. chimanimani* Kupicha, with *O. lancea* Lam. and *O. exasperata* Jacq. (all members of Sect. *Ligustroides* Benth. & Hook. f.) 'could be included in the group of forms akin to the aggregate species *O. chrysophylla* Lam.', which belongs to Sect. *Olea*. They cast doubt on the status of *O. chimanimani* because the identity of one of the paratypes (*Wild* 4575) had previously been published as *O. africana* (Goodier and

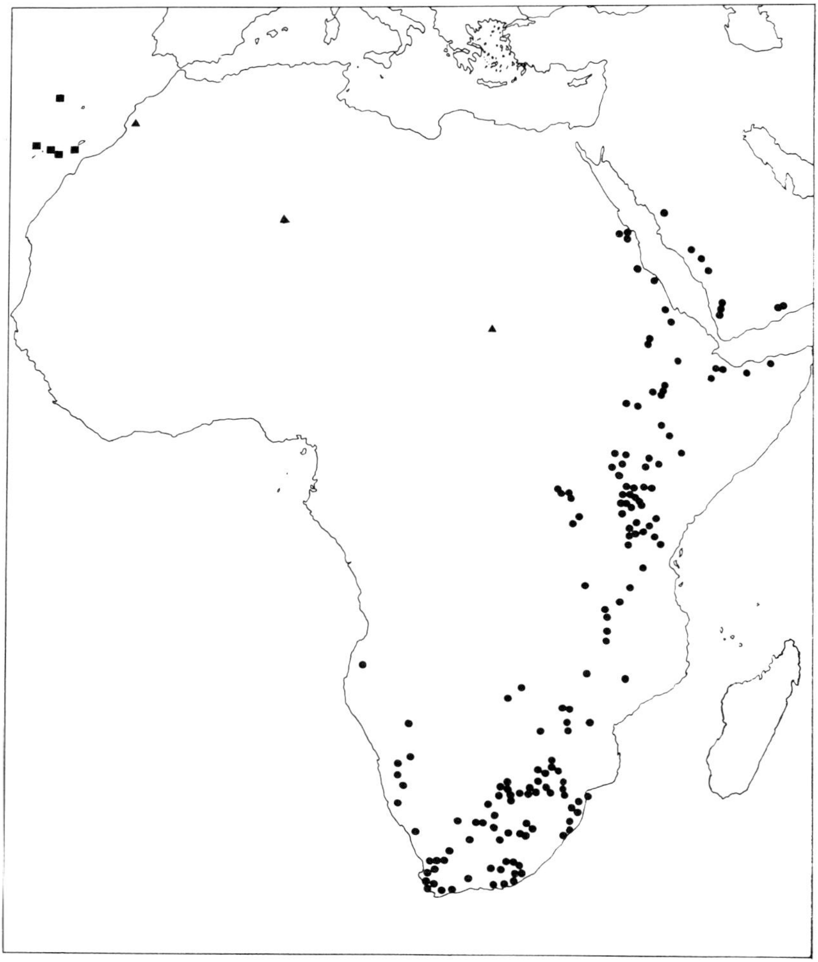

Figure 1. Distribution of *Olea europaea* in Africa. — ▲ subsp. *laperrinei*. — ■ subsp. *cerasiformis*. — ● subsp. *cuspidata*.

Phipps, 1961: 61), but this had been an earlier misidentification and the specimen in the herbarium at Kew possesses neither the dense lepidote indumentum on the underside of the leaves so characteristic of *O. europaea* (including *O. africana*) nor its axillary inflorescences. Nevertheless, it is worth noting that both *O. chimanimani* and *O. europaea* sens. lat. occur in Zimbabwe.

Most recently, Boulos and Hobbs (1986) have employed the name *Olea indica* for the African representative of the *O. europaea* complex. They do not present any new or adequate reasons for maintaining specific rank for this plant or for placing the subspecific name in synonymy. Distribution, even when coupled with fruit size,

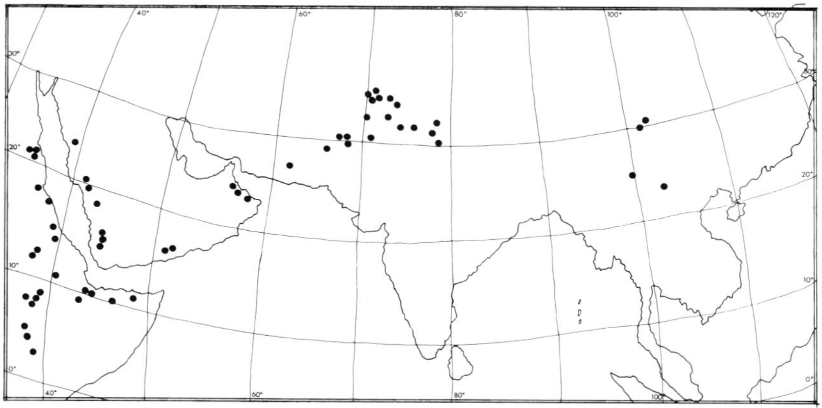

Figure 2. Distribution of *Olea europaea* subsp. *cuspidata* in Asia.

shape and taste are not, on their own, good specific criteria. In fact, on the following page they admit, indirectly by quotation from Zohary and Spiegel-Roy (1975), that fruit size and oil content can vary even within wild and cultivated *O. europaea*, not to mention fruit shape. Furthermore, they are unfortunately misled by Mazzolani and Altamura Betti (whom they cite using their Christian names, Gaspare and Maddalena) into thinking that what they call *O. indica* is 'more closely related to the African element of *Olea* rather than to the Mediterranean *Olea europaea* L.' As we have already pointed out, Mazzolani and Altamura Betti were making comparisons with species in quite different sections, even genera – species quite without the characteristic, dense lepidote indumentum and axillary inflorescences of *O. europaea* sens. lat.

If overall morphology is of little value in distinguishing separate taxa, one wonders whether data from other disciplines might help, but these appear to provide no firm basis for specific differentiation.

The chromosome numbers of all the species of *Olea* that have been counted, whether of the *O. europaea* or *O. capensis* affinities, or even other species, have turned out to be uniformly 2n = 46, except for an early count of n = *c.* 24 (Table 1).

Nor has the wood anatomy of the *Olea europaea* complex helped. Although Benichou (1962) has indicated minor differences between *O. europaea* sens. str. and part of the complex called *O. laperrinei*, they are doubtfully of specific significance and our colleague, Dr David Cutler, has carried out investigations on the wood of several entities and his results have not supported the idea of specific differentiation within this complex (Cutler, personal communication).

Similarly with the micromorphological characters of pollen grains. Although Benichou (1962) indicated that *Olea europaea* sens. str. and *O. laperrinei* might be distinguished by their pollen, a further study reported by Wickens (1976 a: 131–132) indicates that it is impossible with any confidence to separate these two entities on palynological grounds. Furthermore, while Pacini and Vosa (1979) have

Table 1. Chromosome counts recorded within *Olea*.

Author, reference and page	Species	Chromosome number	
		n	2n
Anderson (1931): 57	*O. europaea*	*c.* 24	
	O. capensis (as *O. laurifolia*)		*c.* 48
Sax & Abbe (1932): 39	*O. europaea*		46
Taylor (1945): 347	*O. europaea*		46
Breviglieri & Battaglia (1954): 271	*O. europaea*		46
Battaglia & Breviglieri (1944): 63	*O. europaea*		46
Coutinho (1956): n.v.	*O. europaea*		46, 48
Mehra & Gill (1968): 575	*O. europaea* (as *O. cuspidata*)	23	
Mehra & Bawa (1969): 471	*O. dentata*	23	
	O. salicifolia	23	
Briggs (1970): 127	*O. paniculata*	23	
Mehra (1976): n.v.	*O. dentata*	23	
	O. europaea (as *O. cuspidata*)	23	
	O. paniculata (as *O. glandulifera*)	23	
Bir *et al.* (1982): 592	*O. dioica*	23	
Bedi & Bir (1986): 136	*O. europaea* (as *O. ferruginea*)	23	
	O. paniculata (as *O. glandulifera*)	23	

recognized differences in the pollen exine patterns of several cultivars of the olive, an examination on a broader scale, of pollen from several species throughout the genus (Nilsson, 1988), has shown that, although there is weak differentiation into two groups, the 'variability of various features does not allow clear-cut differences' between species.

In contrast, however, striking evidence for recognizing one broad species is provided by an analysis of the flavonoids present in the genus *Olea*. This shows a uniform pattern of occurrence in *O. europaea* sens. lat. and at the same time emphasizes its distinctness from the rest of the genus. We are most grateful to Professor J.D. Harborne for permitting us to publish the results of his analysis of the flavonoids in *Olea* in Table 2.

It is nevertheless to be expected that over the wide geographical range postulated for the *Olea europaea* complex that some local differentiation will have taken place and, although one cannot with certainty assign every plant to a known geographical type, one can in some cases recognize certain kinds which are characteristic of one particular area. For example, the lepidote indumentum on the underside of the leaves of wild olives from Asia tends to be of a stronger and frequently of a more reddish-brown colour than in East African specimens. The leaves in material from the isolated mountain massifs of the Sahara and from the Macaronesian islands tend to be narrow in proportion to length when compared with those of specimens

Table 2. Flavonoid glycosides in the leaves of *Olea* species (analysis of dried leaves unless otherwise stated).

Taxon	Origin and reference	Regular flavonoids							Sporadic flavonoids					Phenols	
		1	2	3	4	5	6	7	8	9	10	11	12	A	B
O. europaea subsp. *europaea* var. *europaea*	Greece, 1887, *meldreich 859*	+	–	+	+	+	+	+	–	–	–	+	–	+	–
	Greece, 1934, *Hill et al. 2745*	+	+	+	+	+	+	+	+	–	–	+	+	+	+
	Egypt, 1927, *Simpson 5380*	+	+	+	+	+	+	+	–	–	–	–	–	+	+
	Cyprus, 1937, *Syngrassides 1652*	+	+	+	+	+	+	+	–	–	–	+	–	–	–
	Cult. Kew 1970 (fresh leaves)	+	+	–	+	+	+	+	–	–	–	–	–	–	–
var. *sylvestris* (var. *oleaster*)	Greece, 1934, *Hill et al. 2575*	+	+	+	+	+	+	+	+	+	–	–	–	+	+
	Greece, 1926, *Mattfeld 2639*	+	+	–	+	+	+	+	–	–	–	+	–	+	–
	Turkey, 1966, *Tobey 1564*	+	+	+	+	+	+	+	–	–	–	+	–	–	+
	? 1969	+	+	+	+	+	+	+	–	–	–	+	+	+	+
subsp. *cuspidata* (subsp. *africana*)	China, Yunnan, 1924, *Forrest 26007*	+	+	+	+	+	+	+	–	–	–	–	–	+	–
	Egypt, 1933, *Palmer 13*	+	+	+	+	+	+	+	–	+	–	–	+	–	+
	Uganda, 1950, *Eggeling E5929*	+	+	+	+	+	+	+	+	–	–	–	+	–	–
	Kenya, 1960, *Tweedy 2048*	+	+	+	+	+	+	+	–	–	+	–	–	–	–
	Tanzania, 1954, *Hughes 275*	+	+	+	+	+	+	+	–	+	–	–	+	–	–
	Mozambique, 1957, *Mogg 27438*	+	+	+	+	+	+	+	–	+	–	+	–	–	–
	S. Africa, Natal, 1969, *Moll 4701*	+	+	–	+	+	+	+	–	–	–	–	–	–	–
	S. Africa, Cape, 1948, *Rodin 3629*	+	+	+	+	+	+	+	–	+	–	–	–	+	–

Taxon	Locality, date, collector	1	2	3	4	5	6	7	8	9	10	11	12	13	14
subsp. *laperrinei*	Sudan, 1953, *Jackson & Ramsay 2*	+	+	+	+	+	−	+	−	+	+	−	−	+	−
subsp. *cerasiformis*	Madeira, 1937, *Barreto s.n.*	+	+	+	+	+	+	+	−	+	−	−	−	+	−
*Olea brachiata**	China, Hainan, 1928, *Tsang 467*	−	−	−	−	−	−	+	−	−	+	−	−	+	−
	Malaya, Pahang, 1917, *Evans s.n.*	−	−	−	−	−	−	+	−	−	+	−	−	+	−
*O. capensis** subsp. *capensis*	S. Africa, Cape, 1976, *Haynes 1221*	−	+	−	+	+	−	−	−	−	−	−	−	−	−
subsp. *hochstetteri*	Tanzania, 1951, *Greenway & Hughes 8565*	−	+	−	+	+	−	+	−	−	−	−	−	+	−
subsp. *welwitschii*	Tanzania, 1951, *Greenway & Hughes 8564*	−	−		−	+	−	+	−	−	−	−	−	+	−
*O. lancea**	Mauritius, 1966, *Colville-Barclay s.n.*	+	−	−	+	+	−	+	−	−	−	−	−	−	−
*O. paniculata**	India, 1934, *Parker 3306*	+	−	−	+	+	−	+	−	−	−	−	−	+	−
(*O. glandilifera*)	Australia, Queensland, 1929, *Kajewski 1133*	−	+	−	+	+	−	+	−	−	−	−	−	−	−
*O. rosea**	Thailand, 1931, *Kerr 3774*	−	+	−	+	+	−	+	−	−	−	−	−	−	−

*Species which may have additional flavonoid components not present in the :O. europaea *complex* from elsewhere. Also Wickens (1971), after an examination under the scanning electron microscope of lepidote scales from the underside of leaves, found no differences except that the Macaronesian specimen examined had about 19 rays to the scale as compared with 29–31 rays in other representatives of the complex.

We believe that such geographical differentiation as exists in this complex is best expressed taxonomically at the rank of subspecies and we accordingly set out the nomenclature below, disposing of the formidable synonymy involved. It should be noted that many of the species which have been proposed in the past and are, for example, listed in the 'Index Kewensis', would today be treated as horticultural cultivars. These names need to be accounted for and although strictly synonyms of var. *europaea*, they are here listed separately in order not to overburden the more botanical synonymy, as most of them are *nomina nuda*.

TAXONOMY

OLEA L., Sp. Pl. 8 (1753).

Sect. OLEA

Syn.: Sect. *Oleaster* Endl., Gen. Pl. 1: 572 (1838)
Sect. *Euelaea* DC., Prodr. 8: 284 (1844)
Subgen. *Euelaea* (DC.) Pfeiffer, Syn. Bot. 191 (1870)

OLEA EUROPAEA L., Sp. Pl. 8 (1753). Type: Hort. Cliff. Sicc. (lectotype, BM)*.

KEY TO SUBSPECIES

1a. Fruits large, mesocarp thick, fleshy, drupes 2–4 cm long; leaf-blades narrowly-elliptic to elliptic, undersurface silvery, (3–)5–6(–8) cm long, (0.5–)1–1.5(–2) cm broad . . . . subsp. *europaea*

1b. Fruits small, mesocarp fleshy but thin; drupes 0.5–1.2 cm long.
- 2a. Leaf-blades narrowly elliptic to elliptic (to broadly elliptic), undersurface silvery or often reddish, (3–)5–6(–9) cm long, (0.5–1)1–1.5(–3) cm broad; drupes 5–7 mm long . . . . .subsp. *cuspidata*
- 2b. Leaf-blades narrowly elliptic, undersurface silvery, (3–)4–5(–7) cm long, (0.5–)0.6–1 (–1.5) cm broad; drupes 6–7 mm long . . . . .subsp. *laperrinei*
- 2c. Leaf-blades (linear to) very narrowly elliptic, undersurface silvery, (3–)4–6(–8) cm long, (0.3–)0.4–0.6(–0.8) cm broad; drupes 8–12 mm long. . . subsp. *cerasiformis*

subsp. EUROPAEA. Mansfield, Die Kulturpflanze 2: 349 (1939)
Syn.: *O. europaea* subsp. *euromediterranea* Ciferri in L'Olivicoltore 19(5): 96. 1942, *nomen*.

var. EUROPAEA.
Syn.: *O. officianarum* Crantz, Inst. Herb. 2: 379 (1766) Type: *Olea cum fructu ac flore*, Dodoens, Stirp. Hist. Pempt. ed. 2, 821 fig. (1616).
O. gallica Miller, Gard. Dict. ed. 8, *Olea* 1 (1768) Type: Cult. Hort. Chelsea. (BM).
O. hispanica Miller, *op. cit.*, *Olea* 2 (1768) Type: Cult. Hort. Chelsea. (BM).
O. sativa Weston, Bot. Univ. 1: 191 (1770) Type: not indicated.
O. sativa Gaterau, Descr. Pl. Mantaub: 26 (1789) Type: France, not indicated.
O. sativa Hoffmanns. & Link, Fl. Port. 1: 388 (1809) *nom. illegit*,; *O. europaea* subsp. *euromediterranea* series *sativa* Ciferri in L'Olivicoltore 19(5): 96 (1942) *nomen*, contrary to Art. 4 of the Code.

* The specimens in LINN, numbers 20.1, 20.2 and 20.3 are none of them from 'Europa australi', we therefore select the specimen in the Clifford herbarium [*Hort. Cliff. Sic.* 4, Olea 1 (BM)] as lectotype.

O. europaea L. var. *sativa* Lehr, Diss. Bot.-Med. Olea europaea: 21 (1779) Type: as for *O. europaea* L.

O. europaea L. var. *communis* Aiton, Hort. Kew. 1: 13 (1789) Type: Cult. Hort. Kew. (BM-Herb. Banks, n.v.).

O. lancifolia Moench, Meth.: 478 (1794), *nom. illegit.*; Stokes, Bot. Comment.: 27 (1830). *nom. illegit.* [Both based on *O. europaea* L.].

O. pallida Salisb., Prodr.: 13 (1796) *nom. illegit.* [Based on *O. europaea* L.].

O. europaea L. var. *culta* Lam. & DC., Fl. Franç. ed. 3, 3: 497 (1805) Type: France, not indicated, *O. gallica* Mill. included in synonymy.

O. europaea L. var. *sativa* DC., Prodr. 8: 284 (1844); *O. europaea* race *sativa* (DC.) Rouy, Fl. France 10: 224 (1908); *O. europaea* subsp. *sativa* (DC.) Hegi, Ill. Fl. Mitt.-Eur. 5(3): 1936 (1927) [cited as 'subsp. *sativa* (Hoffm. & Link) Rouy']; Type: based on *O. europaea* L.

O. europaea L. var. *linarifolia* Chev. in Rev. Bot. Appl. 28: 10, 20, fig. 3/A [sic, referring to fig. 4A due to transposition of the plate captions] (1948), *nomen*, based on a Wiedemann collection from Anatolia.

This is the cultivated olive; the fruit with a thick, fleshy mesocarp.

var. SYLVESTRIS (Miller) Lehr, Diss. Bot.-Med. *Olea*: 20 (1779); Brotero, Fl. Lusit. 1: 10 (1804).

Syn.: *O. sylvestris* Miller, Gard. Dict. ed. 8, Olea 3 (1768); *O. europaea* race *sylvestris* (Miller) Rouy, Fl. France 10: 223 (1908); *O. europaea* subsp. *sylvestris* (Miller) Hegi, Ill. Fl. Mitt.-Eur. 5(3): 1936 (1927), cited as '(Miller) Rouy'. Type: Cult. Hort. Chelsea. (BM).

O. mucronata Weston, Bot. Univ. 1: 192 (1770). Type: not indicated.

O. europaea L. var. *buxifolia* Aiton, Hort. Kew 1: 13 (1789); *O. sativa* var. *buxifolia* (Aiton) Roemer & Schultes, Syst. 1: 69 (1817); *O. buxifolia* (Aiton) Steud., Nom. 562 (1821), *nom. in syn.*, *non* Miller (1768); *O. lancifolia* var. *buxifolia* (Aiton) Stokes, Bot. Comment. 29 (1830). Type: Cult. Hort. Kew. (BM-Herb. Banks, n.v.).

O. oleaster Hoffmanns. & Link, Fl. Port. 1: 387 (1809); *O. communis* Steud., Nom.: 562 (1821); *O. europaea* var. *oleaster* (Hoffmanns. & Link) DC., Prodr. 8: 284 (1844);

O. europaea forma *oleaster* (Hoffmanns. & Link) Ceballos and C. Vicioso, Estud. Veg. Fl. Forest. Málaga: 245 (1933); *O. europaea* subsp. *euromediterranea* series *oleaster* Ciferri in L'Olivicoltore 19(5): 96 (1942), contrary to Art. 4 of the Code; *O. europaea* subsp. *oleaster* (Hoffmanns. & Link) Zohari in Inst. Pap. 26, Imp. Forestry Inst.: 47 (1951). Type: Portugal, Algarve (B-Willd.? – n.v.).

O. brevifolia Raf., Sylva Tellur. 9 (1838). Type: not indicated.

O. europaea L. var. *phyllireoides* Mazzolani in L'Olivicoltore 18(5): 144 (1941); *nomen*.

This is the 'wild' olive of the Mediterranean region, characterised by its usually shrubby growth, frequently small leaves, and small fruit with thin mesocarp. Spontaneous from the seed of the cultivated plant.

subsp. CUSPIDATA (Wall. ex DC.) Ciferri in L'Olivicoltore 19(5): 96 (1942).

Syn.: *O. indica* Kleinhof ex Burm. f., Fl. Ind.: 6. Mar.–5 Apr. (1768). Type: Java, *Kleynhof* [sic] (holo. G).

O. africana Miller, Gard. Dict. ed. 8, *Olea* 4 (16 Apr. 1768); *O. europaea* L. subsp. *africana* (Miller) P.S. Green in Kew Bull. 34: 69 (1979). Type: Cult. Hort. Chelsea. (holo. BM).

O. buxifolia Miller, Gard. Dict. ed. 8, *Olea* 5 (1768). Type: Cult. Hort. Chelsea. (holo. BM).

O. chrysophylla Lam., Tab. Encycl. 1: 29 (1791); *O. chrysophylla* Lam. var. *euchrysophylla* Chev. in Rev. Bot. Appl. 28: 18 (1948), *nom. inval.* Type: Ins. Mauritiana. (holo. P, n.v., fiche IDC 6207, 432.8).

[*O. europaea* sensu Thunb., Prodr. Pl. Cap. 2 (1794), *non* L. (1753).]

O. europaea L. var. *verrucosa* Willd., Sp. Pl. ed. 4, 1: 44 (1797); *O. sativa* Hoffmanns. & Link var. *verrucosa* (Willd.) Roemer & Schultes, Syst. Veg. 1: 69 (1817), *nom. illegit.*; *O. verrucosa* (Willd.) Link, Enum. Hort. Berol. 1: 33 (1821); *O. chrysophylla* Lam. var. *verrucosa* (Willd.) Chev. in Rev. Bot. Appl. 28: 20 (1948). Type: South Africa, Cape of Good Hope, *Thunberg* (holo. UPS n.v., fiche IDC 5184, 176).

O. similis Burchell, Trav. 1: 172 (1822), *nomen.*

O. asiatica Desf., Cat. Hort. Paris ed. 3: 88 (1829), *nomen.* [Nepal, marked with an asterisk in the enumeration, indicating the presence of a description at the end of the catalogue, but the description does not in fact exist.].

O. sativa Wall., Cat. No. 2817 (1831), *nomen. in synon.*

O. ferruginea Royle, Ill. Bot. Him. 267, t. 65 f. 1 (1835); *O. chrysophylla* Lam. var. *ferruginea* (Royle) Chev. in Rev. Bot. Appl. 28: 18 (1948); *O. europaea* L. subsp. *ferruginea* (Royle) Ciferri in Olearia 3–4: 115 (1950). Type: India, banks of Jumna and of Sutlej, near Turanda (holo. LIV? n.v.; iso.? K).

O. cuspidata Wall. ex G. Don, Gen. Syst. 4: 49 (1837); *O. chrysophylla* Lam. var. *cuspidata* (Wall. ex G. Don) Chev. in Rev. Bot. Appl. 28: 18 (1948). Type: India, Kamaon, Wallich Cat. 2817 (holo. K-W).

O. ferruginea Hort. ex Steudel. Nom. ed. 2, 2: 209 (1842), *nomen.*

O. verrucosa (Willd.) Link var. *brachybotrys* DC., Prodr. 8: 285 (1844). Type: South Africa, near Mooyplats, *Drège* (holo. G-DC).

O. somaliensis Baker in Thiselton-Dyer, Fl. Trop. Afr. 4(1): 18 (1904); *O. chrysophylla* Lam. var. *somaliensis* (Baker) Chev. in Rev. Bot. Appl. 28: 18 (1948). Type: Somalia, *Hildebrandt* 1524 (holo. K).

O. europaea L. var. *nubica* Schweinf. ex Baker, loc. cit. (1904); *O. chrysophylla* Lam. var. *nubica* (Schweinf. ex Baker) Chev., loc. cit. (1948). Type: Sudan, Erkowit Mountain, *Schweinfurth* 249 (holo. K).

O. schimperi Gandoger in Bull. Soc. Bot. Fr. 65: 58 (1918). Type: Ethiopia, near Gennia, *Schimper* ed. II n. 945 (holo. LY? n.v.; iso. K).

O. monticola Gandoger, loc. cit. (1918). Type: Eritrea, Mt. Ira-Nuret, *Pappi* 910 (holo. LY? n.v.).

O. chrysophylla Lam. var. *subnuda* R.E. Fr., Wiss. Erg. Schwed. Rhodesia – Kongo – Exped. 1911–12, 1(2): 258 (1918). Type: Zaire, Kasindi, *Fries* 1856 (holo.? UPS, n.v.).

O. kilimandscharica Knobl. in Notizbl. Bot. Gart. Berlin 12: 200 (1934). Type: Tanzania, Kilimandjaro: *Schlieben* 5065 (holo. B, destroyed; iso. K).

O. subtrinervata Chiov. in Atti R. Accad. Ital., Mem. Cl. Sc. Fis. 11 ('Pl. Nov. Aethiop.') 49 (1940). Type: Ethiopia: Ogaden, Mt. Ellot, *Reghini* 49.

O. chrysophylla Lam. var. *albida* Chev. in Rev. Bot. Appl. 28: 10, 18 & 21 fig. (1948), *nomen* (descr. franc., non lat.). Based on Yemen, *Deflers* 246.

O. chrysophylla Lam. var. *aucheri* Chev., op. cit. 18, 21 fig. 22 (1948), *nomen* (descr. franc. non lat.); *O. aucheri* Chev. ex Ehrend. in Österr. Akad. Wiss., Math.-Naturwiss. Kl., Anz. 97: 156 (1960). Type: Oman, Djebel Akadar, *Aucher–Eloy* 4918.

This is the wild African and Asiatic plant: the fruit is small and has a poorly developed mesocarp, and the leaves are often tinged reddish beneath, especially in Asia.

Under pressure in 1979 to produce a name for the eastern and southern African representative for the treatment of the Oleaceae for *Flora Zambesiaca*, and with a desire to limit the introduction of yet another name to African botanists, foresters, etc. (i.e. *O. indica* Kleinhof ex Burm. fil.), Green proposed the combination *O. europaea* L. subsp. *africana* (Miller) P.S. Green. Unfortunately, Ciferri's combination, based on the Asian *O. cuspidata*, has priority at this rank, and, after a survey of all the numerous collections available at Kew throughout the whole geographical range of *O. europaea sensu lato*, it has proved impossible consistently to distinguish taxonomically between the representatives from Africa and Asia.

subsp. LAPERRINEI (Batt. & Trabut) Ciferri in L'Olivicoltore 19(5): 96 (1942).

Syn.: *O. laperrinei* Batt. & Trabut in Bull. Soc. Bot. Fr. 58: 626 & 672 (1912), as '*laperrini*'; *O. europaea* L. subsp. *laperrinei* (Batt. & Trabut) Ciferri var. *typica* Ciferri, loc. cit. (1942), *nomen*. Type: Algeria, Haggar Mts., *Laperrine* (holo.? AL, n.v.).
O. europaea L. subsp. *laperrinei* (Batt. & Trabut) Ciferri var. *cyrenaica* Ciferri in L'Olivicoltore 19(5): 96 (1942), *nomen*.
O. europaea L. subsp. *laperrinei* (Batt. & Trabut) Ciferri var. *maireana* Ciferri, loc. cit. (1942), *nomen*.

This is the wild olive of the Sahara; the leaves are narrow in relation to their length, and the fruit is small with a poorly developed mesocarp. It is found in three disjunct mountain areas: Jebel Marra in the Sudan, the Hoggar Mountains in southern Algeria and on southern slopes of the Atlas Mountains in Morocco (Wickens, 1976b, p. 127, map 106).

subsp. CERASIFORMIS (Webb & Berth.) Kunkel & Sunding in Monogr. Biol. Canariensis 3: 58 (1972).

Syn.: *O. maderiensis* Hort. Cels. ex Steudel, Nom. 563 (1821), *nomen*.
O. europaea L. var. *maderensis* Lowe in Trans. Camb. Phil. Soc. 4(3): 537 (1838) & Nov. Fl. Mader. 15 (1838); *O. chrysophylla* Lam. var. *maderensis* (Lowe) Chev. in Rev. Bot. Appl. 28: 6 & 20 (1948). Type: Madeira: *Masson* (holo. BM?, n.v.).
O. europaea L. var. *cerasiformis* Webb & Berth., Hist. Nat. Îles Canaries III, 2(3): 162 (1845), *nom. illegit.*
O. chrysophylla Lam. var. *mandonii* Chev. in Rev. Bot. Appl. 28: 20 (1948), *nomen*, as '*Mandon*'. Based on Madeira, *Mandon* Pl. Mad., 174 (P, n.v.).
O. europaea L. var. *canariensis* Willd. ex Ciferri in Olearia 3–4: 121 (1950), *nomen*.

This is the wild olive of the Macaronesian Islands which, like the Sahara plant, has proportionately narrow leaves. It is now much introgressed with the cultivated olive, following the latter's introduction to the islands, but the fruit size in the key above is based on early descriptions. Its occurrence on Gran Canaria has recently been surveyed by Pérez & Parada (1986: 11–13) and by Parada *et al.* (1986: 26–27, 33).

Finally, we list below the published names which have been included in the *Index Kewensis*, of forms which would today be treated as cultivars of *Olea europaea* var. *europaea*. Only a few of them meet the requirements of the International Code of Botanical Nomenclature.

O. cajetana Petagna, Inst. Bot. 2: 19 (1787); *O. sativa* Hoffmanns. & Link var. *cajetana* (Petagna) Roemer & Schultes, Syst. Veg. 1: 69 (1817).
O. europaea L. var. *ferruginea* Aiton, Hort. Kew. 1: 13 (1789); *O. sativa* Hoffmanns. & Link var. *ferruginea* (Aiton) Roemer & Schultes, loc. cit. (1817); *O. lancifolia* Stokes var. *ferruginea* (Aiton) Stokes, Bot. Comment. 29 (1830).
O. europaea L. var. *latifolia* Aiton, loc. cit. (1789); *O. sativa* Hoffmanns. & Link var. *latifolia* (Aiton) Roemer & Schultes, loc. cit. (1817); *O. latifolia* (Aiton) Steudel, Nom. 563 (1821), *nomen in syn.*
O. europaea L. var. *longifolia* Aiton, loc. cit. (1789); *O. sativa* Hoffmanns. & Link var. *longifolia* (Aiton) Steudel, loc. cit. (1821), *nomen in syn.*
O. europaea L. var. *obliqua* Aiton, loc. cit. (1789); *O. sativa* Hoffmanns. & Link var. *obliqua* (Aiton) Roemer & Schultes, loc. cit. (1817); *O. obliqua* (Aiton) Steudel, loc. cit. (1821), *nomen in syn.*

O. amygdalina Gouan, Herbor. 264 (1796), *nomen.*
O. angulosa Gouan, loc. cit. (1796), *nomen.*
O. atrorubens Gouan, loc. cit. (1796), *nomen.*
O. cranimorpha Gouan, loc. cit. (1796), *nomen.*
O. praecox Gouan, loc. cit. (1796), *nomen.*
O. oblonga Gouan, loc. cit. (1796), *nomen.*
O. racemosa Gouan, loc. cit. (1796), *nomen.*
O. sphaerica Gouan, loc. cit. (1796), *nomen.*
O. variegata Gouan, loc. cit. (1796), *nomen.*
O. viridula Gouan, loc. cit. (1796), *nomen.*
O. regia Rozier ex Lam., Encycl. 4: 538 (1798), *nomen in syn.*
O. odorata Rozier ex Lam., op. cit. 539 (1798), *nomen in syn.*
O. alba Lam. ex Steudel, Nom. 562 (1821), *nomen in syn.*
O. oblonga-atrovirens Lam. ex Steudel, loc. cit. (1821), *nomen in syn.*
O. polymorpha Risso ex Schultes, Mant. 1: 75 (1822), *nomen in syn.*
O. lancifolia Stokes, *nom. illegit.* var. *lanceolata* Stokes, Bot. Comment. 28 (1830).
O. angustifolia Raf., Sylva Tellur. 9 (1838).
O. bifera Raf., op. cit. 8 (1838).
O. europaea L. var. *semperflorens* Raf., loc. cit. (1838).
O. verrucosa Raf., loc. cit. (1838).
O. cayana Raf., op. cit. 9 (1838).
O. arolensis Clem. ex Steudel, Nom. ed. 2, 2: 209 (1841), *nomen in syn.*
O. argentata Clem. ex Steudel, loc. cit. (1841), *nomen in syn.*
O. hispalensis Clem. ex Steudel, loc. cit. (1841), *nomen in syn.*

ACKNOWLEDGMENTS

As well as our thanks to Professor Jeffrey Harborne of Reading University, already expressed, we would like to record our gratitude to Dr Charles Jarvis of the British Museum (Natural History) for advice with the typification of *Olea europaea* L., to Frank White of Oxford University for his help and advice with Fig. 1, to our colleagues Nigel Hepper, for information on olives in antiquity, and to Yvette Harvey for drawing Figs. 1 and 2.

LITERATURE CITED

Anderson, A. 1931. Studien über die Embryologie der Familien Celastraceae, Oleaceae und Apocynaceae. Acta Univ. Lund. 27(7): 1–112.

Battaglia, E., and N. Breviglieri. 1955. Microsporogensi regolare ed irregolare in *Olea europaea* L. Caryologia 8: 45–68.

Bedi, Y.S., and S.S. Bir. 1986. Cytological observations on certain woody members of family Oleaceae. New Botanist 12: 135–141.

Benichou, A. 1962. Recherches critiques sur l'olivier de Laperrine. Univ. Alger. Inst. Rech. Sahar. Mém. 6: 1–55.

Bir, S.S., B.S. Gill, and G. Singh. 1982. *In* A. Löve [ed.], IOPB chromosome number reports, LXXVI. Taxon 31: 574–598.

Boulos, L., and J. Hobbs. 1986. Three arboreal species new to the Eastern Desert of Egypt. Candollea 41: 183–191.

Breviglieri, N., and E. Battaglia. 1954. Recherche cariologiche in *Olea europaea* L. Caryologia 6: 2–3 & 271–283.

Briggs, B.G. 1970. Some chromosome numbers in the Oleaceae, Contr. New S. Wales Natl. Herb. 4: 126–129.

Coutinho, L. de A. 1956. Susidios para o estudo cariologico da *Olea europaea* L. Genet. Iber. 8: 3–26.

Goodier, B., and J.B. Phipps. 1961. A revised check-list of the vascular plants of the Chimanimani Mountains. Kirkia 1: 421–466.

Johnson, L.A.S. 1957. A review of the family Oleaceae. Contr. New S. Wales Natl. Herb. 2: 395–418.

Kunkel, G. [ed.]. 1975. Inventario de los recursos naturales renovables de la Provincia de Las Palmas. World Wildlife Fund.

Linnaeus, C. 1753. Species Plantarum. Stockholm. 1200 p. + 31 p. indices & errata.

Maley, J. 1980. Les changements climatiques de la fin du Tertiaire en Afrique: leur conséquence sur l'apparition du Sahara et de sa végétation, pp. 63–86. *In* M.A.J. Williams and H. Faure [eds.], The Sahara and the Nile. A.A. Balkema, Rotterdam.

Mazzolani, G., and M.M. Altamura Betti. 1979. Elementi per la revisione del genere *Olea* (Tourn.) Linn. II - Ciclo di *Olea chrysophylla* Lam. Ann. Bot. (Rome) 37: 127–154.

——, and ——. 1980. Elementi per la revisione del genere *Olea* (Tourn.) Linn. III - *Olea chrysophylla* Lam.: forme affini. *Op. cit.* 38: 15–31.

——, and ——. 1981. Osservazioni aggiuntorie su alcune *Olea* africane. *Op. cit.* 39: 309–315.

Mehra, P.N. 1976. Cytology of Himalayan Hardwoods. Sree Saraswaty Press, Calcutta.

——, and K.S. Bawa. 1969. Chromosomal evolution in tropical hardwoods. Evolution 23: 466–481.

——, and B.S. Gill. 1968. *In* Á. Löve [ed.], IOPB chromosome number reports, XIX. Taxon 17: 573–577.

Newberry, P.E. 1937. On some African species of the genus Olea and the original home of the cultivated olive-tree. Proc. Linn. Soc. 150: 3–16.

Nilsson, S. 1988. A survey of the pollen morphology of *Olea* (Oleaceae) with particular reference to *Olea europaea* s. lat. Kew Bull. 43: 303–315.

Pacini, E., and C.G. Vosa. 1979. Scanning electron microscopy analysis of exine patterns in cultivars of olive (*Olea europaea* L.) Ann. Bot. (London) 44: 745–748.

Parada, V.M., J.D.R. Pérez, and D. Bramwell. 1986. Sobre la vegetación de Gran Canaria. Bot. Macarones. 12 & 13: 17–50.

Pérez, J.D.R., and V.M. Parada. 1986. Distribución de especies significativas para la comprensión de las formaciones boscosas en Gran Canaria (Islad Canarias), I. Bot. Macarones. 12 & 13: 3–16.

Renfrew, J.M. 1973. Palaeoethnobotany. The prehistoric food plants of the Near East and Europe. Methuen, London and Columbia University Press, New York. xviii + 248 p.

Sax, K., and E.C. Abbe. 1932. Chromosome numbers and the anatomy of secondary xylem in the Oleaceae. J. Arnold Arbor. 13: 37–51.

Stefanoff, B., and D. Jordanoff. 1935. Studies upon the Pliocene flora of the plain of Sofia (Bulgaria). Sborn. Bulg. Akad. Nauk. 29: 1–150.

Stojanoff, N., and B. Stefanoff. 1929. Beitrag zur Kenntnis der Pliozänflora der Ebene von Sofia. Z. Bulg. Geol. Ges. 2(3): 3–110.

Taylor, H. 1945. Cyto-taxonomy and phylogeny of the Oleaceae. Brittonia 5: 337–367.

Turrill, W.B. 1951. Wild and cultivated olives. Kew Bull. 3: 439–442.

Tutin, T.G. *et al.* [eds.]. 1972. Flora Europaea 3. Cambridge University Press, Cambridge. 370 p.

Tufnell, O. 1958. Lachish. Oxford University Press, Oxford. 351 p.

Wickens, G.E. 1971. The flora of Jebel Marra (Sudan Republic) and its geographical affinities. Ph.D. thesis, Reading University.

——. 1976a. Speculation on long distance dispersal and the flora of Jebel Marra, Sudan Republic. Kew Bull. 31: 105–150.

——. 1976b. The flora of Jebel Marra (Sudan Republic) and its geographical affinities. Kew Bull. Addit. Ser. 5: 1–368.

Zohary, D., and P. Spiegel-Roy. 1975. Beginnings of fruit growing in the Old World. Science 187: 319–327.

25. FIFTY YEARS OF BOTANICAL RESEARCH IN THE *FLORA IRANICA* AREA (1937–1987)

KARL HEINZ RECHINGER

Beckgasse 22, A-1130 Wien, Austria

ABSTRACT

A report is given on ten botanical expeditions to Iraq, Iran, Afghanistan and Pakistan (West of the Indus) undertaken in the years 1937, 1948, 1956/57, 1962, 1965, 1967, 1971, 1974, 1975 and 1977. It includes itineraries, collecting numbers, botanical discoveries, occasional comments on phytogeography, all of which are interwoven with accounts of various incidents that occurred in my travels. How the idea of writing *Flora Iranica* originated is also outlined together with a short statistical survey of progress to date.

Key words: botanical exploration (1937–1977), itineraries, *Flora Iranica*, Iraq, Iran, Afghanistan, Pakistan.

INTRODUCTION

On May 7th 1987 it was fifty years since I set out on my first journey to Iran (Persia). It might be appropriate to review my ten expeditions to the *Flora Iranica* area and how the idea of writing *Flora Iranica* gradually crystallized.

Friends have asked me time and again how it came about that my botanical interests have been directed to the East. I remember a mute map on the wall of a classroom distracting my attention from what I was being taught at the time. The map showed the blue oceans, green plains and a gradation of brown in the mountains crossed by thick black snakes symbolizing rivers. I realized that most of them ended in oceans; some of them ended, however, in the interior of Asia in deserts. That aroused my curiosity and as a Christmas present I got three volumes of Sven Hedin's 'Von Pol zu Pol', a selection of travel accounts adapted for children, including Hedin's own experiences. I was most fascinated by his accounts on the interior of Asia and I felt that sooner or later I would have to visit this part of the world. I also felt something of the spell of the East when looking east over the Neusiedlersee for the first time in 1922. In 1927–1936 (and 1942) I explored the islands of the Aegean Sea (*Flora Aegaea*, 1943). An attempt to visit the Caucasus in 1929 failed because the visa had been delayed. A plan to go to eastern Turkey in 1931 had to be cancelled because I fell seriously ill. So I had passed the age of 30 before I came to Persia in 1937. Owing to the Second World War most of my expeditions could not be undertaken before I reached the age of 50 in 1957; my last excursion was in 1977, when I was aged 70. A map showing the routes of my journeys is provided in Fig. 5, p. 320.

The first vague idea of synthesizing the available data arose in 1937 after the return of my first journey to Persia. Upon my request, my father Dr Karl

Rechinger (1867–1952) abstracted them in the form of a card index which was later continued by my first wife, Dr Frida Rechinger up to 1951 and extended to include Afghanistan and the North of Iraq. After the death of Josef Bornmüller in 1948 I was entrusted to name more and more collections from the area of Boissier's *Flora Orientalis* (1867–1888), e.g.: Samuelsson's plants from Israel and Syria (S*; Rechinger 1949–1959), Haradjian's plants from Syria (G; dupl. in W; Rechinger 1949–1959), John Frödin's plants collected in Armenia and Kurdistan (UPS; Rechinger 1952), Koeie's and Edelberg's plants from Afghanistan (C; dupl. in W; Koeie & Rechinger 1954–1965) and Walter Koelz's plants from Iran and Afghanistan (US; dupl. in W; Koeie & Rechinger 1954–1965). In between I collected several times in Greece. So I became well acquainted with the plants from the area of *Flora Orientalis*. When Peter Davis started work on *Flora of Turkey* in 1961 I decided to link up to the East from Turkey, accepting the political boundary of Turkey as the western limit of *Flora Iranica*. When accepting the north-south course of the Indus as the eastern boundary, splitting Pakistan into two halves, the *Flora of Pakistan* project was not yet in existence.

Without the untiring help of my second wife Wilhelmina as a driver and assistant collector in the field and as a secretary at home, *Flora Iranica* would not exist.

1937
EXSICCATAE NOS. 1–2599

My first trip to Iran was made possible by the invitation of Dr Erwin Gauba, an Austrian, who had been Professor of Botany at the Agronomic High School at Karaj 45 km west of Tehran since 1933. Gauba was genuinely interested in botanical exploration. Together with Josef Bornmüller he had started to publish an enumeration of plants collected during his excursions around Karaj (Bornmüller & Gauba 1935–1942) and during his vacations he used to make longer excursions (Gauba 1950, 1951, 1952, 1953; 1955).

The cheapest and quickest way to reach Tehran from Vienna in those days was by rail (Warsaw — Kiev — Rostov — Baku), then by boat from Enzeli (Pahlavi) and finally by car to Tehran. I left Vienna with my first wife, Dr Frida Rechinger (née Moser) on May 7th and reached Pahlavi on the morning of May 13th. For my blotting paper the customs officer asked a fantastic sum of import duty which I could not possibly afford. I left the blotters with the customs not knowing that it would take six weeks before I could recover them. Paper of any kind including newsprint was not available in Tehran at this period. Gauba's own supply was extremely scarce so limited collecting around Tehran and Karaj only would be possible. Invaluable time was lost by repeated interventions at the central customs authorities at Tehran.

Rain was pouring down on the Caspian lowlands. The normal road from Gilan to Tehran via Qazvin was blocked by landslides. So we had to try the Chalus-Kandavan road which was not yet officially opened to the public. Crossing Gilan with lowland forests, rice fields and tea plantations on the foothills was most

* Herbarium abbreviations follow *Index Herbariorum*, ed. 7 (1981).

impressive. Traditional houses covered with rice-straw! The inhabitants, balancing two sacks of rice on a long thin pole on their shoulders, could be seen. The mountain slopes were hidden in heavy rain clouds. The new road was underwashed by the river in many places and in others blocked by landslides. A huge block on the road had to be circumnavigated with two wheels of our car on the road and two on the slope. It was late afternoon when we reached the Kandavan Pass at 3010 m. (The tunnel at 2600 m was constructed years later.) Late in the night we finally reached Karaj. My first impression from the arid south slopes of the Elburz was not characteristic. Rain was dripping through the flat roof of Gauba's house. In every room containers of all sorts tried to collect the melodiously tinkling raindrops. At the end of June my belongings had finally been recovered from the main customs office. The start of our expedition to Khorasan had been unduly delayed until June!

On June 29th we finally set out eastwards by way of Firuzkuh — Semnan — Damghan — Shahrud — Sabzevar — Neyshabur to Mashhad. In 1937 this was the main road from Tehran to the East. It was not paved and difficult to use for normal low saloon cars because of repeated crossings of irrigation ditches, stretches of loose sand and mud on the desert borders.

The second day of the journey east of Tehran led to the discovery of the gypsum desert on the southern slopes of the Elburz foothills near Sorkheh west of Semnan. I have repeatedly visited this area which is quite exceptional in its geology (see coloured pictures in *Flora Iranica* 139: t. 417 & 418). *Anthochlamys multinervis* Rech. f., *Astragalus fridae* Rech. f. (Sect. *Proselius*), *A. semnanensis* Bornm. & Rech. f. (Sect. *Leucocercis*), *Euphorbia gypsicola* Rech. f. & Aell., *Gypsophila mucronifolia* Rech. f., *Moltkia gypsacea* Rech. f., and *Nepeta eremokosmos* Rech. f. grow exclusively on gypsaceous soil; most of these species are confined to the Semnan area. *Lomatopodium staurophyllum* (Rech. f.) Rech. f. and *Silene gertraudiae* Melzheimer grow on gypsum as well as on serpentine soil; both are known from Semnan as well as from serpentine soil in Khorasan.

We spent several weeks at Mashhad during July and various excursions were undertaken radiating from Mashhad. Kuh-e Bezg, not visited by any botanist since Aitchison in 1879, was climbed from July 5th to 7th. We went up from Torbat-e Heydariyeh not far from the Afghan border. It has loose stands of *Juniperus excelsa* M.-B. at higher elevations. As our visit was in the summer we found several new thistles, namely, *Cousinia chrysandra* Bornm. & Gauba, *C. euchlora* Bornm. & Rech. f., *C. rechingerorum* Bornm. *C. tetanocephala* Bornm. & Gauba as well as *Echinops arachniolepis* Rech. f. We also found the labiate *Hymenocrater platystegius* Rech. f. and the umbellifer *Trachydium vesiculoso-alatum* Rech. f.

On July 10th and 11th the serpentine hills around Robat Safid were explored on the Mashhad — Torbat-e Heydariyeh road. Here an especially interesting plant was found: the new caryophyllaceous genus *Diaphanoptera* Rech. f. with its type species *D. khorasanica* Rech. f. This species has never been recollected; all efforts by myself and several other botanists have been in vain but subsequently five more species of this genus related to *Acanthophyllum* C.A. Meyer have been discovered, all of them endemic to Afghanistan, NE Persia and Turkmenistan.

Jurinea pungens Boiss., the only species of this genus with rigid leaves ending in a terminal spine, was based on a single specimen without locality collected by

Figure 1. 1. S.M.H. Jafri and S.I. Ali, Karachi, Pakistan, 6 April 1965. — 2. Esfandiar Esfandiari, Evin, Tehran, Iran. 29 June 1971. — 3. S.M.A. Kazmi, Peshawar, Pakistan. June 1965. — 4. Per Wendelbo from Bergen, Norway and 1974–1978 botanical adviser of Ariamehr Botanical Garden, Iran. — 5. M. Assadi, Bazman, Baluchistan, Iran. May 1977. — 6. M. Iranshahr, Azerbaijan, Iran.

Jenisch. It was found to be dominant at Robat Safid over considerable areas, growing exclusively on serpentine and has not been found elsewhere. *Sclerorhachis* (Rech. f.) Rech. f., a genus of four species that I had separated from *Anthemis* L., is endemic to Khorasan, Afghanistan and Turkmenistan. It is represented at Robat Safid by its tallest and most widely distributed species *S. platyrachis* (Boiss.) Podlech which shows a pronounced predilection for serpentine. Other species confined to this area are *Cousinia meshhedensis* Bornm. & Rech. f., *C. trachyphyllaria* Bornm. & Gauba, *Scorzonera microcalathia* Rech. f., and *Ferula xylorhachis* Rech. f., the only *Ferula* with indurated rosette leaf bases.

Another interesting excursion from Mashhad from July 13th to 15th led from Quchan into the Kopet Dagh on the Turkmenian (USSR) frontier to the Alamli and Allah Akbar passes. The northern slope facing the Aralo-Caspian desert plain is very steep and the view of the unlimited Turanian desert plain is most impressive. In spite of the advanced season and heavy grazing the limestone rocks furnished several new species: *Cousinia dipterocarpa* Bornm. & Rech. f., *Graellsia integrifolia* (Rech. f.) Rech. f., *Johreniopsis stricticaulis* (Rech. f.) M. Pimenov, *Jurinea kopetensis* Rech. f., *Onobrychis ptychophylla* Širj. & Rech. f. and *Rubia rechingeri*

Ehrendf. Above Quchan the peculiar looking *Centaurea galactochroa* Rech. f. and *Acantholimon kutschanense* Rech. f. & Schiman-Czeika were found.

The last trip from Mashhad was from July 25th–27th to Bojnurd and northwest in the direction of Tappeh-ye Moraveh. The plan was to reach the easternmost fringes of the Caspian forests. The car, however, no longer in satisfactory condition when we left Tehran and by its construction certainly unfit for travelling off the main roads, was being repaired for half the time during our stay at Mashhad and, in addition, broke down several times on each excursion. The driver turned out to be a heavy opium-addict with all its consequences. Unfortunately, Dr Gauba was neither a good organizer nor acquainted with the technical intricacies of an engine. So we had to give up and, after sampling the vegetation west of Bojnurd, comparatively rich in shrubs such as *Paliurus*, *Crataegus*, *Rhamnus sintenisii* Rech. f. and the dried-up remains of a rich herbaceous undergrowth, we retreated with difficulty the more than 300 km distance back to Mashhad.

Because of the very limited capacity of Gauba's car we had to send two wooden boxes containing my plants collected in Khorasan by lorry to Tehran. They arrived weeks later when we had already left for Europe. Our camping equipment had been sent in advance from Mashhad to Shahrud from where we wanted to cross the Elburz Range by mule to Gorgan. This shipment too had not arrived in time so we continued by car to Semnan and went to Firuzkuh, crossed the Elburz by the Gaduk Pass, went down to the Talar Valley, where *Trifolium mazandaranicum* Rech. f. and *Asperula mazandaranica* Ehrendf. were found, and back through Mazandaran, the Chalus Valley and the Kandavan Pass to Karaj. In the Chalus Gorge the car gave up completely and had to be towed to Karaj by a charcoal lorry, with me sitting on top of the charcoal load.

It was not possible to travel back to Vienna by the shortest way through the USSR because in those days transit passengers were not allowed to take along anything written (diaries!) nor photographs. By the way, one of the peculiarities of the Reza Shah regime was that photography of 'old' (traditional) objects and people dressed in their traditional costume was strictly prohibited. Female persons were not allowed to wear the Chador (the veil hiding the face) – exactly the opposite situation to that prevailing now.

We hired two seats in a taxi to Baghdad and arrived there on August 16th. Here we had to borrow money to continue in a cheap Arabic bus across the desert via Rutba, Damascus, Antilebanon and Lebanon to Beirut during the hottest days of the year. In Beirut the money from Europe reached us at last. We could then relax on an Italian steamboat from here via Haifa — Cyprus — Brindisi to Trieste, landing on August 30th. From there we went by train to Austria. Part of my collections reached Vienna in December 1937. The Khorasan boxes arrived more than one year later!

In spite of time-consuming (weeks!) problems with the Persian customs authorities with my collecting equipment, a financial crisis at the end of the journey (which had been undertaken by modest private means), and in spite of temporarily failing health, the results of my first expedition to Persia were satisfactory. Most of the new taxa discovered have been published in Feddes Repert. (1940 and 1941) and a complete enumeration of the specimens collected appeared in Ann. Naturhistor. Mus. Wien 50 (1940, as 1939), 51 (1941), 53 (1943, as 1942), 55 (separately issued

1944, reprinted 1947), 56 (1948) and 57 (1950). An illustrated sketch of the vegetation appeared in Karsten & Schenk, Vegetationsbilder (1939). My lecture 'Grundzüge der Pflanzenverbreitung in Iran' presented at the Stockholm Congress 1950 (published in 1951) was based on this and the 1948 journey.

A special feature of the results of my expedition is that the mycologist Dr Franz Petrak, who settled at Vienna in 1938, searched carefully for parasitic and saprophytic fungi through my whole phanerogamic collections from 1937 and 1948. This tremendous labour resulted in a series of publications in Ann. Naturhistor. Mus. Wien 50: 414–518 (1940, as 1939) and in Sydowia 3: 135–148, 268–332 (1949) thus laying the foundation for the knowledge of lower fungi for Persia and adjacent countries. He described 7 new genera and 51 new species taking into account fungi recovered from relevant oriental phanerogamic collections from the Vienna Herbarium.

1948
EXSICCATAE NOS. 2600–7550

On April 3rd 1948, only about three years after the end of the Second World War, encouraged by Dr E. Esfandiari and Dr A.A. Azizi, I set out on a second expedition to Iran accompanied by Dr Frida Rechinger. Starting from Geneva, Tehran was reached by air via Paris, Tunis and Lydda. Meanwhile, Dr Gauba had emigrated to Canberra, Australia. Dr E. Esfandiari, Director of the Plant Pests and Diseases Institute of the Ministry of Agriculture, now at Evin, Tehran, offered hospitality and the use of his house as our headquarters and also offered transport for a 'big round trip'.

After our arrival at Tehran on April 5th it took two weeks to accomplish all the formalities to obtain travel-permits, and collect the necessary letters of introduction to local authorities and persons of influence. During this wait we made a few excursions which yielded only some early flowering plants since the vegetation in the north of Iran developed quite late.

On April 19th we left Tehran with Dr Esfandiari and Paul Aellen in an old military high-wheeled 'command-car' of American origin, furnished with a trailer for luggage. The car served its purpose but the trailer had to be abandoned at Kerman. It was replaced by a lorry, which however very often went its own ways and caused many inconveniences. South of Yazd towards Kermanshahan the spring vegetation with its many annuals was in flower. Kerman was reached in heavy rain on April 23rd. In order to reach the south coast one had to cross several passes between Baghin, Mashiz and Sirjan (Saidabad) between 2300 and 2500 m in altitude. The sight of the peaks more than 4000 m high still deep in snow is very impressive. Few plants were in flower; among them were several species of *Eremurus*. Sirjan (1700 m) is situated in a partly sandy, partly halophytic depression. An exploration at a more advanced season would certainly be worthwhile. Another pass, Chah-e Chaghak (1900 m) had to be crossed before the 'garmsir' (Stapf 1888–1889), the hot region bordering the Persian Gulf, was reached. From Hajiabad and Tarom, at 900 m downwards, several characteristic southern shrubs appear, e.g.: *Acacia hydaspica* J.R. Drumm. ex Park., *A. oerfota* (Forssk.) Schweinf., *Calotropis procera* (Willd.) R. Br., *Dodonaea viscosa* L.,

Ebenus stellata Boiss., *Euphorbia larica* Boiss., *Gymnocarpus decander* Forssk., *Ochradenus baccatus* Del., *Stocksia brahuica* Benth. and *Taverniera cuneifolia* (Roth) Arn. accompanied by the grasses *Chrysopogon aucheri* (Boiss.) Stapf and *Cymbopogon olivieri* (Boiss.) Bor. There were also many herbs and subshrubs, such as: *Anvillea garcini* (Burm.) DC., *Blepharis persica* (Burm.) O. Kuntze, *Cometes surattensis* Boiss., *Diceratella canescens* Boiss., *Forskohlea tenacissima* L., *Gaillonia aucheriana* Decne. ex Boiss., *G. calyptera* (Decne.) Ehrendf., *Gastrocotyle hispida* (Forssk.) C.B. Clarke, *Grantia aucheri* Boiss., *Lotononis platycarpos* (Roth) Arn., *Pimpinella stocksii* Boiss., *Platychaete aucheri* Boiss., *P. glaucescens* Boiss., *Pycnocycla nodiflora* Decne., *Rumex cyprius* Murb. and *R. vesicarius* L.

Five new species were discovered, namely, *Acantholimon austro-iranicum* Rech. f. & Schiman-Czeika, *Anthemis mirheydari* Iranshahr, *Convolvulus argyracanthus* Rech. f. et al., *Lotus laricus* Rech. f. et al. and *Silene austro-iranica* Rech. f. et al. The occurrence of these 'garmsir' species coincides with the appearance of *Phoenix dactylifera* L.

We returned to Kerman on the same track, quite difficult in 1948, today replaced by an asphalt road. We left Kerman on May 6th and proceeded on the main road to Bam. Here a relative of Dr Esfandiari offered us one of the most memorable headquarters imaginable: a large house in classical oriental style set in the midst of a palm-grove. Bam is situated on the southwest corner of the Lut desert basin. From here we crossed via Darzin the Kuh-e Jamal Barez which is the southwestern spur projecting from the great Kerman mountain knot and separating the Lut Basin from the Jiroft (Jaz Murian) Basin. On the ridge near the small village of Deh Bakri, at 2100 m in loose stands of *Acer monspessulanum* L. subsp. *persicum* (Pojark.) Rech. f. et al., *Acantholimon chlorostegium* Schiman-Czeika, *Isatis pachycarpa* Rech. f. et al., *Ranunculus papyrocarpus* Rech. f. et al. and *Silene austro-iranica* Rech. f. et al. were found. On the descent to the south into the hot basin many southern species seen on the way to Bandar Abbas were recollected. On May 11th we left Bam and went on the main road along the southern border of the Lut to Zahedan. From here, an excursion south to Khash and further across the Karevandar Hills to Iranshahr was undertaken; the first half of the way furnished *Euphorbia gedrosiaca* Rech. f. et al., *E. multifurcata* Rech. f. et al. and *Pterocephalus gedrosiacus* Rech. f. et al. We discovered the new monotypic genus *Karvandarina aphylla* Rech. f. et al. (Compositae-*Cynareae*) during the second half as well as *Helianthemum cinereo-flavescens* Rech. f. et al. (which in *Flora Iranica* 110 was erroneously not clearly separated from *H. stipulatum* Forssk.), *Heliotropium multiflorum* Rech. f. et al. and *H. nodulosum* Rech. f. et al.

From Khash we planned to ascend the volcano Kuh-e Taftan; this plan had to be abandoned because I suffered a very serious attack of dysentery. From the foot of Kuh-e Taftan, *Linaria michauxii* Chav. subsp. *robusta* Rech. f. et al. and *Heliotropium taftanicum* Rech. f. et al. have been described. We made a loop around Kuh-e Taftan and returned to Zahedan. We visited Zabol-e Sistan situated in the Hamun basin in which the Hilmand River from the Hindu Kush ends. The desert stretch on the way to Zabol and from Zabol on to Birjand evidently had seen no rain for years. Nothing at all was in flower! In Zabol, except for a few weeds, no plants were collected because the Hilmand had flooded the whole area. In spite of regularly taking preventive medicine I had acquired malaria, which caused

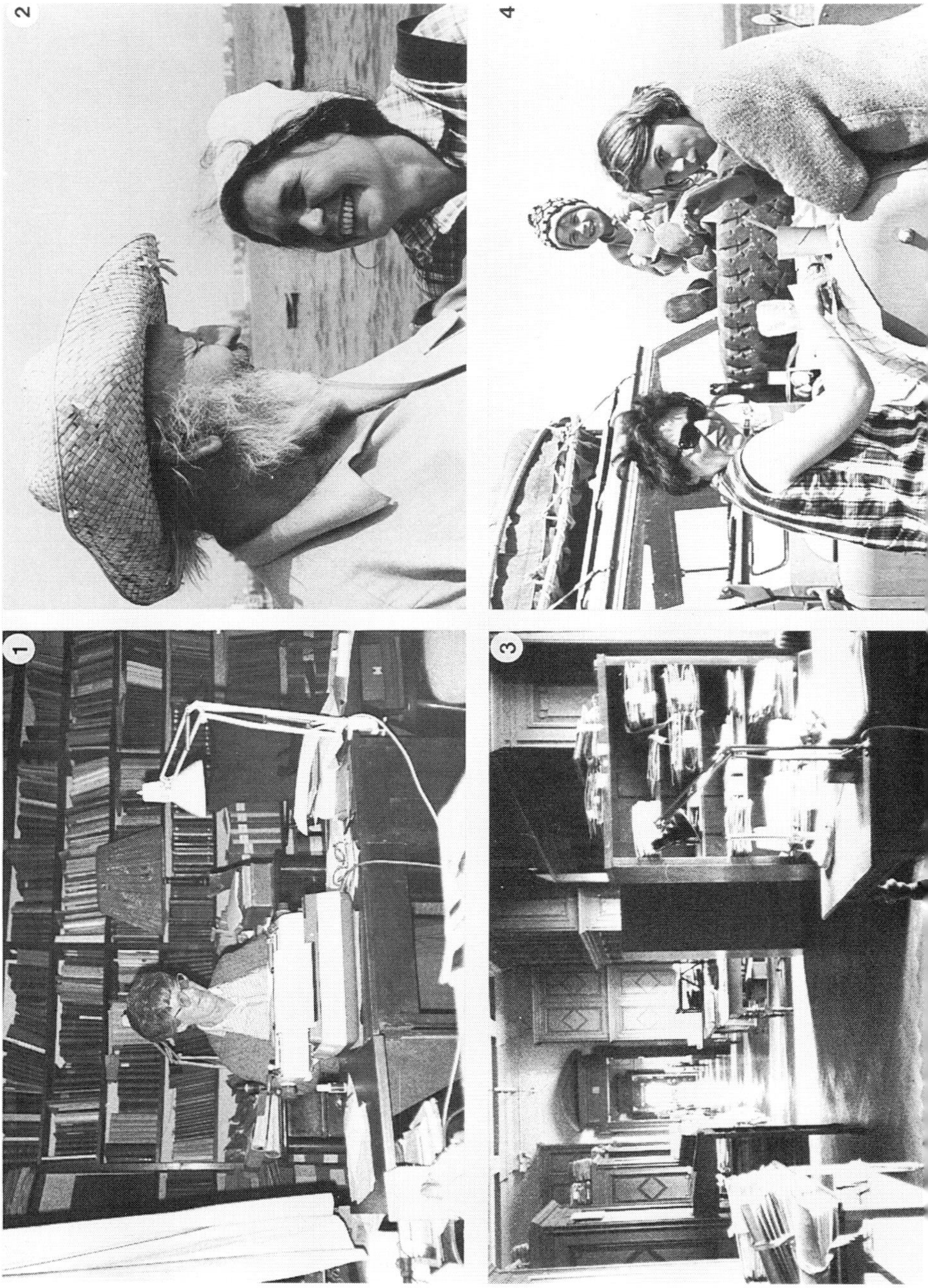

Figure 2. 1. My wife, Wilhelmina Rechinger at home – typing *Flora Iranica* manuscripts. — 2. Wilhelmina and I at Thessaloniki, Greece. 7 August 1965. Photo Jennifer Lamond. — 3. The Herbarium, Natural History Museum, Vienna, where the complete set of my specimens from the *Flora Iranica* area is conserved. — 4. Wilhelmina, Karl Björn and Lilian having lunch between Bojnurd and Mashhad, Iran. 20 April 1967.

recurrent attacks of fever and spells of weakness during the remainder of the 1948 journey and for years after.

At Birjand, we were again the guests of a big landowner for a few days. The most memorable feature of my stay was a beautiful carpet 15 × 10 metres. In my fevered dreams I was in the central medallion of the carpet which revolved around me. We left Birjand on May 26th. Development of the vegetation improved from Qayen to the north. Revisiting the serpentine hills at Robat-e Safid furnished no sensational novelties. Mashhad was reached the following day. As in 1937, Mashhad was our headquarters for a couple of weeks.

The first excursion on May 30th was into the southern parallel range, the Kuh-i Binalud, and into the Darreh Abshar (Waterfall Valley) by car as far as Akhlamad. From there we went by donkey and by foot into the steep valley with narrow gorge-like sections strewn with limestone rocks, on which huge hard cushions of *Gypsophila aretioides* Boiss. and soft ones of *Dionysia tapetodes* Bunge grew. Several hundreds of species were collected, among them five new ones: *Euphorbia chasmophila* Rech. f. & Aell., *Ferula flabellifolia* Rech. f. & Aell., *Linaria melanogramma* Rech. f. & Aell., *Salvia chloroleuca* Rech. f. & Aell. and *Sisymbrium integrifolium* Rech. f. & Aell.

The second excursion (June 2nd–3rd) was into the Kopet Dagh via Chenaran and Quchan to the Alamli pass as in 1937 but 6 weeks earlier. The season was rather cool and humid. Hundreds of species including dozens of Astragali were collected. Four new species were found: *Acantholimon gorganicum* Mobayen, *Aethionema apterocarpum* Rech. f. et al., *Astragalus orthocarpioides* Širj. & Rech. f. (Sect. *Astragalus*) and *Euphorbia aellenii* Rech. f.

From June 7th–10th an expedition into the Hazar Masjed, the highest point of the southeast Kopet Dagh, was undertaken. Special permits and a military escort were necessary. The escort was a nuisance. They were evidently afraid of the Kurdish tribesmen. The escort tried to rush us as quickly as possible through the most interesting collecting grounds and they forced the poor inhabitants of the villages to share their meals with them without any recompense. We three (Esfandiari remained behind at Mashhad) went by car to Tuz, hired several horses and, by way of a series of valleys and limestone gorges between the villages of Ardak, Gash and Talqur, reached the high plateau. Vegetation in the valleys was at its best. *Acanthophyllum chloroleucum* Rech. f. & Aell., *Campanula khorasanica* Rech. f. & Aell., *Ferula latisecta* Rech. f. & Aell., *Graellsia integrifolia* (Rech. f.) Rech. f., *Hesperis khorasanica* Rech. f. & Aell. and *Psychrogeton aellenii* (Rech. f.) Grierson were discovered. On the undulating high plateau, there were huge camps of black tents inhabited by Kurdish nomads. Their innumerable flocks of sheep and goats had already grazed the vegetation down to zero before it could come into flower. On the fourth day, while riding along a river cut deep into the loess, the soil gave way and I fell with my horse into the river. I had difficulties in freeing myself from the struggling horse. The shock was great but no serious damage had been done. However, several rain showers had made the soil very slippery and I preferred to walk back to Tuz where the car was waiting for us.

On June 13th we left Mashhad and reached Shahrud via Neyshabur and Sabzevar. Owing to the cool, rainy season, annuals flowered everywhere in profusion.

On June 17th we went up to the Khush Yaylaq pass (Khash Aylan); *Acantholimon embergeri* Mobayen and *Cousinia esfandiarii* Rech. f. & Aell. were found. During the following days we went west on the main road through the Caspian lowland by Gorgan to Bandar Gaz and Babolsar. *Alcea gorganica* (Rech. f. et al.) Zohary was discovered. Then we went up through the Talar Valley. On June 20th near Sorkhabad I found *Rumex alveolatus* Losinskaja, *R. chalepensis* Mill., *R. crispus* L., *R. dentatus* L. and *R. pulcher* L. subsp. *divaricatus* (L.) Murb. growing together with four new hybrids – *R. alveolatus* × *chalepensis*, *R. alveolatus* × *crispus*, *R. alveolatus* × *pulcher* subsp. *divaricatus* and *R. chalepensis* × *dentatus*. That same evening we reached Tehran.

The days between June 21st and July 12th were spent in Tehran. I had to undergo medical treatment against malaria and dysentery.

The three thousand numbers of specimens collected on the round trip by Aellen, Esfandiari and me, bearing my collecting numbers, were sorted out at Tehran and divided into three sets. The first, including unicates, went to Vienna (W); the second went to Evin (Tehran) where it formed the nucleus of the Evin herbarium (IRAN); and the third set went to P. Aellen, Basel, and is now permanently deposited at Geneva (G-AELL).

From July 13th to 16th we took a trip by rail to Bisheh accompanied by Esfandiari's assistant Manouchehri. Bisheh is situated in the limestone gorge of the Dez River in Luristan. Here M. Koeie had collected in the early spring of 1937. The only access to Bisheh is by rail; there is no road, not even a footpath, leading to the settlement which consists only of a few railway-workers' huts. Even though no new species were discovered, this excursion into one of Iran's areas richest in endemics was worthwhile. In rock crevices where water is permanently dripping and depositing sinter, the endemic *Primula gaubaeana* Bornm. (sect. *Floribundae*) grows. I was able to secure seed for Sir William Wright Smith's unique collection of living primulas at Edinburgh.

On July 18th we went on our own by rail to Shahrud. With the help of Dr Kaveh Farmanfarmajan, headquarters were established at the small village of Nokarman, situated at 2000 m on the south slopes of the botanically unexplored Shahvar Kuh. The surroundings of the village were explored. After two days, three donkeys and two drivers were hired to carry presses, tents and provisions and the ascent started. The south slope furnished: *Cousinia decumbens* Rech. f., *C. irritans* Rech. f., *C. nekarmanica* Rech. f., *C. shahvarica* Rech. f., *Ferula serpentinica* Rech. f., *Gypsophila xanthochlora* Rech. f. and *Mattiastrum pygmaeum* Rech. f. At higher elevations at around 3500 m I found *Myopordon hyrcanum* (Bornm.) Wagenitz, now the only existing specimen as the type was destroyed in Berlin during the war.

On the third day we descended to the tree line on the northern slope and camped near to a huge shepherds' camp. From there I climbed down into a narrow gully with a rich vegetation of tall herbs. Fresh bear droppings were seen. Next morning I climbed up the steep limestone slope on the other side of the shepherds' camp. The slope was covered with dense low carpets of *Juniperus sabina* L. and inaccessible to the flocks. Alarmed by the shouting of the shepherds I realized that a mother bear accompanied by two cubs was approaching the camp and was being driven back by the shepherds. I could watch the mother bear stand up repeatedly on her hind legs and defeating the dogs which were attacking her. Thus she drew full attention to herself and gave the cubs time to disappear and take shelter between the shrubs. Finally all three bears disappeared into the gorge.

Saponaria bodeana Boiss. with bright red flowers was rediscovered here. It was only known without precise locality from a fragment. I have unnecessarily redescribed it under the name *S. ignea* Rech. f. Its identity was established years later at the Boissier herbarium (G). Real novelties were: *Delphinium elbursense* Rech. f., *D. gorganicum* Rech. f., *Erigeron ursinus* Rech. f., *Leontodon stenocalathius* Rech. f. and *Scrophularia gorganica* Rech. f.

On July 28th Nokarman was reached again, after crossing the main ridge farther east. The view south over Shahrud and the broad, now completely dry riverbed far into the desert was breathtaking.

On August 1st Shahrud was revisited and in the night from August 2nd to 3rd we went by train to Tehran.

On August 6th we left Tehran by bus. This time, again accompanied by Mr Manouchehri, we crossed the Kandavan Pass and went down north to a place called Pol-e Zanguleh (small bell bridge). After a day's collecting on the humid northern slope of the Elburz range we organized a small caravan of one horse and three mules. We went east by the coal mines of Elika (Ileka), Kamarband (2500 m), the Naftab pass (3200 m), and Mount Uloj (3400 m), then turned north to go down to Mazandaran via Sanus and Kinch mainly through broadleaved mixed forest to Dasht-e Nasir, an enclave of *Cupressus sempervirens* L., and from here to the lower Chalus Valley. The botanical harvest was rich and varied on account of the different altitudes, exposures and type of soil. *Saxifraga mazandaranica* Rech. f. and *Centaurea kandavanensis* Wagenitz were found near Pol-e-Zanguleh. On Mount Uloj, *Alyssum polycladum* Rech. f., *Arabis rimarum* Rech. f., *Asyneuma virgatum* (Labill.) Bornm. subsp. *mazanderanicum* (Rech. f.) Damboldt and *Ziziphora clinopodioides* Lam. subsp. *elbursensis* Rech. f. occurred. The lower regions in Mazandaran furnished *Centaurea aziziana* Rech. f.; between Sanus and Kinch in the Chalus Valley, *Leutea nematoloba* (Rech. f.) M. Pimenov grew at 1000 m. *Acantholimon ophiocladum* Rech. f. & Schiman-Czeika was discovered on the Kandavan north slope

On August 12th we returned to Tehran. One week had to be spent in Tehran for medical treatment, packing herbarium specimens, and preparing for the homeward journey. With the help of Dr E. Esfandiari I obtained permission from the Ministry of Agriculture to take duplicates from Gauba's herbarium at Karaj for the Naturhistorisches Museum, Wien (W). These were to replace Gauba's duplicates which had been sent to W before the war and which had been destroyed. We spent the days from August 19th to September 2nd as guests of the Department of Agriculture at Karaj and managed to secure approximately 2000 numbers. In between we made small excursions around Karaj collecting late-flowering and/or fruiting species not seen in 1937. Two new species, *Echinops elbursensis* Rech. f. and *E. keredjensis* Rech. f. were found. It took one more week at Tehran to obtain our visa, complete customs formalities, pack the herbarium specimens, etc.

On September 9th we left by Air France via Damascus, Rome, and Paris to Geneva and from there by train back to Vienna. No enumeration of the plants collected in Iran in 1948 has ever been published. New species have been published by me and other authors in a set of 40 papers in various periodicals with the

Figure 3. 1. Jennifer Lamond photographing *Euphorbia caducifolia* Haines between Wad and Bela, Baluchistan, Pakistan. 2 April 1965. — 2. Knut Lindberg with *Ferula narthex* Boiss., between Kabul and Ghazni, east Afghanistan. 29 June 1962. — 3. Wilhelmina and myself with guide on the Kuh-i Taftan, Baluchistan. 3 May 1977. — 4. The Landrover in Baluchistan, Pakistan. April 1965. — 5. Landrover in distress being pushed by Evan Guest, Wadi at-Tib, 70 km north of Amara, Iraq. 28 March 1957. — 6. The Landrover on the road from Panjao to Shahrestan, central Afghanistan. 1 July 1967.

subtitle: '*Rechingeri Iter Iranicum secundum*'. A complete list of these papers dealing occasionally also with specimens from other collectors has been published in Sitzungsberichte der Math.-Nat. Klasse der Österr. Akad. Wiss. 98: 7–9 (1961).

1956–1957
EXSICCATAE NOS. 8000–15940

Early in 1956 I was asked by Nicholas Polunin, then Professor of Botany at Baghdad, whether I would like to spend one year as a visiting professor at the College of Arts and Science in Baghdad, Adhamiya, which was to be transformed into a university. After some hesitation I agreed in view of the possibility to explore the northern mountainous (Kurdish) part of Iraq which was included in the planned *Flora Iranica*. I obtained one year's leave of absence without pay from Vienna. I left Vienna by train in the evening of October 3rd 1956 accompanied by my second wife Wilhelmina (née Goedemans) under rather dramatic circumstances. I realized a few minutes before half past seven in the evening that from October 3rd onwards the winter schedule of the railways was in force and that the Orient Express for economic reasons was combined from Vienna to Bruck an der Mur with the Roma Express and thus should leave Vienna an hour earlier, i.e. at seven-thirty in the evening instead of eight-thirty. We bribed a taxi-driver to disregard all traffic lights and arrived with our piles of luggage on the platform at the very moment when the red terminal (tail) light of the train was disappearing in the dark. It was clear that we had to catch the train by all possible means as the sleeping car reservations to Baghdad had been booked and paid for in advance. Rain was pouring down. We found another taxi-driver, one able and willing to compete with the train and we arrived at Semmering Pass station about 100 km south of Vienna five minutes before the train! As the train stops here for only a minute, we had to push our eleven pieces of luggage into the nearest coach – labelled destination Roma – at the front of the long and overcrowded train. Within 30 minutes we had to carry everything to the very last coach – the sleeper to Istanbul – before the train was split at Bruck. We managed and then had ample time to relax before reaching Istanbul on 5th October and Baghdad on 11th.

At Baghdad it was revealed to me that besides a moderate amount of teaching I was expected to lay the foundation of a University herbarium, collect herbarium material in 50 sets as exsiccatae for exchange in order to build up the herbarium, and write a concise handbook of the *Flora of Iraq*. This was a rather unrealistic programme in view of the lack of a botanical library and herbarium, no vehicle being available for excursions and my time being limited to one year! After several discussions, my proposal to write a *Flora of Lowland Iraq* was accepted. Co-operation with Dr Ali al-Rawi and Mr Evan Guest of the National Herbarium at Abu Ghraib belonging to the Department of Agriculture was established, in spite of the fact that a long-term project of Abu Ghraib and Kew to write a critical flora of Iraq had already been launched. During the winter months I worked three days per week at the Abu Ghraib Herbarium and in my spare time at home extracted literature for the Lowland Flora. The book was completed during the following years in Vienna and published in 1964. We rented a small house with a garden as our home in Adhamiya within reasonable walking distance from the College. I made a few excursions around Baghdad in October and tried to arouse my students' interest in plants. In November the University was closed for several weeks because of students' strikes. So I was able to join Dr al-Rawi and Evan Guest on a major expedition to the Southern Desert from November 4th–10th with the following itinerary: Baghdad — As-Samawah — As Salman — An Nasiriyah — Ur — Al Busayyah — Samah (Neutral Territory) — Al Busayyah — Khidr al-Moi — Al Batin — Hawr al Hammar — Ur — Baghdad. I became acquainted with the colourful autumnal aspect of the desert flora, Chenopodiaceae dominating. Many species collected on this excursion had not previously been known from Iraq and very little fruiting material was available. *Anabasis lachnantha* Aell. & Rech. f., *Convolvulus oxyphyllus* subsp. *oxycladus* Rech. f. as well as *Salsola pseudonitraria* Aell. & Rech. f. were discovered.

A smaller excursion followed on November 20th–21st to Al Mahmudiyah — Karbala — Shithathah — Bakhr al Milh — Ar Rahhaliyah — Ramadi and Fallujah. The Suez crisis had its repercussions on our lives. The arrival of part of our belongings, sent by freight from Vienna via Trieste and Beirut, was delayed for weeks. All foreigners had to report to their embassies for evacuation but gradually the situation calmed down.

The winter 1956–57 in Baghdad was comparatively cold and dry. The fireplace in our sitting room was not working as the chimney was far too narrow. We had to rely on our 'Aladdin', a portable kerosene stove which created a homely atmosphere with a singing tea-kettle. In the second half of January, however, roses started to flower in our little garden, and we could enjoy our lunches in the open. Towards the end of February a hot spell ended in thunderstorms and during a couple of days rain was pouring down. Cascades of rainwater came from the roof and, as the pavement at the entrance was sloping towards the house, the surplus water from the garden flooded the interior. The walls were soaked up to 1.5 metres. Frogs appeared in the garden from out of nowhere and gave polyphonic concerts by their guttural trilling. It was weeks before the house was dry again. Feathery groups of saline crystals emerged from the cracks in the floor and the roof. Weeds in the palm-groves started to flower and I could try to explain the characters of some major plant families to my students with living plants on hand.

I sketched programmes for expeditions for the spring and summer into botanically unknown or little known parts of Iraq including several sections of the deserts and Persian foothills as well as the Qandil Range, second in altitude only to the Algird Dagh (Mount Helgurd). After long discussions, Abu Ghraib as well as the Adhamiya University College agreed to my proposals.

From March 15th–20th 1957 a students' excursion to Basra by rail took place. I camped with Wilhelmina behind the curtain of a small theatre while the students occupied the auditorium. We visited the island of Maqil, the Shatt al-Arab down to Fao, the swamps near Qurna, Al Madinah and Hwair, the Jabal Sanam near Hwair and some oil-wells. In the evening of May 21st Wilhelmina returned with the students to Baghdad. Next day Evan Guest and Ali al-Rawi arrived with two Landrovers and from March 22nd–25th we made a short excursion into the southern Desert: Az Zubayr — Umm Qasr — Jabal Sanam — Safwan — Jarishan — Twai al-Hashaish and back again to Basra. The spring aspect of the deserts – shades of pale green changing to deeper green in the wadis – was surprisingly different from the dull shades of grey-yellow and brown of the autumn one. The crucifer *Horwoodia dicksoniae* Turrill, so far only known from Kuwait, was discovered at several places; it is strictly confined to deep sand. The subshrubby composite *Rhanterium epapposum* Oliv., reported so far from the Southern Desert of Iraq, without locality, was found to be dominant over large sandy areas of the Southern Desert along the Kuwait border.

From March 26th–30th there followed an excursion to the Persian foothills north of Al Amarah: Basra — Qurna — Al Amarah — Shatt at-Tib — Al Amarah — Basra and back to Baghdad. This and the following excursions into the so-called Persian foothills which form the political frontier between Iraq and Iran proved that many lowland Saharo-Sindian species and Iranian highland species find their natural limits of distribution here and several endemics are evidently confined to this region. Therefore, I decided to include the foothills in my treatment of the Lowland Flora of Iraq as well as the Iranian highland flora. *Eremurus rechingeri* Wendelbo, related to *E. kopetdaghensis* M. Popov ex B. Fedtsch. from northern Iran and Turkmenistan, grows here as well as *Koelpinia chrysoglochis* Rech. f., a widely distributed species hitherto not recognized, now known from Algeria, Tunisia, Iraq and southern Iran. Furthermore, we found *Ergocarpon cryptanthum* (Rech. f.) Townsend, a monotypic umbelliferous genus.

Figure 4. Frida Rechinger with Kurdish nomads on the Hazar Masjed, Khorasan. June 1948. — 2. Per Wendelbo (left) and Ian Hedge at Doab, north Afghanistan. 13 June 1962. — 3. Jennifer Lamond, Wilhelmina and myself in Khinjan Valley, north Afghanistan. 1 July 1965. — 4. The staff of Evin, Wilhelmina and myself, from left to right: Fereidoun Terme, Mrs Parvine Partovi-Fahar, Mousa Iranshahr, Eyoulla Behboudi, myself and Wilhelmina, Miss Farideh Matine and Ahmad Zargani.

On April 4th–5th a short but most rewarding excursion across the Jabal Hamrin, a long range of hills, to the Persian frontier near Khanaqin was undertaken. Other visits to the foothills followed on April 10th–14th in a southeasterly direction to the region of Al Kut, Chabbaicha and Badrah.

The most interesting foothill journey was on June 2nd–6th to Balad Ruz, Mandali and Badrah. Here we found *Acanthophyllum khuzistanicum* Rech. f., *Anthemis hamrinensis* Iranshahr, *Centaurea alveicola* Rech. f., *Convolvulus catenifiorus* (Rech. f.) Sa'ad, *Echinops fraudator* C.C. Townsend, *E. tenuisectus* Rech. f., *Mattiastrum densum* Rech. f. & H. Riedl and *Phlomis polioxantha* Rech. f.

From April 21st to 24th another expedition to the Southern Desert followed: As Samawah — As Salman — Ansab — Aidaha — Ash Shabicha — Wadi al-Khirr — Karbala — Baghdad. *Atractylis arabica* Rech. f., *Pulicaria guestii* Rech. f. and *Reseda patzakiana* Rech. f. were discovered.

On May 19th I travelled with Wilhelmina by the Nairn bus to Beirut in order to fetch a Simca Chatelaine I had ordered from Paris. The car dealer's representative turned out to be a descendant of the ominous botanist Rafinesque-Schmaltz! In Beirut the American University harbouring the Post Herbarium (ca. 200 small parcels), the Catholic priest Père Mouterde – the author of the *Flore du Liban et de la Syrie* – and Mr H. Pabot, who had spent years as a FAO representative in various parts of the East, were visited. Fortunately the herbaria of the two last-named botanists were acquired by the Conservatoire Botanique de Genève immediately before the disastrous political troubles started in Lebanon. Wilhelmina's first experience as a driver – she had acquired her driving licence in Vienna just before our departure – was to drive across the Lebanon, Antilebanon and the Syrian Desert to Baghdad, where we arrived on May 26th. Botanically the journey was favoured by the heavy rains, which had been delayed this year.

From June 6th–9th I went for the last time with Evan Guest and Ali al-Rawi to the desert, this time into the hardly known area up to 100 km west of Rutbah. *Ephedra transitoria* H. Riedl and *Scorzonera rawii* Rech. f. & Guest were found. Owning a private car gave us a certain amount of independence.

The first major enterprise was to drive north to Kirkuk and then west across the lower Kurdish hills to Chamchamal — Sulaymaniyah — Dukan — Halabja and Tawila, on to the slopes of the Avroman Mountains on the Persian frontier, which had not been visited since the days of Haussknecht in 1866–1869. The three days around Tawila were rewarding: *Astragalus octopus* Townsend, *A. tawilicus* Townsend, *Pimpinella haussknechtii* Rech. f. & H. Riedl, *Veronica anagalloides* Guss. subsp. *heureka* M.A. Fischer and *V. avromanica* M.A. Fischer were discovered. Above 2000 m the vegetation was hardly developed, whereas back in Baghdad we had to cope with the summer heat: air temperature up to 45°C outside and 40°C inside, which is practically the limit one can endure. The beds had been transferred to the flat roofs which were surrounded by walls for the sake of privacy. When the air temperature before sunrise occasionally dropped below 30°C one started to freeze! Between excursions I had to go gradually through the plants collected so far to label and sort out the set for the university college which finally amounted to nearly 4000 numbers. I do not know whether it has ever been consulted, or whether it still exists.

The next excursion was from July 1st–14th, together with Ali al-Rawi, to Mosul

— Zakho — Sharanish — Jabal Khantur — Marsis — Mount Zawita (2100 m) — Mosul — Dihok — Zawita Village — Suwara Tuka — Sersang — Amadiyah — Kirkuk and Baghdad. The Kurdish limestone ranges together with the adjacent ones on Persian territory definitely form one of the richest centres of more or less local endemics, which include geographically isolated plants like *Pelargonium endlicherianum* Fenzl and the rubiaceous *Wendlandia ligustroides* (Boiss. & Hohen.) Blakelock. The dominant tree of these districts is still *Quercus brantii* Lindl., nowadays (especially at lower elevations) largely destroyed. Many of these endemics inhabit the loose stands of oaks, others are confined to rock crevices. Labiatae, including *Pentapleura subulifera* Hand.-Mazz., and Compositae are well represented. Another feature is the radiation of many Mediterranean plants into the area, e.g., annual *Trifolium* species. The mountains around Sharanish (1800 m) furnished ten new species: *Anthemis kurdica* Iranshahr, *Carlina kurdica* Meusel & Kästner, *Carthamus curdicus* Hanelt, *Cyclotrichium longiflorum* Leblebici, *Echinops amoenus* Rech. f., *Parietaria rechingeri* Chrtek, *Pimpinella kurdica* Rech. f. & H. Riedl, *Scutellaria porphyrantha* Rech. f., *Staehelina kurdica* Merxm. & Rech. f. and *Teucrium chasmophyticum* Rech. f., *Cousinia acanthophysa* Rech. f. on Mount Zawita, *Tragopogon vaginatus* M. Ownbey & Rech. f. at Chamchamal and *Astragalus dendroproselius* Rech. f. at Suwara Tuka were also found.

On July 15th the lease of our house in Adhamiya expired. We had to dispose of those belongings which we did not want to take home to Vienna. The daily temperatures were between 45°C and 49°C. Most Europeans had already left. We lost considerably on selling furniture but funnily enough, we gained on the old railway-sleepers meant for heating in the useless fireplace! More than 50 corrugated cardboard boxes had to be found in the bazaar for packing and air-freighting herbarium specimens to Vienna.

On July 23rd the last excursion began from which we returned to Baghdad on August 15th. The itinerary was: Baghdad — Jalaula — Kirkuk — Sulaymaniyah — Dukan — Ranya — Sharidan — Pushtashan — Mount Qandil — Pushtashan — Darband — Mirza Rustam — Dukan (with 2 days of rest) — Kirkuk — Arbil — Shaqlawa — Rawandiz — Bersorin Gorge — Darband — Nowanda Valley — Algird Dagh (Mount Helgurd) — Haji Omran — Darband — Rawandiz Gorge — Shaqlawa — Arbil — Kirkuk — Baghdad. At Ranya we experienced the highest temperature - 53°C inside a house! Starting from approximately 300 m one had to cross several ranges of increasing altitude up and down - no shade, no water and most of the vegetation of the derelict oakwoods dried-up. We tried to reach Pushtashan before nightfall, but it went dark and one of the mules stumbled and hurtled down the slope, so we had to camp. After two days, partly on mule-back, partly by foot, we reached Pushtashan (1100 m), situated in the shade of walnut trees and with an ample supply of fresh water. Here *Alcea macrocarpa* I. Riedl and *Myosotis kurdica* H. Riedl were found. The following day we went on crossing a belt dominated by spiny Astragali, including *A. chionobiiformis* Townsend and *A. pushtashanicus* Townsend and thistles including *Cousinia qandilica* Rech. f., *Crepis kurdica* Rech. f., *Onosma qandilicum* H. Riedl and *O. rechingeri* H. Riedl as well as *Satureja metastasiantha* Rech. f. were also discovered. Above the spiny belt followed one appearing yellow from a distance, which was composed mainly of *Prangos pabularia* Lindl. and other tall Umbelliferae and the violet-coloured *Vicia gregaria* Boiss. & Heldr. In the evening we reached Goam-e Kirmosoran (Kermasur), a small lake situated at ca. 3200 m. Here we camped, and climbed the mighty limestone Qandil (3500 m) the next day. The snow belt begins above 3200 m – much moving scree, few but interesting high mountain species were collected but no new species discovered. After another night at the lake we returned by a different path to Pushtashan where we spent the night on the flat roof of a house, returning the next day to Ranya and on by car to Dukan. Here we stayed two days in the comfortable resthouse and dried our plants. On August 7th and 8th we drove via Shaqlawa to Rawandiz. The dramatic narrow limestone gorges of Rawandiz and Bersorin, though visited before by various botanists, held three new species and one new subspecies as surprises: *Echinops faucicolus* Rech. f., *Picris strigosa* M.-B. subsp. *kurdica* Lack, *Pimpinella brachyclada* Rech. f. and *Stachys nephrophylla* Rech. f.

Though separated only by the Rawandiz Gorge (cut by the famous Hamilton Road connecting North Iraq with Persian Azerbaijan), one has to make the detour by Arbil in order to reach the Helgurd from Qandil. The exact altitude of the two mountains is not known to us. Qandil is believed to measure ca. 3500 m and the Algurd Dagh (Mount Helgurd) ca. 3700 m. Because of the great differences in temperature and pressure, climbing these mountains is very exhausting. On August 9th we reached Haji Omran after having found *Echinops pachyphyllus* Rech. f. near Rayat. At Haji Omran *Echinops rectangularis* Rech. f. and *Tragopogon rechingeri* Ownbey were discovered. The Algurd Dagh, at ca. 3700 m the highest mountain in Iraq, formed by its lofty pyramidal shape and its dark, mainly igneous rocks quite a contrast to the bleak, broad Qandil. Only about half of the high mountain species occur on both mountains. In spite of the visits of other botanists in earlier days the ascent through the Nowandah Valley furnished several new species: *Astragalus helgurdensis* Townsend, *Cirsium karduchorum* Petrak, *Leutea rechingeri* (Leute) M. Pimenov, *Rosularia rechingeri* Jansson and *Scorzonera tenax* Rech. f.

Back in Baghdad on August 15th the last and most distressing period of our stay in Iraq began. Air temperatures reached between 45°C and 51°C every day. Settling accounts with the University College and the tax office, applying for a transfer permit and exit visa for Iraq, a transit visa for Turkey and Greece, arrangements with the shipping agents, etc., would have taken one week in winter. In summer it took two weeks as most of the office 'bosses' were on leave and replaced by inefficient substitutes. As a matter of fact we managed to leave Iraq at Khanaqin 5 minutes before the customs-licence for our car expired! On the edge of our nerves we slept the rest of the night of the 27th to 28th on the roof of a small hotel.

We crossed the northwest of Persia between August 28th and September 3rd by the route Qasr Shirin — Karand — Shahabad — Kermanshah — Taq-i Bustan — Sanandaj — Hoseynabad — Divandarreh — Saqqez — Miandoab — Bonab — Khanian — Tabriz. Near Bonab and in the Atchi Chay Valley near Tabriz, interesting late flowering halophytes were collected. We went on via Marand and Khvoy to Maku, where we left Iran on September 4th. 'Off-season'-botanizing in Kurdistan and Azerbaijan furnished *Centaurea intricata* Boiss. subsp. *kermanshahensis* Wagenitz, *C. pterocaula* Trautv. subsp. *iranica* Wagenitz, *Echinops pachyphyllus* Rech. f., *Ferulago phialocarpa* Rech. f. & H. Riedl, *Heliotropium gypsaceum* Rech. f. & H. Riedl and *Johreniopsis oligactis* (Rech. f. & H. Riedl) M. Pimenov.

We entered Turkey from the east through the gate of the new customs-office forming a frame for the impressive Ağri Dagh (Mount Ararat). We travelled west via Doğubayazit — Ağri — Tahir Pass (2500 m) — Horasan to Erzurum. At the Tahir Pass many plants were still in flower or fruit. *Rumex gracilescens* Rech. f. was discovered, a rather isolated species probably related to a Central-Asiatic group. It is now known from a second locality in Turkey and from Khan Hasan Saralan in Persian Kurdistan. From Erzurum we went on via Erzincan — Zara — Sivas — Yozgat to Ankara. Here Wilhelmina fell ill with influenza and we had to stay for four days. I made a one-day excursion to Tuz Gölü to compare the Central-Anatolian halophytes with the Azerbaijanian. On my return to the Berlin Hotel in Ankara a catastrophe had happened. A thunderstorm in the hills north of the city had caused a sudden flooding of the river below the hotel, carrying away cars, fragments of houses and trees. On September 13th we left Ankara for Bursa, paid a short visit to Ulu Dagh, visited Prinkipo from Istanbul and left Asia via Edirne on September 18th after $11\frac{1}{2}$ months. We drove through northern Greece to Thessaloniki and south to Athens. Here we visited the Greek amateur botanist C. Goulimis. On September 25th we took a ferry from Piraeus to Venice and reached Bad Aussee in Styria on 26th to visit my mother. On October 3rd we returned to Vienna after exactly one year and a day!

1962
EXSICCATAE NOS. 16000–19890[1]

On May 20th 1962 I left Vienna by air – this time alone – and reached Kabul by Aeroflot via Moscow and Tashkent the next morning. Possibilities of part-time collaboration had been discussed with Ian Hedge, Edinburgh, Per Wendelbo, Göteborg and Dr Knut Lindberg, a medical doctor and zoologist from Lund. All three had reached Afghanistan before me and had left Kabul for expeditions by the time of my arrival. No Landrover in working condition was available for hire. The only one offered was being repaired and not expected to be ready within a fortnight. To fill the time I decided to join Hedge and Wendelbo who had left for the Maimana area. I went in a small shaky Afghan aeroplane across the Hindu Kush to Kunduz on May 25th and on by lorry via Tashkurghan to Mazar-i Sharif. There I suffered an unusually heavy attack of dysentery. I lost several days, then hired a seat in a taxi and went on west via Balkh — Akcha — Shibarghan to Andkhui, where the road approaches the sand dunes bordering the Kara Kum desert south of Andkhui. There was a Landrover stranded with two men waving for help – evidently a Russian and an Afghan. The 'Russian' turned out to be Per Wendelbo and the 'Afghan' Ian Hedge! They were very satisfied with their collecting especially in the Belchiragh Gorge which was full of endemics. I transferred my belongings to their ruined Landrover and we shared a few cans of

[1] Up to and including my first Afghanistan journey (1962) I used parallel sets of numbers for plants from Greece and from the Flora Iranica area. From no. 19891 onwards there exists only one set of consecutive numbers, used without regard to the origin of the plants.

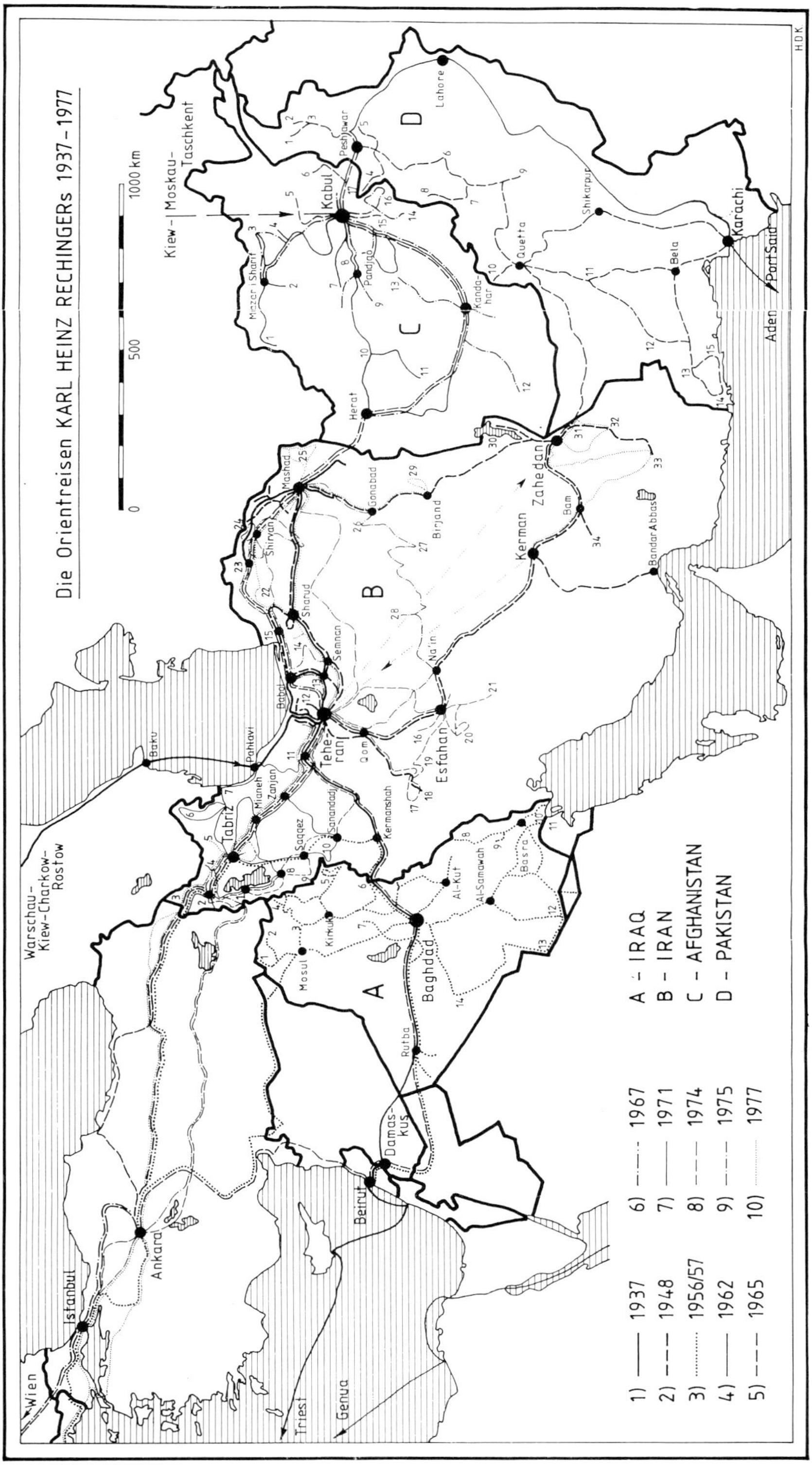
Die Orientreisen KARL HEINZ RECHINGERs 1937–1977
0
500
1000 km
1) —— 1937
2) – – – 1948
3) 1956/57
4) —— 1962
5) – – – 1965
6) –·–·– 1967
7) —— 1971
8) – – – 1974
9) –·–·– 1975
10) 1977
A - IRAQ
B - IRAN
C - AFGHANISTAN
D - PAKISTAN
Wien
Triest
Genua
Istanbul
Ankara
Beirut
Damaskus
Rutba
Baghdad
Mosul
Kirkuk
Al-Kut
Al-Samawah
Basra
Warschau-Kiew-Charkow-Rostow
Baku
Tabriz
Pahlavi
Mianeh
Zanjan
Saqqez
Sanandadj
Kermanshah
Teheran
Qom
Esfahan
Na'in
Babol
Semnan
Sharud
Shirvan
Mashad
Gonabad
Birjand
Kerman
Bam
Zahedan
Bandar Abbas
Herat
Kandahar
Pandjao
Mazar i Sharif
Kabul
Kiew-Moskau-Taschkent
Peshawar
Lahore
Quetta
Shikarpur
Bela
Karachi
Port Said
Aden
HDK

Figure 5. THE ITINERARIES (1937–1977) OF KARL HEINZ RECHINGER IN THE ORIENT
Map scale 1: 9,000,000
Basic Maps:
Haack Hausatlas, 3. Aufl. Gotha-Leipzig 1974, K 191 Vorderasien (1: 9,000,000).
Haack Grosser Weltatlas, Gotha-Leipzig 1968, K 34 Iran, Pakistan, Afghanistan (1: 9,000,000).
The first 6 itineraries have been prepared according to K.H. RECHINGER, Botanische Orientforschungen 1937–1967 and '*Flora Iranica*', Bustan 9: 29–32 (1968), the remainder from handwritten itineraries.
The following localities are indicated by corresponding numbers.

A – IRAQ

1 Zawitah, 2 Amadiyah, 3 Arbil, 4 Hajji Umran, Arl Gird Dagh, 5 Sulaimaniyah, 6 Khanakin, 7 Djabal Hamrin, 8 At Tib, 9 Al Madina, 10 Fao, 11 Djabal Sanam, 12 Ansab, 13 Al Diumaiyma, 14 Nukhaib.

B – IRAN

1 Rezaiyeh, 2 Khvoy, 3 Bazargan, 4 Marand, Zonus, 5 Kiyamaki Protected Region, 6 Dasht-e Moghan, 7 Astara, 8 Mahabad, 9 Baneh, 10 Marwan, Kuh-e Chehel Chasmeh, 11 Qazvin, 12 Kojur, Nur, Amol, 13 Firuzkuh, 14 Fulad Mahalleh, 15 Gorgan, 16 Mooteh Protected Region, 17 Khorramabad, 18 Bisheh, 19 Oshtoran Kuh, 20 Shahr Kord, Borujen, 21 Semirom, 22 Mohammed Reza Shah National Park, 23 Bojnurd, 24 Kopet Dagh, 25 Sarakhs, Jannatabad, 26 Bejestan, 27 Deyhuk, 28 Khvor, 29 Ahangaran, Zaidan, 30 Zabol-i Sistan, 31 Mirjaveh, 32 Kash, 33 Iranshahr, Bampur, 34 Sabzvaran.

C – AFGHANISTAN

1 Andkhui, 2 Akupruk, 3 Kunduz, 4 Pul-i Khumri, 5 Andarab, 6 Kandai, 7 Band-i Amir, 8 Bamian, 9 Sharestan, 10 Qala Shahrak, 11 Parjuman, 12 Bhagat, 13 Orozgan, 14 Urgun, 15 Gardez, 16 Khost, 17 Jalalabad.

D – PAKISTAN

1 Ushu, 2 Utror, 3 Kalam, 4 Kurram, 5 Attock, 6 Dera Ismail Khan, 7 Fort Sandeman, 8 Wana, 9 Fort Munro, 10 Khojak Pass, 11 Kalat, 12 Panjgur, 13 Turbat, 14 Gwadar, 15 Pasni.

---

beer cooled in the wind with the help of soaked handkerchiefs. All efforts to repair their Landrover were in vain. We had to wait until the following night. There came a bus which towed us to Andkhui, where the tank of a ruined lorry was fixed on the roof of the Landrover and with more petrol dripping down onto the vehicle than into the engine, we reached Mazar-i Sharif after two days. From there a mechanic with a new petrol pump was ordered from Kabul by telephone. We used the spare days at Mazar-i Sharif for excursions into the Balkh Gorge near Chashmeh Shafal and also to Akupruk. On the way to Kabul we made a side excursion from Doab into the Ajar Valley (King's Valley), most picturesque with colourful rocks and bizarre erosion forms. *Cephalorhizum pachycormum* Rech. f., *Allium pogonotepalum* Wdb., *A. spirophyllum* Wdb. and *Nepeta rechingeri* Hedge were discovered.

After a few excursions from Kabul I undertook together with Dr Knut Lindberg an expedition with the following itinerary: June 25th to July 8th: Kabul — Ghazni — Karabagh — Sang-i Masha — Kotal-e Kechru — Miradina — Dasht-i Nawar — Okak — Lake Djalgat Shahr Kawat — Sabzab — Behzud — Kabul. A gypsum area at the pass near Loman furnished *Scrophularia elegantissima* Rech. f. & Wdb., *Saponaria pachyphylla* Rech. f. and *Trigonella pycnotricha* Rech. f. The Dasht-i Nawar is a wide basin at ca. 3000 m altitude with a flat saline lake in its centre. It is surrounded by mountains mostly more than 4000 m high, at the snow covered northeast side possibly more than 4500 m. At the time of our visit the lake was populated by immense flocks of flamingos. On the flats northeast of the Dasht-i Nawar hard hummocks of several species of *Acantholimon* are dominant over wide areas.

The foothills near Okak at 3000 m furnished interesting new taxa: *Mesostemma kotschyanum* (Fenzl) Vved. subsp. *afghanicum* Rech. f., *Erysimum ghaznicum* Cullen & Rech. f., *Mattiastrum polyanthum* Rech. f. & H. Riedl, and *M. cynoglossoides* Rech. f. & H. Riedl. In a narrow limestone gorge north of the Dasht-i Nawar towards Behzud the new caryophyllaceous genus *Pentastemonodiscus* Rech. f. (*P. monochlamydeus* Rech. f.) and the dwarf, taxonomically isolated *Steptorhamphus pumilus* (Rech. f. & Tuisl) Tuisl were discovered. In return for a few days of his hospitality the governor of Behzud begged us to take along on the rear running board of my vehicle a criminal and his guard to be delivered at Behzud jail. The prisoner acted at the same time as our guide and we shared our meals with him.

From July 12th to 17th I visited Bamian and the lakes of Band-i Amir in central Afghanistan with K. Lindberg. Bamian is well known for the huge Buddha statues over 50 m high carved in the rocks. The blue and green lakes of Band-i Amir form their own natural dams by the sediments of their continuous overflow. These white walls show a strong contrast with the deep, blue water. Though visited by many botanists and tourists before, these localities still revealed several new species one of which was *Turaniophyton codringtonii* (Rech. f.) Podlech.

Back to Kabul. Lindberg's Russian jeep and my Landrover had been repaired in the meantime and we could finally set out in two vehicles on the long-planned 'Central Road', the crossing of central Afghanistan from Kabul to Herat had not been previously completed by a botanist before. Two vehicles were compulsory as for long distances no regular supply of petrol, drinking water and food could be expected. We left Kabul on July 22nd and reached Panjao via Maidan and the Unai Pass. Panjao is situated on one of the tributaries of the Hilmand River. From there we crossed several passes at altitudes between 2800 and 3200 m and across undulating high plateaux before reaching the upper tributaries of the Hari Rud. This central part of the Afghan Hindu Kush is called Hazarajat after its nomadic (Mongolian) inhabitants, the Hazaras. During the summer, additional nomadic tribes from all directions converge here, erecting colourful cities of tents and 'yurts'. Locally, dry farming appears to be successful at altitudes up to 3400 m, which points to comparatively high humidity in early summer. For the botanist, however, this region is rather disappointing. The only remnants of the native vegetation are plants unpalatable to cattle: *Leucopoa karatavica* (Bunge) Krecz. & Bobr., *Salvia rhytidea* Benth. and several species of *Eremurus*. Everything else is used as fuel. Beyond several passes near Lal-i Sar Jangal, the upper tributaries of the Hari Rud are reached. The valley appears to be very dry at this advanced season. West of Chakcharan the track leaves the Hari Rud valley to avoid narrow gorges and crosses the mountain range south of and parallel to the river on the Shutur Khan Kotal. Here *Koeiea ghorana* Rech. f. and *Acantholimon ahangarense* Rech. f. & Schiman-Czeika were discovered. Instead of continuing west towards Qala Sharaq we turned further south to Taiwara. The landscape changes: the undulating plateau dissolves into numerous steep and parallel limestone crests running northeast to southwest.

The province of Ghorat is probably one of the botanically most promising regions of Afghanistan. Access, however, is difficult. Until late spring access from the North is blocked by snow whereas access from the South is impossible as the gorges separating the ridges are blocked by meltwater. Thus only a few summer months are left for botanical exploration. Going southwest from Taiwara we reached the small village of Parjuman and explored the limestone gorge below and the steep limestone crest south of the village up to 2900 m. The few days spent in Ghorat at the end of July furnished an impressive list of new species: 9 Cousinias belonging to various sections and all described by me: *C. araneotexta*, *C. chaetolepis*, *C. ghorana*, *C. infundibularis*, *C. leptomera*, *C. malacophylla*, *C. miserabilis*, *C. parjumanensis*, *C. tricolor*; and 6 Acantholimons jointly described with Schiman-Czeika: *A. ahangarense*, *A. carinatum*, *A. catenatum*, *A. ghoranum*, *A. schizostegium*, and *A. stereophyllum*. Other genera: *Astragalus caroli-henrici* Deml (Sect. *Aegacantha*), *Amygdalus jugata* Browicz, *Lotus corniculatus* L. subsp. *fruticosus* Chrtkova-Žertova, *Ferula xanthocarpa* Chamberlain, *F. ghorana* Rech. f., *F. trachelocarpa* Rech. f., *Dionysia balsamica* Wendelbo & Rech. f., *D. saponacea* Wendelbo & Rech. f., *Eremostachys ghorana* Rech. f., *Scutellaria ghorana* Hedge, *Salvia rechingeri* Hedge, *Echinops ghoranus* Rech. f., *E. hololeucus* Rech. f., *Scariola leucoclada* (Rech. f. & Tuisl) Rech. f. and a new monotypic umbelliferous genus *Mastigosciadium* Rech. f. based on *M. hysteranthum* Rech. f. & Kuber.

On August 2nd we left Parjuman and returned by the same difficult track to the spot where we had left the 'Central Road' and went to Qala Sharak. Here I said goodbye to my companion Knut Lindberg. He went southwest to Shindand on the Herat-Kandahar road by way of Tulak and Farsi. I continued west across the mountain range to Chisht. For many kilometres a narrow gorge served as a track. Then, above Chisht, the Hari Rud valley was reached. The wooden bridge across the Hari Rud was in an unconventional state. It consisted of strong longitudinal beams with thinner loose crossbeams but the latter were laid for only half the length of the bridge. Any vehicle that wanted to cross the river had to proceed to the middle of the bridge and had to wait there until a few men took up the loose beams from behind and placed them in front of the vehicle. I preferred to walk. The missing crossbeams had evidently been stolen.

From Chisht onwards a proper road to Herat exists. It follows the gradually broadening Hari Rud valley. The influence of the summer north 'Bad-i-sad-o-bist-roz', which literally translated means 'wind of the 120 days' increases. The air is dim from dust. In the hot valley, vegetation had dried up completely. On a sheltered north facing rock I discovered an intriguing green spot. I crossed the river, half-wading, half-swimming, and discovered the second locality of *Dionysia balsamea* Wendelbo & Rech. f. Petrol was getting scarce. The driver let the shaky car run downhill in neutral at full speed. We passed the night at Obeh and bribed three lorry drivers to sell us two gallons of petrol each at fancy prices. We went up to Chashmeh Obeh on the south slope of the Paropamisus. Here I discovered *Calyptrosciadium polycladum* Rech. f. & Kuber, a new monotypic umbelliferous genus forming large tufts with hundreds of fine, repeatedly branching stems standing out like yellow veils against the sky. In the little thermal spring of Chashmeh Obeh I had my first hot bath since I left Vienna on May 20th. I reached Herat the same evening. Our East-West crossing of the Hindu Kush came to its end. It took six more days to return to Kabul via Farah and Kandahar as the vehicle had to be repaired time and again. In 1962 the new main road was still under construction and the old track was evidently not being repaired any more.

On August 20th I left Kabul by car via Sarobi and Jalalabad to Torkham. Because of political tension there was no normal traffic across the border to Pakistan. I had to walk across the no-man's-land at the frontier. The British Consul at Kabul had kindly agreed to take the wooden boxes containing my herbarium specimens to the High Commissioner at Peshawar. On the Pakistan side no normal vehicle was available to take me on to Landi Kotal on top of the Khyber Pass from where a bus service to Peshawar existed. After many hours of discussing and waiting a wreck of a jeep appeared. The radiator was out of action and was replaced by a small half-blind boy sipping water from a tin and spraying it by mouth on the engine thus replacing the cooling system as we wound up to the Khyber Pass. Repeatedly repairing the jeep en route took hours. I reached Peshawar finally in the evening. A night at the Dean's Hotel with luxurious meals and drinks and an air-conditioned room made up for the hardships and the heat of an eventful day.

From August 22nd to 25th I made an excursion by car via Malakand, Saidu Sharif and Madyan to Kalam in Swat. Forests up to about 3500 m, glaciers, the continual gurgle of the rivers and a dense cover of lush green ground vegetation, overtopped by peaks of up to 7000 m high, formed an impressive contrast to several months of field work in arid Afghanistan. This was my first acquaintance with the Sino-Himalayan Flora. When putting my plants into the press in front of one of the small resthouses I realized similar activities were going on in a neighbouring hut; the fellow botanist turned out to be E. Nasir from Rawalpindi. Nasir took me to Rawalpindi the next day and I was invited to spend the night in R.R. Stewart's room as he had left for America. I returned by train to Peshawar. I managed to collect all my belongings on 26th and to get them aboard the 'Khyber Mail' to reach Karachi in the early morning of 28th. I left Karachi August 29th by the 'Circassia' (Anchor Line) and reached Suez via Aden and Port Sudan on September 7th. I spent a few days in Egypt, left Port Said on 11th and reached Vienna via Alexandria, Piraeus, Naples and Genoa on September 22nd.

1965

EXSICCATAE NOS. 27000–32999

At the end of 1963 the first three parts of *Flora Iranica* were published. In preparing the various treatments it appeared that there was a lack of specimens from the east (more especially the southeast) of the *Flora Iranica* area. Connections had been established with Dr S.M.A. Kazmi, Peshawar, who had recently graduated from Munich. I invited him to Vienna and the framework of a project covering most of Pakistan west of the Indus was worked out. A Landrover was acquired by concentrated effort on the part of the Austrian Ministry of Education, the Verein der Freunde des Naturhistorischen Museums and the Hammer-Purgstall Gesellschaft. A fixed sum per km, petrol, maintenance and a set of new tyres would have to be paid by the user. The plan of the expedition comprised 5 months including hundreds of miles of cross-country driving. Fortunately, Jennifer Lamond of Edinburgh was prepared to join us and share the driving with my wife, Wilhelmina.

We set out from Vienna on March 7th 1965 via Graz — Zagreb — Belgrade — Istanbul — Ankara — Iskenderun — Latakia — Tripolis — Beirut — Damascus — Deraa — Rutba — Baghdad. By this detour we avoided the high passes of eastern Turkey which, according to our information, were often blocked by snow in March. On March 21st we reached the *Flora Iranica* territory at Qasr Shirin. The Persian frontier officer kindly invited us to his house to take part in his Nowruz (Persian New Year) party. The remarks scribbled by him in our passports made his colleagues smile when we left the country on the Pakistan border a few days later! We reached Tehran on the evening of March 22nd. On the 24th we took the main road via Qom — Esfahan — Yazd and Kerman. Due to the New Year festivities it was impossible to get normal accommodation. We camped on the south border of the Lut desert east of Bam on the night of 26th and 27th. Jennifer managed to create a homely atmosphere by reading to us 'Winnie the Pooh'. Next morning I went botanizing in a humid depression sampling several different *Tamarix*. One of them, *T. kermanensis* B.R. Baum, turned out to be the first of many new species to be discovered during this expedition. Few spring plants flowered along the road by Nosratabad and Zahedan to the southeast boundary of Iran, which was reached on 28th near Nok Kundi. From Karachi a few small excursions were undertaken, partly with Dr S.I. Ali and his collaborators. We were kindly allowed the use of the University guest-house as our headquarters. In return I was asked to improvise a lecture on the phytogeography of the *Flora Iranica* area for the students – not an easy task without any pictures or maps and with hardly any knowledge of Latin plant names on the side of the audience. The memorable fact about this lecture was that before it began a girl student recited with a singing voice several suras from the Koran.

The following weeks from 8th to 23rd April were devoted to the exploration of Makran. Dr Kazmi, whom we met at Karachi, proved to be a kind and helpful companion. He acted as an interpreter and his dealing with the indigenous population was very valuable. Travelling in Makran was hampered by lack of roads, water and petrol. Many times our vehicle was stuck in mud or sand or in water courses swollen by sudden showers.

Only scattered bibliographical references from Makran have been published so far and herbarium material in local as well as in European herbaria is extremely scarce. The mountain ranges bordering the Iranian Highlands in the east fan out gradually south of Quetta towards the Arabian Sea. The main features of the barren landscape of Makran – the land in southwest Pakistan west of the Indus – are ranges of low hills composed of soft, geologically young materials cut by numerous valleys broadening towards the coast and weathering sometimes into grotesque forms. Vegetation is rather sparse, composed of 'subtropical', not 'Iranian' elements, partly Saharo-Sindian, partly 'Nubo-Arabian' in connection. Few novelties were discovered: *Saponaria macranica* Rech. f., *Lotus compactus* Chrtkova-Žertova, *Heliotropium remotiflorum* Rech. f. & H. Riedl, *Lycium macranicum* Schönb.-Tem., *Argyrolobium pulvinatum* Rech. f., *Psammogeton lamondiae* (Rech. f.) Townsend and *Medicago macranica* Heyn growing on the flat-topped peninsula Ras Nuh near Gwadar are the most remarkable. *Pseudolotus* Rech. f. (*P. makranicus* (Rech. f. & Esfand.) Rech. f), a monotypic genus described from Persian Makran, was also found in various localities in Pakistani Makran.

Then followed a survey of the interior hilly section of southern Baluchistan. Quetta and Karachi had to be revisited in order to replenish equipment, food and petrol supplies and to repair the Landrover. Based mainly in Quetta, a couple of excursions were undertaken between May 6th and 20th including Loralai, Ziarat, Sanjawi, Harnai, Gumbaz, Shinghar and Fort Monro in the Sulaiman Range. Travelling in central and northern parts of west Pakistan is much easier than in the south as far as water and petrol supply is concerned. But owing to political tensions between Pakistan and India, the authorities became more and more restrictive in granting travel permits for tribal territories. An attempt to penetrate into South Waziristan failed. The military authorities had declared the road as 'operational' because of tensions with the nomads and we had to return to Quetta.

The highest mountains of Baluchistan are situated east of Quetta. They reach more than 3300 m, consist mainly of limestone and are largely covered by *Juniperus excelsa* M.-B. Pending a detailed phytogeographical analysis of Baluchistan it is evident that the 'Iranian' element dominates at higher elevations: spiny Astragali, *Cousinia*, *Eremurus*, *Artemisia*; *Onobrychis cornuta* (L.) Desv. subsp. *cornuta* is replaced in Baluchistan as well as in eastern Afghanistan by the locally dominant subsp. *leptacantha* Rech. f. Among the new species discovered are: *Saponaria subrosularis* Rech. f., *Cerasus rechingeri* Browicz, *Heliotropium rechingeri* H. Riedl, *Phagnalon pycnophyllon* Rech. f., *Echinops sulaimani* Rech. f., *Cousinia quettensis* Rech. f., and *Scorzonera koelpinioides* Rech. f. Two new endemic genera have also been described from here: *Sulaimanya otostegioides* (Prain) Hedge & Rech. f. based on *Moluccella otostegioides* Prain and *Hyalochaete modesta* (Boiss.) Dittrich & Rech. f. based on *Jurinea modesta* Boiss. The limestone gorge above Harnai furnished *Sida quettensis* I. Riedl, *Tephrosia rechingeri* Ali and *Heliotropium ulophyllum* Rech. f. & H. Riedl.

The authorities permitted us to cross the northern section of the Sulaiman Range from Fort Sandeman to Dera Ismail Khan with an escort of 25 militia. There we were stuck for 3 days because of the road to the north towards Bannu and Kohat being flooded by tributaries of the Indus. Botanizing along the main road we reached Peshawar on May 24th. Thanks to Dr Kazmi's hospitality we were able to establish our headquarters at his home.

Some excursions from Peshawar in various directions proved rewarding. A major one to Kurram was a failure as the travel permits granted by the civil authorities in Peshawar were not acknowledged by the military ones at Parachinar. We were confined to the resthouse for two days and on the third day ordered to return to Peshawar. No permits were granted for visiting the Kabul Gorge and Chitral. The only promising major excursion possible was to revisit Swat at an earlier season (June 1st–6th); a side-trip to Shangla was also very rewarding.

Swat, though botanically rather well known, yielded three new species: *Astragalus parkeri* Deml (Sect. *Aegacantha*), *Erigeron cedretorum* Rech. f. and *Myosotis albicans* H. Riedl. The vegetation in lower and middle Swat was well developed. Higher up and around Kalam not many plants were seen in flower, far fewer than at the end of August 1962. The sight of the forests and the torrents with the high glacier peaks as background was as impressive as ever.

The attitude of Pakistani authorities against foreigners trying to visit tribal territories deteriorated. Even the intervention of Mr Ali Akhtar, Dr Kazmi's father, at the central authorities at Lahore had no result. So I decided to leave Pakistan and spend the remainder of my leave of absence in Afghanistan.

On June 17th we left Peshawar, said goodbye to Dr Kazmi, crossed the Khyber Pass into Afghanistan and reached Kabul the same evening. We made several small excursions from Kabul and tried in vain to rediscover the enigmatic caryophyllaceous genus *Kabulia*, the type of which, *K. akhtarii* Bor & Fisch., was named after Dr Kazmi's father. As usual many days were lost by the formalities necessary to obtain residence- and travel-permits for Afghanistan. My main task was to link our Afghan activities with those in Pakistan as closely as possible. We visited the Altimur (Tera Kotal) and the Sata Kandao-pass, Shinkai, and the areas near Gardez and Khost. On the Peiwar Kotal we stood exactly on the Pakistan border – without any further formalities. Another trip took us back to Jalalabad and into southern Nuristan into the lower parts of the Pech and Kunar valleys.

Since my last visit to Afghanistan the road across the Salang Pass had been opened to the public. It is the easiest way to reach the 'nival' region of the Hindu Kush. Phytogeographically and geologically it forms an impressive transect. We were rewarded by the discovery of *Corydalis griffithii* Boiss. subsp. *salangensis* Wendelbo, *Cousinia immitans* Rech. f., *C. porphyrostephana* Rech. f., *Erysimum salangense* Polatschek & Rech. f., *Mattiastrum formosum* H. Riedl and *Reseda aucheri* Boiss. subsp. *afghanica* Rech. f. The Salang Pass has been visited by few botanists before and by many after us and it soon became a favourite haunt of the many Europeans stationed at Kabul at that time. They crossed the pass to reach the rich trout-fishing grounds in the north. Between excursions we even took part in the vivid international social scene and musical activities in Kabul.

On July 21st all formalities for the homeward journey were completed and we reached Kandahar the same day. Via Herat, Mashhad and Bojnurd we arrived at Tehran. A few days there with official visits to the Ministry of Agriculture, and the Austrian Embassy, and a luxurious reception at the Tehran 'Hilton' formed impressive contrasts to our prolonged spartan field-life. As guests of the Austrian cultural institute we really enjoyed our days in Tehran. On July 30th we continued our homeward journey via Tabriz — Maku — Bazorgan — Erzurum — Aşkale — Kop Dagh — Bayburt — Zigana Pass — Trabzon — Samsun to Istanbul. Here on August 10th we said goodbye to Jennifer Lamond, the friendliest and most reliable companion one can imagine. She had shared all the hardships and adventures with us for five months. Jennifer travelled home by rail from Istanbul. We drove home the short way via Sofia and Belgrade and reached Vienna on August 16th. This was my first expedition undertaken all the way from Vienna by Landrover. On the whole it had served its purpose and many areas would not have been accessible to us without it. But the vehicle was not nearly as robust as we had hoped. Many irreplaceable days had been lost through repairs.

The botanical result was very satisfactory. Ample herbarium material from the least known and in part botanically unexplored section of the *Flora Iranica* area was brought home, and I had gained insight into the intricacies of the phytogeographical relations of the southeastern corner of this region. It is important when travelling in these countries to have an emergency programme on hand or to develop one at short notice in case the original plan cannot be carried out.

1967
EXSICCATAE NOS. 33000–37795

In view of the important results of the 1965 expedition and the Landrover being still available I planned a new journey. The main purpose was to explore the spring flora of Afghanistan and to visit those regions from which no herbarium specimens had been available so far. This time my wife Wilhelmina alone was the driver as she had gathered enough experience with the tricky Landrover on the 1965 trip. We also took our two children, Lilian (15 years) and Karl Bjorn ($4\frac{1}{2}$ years) along. They lived with the family of a teacher at the Austrian school of music in Kabul. The elder sister took care of her little brother during the periods when their parents set out on expeditions from Kabul. My daughter still remembers changing and drying blotters for piles and piles of plants during our absence!

We left Vienna on April 4th 1967 and took the direct road (already well known to us) to Graz where we delivered the manuscript of Lieferung/fascicle 57 (Cruciferae) of *Flora Iranica* to the publisher. Then on via Belgrade — Sofia — Istanbul — Ankara — Erzurum — Tabriz to Tehran where we arrived on April 14th. On April 9th I remembered the 100th anniversary of my father's birth, and on April 12th at Erzurum we celebrated our daughter's 15th birthday. So far everything had gone well. In many places crocuses and other early 'bulbs' started to flower – a new experience for me as I had spent my only oriental spring (1957) in the deserts of Iraq. In eastern Turkey the weather gradually got worse. The highest pass, Tahir (2500 m), was kept open by the Turkish roadmen but the snow walls on both sides of the road reached up to 3 metres. It soon started to snow at high altitudes and to rain at lower elevations. The road from the Persian border to Tabriz and Mianeh was still under construction. Lorries and buses collided or went off the slippery tracks; rivers swollen by melting snow had to be forded. Most obstacles could be overcome by our Landrover but once we got stuck by an invisible rock in the middle of a deep opaque river. The ice cold water started to run into the vehicle which had to be lifted by a dozen men over the obstacle. We reached Tehran on April 14th and continued on 17th via the Haraz Valley to Babol Sar and along the Caspian by Sari — Behshahr — Gorgan to Shahpasand. Early spring flowers of the forest clearings were in full bloom. Further east beyond the Golestan forest, which represents the easternmost extension of the humid Caspian vegetation, patches of various tulips and of the impressive pale yellow-flowered *Fritillaria raddeana* Regel were seen. More than in other cases the outward journey added considerably to the botanical results. We continued eastwards via Mashhad to Herat and along the main road by Farah Rud — Dilaram — Girishk to Kandahar. Spring annuals on the borders of Dasht-i Margo and Registan were at their best. Occasional showers turned into continuous rain beyond Ghazni and during the last hours before reaching Kabul on April 25th the windscreen wipers had to be moved by hand. Rain continued during the first days in Kabul where we completed the usual formalities, organized our headquarters and obtained the necessary permits.

During the last few years, a team of German visiting professors had been established at Kabul University. The botanists of this 'Deutschen Partnerschaftsteams' of the Faculty of Science, University of Kabul were in turn: O. Anders, S.-W. Breckle, H. Freitag and D. Podlech. This resulted in an intensification of botanical research in Afghanistan not previously anticipated. In one way these botanists were in a more favourable position than travelling botanists from abroad as they could study plants all the year round. In another respect their position was difficult as the main vacation fell in the winter and these weeks could hardly be used for botanical fieldwork. In many cases their excursions had to be carried out during long weekends or short vacations. They have all contributed considerably to *Flora Iranica* either through their collections or as collaborators. Prof. Podlech and Prof. Freitag assisted in both ways.

When I arrived at Kabul I took up contact with Prof. Freitag, who only recently had started his activities as a member of the above-mentioned team. He was keen to familiarize himself with the flora of Afghanistan and we undertook some excursions together.

Page proofs of *Flora Iranica* Lieferung 48 (Boraginaceae) had been waiting for us in Kabul. They had to be read (within a week!) before setting out on the first major excursion from May 3rd to 11th across the Salang Pass (still deep in snow) and down to the Khinjan valley to Pul-i Khumri. During the following days the spring flora of the northern plains and foothills from Pul-i Khumri — Kunduz — Tashkurghan — Mazar-i Sharif to Haibak was explored. The result was more than satisfactory.

Here follows a list of the most characteristic, mostly Turanian spring annuals growing on loess: *Acanthocephalus amplexifolius* Kar. & Kir., *A. benthamianus* Regel & Schmalh., *Cousiniopsis atractyloides* (Winkl.) Nevski, *Crepis pulchra* L. subsp. *turkestanica* Babc., *Cymbolaena griffithii* (I. Gray) Wagenitz, *Epilasia mirabilis* Lipsch., *Koelpinia macrantha* C. Winkl., *K. tenuissima* Pavl. & Lipsch., *Microcephala lamellata* (Bunge) Pobed., *Drepanocaryum sewertzowii* (Bunge) Fedtsch., *Hypogomphia turkestana* Bunge, *Lallemantia baldshuanica* Gontsch., *Medicago lanigera* C. Winkl. ex Fedtsch., *Trigonella grandiflora* Bunge, *T. monantha* C.A. Mey. subsp. *geminiflora* (Bunge) Rech. f., *Consolida leptocarpa* Nevski, *Aphanopleura leptoclada* (Aitch. & Hemsl.) Lipsky, *Valerianella cymbicarpa* C.A. Mey., *V. oxyrrhyncha* C.A. Mey., *V. szowitsiana* Fisch. & C.A. Mey., *V. turkestanica* Regel & Schmalh.

Between May 12th and 15th we collected the spring flora of the Tang-i Gharu and the Kabul Valley all the way to Sarobi and beyond.

From May 17th–28th we drove to Kandahar and on to Girishk. This time we joined forces, with Prof. Freitag driving his own vehicle. The idea was to penetrate along the lower Hilmand southwest into the Registan desert and to reach the Chagai hills, an area so far botanically unknown. Two vehicles at least are needed for desert expeditions for safety's sake as well as in order to carry sufficient supplies of water and petrol. Air temperature reached 42°C at noon. The route was Lashkargah (Bhost) — Darweshan — Safar — Bhagat. We found *Anthemis kandaharica* Iranshahr, *Cithareloma registanicum* Rech. f., *Launaea platyphylla* Rech. f. and *Psammogeton registanicus* Rech. f. It was impossible to cross the deep sand dunes south of Bhagat, and we were forced to return to Kandahar. From there we tried to drive northwest along the Arghandab reservoir to Tirin, Orozgan and Malestan to Sang-i Masha. On a high pass our vehicle was stuck in the mud and Freitag's vehicle a few kilometres further on was unable to ford the river. Our vehicle was rescued by Freitag's driver and with the help of nomads using stones and a stolen telegraph pole as a lever. We returned to Kandahar; Freitag continued immediately to Kabul. We stayed in Kandahar and made a short but very successful reconnaissance trip directly into the Registan desert south of Kandahar. Here *Astragalus registanicus* Rech. f. (Sect. *Eremophysa*), *Heliotropium leucocladum* H. Riedl, *H. arenicolum* Rech. f. & H. Riedl and the new umbelliferous genus *Registaniella hapaxlegomena* Rech. f. were found. On May 28th we returned to Kabul. Though on both excursions from Kandahar the geographical goal (the Chagai hills) could not be reached, both trips yielded good botanical results. Before our visit hardly anything had been known of the sand desert flora of south Afghanistan. In narrow rock crevices between Tirin and Orozgan, the peculiar new monotypic fumariaceous genus *Cryptocapnos chasmophyticus* Rech. f. was discovered.

From June 7th–10th we made a trip in the province of Paktia. The route was as follows: Kabul — Logar — Gardez — Khost — Jakubi — Narai Kotal — Kurram Valley between Chakmani and Ahmad Khel — Khost — Matun Valley — Domandi — Khost — Urgun — Gardez — Kabul. This trip was arranged to learn more about the boundaries of the Sino-Himalayan type of forest influenced by the monsoon and to study the botanically little known area between Khost, Urgun and Gardez. In this case we were forced by the Kabul authorities to take along a young man originating from the Khost area. I was expected to teach him something of the flora and of the technique of making herbarium specimens. He was expected to act as an interpreter, a local guide and an assistant. He turned out to be not interested in any way. Occupying the back seat of the vehicle he used to stretch out his naked feet either between us (sitting in front) or out of the window. On a very steep slope when the Landrover was in danger of rolling back we asked him to put stones against the wheel. He politely asked blankly, 'Yes, should I put the stone in front or behind the wheel?' He was proud of knowing some German. Actually it was a single word of very bad language, 'Arschlochmensch'. He used it on every occasion. Apparently he had learned it from one of the German foresters working at Kotkai. After a few days we made him visit his relatives and picked him up as many days later as possible.

The sight of the destruction of the beautiful Deodar forests was distressing. The trees were cut and the wood carved away so that quadrangular 2.5 metre long blocks remained. These were loaded on camels, two by two, carried down to the valleys and exported to Pakistan. Incredible quantities of wood were thus wasted.

One of the most interesting localities visited on this excursion was the sandy gypsiferous area north of Urgun where we found *Saponaria stenopetala* Rech. f. Unfortunately we had both acquired heavy dysentery and several days were lost before we could continue our excursions.

Our next trip was in central Afghanistan, northwest and south of Panjao, in the Deh Kundi area from June 20th–July 4th. We travelled by the following route: Kabul — Maidan — Siah Sang — Behzud — Panjao — Shahtu — Shahtu Kotal — Kotal-e Naiak — Sauzak Sumaj — Serdak — Sad Bark Pass — Panjao — Waras Pass — Khonak Pass — Shahrestan area — Panjao — Behzud — Kabul.

This was the only time that we decided to engage an experienced native driver as we planned to penetrate into roadless areas which to my knowledge had not been visited by botanists before. This decision proved to be wrong as it resulted in damage to our vehicle and considerable loss of time. Up on the well-known road to Panjao and north across the Shahtu Kotal everything went well. Northwest from Naiak we tried to penetrate into the Kuh-i Hisar, an area comparatively rich in water. Returning to Panjao at the north foot of the Shahtu Kotal our driver broke the long axle of the Landrover by incautious gear-shifting. We had to stay for five hot days and five cold nights in our little tent pitched on a flat lentil-field at 3000 m. Our driver was sent to Kabul at the first opportunity, a lorry that passed the next day. I had ample time for botanizing in the surroundings. One morning a caravan passed. The tribal head – a one-legged invalid mounted on a beautiful white horse – stopped and invited us for tea at his camp on top of the pass. The following night peculiar noises woke us. A heavily armed man opened our flimsy tent and said '*pul, pul*!', which means 'money, money'. We tried to keep calm and smiling. Wilhelmina opened a conversation, explaining that we were penniless as our vehicle had broken down and our driver had taken all our money with him to buy new spare parts in Kabul. Moreover she told him that we were on friendly terms with his one-legged boss who was expecting us in his camp next day. This made the intruder quickly disappear. We took the last sips from our brandy-bottle and tried to sleep again. Next night there was a sudden noise of gurgling water. Apparently our tent was situated near the centre of the irrigation system for the surrounding fields of the village of Shahtu and was blocking the normal flow of water. With the help of torches and ice-axes we diverted the imminent rivulet. Next morning the problem was solved by a friendly discussion. On the sixth morning after his departure for Kabul our driver reappeared with a new long axle. He installed it within a few hours, lying on his back on the rocky soaked soil where the accident had happened. We returned over the Shahtu Kotal to Panjao, not before drinking tea with the tribal chief and with a young French couple collecting butterflies. Their vehicle had been pushed up the pass manually by a group of co-operative nomads.

We went back to Panjao and, on a miserable very seldom used track, turned south crossing the Waras Pass (3130 m) and the Kunuk Pass (3300 m) to Shahrestan, another botanically unexplored area. *Delphinium bulbilliferum* Rech. f., the only *Delphinium* developing bulbils in the leaf axils, was discovered. In rock crevices I found *Psychrogeton shahrestanicus* Rech. f.

Shahrestan, at an elevation of 2200 m and situated in a rather arid valley on tributaries of the Hilmand river open to the south, is surrounded by a landscape dominated by pale granite-like rocks weathering into bizarre blocks and fine sand. This area furnished several interesting new species such as *Scrophularia horizontalis* Rech. f., *Jurinea shahrestanica* Rech. f., *Cousinia insignis* Rech. f. and *Astragalus shahrestanicus* Podlech & Deml. (Sect. *Aegacantha*). To the west of Shahrestan towards Deh Kundi shale slopes were covered by thousands of the new species *Gypsophila myriantha* Rech. f., a tall plant with spinescent basal leaves and ample panicles of hair-like branches with minute flowers waving in the wind like a wheat-field. Other new species were *Sclerorhachis polysphaera* Rech. f. and *Eryngium rechingeri* Tamamschian & M. Pimenov.

We had planned to go south from Shahrestan via Gizao and to link up with the expedition to Orozgan earlier in the year. This proved impossible as the track had been washed away in the spring. So we had to return to Kabul. The remainder of the time was used to revisit various promising areas seen on earlier occasions such as the Ghorband-Shibar Pass, Ajar Valley — Bamian — Doab — Doshi, the Dasht-i Nawar area, the Salang Pass region climbing up from the north to the lake (3700 m) and by an excursion across the Kabul River gorge to Jalalabad and to Peshawar. Revisiting the Dasht-i Nawar was especially rewarding; a very limited sandy area at about 3000 m altitude has no fewer than four endemics: *Astragalus antheliophorus* Deml (Sect. *Aegacantha*), *Fibigia compacta* Rech. f., *Cousinia ammophila* Rech. f. and *Piptatherum rechingeri* Freitag. On different habitats two perennial violet-flowered Trigonellas are endemic: *T. ionantha* Rech. f. and *T. pycnotricha* Rech. f.; also *Mattiastrum polyanthum* Rech. f. & Riedl and *M. cynoglossoides* Rech. f. & Riedl, all collected only once by me. We also found *Nepeta nawarica* Rech. f., *Acantholimon acanthobryum* Rech. f. & Schiman-Czeika and *A. nawaricum* Rech. f. & Schiman-Czeika. We failed to find again the peculiar endemic cruciferous genus *Cyphocardamum aretoides* Hedge, only once collected in this area in 1965 by the Austrian entomologist Kasy.

The result of the 1967 Afghanistan expedition was three new genera: *Kandaharia* Alava, *Registaniella* Rech. f. (both Umbelliferae) and *Cryptocapnos* Rech. f. (Fumariaceae) and approximately 55 new species including five Cousinias.

We left Kabul on August 6th for good and went on the usual route, Mashhad — Tehran — Erzurum. Here we took the northern road through Turkey via the Kop Pass — Bayburt — Gümüşhane — Trabzon — Giresun — Sivas — Ankara and on to Istanbul — Alexandroupolis — Thessaloniki. We spent a couple of days bathing and relaxing in the Aegean Sea and reached Vienna on September 4th.

1971

EXSICCATAE NOS. 39101–44439

After three expeditions in Afghanistan and Pakistan I felt it necessary to fill some more gaps in my knowledge, this time of the western regions of the *Flora Iranica* area. So the seventh journey was devoted to the west, more especially Gilan, Azerbaijan and Kurdistan. When working through several plant families it became evident that the Armenian-Azerbaijanian mountain knot, where the ranges bordering the Anatolian and Iranian highlands converge, is an important centre of endemism. So we, Wilhelmina and I, this time again together with our Scottish friend and experienced companion Jennifer Lamond, set out by Landrover – for the last time – from Vienna on April 27th and reached Tehran by the short route through Yugoslavia, Bulgaria and Turkey. Following the kind invitation of Dr Esfandiar Esfandiari, close collaboration with the Evin Herbarium (Plant Pest and Diseases Research Institute) was established under his direction. From May 11th to June 22nd Dr M. Iranshahr in an official Landrover accompanied us, succeeded by Dr F. Terme from June 29th to August 1st. Three independently numbered sets of specimens, mainly identical, were collected. Mine went to Vienna (W), Jennifer's to Edinburgh (E) and Iranshahr's and Terme's to Tehran-Evin (IRAN). When on occasional separate excursions the main finds were shared by the three participants.

From May 11th to 25th we collected mainly in the Gilan lowland and on the humid slopes from Astara to the Heyran Pass. We then travelled from Ardabil along the Talysh border across the Dasht-e Moghan to Ahar and Tabriz. Comparatively few plants had so far been reported from the Dasht-e Moghan, a low arid depression bordered in the north by the Araxes (Aras) river. Our visit could not add very much – with the exception of *Anthemis moghanica* Iranshahr – as an unusual drought had prevailed during winter and spring.

From May 25th to the end of June we hired rooms in a small house in Tabriz as our headquarters. Long and short excursions were undertaken in various directions from Tabriz. I could not take part in two of them because of illness. Jennifer Lamond and Dr Iranshahr visited the Kuh Anguran between Manjil and Zanjan and the Kaflan Kuh near Mianeh from May 30th until June 2nd. Together with F. Terme, Jennifer visited Kuhha-ye Savalan and the Qara Dagh mountains. We all explored together the Qotur Valley up to the Turkish frontier and the area around Zonus east of the road from Tabriz to Jolfa.

In the beginning of June we went to the western side of Lake Urmia (Rezaiyeh) to the town of the same name and to the Turkish frontier near Sero and on into Kurdistan via Balanesh, Rashkan and Heydarabad to Mahabad. From there we returned to Tehran for a couple of days.

On June 29th we went with Terme again to Kurdistan via Qazvin — Takestan — Zanjan to Bijar (Kuh Hamzeh Arab) — Divandarreh — Hoseynabad — Sanandaj (excursions to the Ariz and Salavatabad passes) — Marivan — Chehel Chashmeh mountains — Saqqez and on to Tabriz.

From July 14th to 16th I returned with Wilhelmina from Tabriz via Mianeh to the Gilan forests between Khalkhal (Herowabad) and Assalem.

On July 26th and 27th Terme visited the Kiyamaki Dagh area south of Jolfa and on July 29th the Kuh-e Mishab Dagh near Jam.

On August 1st my three companions climbed the Ghogeh Dagh near Bazorgan on the Turkish frontier. This day added not only a few new records to the Iranian as well as to the Turkish flora but also furnished a few new species, e.g. *Gypsophila bazorganica* Rech. f. and *G. wilhelminae* Rech. f.

The next day we left Iran and went on the straight road across Turkey to

Istanbul and along the northern Aegean coast to Thessaloniki. Here we said goodbye to Jennifer, who took the train homeward. I spent a few days with Wilhelmina in Greece before returning home.

The results of the 1971 trip to Persia were satisfactory in spite of my temporary illness, thanks to the friendly co-operation of all participants. Considerable herbarium material of many polymorphic genera was collected, e.g. of *Cousinia* Sect. *Cynaroides*, *Astragalus*, *Acantholimon*, *Crepis* and others. I got better ideas about their intricate and closely knit patterns of distribution (see, e.g., K.H. Rechinger, 1986, with distribution maps of *Cousinia*). Also from a more general phytogeographical point of view, the visits to the boundaries of the Hyrcanian, more humid type of forests and the Zagrosian (Kurdish-Luristanian) type so rich in endemics, were extremely interesting. Here follows a list of the new taxa found in 1971: *Acantholimon laxiusculum* Rech. f. & Schiman-Czeika, *A. wilhelminae* Rech. f. & Schiman-Czeika, *Alcea crassicaulis* I. Riedl, *A. wilhelminae* I. Riedl, *A. xanthochlora* I. Riedl, *Alchemilla subpotentilla* Fröhner, *Anthemis atropatana* Iranshahr, *A. moghanica* Iranshahr, *Campanula lamondiae* Rech. f., *Cousinia bijarensis* Rech. f., *C. millefontana* Rech. f., *C. qaradaghensis* Rech. f., *C. stroterolepis* Rech. f., *C. wilhelminae* Rech. f., *Crepis sancta* L. subsp. *azerbaijanica* Rech. f., *Dianthus longivaginatus* Rech. f., *Eremostachys azerbaijanica* Rech. f., *Gypsophila bazorganica* Rech. f., *G. wilhelminae* Rech. f., *Heracleum rechingeri* Mandenova, *Lactuca azerbaijanica* Rech. f., *Myosotis anomala* Rech. f., *Nepeta scrophularioides* Rech. f., *Scrophularia azerbaijanica* Grau, *S. rechingeri* Grau, *Scutellaria bornmuelleri* Hausskn. & Bornm. subsp. *mianensis* Rech. f., *S. theobromina* Rech. f., *Vicia rechingeri* Chrtkova-Žertova.

1974
EXSICCATAE NOS. 46000–50093

During the ten years from 1964–1974, main road conditions in Turkey and Iran had gradually improved. As the Landrover was not available any more, we decided to go east by our private car, a Volvo-Combi, on May 9th. (This was also true for the 1975 and 1977 journeys to Persia.) We used the Volvo also within Persia as far as possible. For some more hazardous enterprises off the main roads we depended on the help of the 'Evin' and 'Ariamehr' institutions. Thanks to the goodwill of Mr Eskandar Firouz and Dr Esfandiar Esfandiari and the co-operative spirit of the various drivers everything went well. In some expeditions during my 1974, 1975 and 1977 travels my oldest botanical friend (since 1927), Dr Jany Renz from Basel, Switzerland, took part with his private Range Rover. In many cases this enabled us to establish base camps at high elevations from where he used to climb the highest peaks and to contribute specimens from localities inaccessible to other participants.

We went to Tehran again by the short way through Yugoslavia, Bulgaria and Turkey and reached Tehran on May 16th. Under the administration of the Department of Environment a second botanical centre had been created independent from the Evin Herbarium: the Ariamehr Botanical Garden with a herbarium situated about 30 km west of Tehran towards Karaj. Per Wendelbo, specializing on Liliiflorae and Primulaceae, had been appointed as a botanical adviser. The administrative director was Mr Chaichi, a businessman without any understanding for plants or for nature at all. This institution was then under the Department of Environment and Nature Conservation. (It is now called the Botanical Institute of Iran and is under the adminstration of the Forest Department.)

The head of the Department of Environment and Nature Conservation was Mr Eskandar Firouz, the creator and organiser of 54 National Parks and Protected Areas in Iran (E. Firouz, 1974). Even though primarily created for the protection of animals, these areas are of vital botanical interest. Most of them are well placed from a phytogeographical point of view, the foremost being the former Mohammad Reza Shah National Park (now called 'Golestan') east of Gorgan situated at the eastern boundary of the Caspian forests. Mr Firouz showed an active interest in my *Flora Iranica* activities. I obtained permission to collect herbarium specimens in the protected areas. In return a series of publications giving inventories of plants was planned. For the two largest areas, the Kavir Protected Region and the Touran Protected Region, these lists have been published (K.H. Rechinger with contributions by Per Wendelbo, 1976 and 1977.) The material for a third one on the Mohammad Reza Shah National Park has been collected by Per Wendelbo and his Iranian collaborators and by me, but the untimely death of Per Wendelbo and the revolution in Iran have prevented organizing and editing of the material and notes. By visiting several of these protected areas it became evident to me what an amazing power of recovery there is in plant life in arid regions. In some places unprotected sites completely grazed could be observed side by side with areas fenced in for five to ten years. Prolonged seed dormancy of many xerophytes evidently helps recovery. The only long-term solution (if any) to the problem of overgrazing would be to allow the flora recovery periods of several years and to shift grazing rights from one region to another in good time. Such a procedure would give far better results from an economic viewpoint than the introduction of alien species.

The programmes of my 1974, 1975 and 1977 journeys to Persia have in part been influenced by the exploration of several of these protected areas. Mr Firouz supported my work by temporarily putting at our disposal a four-wheel drive car equipped with driver, and by permitting us to use the local centres of the Environment Organization. His help is here gratefully acknowledged.

The time from May 22nd to 26th was devoted to the Kavir Protected Region situated at the northern border of the Kavir Desert. Our friend, M. Iranshahr, collecting for the Evin Herbarium, accompanied us. One of the most impressive features was the sand dunes in the northeast of the area with saxaul (*Haloxylon aphyllum* (Minkw.) Iljin, *H. persicum* Bunge ex Boiss.). Here *Chamaesphacus ilicifolius* Schrenk, *Chrozophora gracilis* Fisch. & C.A. Mey. and *Salsola aptera* Paulsen, all Turanian elements, were discovered as new to the flora of Persia. A new species, *Heliotropium kavirense* H. Riedl, was also found (Rechinger & Wendelbo, 1976).

We crossed over south of the Daryacheh-ye Namak, the salt lake east of Qom, to Ardestan and went up by road to the Kuh-i Karkas (Kargiz) at 2880 m, then by Moorcheh Khvort to Esfahan, where our headquarters were established at the local branch of the Plant Pests and Diseases Institute. In the Kolah Ghazi National Park 25 km southeast of Esfahan we found *Malabaila isfahanica* Alava. In the Mooteh Protected Region (33°37′N, 50°47′E), *Cousinia mutehensis* Rech. f. and *Ferula kashanica* Chamberlain were collected. *Stachys tomentosa* (Benth.) Rech. f. is now being interpreted as the hybrid *S. inflata* × *lavandulifolia* on the basis of new material. Between Borujen and Shahr Kord, *Haplophyllum rechingeri* C.C. Townsend was found. The Tang-e Sayad Protected Region 32 km east of Shahr Kord furnished *Cousinia tenuiramula* Rech. f. and *Tragopogon bakhtiaricus* Rech. f. At Esfahan, Dr Jany Renz (Basel) and Dr R. Alava (Turku), both active collaborators of *Flora Iranica*, joined our group. On June 5th we set out on a major excursion from Esfahan to the south by Shahreza and Kohruyeh to Semirom. From here we visited the Pashma Kuh, so far not botanically explored. Though heavily grazed, three new species were discovered in rock crevices between 2850 and 3000 m: *Campanula hermannii* Rech. f., *Andrachne merxmuelleri* Rech. f. and *Dionysia iranshahrii* Wendelbo. The Kuh-e Surmandeh, nearly 3800 m high and hitherto also botanically unknown, yielded *Scrophularia flava* Grau forming hybrid populations with *S. glauca* Decne. ex Benth., and also *Dracocephalum surmandinum* Rech. f. and *Silene caroli-henrici* Melzh.

On June 10th we left Isfahan for good and went by Damaneh to Daran where we found *Jurinea prasinophylla* Rech. f. and *Scorzonera helodes* Rech. f., then on to Azna and Dow Rud in Luristan. Since my visit in 1948, Dow Rud had developed into a local traffic and industrial centre dominated by a concrete factory spitting out formidable clouds of dust and smoke day and night. Nevertheless, we pitched camp in the courtyard of the local Environment Office and made excursions to the entrance of the Dez River Gorge and to Khorramabad. We continued across the remnants of the *Quercus brantii* forests towards Sefid Dasht and made major excursions to Aligudarz, the Khali Kuh and the Oshtoran Kuh. On June 18th we returned to Tehran via Arak and Saveh.

From June 19th to 26th we stayed at Tehran, reading *Flora Iranica* proofs, making smaller excursions and working in the Evin and Ariamehr herbaria.

On June 29th we left Tehran and travelled via Takestan — Hamadan — Sanandaj — Divandarreh and Saqqez to Rezaiyeh. From here we tried to cover as much as possible of the area between Lake Urmia and the Turkish frontier by car into the valleys and by foot on the high mountains. The Chalil Kuh (3200 m) was climbed by Jany Renz and Wilhelmina from different sides. *Centaurea ochrocephala* Wagenitz, *C. pseudoscabiosa* Boiss. & Buhse subsp. *armata* Wagenitz, *Scorzonera renzii* Rech. f. and *Scrophularia kurdica* Eig subsp. *glabra* Grau were found here. Interesting remnants of the flora of lower elevations rich in endemics were found near Balanesh and Oshnoviyeh. Hospitality was offered by the Agricultural College of Rezaiyeh. Between excursions I tried to put the local herbarium in order and to advise the assistants on how to make decent herbarium specimens and how to label them. Excursions were undertaken in various directions: to the south in order to link up with former activities in Kurdistan, and northwest again to the Qotur valley where Jany Renz and Wilhelmina climbed the Kuh Kani Ziarat (3200 m). Thunderstorms in summer occur rather regularly in west Azerbaijan and make summer vegetation more lush than in most other parts of Iran. Hitherto very little herbarium material had been available from this part of the *Flora Iranica* area. Novelties found included: *Amygdalus pabotii* Browicz, *Scrophularia kurdica* Eig subsp. *glabra* Grau, *Centaurea ochrocephala* Wagenitz, *Cousinia sardashtensis* Rech. f. and *Picris strigosa* M.-B. subsp. *macrotricha* Lack.

On July 23rd we left Persia at Sero with Jany Renz driving his own vehicle. We passed through Yüksekova — Başkale — Reşadiye — Tatvan (south of Lake Van) — Muş — Bingöl — Elaziğ — Gürün — Kayseri — Ürgüp — Göreme — Şereflikoçhisar — Ankara — Lake Abant — Istanbul — Sofia — Belgrade and reached Vienna on August 1st. Jany Renz left us in Zagreb to drive home via Italy.

1975
EXSICCATAE NOS. 50100–54020

Encouraged by the good results of the previous year's journey and by the good connections established at Tehran, we left Vienna again in our Volvo on April 3rd 1975; we drove the short way across our usual route – Yugoslavia, Bulgaria and Turkey – and reached Tehran on April 11th. This time our headquarters were at Per Wendelbo's flat in Tehran-Yussufabad. In return for the support offered by Mr Firouz, every spare minute between excursions was devoted to naming the recently collected herbarium material at the Ariamehr herbarium. In the meantime Dr Esfandiari had retired from the directorship of the Evin institution and had been provisionally replaced by Dr Sharif. One of my main plans was a trip round the Kavir Desert. A Landrover, with our friend Iranshahr and a driver, was delegated from the Evin side and another car with an experienced driver from Mr Firouz. On Firouz's request the Kavir round trip was combined with an exploration of the flora of the Touran Protected Area (the largest of the protected areas) situated in the north of the Kavir Desert between 34°45′–36°30′N and 55°–57′E.

After a few minor excursions we left Tehran on April 24th. On our way we paid a short visit to the gypsum area rich in local endemics discovered on my first expedition to Persia in 1937. I proposed the protection of this area as well as of the serpentine area of Robat Safid in Khorasan – as far as I know in vain. We reached the Touran Protected Area (called after a village Turan and not after the Turanian desert plains east of the Caspian) on April 25th and spent 6 days in the area. As was the case with the plants of the Kavir Protected Area, the first set of plants is in W, a duplicate set at Ariamehr and a practically identical but separately numbered set collected by Iranshahr is at Evin. (For details see K.H. Rechinger, 1977). No new species were discovered. The most interesting finds were those new for Persia: *Heliotropium acutiflorum* Kar. & Kir., *Horaninovia ulicina* Fisch. & C.A. Mey. (discovered by Per Wendelbo), *Cithareloma lehmannii* Bunge, *Isatis trachycarpa* Trautv., *Smirnovia turkestana* Bunge – all Turanian psammophytes, and *Paracaryum calathicarpum* (Stocks) Boiss. and *P. platycalyx* Rech. f. & H. Riedl, the latter two known previously from Afghanistan and Pakistan.

From the Touran Protected Area we proceeded east to Doruneh and Kashmar and from there through the hills called by geologists 'coloured mélange', north to Neyshabur and Mashhad in order to have the vehicles repaired and to buy food.

It became evident already on this first section of the Kavir round trip that prolonged drought had hindered the development of spring vegetation. Thunderstorms with heavy rain occurred unusually late in the season and continued throughout our expedition. This made the crossing of low mud flats difficult and hazardous. On several occasions one vehicle had to pull out the other which was stuck in the mud. In the hills bordering the desert vegetation had been grazed to the ground and had not yet recovered on account of the delayed rain. The hills immediately north of Kashmar were an exception where *Scrophularia valida* Grau was discovered but literally no plants at all were seen farther north towards Neyshabur except on sand. We left Mashhad on May 7th, revisited the serpentine hills near Robat Safid – again extremely dry and overgrazed – and proceeded to Torbat-e Heydariyeh and southwest to Faizabad, where we had been invited by a descendant of the Kajar Dynasty, a big landowner residing in an impressive old but half-dilapidated house. This gave us welcome shelter during the night when heavy rain was pouring for hours. We had to wait one day before we could proceed to Juymand (Gonabad) via Bejestan — Ferdows — Boshruyeh — Deyhuk — Asfak and Tabas. During these days we passed several caravanserais dating back to the time of Shah Abbas who had the buildings erected ± 40 km apart. We were fortunate. Only a few years later Tabas built in classical style from bricks dried in the sun, with all the charm of the old Orient and bedded in lush green gardens, was entirely destroyed by an earthquake.

On dry hills the small and taxonomically isolated *Astragalus mirus* Širj. & Rech. f. (Sect. *Mirae*) (Syn. *Dorycnium calycinum* Stocks), described from Baluchistan and found once in southwest Afghanistan, and the rare *A. durandianus* Aitch. & Baker were growing. In crossing a stretch of sand dunes one of the vehicles broke its long axle; exactly the same misfortune as had happened to us in central Afghanistan eight years before. All attempts to replace the axle by one taken from a wrecked vehicle were in vain. So we had to go round half of the Kavir back to Tehran by front-wheel drive only. West of Boshruyeh, *Cousinia raphiostegia* Rech. f. was discovered. On May 14th we left Tabas. In certain areas rain water filled every depression of the desert. Looking into the distance, waterflats glittering in the sun could hardly be distinguished from saline flats, the latter devoid of any vegetation. The track to Robat-e Qarehbil and Robat Khan leads through a dry valley between rocky barren hill ranges. Many recent parallel landslides filling in the valleys on the steep slopes seem to be due to earthquakes, frequent in this area. At Posht-e Badam we obtained permission to put up our camp beds in a small dirty school room as threatening rain clouds gathered again. The next morning we had difficulties in finding the track leading by Baiazeh to Khvor (Khur). Here we camped on the flat roof of the local centre of the Department of Agriculture. The night was clear, lightning in the distance far to the north illuminated the scene, but the mixture of kitchen and lavatory smell, opium smoke and insecticides from the back of the building was rather unpleasant. Later on we found out that people had been smoking below in a room where hundreds of litres of highly inflammable high octane petrol for a helicopter were stored. This aircraft was used to spray insecticide at intervals on the palm groves. From Khvor we went along a glittering white salt desert looking like a glacier, without any vegetation. We turned west and finally reached the Rig-e Jin (Sand of Ghosts), probably the highest sand dunes in Persia, with: *Agriophyllum minus* Fisch. & C.A. Mey., *A. latifolium* Fisch. & C.A. Mey., *Astenatherum forskalii* (Vahl) Nevski, *Cithareloma registanicum* Rech. f., *Heliotropium arguzioides* Kar. & Kir., *H. transoxanum* Bunge, *Oligochaeta minima* (Boiss.) Briq., *Smirnowia turkestana* Bunge, *Spirorhynchus sabulosus* Kar. & Kir. and *Stipagrostis pennata* (Trin.) de Winter.

We continued west by Chupanan and Anarak to reach Nain the following day. Thus the more adventurous part of the Kavir round trip, leading in part through botanically unexplored country, was completed. Skimming along the road we reached Tehran on May 18th via Kashan and Qom. On this trip M. Iranshahr had been our companion.

Ten days of rest, of working in the Evin and Ariamehr herbaria and of minor excursions around Tehran followed. On May 28th I left Tehran with Wilhelmina in our Volvo. We took the road to Garmsar, revisited the gypsum area near Semnan and risked another transect through the Elburz range by Shahmirzad (sampling the Parvar Protected Region) and Fulad Mahalleh to Sari in Mazandaran. *Onobrychis mazandaranica* Rech. f. and *Anthemis altissima* L. var. *discoidea* Iranshahr were found in loose stands of *Carpinus*. We went on to Gorgan and stayed a couple of days at Tang-e Rah, Tang-e Gol and Almeh in 'Golestan', the Mohammad Reza Shah National Park mentioned before. A special treat awaited me near the western entrance of the reservation. In contrasting adjacent habitats, five different species of *Rumex*: *R. alveolatus* Losinskaja, *R. crispus* var. *strictissimus* Rech. pat., *R. pulcher* L. subsp. *anadontus* (Hausskn.) Rech. f., *R. conglomeratus* Murr. and *R. sanguineus* L., occurred and most of the possible hybrids were found. Another interesting feature in the same area was a south facing slope harbouring no less than 13 mediterranean elements, namely: *Caucalis platycarpos* L., *Convolvulus cantabrica* L., *Cynoglossum creticum* Miller, *Medicago rigidula* (L.) All., *Pallenis spinosa* L., *Petrorhagia prolifera* (L.) Ball & Heywood, *Securigera securidaca* (L.) Deg. & Dorfl., *Sideritis montana* L., *Teucrium polium* L. var. *tonsum* Stapf, *Trifolium angustifolium* L., *T. scabrum* L., *Vicia amphicarpa* Lam. and *V. lutea* L.

The 'dasht' (level plain) in the east of the protected area near Robat-e Qarehbil furnished some interesting plants, e.g. *Cousinia qarehbilensis* Rech. f. and *Diaphanoptera stenocalycina* Rech. f. & Schiman-Czeika, the sixth species of this small caryophyllaceous genus endemic to Turcomania, northeast Persia and Afghanistan, described by me in 1940; each species is known from a single locality or restricted to a very limited area. The Caspian forests have their eastern boundary within the reservation and are largely in good condition in spite of the fact that the main road to Afghanistan, Pakistan and India cuts right through the reservation. At higher elevations, stands of *Juniperus excelsa* M.-B. are well preserved. At Tang-e Gol our friend Jany Renz arrived with his Range Rover and Terme from Evin arrived in an official car. We left the Volvo at Tang-e Gol and shifted into the Range Rover to go east to Shirvan. From there we turned north into the Kopet Dagh. It was a rather hazardous drive through narrow flooded gorges and on extremely steep and slippery slopes via the villages of Lujali, Bik, Reshvanlu and Namanlu to the Takht-e Gulul, the centre of the Gulul Sarani Protected Area. This reservation included a high plateau at ca. 2000 m altitude and the hills north of it along the Turkmenian (Soviet) border. Grazing and woodcutting is strictly forbidden in this area and armed guards defend it against intruding shepherds, with the result that within a few years, the vegetation had recovered considerably. One day upon returning from an excursion to the guards' camp, where we had pitched our tents, a man encrusted with mud and dust turned up on a heavy motorbike. He had a letter from the Imperial Court inviting me to an audience with Her Majesty the Empress Farah Diba. She had heard about me as being the author of *Flora Iranica*. The Iranian Embassy had tried to contact me in Vienna but soon found out that I had left for Tehran. Upon enquiry in Tehran, they found that I had left for the Gulul Sarani Protected Region. The poor man on the bike had two days and nights of hard riding before he could catch up with me. It was clear that we could not be back in time to keep the appointment. So upon our return to Shirwan, Wilhelmina had great trouble to reach the Imperial Court by telephone and to convince the officials at the other end of the line of the impossibility to comply with the Imperial order and to arrange another date. Finally it was postponed until June 23rd. We could now continue our journey. So we went to Bojnurd, took a narrow road, crossed the northern of the two mountain ranges forming the eastern extension of the Elburz range to Esfarayen, visited the Dasht-e Jahan Protected Area, crossed the valley between the two ranges used by the Tehran-Mashhad railway line and went on south across the Yoghatay Range north of Sabzevar. To my knowledge no plants had so far been recorded from this area. Except for a few interesting plants in the Shah Jahan Reserve there was not much to collect – the hills were dry and grazed and the valley cultivated. The Yoghatay Range, however, consisting of serpentine rock, was worthwhile! *Cousinia bienertii*, described by Bunge and not found again since, was rediscovered, along with *C. sabzevarensis* Rech. f. and *Trigonella subenervis* Rech. f., both new to science. From Sabzevar we made a loop east, crossed the low ranges again and returned via Soltanabad — Abdollah Givi — Quchan to Bojnurd. Then we collected our Volvo again at Tang-e Gol and returned on June 19th to Tehran.

From Tehran, several excursions were undertaken to the Kuh Dashteh, the Kandavan Pass and to Lashkarak, where Mr Firouz had invited us to his marvellous summer resort. On June 23rd the audience with Empress Farah Diba took place. First of all I had to dress up. I had to get a dark suit. A colourful Baluchi necktie was presented to me by Mr Dalaki, one of Mr Firouz's collaborators, and a white shirt was bought in the bazaar, while gold cufflinks were borrowed from Jany Renz. Mr Firouz accompanied me. The interview was in the Niavaran Palace, a modern but impressive building amidst a park of beautiful old trees close to the old Kajar Palace. On a small table in the library tea was served. I was asked to report on my botanical activities in Iran, on the plan and progress of the Flora, on its implications on nature conservation, on my ideas about how to limit overgrazing, etc. The interview lasted about 40 minutes. My impression of the Empress was that of a highly cultured personality with a remarkably wide background and seriously concerned about her country, its population and also its natural history. Conversation was in fluent English. I could not help noticing, in spite of a polite conventional smile, shadows in her expression as if she had some forebodings of the impending tragedy that would fall on the Shah's family and the whole regime.

During the last hot days of June in Tehran we were exposed to an unexpected wave of publicity. Journalists, representing various Persian newspapers and magazines, competed in interviewing me, and *Flora Iranica* was officially presented to the public. We had to act 'botanical field work' for television and learned the feelings of a person exposed to flashlights. At the beginning the camera team was not at all happy with our performance but as soon as I started to make fun of it they were highly satisfied. As far as I know the whole fuss was for nothing.

The homeward journey took place between June 30th and July 10th via Sero — Yüksekova — Hoşap — Gevaş — Reşadiye (south of Lake Van) — Muş — Bingöl — Elaziğ — Malatya — Ankara — Istanbul — Alexandroupolis — Thessaloniki. On July 9th we said goodbye to our wonderful companion Jany Renz. Next day we reached Vienna.

1977
EXSICCATAE NOS. 54400–57650

The next visit to Persia had originally been planned for 1976. It had to be postponed to the following year because on May 20th 1976 the doctorate *honoris causa* was to be conferred on me by the University of Lund, Sweden. This great festivity took place in the Roman cathedral of Lund. Lund has a special significance for me because in 1920, after the First World War, I had been invited by Prof. Svante Murbeck, an old friend of my father, to stay with him as a 'Kriegskind' and to be properly fed. I earned my pocket money by mounting herbarium specimens for the Lund herbarium. With this money I could pay for my homeward journey and a trip to Transylvania in 1922. It was a peculiar feeling to be in Lund 56 years later in the centre of a big ceremony amidst my Aegean friends Runemark, von Bothmer, Snogerup and many others.

On March 30th 1977 we left Vienna for the third time in our Volvo. At Graz the manuscripts of several families for *Flora Iranica* and the page proofs of Compositae-*Lactuceae* were delivered to the publisher. Beyond Marburg (Maribor) we were blocked by a snowstorm and had great difficulties in returning to Marburg. Late next morning the weather had improved and we could continue via Zagreb — Belgrade — Nis — Thessaloniki — Alexandroupolis — Tekirdagh — Ankara — Sivas — Erzurum — Tabriz to reach Tehran on April 7th. Close co-operation with the Department of Environment was established. At the Ariamehr Botanical Garden Prof. Hans Runemark had succeeded Per Wendelbo as a botanical consultant. After a heart attack in autumn 1976 Wendelbo had left Iran for Europe to undergo an operation at home. The time in between my expeditions to Baluchistan, Khorasan and Azerbaijan was spent in the Ariamehr herbarium.

From April 23rd to May 5th an expedition to Baluchistan was organized. Mr Assadi, a young ambitious man, trained by Per Wendelbo and working as an assistant at the herbarium at Ariamehr, was delegated to be our companion. He is genuinely interested in plants and within a short time built up a considerable general knowledge of the Persian Flora. A 'Wagoneer' car with our heavy luggage was sent ahead to Zahedan. We – Wilhelmina and myself – went by air to Zahedan on April 23rd. On the same day we continued by car towards Mirjaveh, crossed west at the north foot of the Taftan and returned on the Khash road to Zahedan. Although early in the season the weather was rather hot and dry. The next day we went west on the main road to Nosratabad close to the southeast corner of the Lut Desert. My main concern was to reach the big sand dunes on the west side of the Lut. This was, botanically, a complete failure. There was no vegetation at all on the sand dunes. This situation was probably due to the fact that for several years there had not been any rain, neither in winter nor in early spring. Rains – if ever – came at a time when the sun's rays caused strong evaporation; thus they are not beneficial to plant growth. In addition, the 'Wagoneer' was in a poor condition and unfit for prolonged desert travelling and the assistants of the local Environment Office were also not co-operative. Fortunately I had prepared an emergency programme, i.e. visits to the Kuh-e Taftan and the Kuh-e Bazman, two highly isolated volcanic mountains in Baluchistan, both botanically not well known. We left Zahedan on April 25th for Khash. About 100 km south of Zahedan, at an altitude of about 1750 m, on the borders of neglected fields was the only locality where – evidently due to local rain – annuals were still in good condition. We reached Khash and took a small track leading up to the south slope of the Kuh-e Taftan to a small village called Torshab where we pitched our tents. From there, the three of us climbed the slopes as high as possible (2300–2700 m) in different directions. In general the slopes were very dry and grazed by goats. Still, a few interesting plants could be secured, e.g.: *Amygdalus browiczii* Freitag, *Convolvulus turrillianus* Parsa, *Cousinia gabrielae* Bornm., *Hyoscyamus malekianus* Parsa, *Nepeta juncea* Benth. subsp. *desertorum* Bornm., *Polygonum olivascens* Rech. f. & Schiman-Czeika, *Pterocephalus kermanensis* Bornm., *Salvia sharifii* Rech. f. and *Trachyspermum gedrosiacum* (Bornm.) Hedge, Lamond & Rech. f.

The next day we returned to Khash and took the road along the south and east slopes of the Taftan towards Mirjaveh. On April 28th we turned west and tried to reach Bazman, crossing a saline depression, an area not yet visited by any botanist. This was locally quite difficult. We camped amongst *Pistacia atlantica* Desf. subsp. *mutica* (Fisch. & C.A. Mey.) Rech. f., *Stocksia brahuica* Benth. and *Withania coagulans* (Stocks) Durral. Two cars passed on the small desolate track. Our companions took them to be smugglers. About 20 km before the village of Bazman we reached the main road. Near Bazman we put our camp beds amidst a date palm grove close to the only watering place, the mouth of a 'qanat', one of the subterranean aqueducts often several kilometres long. Among the plants to be pressed were three different species of *Ferula* – each trying to stink more than the other!

On April 30th we went by car to Bampur and Iranshahr in order to gather information about the access to Kuh-e Bazman. On May 1st we explored the lower slopes of Kuh-e Bazman, all dark volcanic soil with sparse vegetation. We found *Karvandarina subaphylla* Rech. f. This Compositae-*Cynareae* monotypic genus had been discovered near Karevandar in 1948. It proved to be widely distributed and locally plentiful on barren dark volcanic soil though not easily visible in full sunshine on account of the thread-like leafless stems and small heads. Once *Salvia mirzayani* Rech. f., described from Baluchistan in 1952, then discovered in southeast Arabia, was encountered, alas not flowering. In some valleys the dwarf palm *Nannorrhops ritchiana* (Griff.) Aitch. was frequent. In Pansare, an accumulation of tortoise-like huts constructed from palm leaves, six camels were hired with Assadi's help. The camel expedition lasted from May 2nd to May 4th. It was our first and only one. As there is absolutely no water supply, neither in the desert at the foot of the mountains nor on their east slope, and as considerable distances had to be covered, travelling by camel (with an air temperature of 37–40°C and without any shade) is still the only way. We soon got used to the rolling gait of the camels on the flat desert. Going downhill the gait is transformed into a short and hard staccato – far less pleasant. Through sitting astride and on top of the plant presses and water containers our legs became stiff and aching. We reached an altitude of about 2000 m by camel, and camped on the soft sandy bottom of a sheltered gorge. Our camel drivers baked their bread in the sand.

The flora of the Bazman slopes is not rich but interesting. Plants such as *Centaurea gabrielae* (Bornm.) Wagenitz, *Heliotropium oliganthum* Rech. f., Aell. & Esfand., *H. taftanicum* Rech. f., Aell. & Esfand., *Scorzonera longipapposa* Rech. f., *Teucrium gabrielae* Bornm., *Trachyspermum gedrosiacum* (Bornm.) Hedge, Lamond & Rech. f. were found at higher altitudes. Lower down we found: *Convolvulus turrillianus* Parsa, *Danthoniopsis stocksii* (Boiss.) C.E. Hubbard, *Gaillonia calycoptera* Jaub. & Spach, *G. eriantha* Jaub. & Spach, *Heliotropium serpentinicum* Rech. f., *Karvandarina aphylla* Rech. f., *Linaria campyloceras* Rech. f. and *Salvia mirzayani* Rech. f. Assadi climbed up on foot and reached ca. 3000 m. To my knowledge the only plants known before from Kuh-e Bazman were the ones collected by the Austrian medical doctor and geographer Alfons Gabriel and his wife Agnes, an account of which had been published by Bornmüller (1939). On May 5th we left Bazman by car, crossed the southern extension of the Lut desert towards Bam which had very sparse vegetation and took the main road to Kerman. In contrast to further south, the vegetation, including many annuals, was found in full flower around Mahan near Kerman. On May 7th we returned to Tehran by air. The sight of the desert interior which I had crossed so many times by road was very impressive from the air. The southeastern branch of the Trans-Iranian railway had recently reached Yazd and the track on to Kerman was under construction.

After two busy weeks at the Ariamehr herbarium we set out in our Volvo. The night before, the rear window had been taken out of the car but not broken. Someone evidently wanted to get hold of the frame! This time we took the southern route to Semnan. The weather was cool with repeated rain showers and thunderstorms. On our way east we visited the pass Abr northeast of Shahrud, a botanically rather well-known locality situated on the range which separates sharply the dry interior Iranian from the humid Caspian region. Climatically this situation is parallel to the Cordillera Cantabrica in northern Spain or to certain parts of the ranges bordering South Africa. Approaching from Shahrud the incision on the arid south slope is marked by a lush green tongue of vegetation due to the overboiling of fog from the north under certain weather conditions ('Abr' in the Farsi language means 'fog'.) We then crossed over to the Caspian side further east by the Khush Yailaq to Gorgan and on to Tang-e Gol in the Golestan Reserve where Jany Renz arrived by Range Rover. We continued to Mashhad. Here Hans Runemark joined forces. He arrived by air and the ominous blue 'Wagoneer' of the Ariamehr Botanical Garden arrived by road. Together we continued to Sarakhs via Mozduran in the northeastern corner of Khorasan right on the Soviet border (actually in the Turanian lowlands) slightly beyond the low hills forming the southeastern extension of the Kopet Dagh. My hope was to find representatives of the Turanian sand-desert flora but everything had been grazed.

Pistacia vera L. is frequent on the hills at the westernmost extension of its natural north-Afghan area. At the time of publication of the Anacardiaceae in *Flora Iranica* Lieferung 63 (1969) no exact locality on Persian territory was known. The following remarkable plants were found: *Amberboa bucharica* Iljin, *A. turanica* Iljin, *Chesneya kopetdaghensis* A. Boriss., *Cousinia arctotidifolia* Bunge, *C. linczevskyi* Juz., *C. monocephala* Bunge, *Hymenocrater platystegius* Rech. f., *Jurinea catharinae* Iljin, *Lycium kopetdaghense* A. Boriss. and *Pistacia vera* L.

As the northeast corner of Khorasan was botanically not yet known we tried to follow the Soviet and later the Afghan frontier as closely as possible. This turned out to be difficult because of problems in fording rivers swollen by rain and repeated checks of our papers by the frontier guards. During the days from May 26th–29th we proceeded from Salehabad to Jannatabad and crossed the low range forming the link between the Paropamisus Range on the Afghan side of the Hari Rud (Tedschen) and the Kuh-i Binalud on the Persian side. Between Jannatabad and Malu, *Cousinia lignosissima* Rech. f. was discovered, forming 50 cm high hummocks thick and woody at the base. The 'dasht' south of the range was completely flooded by the repeated rain showers and we took refuge in the school of the coal-mining village of Maidan-e Gulan. We then proceeded slowly across the soaked loamy depressions and numerous gullies filled with chocolate-coloured roaring water courses. Fording the main river before reaching Torbat-e Sheikh Jam was only possible with the help of a tractor. We returned to Mashhad on the evening of May 28th where we dried ourselves and the blotters.

On May 29th we went south from Mashhad on the Birjand road and once more combed the serpentine hills near Robat Safid, the four of us radiating in different directions hoping to re-collect *Diaphanoptera khorasanica* Rech. f., the type species of the genus which had been discovered here by me in 1937. All our day-long attempts to find it were in vain. But many other plants were found in flower due to the rain: *Mattiastrum crista-galli* Rech. f. & H. Riedl, *Onobrychis amoena* Pop. & Vved., *Scorzonera microcalathia* (Rech. f.) Rech. f. and *Silene gertraudiae* Melzh.

We passed a restless night in the only dirty hotel at Torbat-e Heydariyeh, the four of us on deeply concave beds, each one creaking in a different tune. The next morning we went on to Qayen, a quiet little place where we found pleasant accommodation in the clean little 'Inn', one of a chain of official resthouses partly still under development, scattered along the main roads all over the country. The sandy road from Qayen to Birjand crosses the rounded hills, 1900–2000 m high, to which the name of 'Kuhistan' was applied by A. Gabriel. Few plants have been reported from this area which is comparatively easily accessible. At Birjand we contacted the local Environment Officer and he supplied a man to guide us to the Kuh-e Ahangaran on the east border of the 'Kuhistan hills', so far not visited by botanists. On June 1st we went 95 km east to Assadabad where recently a sugar factory had been built. We continued 'roadless' to the village of Ahangaran where we pitched our tents. *Cousinia fabrorum* Rech. f. was found in two places. This plant, as well as *C. lignosissima* discovered a few days earlier belong to the small peculiar section *Neurocentrae* Bunge confined to the arid hills bordering the Iranian and Afghan deserts (for distribution maps see K.H. Rechinger, 1986). Next morning the indefatigable Jany Renz climbed the highest ridge, about 2800 m, extremely difficult because of loose rolling stones on steep limestone plates. He discovered *Silene renzii* Melzh. and *Hedysarum renzii* Rech. f. *Sclerorhachis leptoclada* Rech. f., the fourth species of this genus of Compositae-*Anthemideae*, endemic to the *Flora Iranica* area, was discovered along the road between Assadabad and Birjand. On June 5th on the way back to Tehran three side excursions were made. From Mashhad I went with Jany Renz to Zoshk, a valley in the Kuh-i Binalud cut deep into schistose rocks. We found several species including: *Dianthus binaludensis* Rech. f., *Sisymbrium integerrimum* Rech. f. & Aell. and *Stroganowia affghana* (Boiss.) Pavlov.

From Shahrud the Abr Pass was revisited to collect fruiting specimens of some species collected a few weeks earlier in flower. A short trip undertaken on June 7th from Damghan north into the eastern Elburz was most rewarding as we found: *Bufonia koelzii* Rech. f., *Colutea porphyrogramma* Rech. f., *Gypsophila xanthochlora* Rech. f., *Rosa* × *kopetdagensis* Mef. (= *R. persica* Michaux ex Juss.* × *hemisphaerica* J. Herrmann), *Stelleropsis antoninae* Pobed.

We reached Tehran on June 8th. The wind had turned to the south. Air temperature was 41°C! Saying goodbye to Mr Firouz at his office, he said: 'You can rely on me as long as I am sitting here'. Only a few weeks later he was removed from his position due to a political intrigue preceding the revolution!

Between June 13th and 21st we went again with Jany Renz to Azerbaijan. From Mianeh we went to Siah Chaman (Karachaman) and crossed over to Maragheh and along Lake Urmia (Rezaiyeh) to Tabriz, thus encircling Mount Sahand. We spent a few days in the Kiyamaki Protected Region up to 3200 m, on the west end of the Qareh Dagh and northwest up to Sahand beyond the village of Kandavan (not to be mistaken for the homonymous Kandavan Pass northwest of Tehran). Here people live in caves at an elevation of ca. 2000 m. For our way home we chose the road from Bazorgan northwards along the foot of the Ağri Dagh (Ararat) to Kağizman and through the Aras (Araxes) valley to Erzurum, thus avoiding the 2500 m high Tahir Pass which was becoming more and more blocked by heavy traffic every year. From Ankara we took the short cut via Bursa, crossed the Dardanelles on a small ferry and via Thessaloniki went on to Athens by ferry from Patras to Ancona, and across northern Italy and through the Mont Blanc tunnel to Geneva, returning to Vienna in September.

Only one year later (in 1978) came the revolution in Iran, the outbreak of the unfortunate war between Iran and Iraq and the invasion of Afghanistan by the USSR. The period of easy travelling in Iran and Afghanistan had roughly coincided with my age of 50 to 70 years. Many gaps in the knowledge of the *Flora Iranica* area have been filled but much still remains to be done, however, most of

* *Rosa persica* has been separated by many authors as a separate genus: *Hulthemia*.

it in areas too hard of access for an old man. Younger native botanists are taking over as the contents of the recent numbers of the Iranian Journal of Botany clearly show.

An extremely well-balanced period of my life had thus come to its end. From the year 1957 onwards, periods of intensified working at European herbaria, alternating with field work, meant each time relaxing from the one by doing the other and thus never getting tired.

FLORA IRANICA FACTS AND FIGURES

To date (1987) 162 parts of *Flora Iranica* have been published, beginning in 1963. Most parts contain a single plant family, either large or small. Rosaceae comprise two parts. Of Caryophyllaceae, Liliaceae and Scrophulariaceae one part each is published, the second is due. The Compositae have been published in six parts, the seventh and final one is in press. Of Papilionaceae two parts have been published, three more dealing with the estimated ca. 1200 species of *Astragalus* are due. *Astragalus* will be the largest genus in *Flora Iranica*.

The following genera with more than 100 species are: *Cousinia* 378, *Acantholimon* 164, *Silene* 140, *Allium* 139, *Oxytropis* 115, *Nepeta* 107.

The following families have more than 100 species: Papilionaceae (incl. ± 1200 Astragali) ca. 2414 species, Compositae 1607, Gramineae 758, Labiatae 541, Umbelliferae 515, Cruciferae 470, Caryophyllaceae 460, Boraginaceae 363, Rosaceae 268, Plumbaginaceae 193, Alliaceae 141, Polygonaceae 122, Euphorbiaceae 118, Malvaceae 108.

Umbelliferae are unique in the fact that approximately half of the 140 genera are either monotypic or represented in the *Flora Iranica* area by a single species.

Up to and including part 162 (Umbelliferae) there are 7394 pages of text dealing with 7855 species, 3947 black & white plates and 196 coloured plates. If both parts currently in press, No. 163 (Caryophyllaceae III) and 164 (Compositae VII) are added, 8355 species which belong to 1326 genera have been treated.

It is difficult to estimate the number of species accounts of which still remain to be written; *Astragalus* will have ± 1200 species (according to Prof. D. Podlech). The remaining families will have ca. 500–700 species so the total number of species in the Flora will be more than 10,000. Thus the original estimate of ca. 9000 species was too low. In comparison, *Flora Europaea* has 11,557 species and *Flora of Turkey* just under 8800, with 8576 native.

ACKNOWLEDGMENTS

I am indebted to the following for their contribution and help: Dr Hans Dieter Knapp, Waren-Müritz, DDR for the route-map, Dr Ernst Vitek, Wien, for adapting and mounting the photographs and Dr D.P. Winstanley, Brentwood, Essex, UK for the translation of the announcement of *Flora Iranica*, and Dr Kit Tan, Edinburgh, for editing the text.

The National Science Foundation, USA, the Hammer-Purgstall Gesellschaft, Wien, Die Freunde des Naturhistorischen Museums, Wien and the Fonds für wissenschaftliche Forschung, Wien have provided grants-in-aid (since 1973) for individual contributors to *Flora Iranica*.

The Austrian Ministry of Education subsidised my 1962 and 1965 expeditions to Afghanistan and Pakistan and those to Iran in 1967 and 1971.

The Schweizerische Nationalfonds supported my three visits to the Conservatoire Botanique de Genève between 1978 and 1980, which were mainly devoted to the elaboration of No. 150 Labiatae, published in 1982.

ANNOUNCEMENT OF *FLORA IRANICA*

Extract from 'Die Flora Iranica' by Karl Heinz Rechinger, Adeva Mitteilungen 20: 7–14 (September 1969). Akademische Druck – u. Verlagsanstalt, Graz, Austria.

The territory covered by the Flora amounts to just under 3 million square kilometres, an area roughly equal to that of Great Britain, France, Germany and the Scandinavian countries.

Such an enormous area might be expected to yield a large number of species. The number has in fact been estimated at some 9000. Furthermore, there are extreme climatic contrasts; whereas the humid districts on the shores of the Caspian Sea enjoy an annual precipitation of 1500–2000 mm, the interior of the deserts of Kawir and Lut have practically no rain in some years. However, it is an astonishing fact that steppes or semideserts with a sparse cover of vegetation often provide homes for unexpectedly large numbers of species. Comparisons with adjacent desert territories, especially in the west and south, namely the Saharo-Sindian desert zone, which because of its lower altitude has a different climate characterized mainly by higher winter temperatures, have shown that the number of species in the Iranian highlands is several times greater than those found in any area of equal size selected from the Saharo-Sindian desert zone. Certain studies, which cannot be dealt with in detail here, have led to the conclusion that the Iranian highlands constitute an area in which a flora adapted to severe aridity has evolved and survived. A considerable number of genera and thousands of species are endemic to the Iranian highlands, in other words they are found there and nowhere else. This fact means that the Iranian flora, besides being of great practical interest, is an extremely rewarding subject for scientific study.

The immense practical importance of a Flora – a published description of the plants of a country – needs no emphasis. The Iranian highlands and their peripheral mountain ranges are the original homeland of the wild ancestors of many cultivated plants, such as cereals and fruits, and there is growing interest in the possibility of using them for crossbreeding with the object of improving the strains at present in cultivation. Natural forests and woodlands are restricted to narrow areas where favourable conditions exist, and in many districts deforestation has had catastrophic effects. The task of combating deforestation is never easy, and where woodland is struggling to survive at the extreme edge of its climatic range, it may be almost impossible. Even in treeless, unwooded districts suitable for use as pastureland, there are many areas which have suffered serious damage from over intensive grazing. Where the soil surface is denuded of its natural vegetation the ultimate result is excessive erosion by wind and water. Effective measures for control of pests such as the migratory swarms of locusts, and also of plant diseases, is feasible only on a foundation of detailed knowledge of the host plants on which they live and their geographical distribution.

What has just been said does no more than touch upon a few of the salient problems with which developing countries are confronted, problems which cannot possibly be solved without detailed knowledge of the indigenous plants of the region. There are numerous costly projects which international bodies such as UNESCO and FAO have undertaken in this and similar spheres, but until such time as '*Flora Iranica*' has been completed there can be no proper basis on which they can operate, as is all too clearly evident from the constantly increasing inflow of enquiries and plant specimens sent for identification.

The framework or plan on which '*Flora Iranica*' is based can be briefly outlined. As soon as the text for each family of plants has been completed it is published as a separate issue, and is hence immediately available for use. These instalments or parts are published as they become ready, irrespective of the sequence of families in the systematic classification. When the whole work is complete, users will be able to have them assembled and bound in whatever order they wish – in keeping with the taxonomic classification which they prefer, in alphabetical order or in chronological order by date of publication. For each family there is a key covering all the genera, and for each genus a key to all the species. In addition to the valid scientific name, the principal synonyms are listed, i.e. those names which are no longer valid under the current rules of nomenclature but under which the species from the Iranian floristic area may be cited in literature. There is a concise description which emphasises the most important characters of the species and hence enables the user to check the identification which he or she has reached with the aid of the key. For each species the 'type' is cited, i.e. the original herbarium specimen from which the species in question was first described and on which its name is based. Then follows a list of all the localities in the *Flora Iranica* region in which the species has been found. These localities are grouped according to landscape or habitat type. Some collectors' names are abbreviated. Collectors' numbers are given when available. Specimens which have been examined by the author are indicated by exclamation marks. The external distribution of the species is given, followed where necessary by critical comments and observations on variability, etc.

LITERATURE CITED

Bornmüller, J., and E. Gauba. 1935–1942. Florulae Keredjensis Fundamenta. Feddes Repert. 30: 73–124 (1935–1936); *op. cit.* 41: 297–344 (1937); *op. cit.* 47: 52–80 and 121–137 (1939); *op. cit.* 49: 253–272 (1940); *op. cit.* 50: 365–376 (1941); *op. cit.* 51: 33–48, 84–112 and 209–239 (1942).

Firouz, E. 1974. Environment Iran. Tehran.

Freitag, H. 1986. Notes on the distribution, climate and flora of the sand deserts of Iran and Afghanistan. Proc. Roy. Soc. Edinb. 89B: 135–146.

Gauba, E. 1950–1953. Botanische Sammelreisen in der persischen Dattelregion. Ann. Naturhistor. Mus. Wien 57: 42–52 (1950); *op. cit.* 58: 13–32 (1951); *op. cit.* 59: 119–134 (1952–1953).

——. 1955. Ein Besuch der Kaspischen Wälder Nordpersiens. Ann. Naturhistor. Mus. Wien. 60: 60–76.

Koeie, M., and K.H. Rechinger. 1954–1965. Symbolae Afghanicae. I. Labiatae. Det Kongelige Danske Videnskabernes Selskab, Biologiske Skrifter 8 (1): 1–80; II. Compositae, *op. cit.* 8 (2): 1–215 (1955); III. Leguminosae, *op. cit.* 9 (3): 1–208 (1957); IV. Familiae variae, *op. cit.* 10 (3): 1–194 (1958); V. Familiae variae, *op. cit.* 13 (5): 1–267 (1963); VI. Familiae variae, *op. cit.* 14 (4): 1–91 (1965).

Petrak, F. 1940 [as 1939]. Fungi. *In* K.H. Rechinger, Ergebnisse einer botanischen Reise nach dem Iran, 1937. Ann. Naturhistor. Mus. Wien 50: 414–521.

Rechinger, K.H. For full bibliography see Ann. Naturhistor. Mus. Wien 75: 5–16 (1971) and Proc. Roy. Soc. Edinb. 89B: 3–5 (1986).

——. 1939. Vegetationsbilder aus dem nördlichen Iran. *In* G. Karsten and H. Schenck, Vegetationsbilder 25. Reihe, Heft 5.

——. 1940–1941. Plantae novae Iranicae I. Feddes Repert. 48: 33–48 (1940); II., *op. cit.* 48A: 121–167 (1940); III, *op. cit.* 50: 255–262 (1941).

——. 1940–1950. Ergebnisse einer botanischen Reise nach dem Iran, 1937. I. Ann. Naturhistor. Mus. Wien 50: 410–536 (1940, as 1939); II, *op. cit.* 51: 374–428 (1941); III, *op. cit.* 53: 340–357 (1943, as 1942); IV, *op. cit.* 55: 265–295 (1946); V, *op. cit.* 56: 200–245 (1948); VI, *op. cit.* 57: 53–70 (1950).

——. 1943a. Flora Aegaea. Denkschr. Akad. Wiss. Wien, Math.-Nat. Kl. 105(1): 1–924.

——. 1943b. Neue Beiträge zur Flora von Kreta. Denkschr. Akad. Wiss. Wien, Math.-Nat. Kl. 105(2): 1–184.

——. 1949–1960. Rechingeri Iter Iranicum secundum. Published in various periodicals. For a complete list see Sitzungsber. Math.-Nat. Kl. Österr. Akad. Wiss. 98: 7–9 (1961).

——. 1949–1959. Reliquiae Samuelssonianae, I. Arkiv for Botanik Ser. 2, 1(5): 301–327; II. Zur Flora von Cypern, *op. cit.* Ser. 2, 1(9): 413–436 (1950); III. Plantae novae Syriacae, *op. cit.* Ser. 2, 1(14): 505–512 (1951); IV, Zur Flora der Türkei, *op. cit.* Ser. 2, 2(15): 513–543 (1951); V, Zur Flora von Palaestina und Transjordanien, *op. cit.* Ser. 2, 2(5): 271–455 (1952); VI, Zur Flora von Syrien, Libanon und den angrenzenden Gebieten, *op. cit.* Ser. 2, 5(1): 1–488 (1959).

——. 1950. Grundzüge der Pflanzenverbreitung in der Ägäis, I. Vegetatio Acta Geobotanica 2: 55–119; II, *op. cit.* 2: 239–308; III, *op. cit.* 2: 365–386.

——. 1951. Phytogeographia Aegaea. Denkschr. Akad. Wiss. Wien, Math.-Nat. Kl. 105(Abt. 2): 1–208.

——. 1951. Grundzüge der Pflanzenverbreitung in Iran (Rechingeri Iter Iranicum secundum Nr. 16. Verh. Zool. Bot. Ges. Wien 92: 181–188 = Flora and Vegetation of Iran. Proc. of the VII Intl. Bot. Congress, Stockholm 1950 (1954).

——. 1952. Pflanzen aus Kurdistan und Armenien gesammelt von Prof. John Frödin. Symb. Bot. Upsal. 11(5): 1–56.

——. 1961–1972. Notizen zur Orient-Flora. Anz. Math.-Nat. Kl. Österr. Akad. Wiss., in volumes *98–109*.

——. 1963→. Flora Iranica. Akademische Druck- u. Verlagsanstalt, Graz.

——. 1964. Flora of Lowland Iraq. Cramer, Weinheim. 746 p.

——. 1977. Plants of the Touran Protected Area, Iran. Iran. J. Bot. 1: 155–180.

——. 1979–1986. Florae Iranicae Praecursores 1–68. Plant Syst. Evol., in volumes 133–153.

——. 1986. *Cousinia:* morphology, taxonomy, distribution and phytogeographical implications. Proc. Roy. Soc. Edinb. 89B: 45–58.

——, and P. Wendelbo. 1976. Plants of the Kavir Protected Region, Iran. Iran. J. Bot. 1: 23–56.

Stapf, O. 1888–1889. Der Landschaftscharakter der persischen Steppen und Wüsten. Österr.-Ungar. Revue, N.F. 4: 227–251, 348–366 (1888); 5: 51–62, 155–165 (1889).

INDEX